ABRAHAM ROBINSON

Abraham Robinson (1918–1974)
from a portrait by Rudolph F. Zallinger

Abraham Robinson

The Creation of Nonstandard Analysis
A Personal and Mathematical Odyssey

JOSEPH WARREN DAUBEN

PRINCETON UNIVERSITY PRESS, PRINCETON, NEW JERSEY

Published by Princeton University Press, 41 William Street,
Princeton, New Jersey 08540
In the United Kingdom: Princeton University Press,
Chichester, West Sussex

Library of Congress Cataloging-in-Publication Data
Dauben, Joseph Warren, 1944–
Abraham Robinson : the creation of nonstandard analysis : a personal and mathematical odyssey / Joseph Warren Dauben.
p. cm.
Includes bibliographical references and index.
ISBN 0-691-03745-0
1. Robinson, Abraham, 1918–1974. 2. Mathematicians—Germany—Biography. 3. Nonstandard mathematical analysis. I. Robinson, Abraham, 1918–1974. II. Title.
QA29.R57D38 1995
510'.92—dc20
[B] 94-32715

This book has been composed in ITC New Baskerville

Princeton University Press books are printed on
acid-free paper and meet the guidelines for permanence
and durability of the Committee on Production
Guidelines for Book Longevity of the
Council on Library Resources

Printed in the United States of America

10 9 8 7 6 5 4 3 2 1

TRIUMVIRATUS

Garrett Birkhoff
I. Bernard Cohen
Dirk J. Struik

And with special thanks to
Hannah Katzenstein,
who first showed me the historic beauty of Jerusalem,
that *Stadt aus Gold*,
from the walls of the Old City on a crisp
winter's day in January of 1982.

ERRATA

The transfer of electronic files between compositor and printer inadvertently resulted in the improper conversion of certain characters in the Symbols font. Consequently, during printing several symbols were either omitted or replaced by different characters.

PAGE	FOR	READ
185, line 12	" -t heorem,"	"π-theorem,"
208, line 1	b_{nk} 0, R obinson	$b_{nk} \geq 0$, Robinson
line 2	C, c_{nk} 0, such	C, $c_{nk} \geq 0$, such
line 30	$J_0 \subseteq J_1 \subseteq \ldots \subseteq J_{2k+1}$ k 0,	$J_0 \subseteq J_1 \subseteq \ldots \subseteq J_{2k+1}$ $k \geq 0$,
214, line 20	commutative field , and if ev ery	commutative field Σ, and if every
line 21	extension of is a lso a zero of p,	extension of Σ is also a zero of p,
219, line 20	valuation j	valuation φ
236, line 32	index r	index ρ
last line	*for some r and* $q_i(x_1, \ldots, x_n)$, $p^{\rho} = \quad q_i p_i$.	*for some* ρ *and* $q_i(x_1, \ldots, x_n)$, $p^{\rho} = \Sigma\, q_i p_i$.
275, line 21	defined in K, k 0,	defined in K, $k \geq 0$,
285, line 10	such as $(\ x_n - x_n - 1)y_n$.	such as $\Sigma(x_n - x_{n-1})y_n$.
line 35	$(\ x_n - x_{n-1})y_n$."[141]	$\Sigma(x_n - x_{n-1})y_n$."[141]
359, line 28	$p(z) = z + a_2 z^{n}2 +$	$p(z) = z + a_2 z^{n_2} +$
430, line 25	However, N^F N.[85]	However, $N^F \neq N$.[85]
line 33	P^F N^F	$P^F \neq N^F$
note 85	Consequently, N^F N.	Consequently, $N^F \neq N$.
439, line 25	p 0, f or which she established the	$p \neq 0$, for which she established the

IN ADDITION:

89, figure	flopped image	reversed image
167, note 99 (first line)	Henken	Henkin
255, line 32	Robinsohn's	Robinsohns'
263, line 30	constructability, constructable	constructibility, constructible
note 64	Constructability	Constructibility
331, figure caption	Robinson at the International Congress for Logic, Methodology, and Philosophy of Science, Jerusalem, 1964.	Conference on Functional Analysis, Jerusalem, 1961.
492, figure	flopped image	reversed image
531, line 31	Schmeiden	Schmieden

CONTENTS

FOREWORD

BY BENOIT B. MANDELBROT

ABRAHAM ROBINSON'S LIFE was extraordinary in many ways, if only because his professional work spanned three significant fields: airplane design, symbolic logic, and mathematical analysis. His untimely death was a blow to many people worldwide, and I was one of them; in fact, Robinson and I had hoped to work together on a presentation of fractals in terms of nonstandard analysis (unfortunately, he was on sabbatical when I spent a term at Yale in 1970). This direct personal link, in addition to both my friendship with Renée Robinson and being the Robinson Professor at Yale, made me treasure the invitation to contribute this Foreword. But I did not accept until I had had the opportunity to read the book.

I was fascinated, and I hasten to congratulate Joseph Dauben for having produced a masterful and beautiful work. Very few mathematicians' stories justify a full biography or autobiography. In fact, most mathematicians would not want one, and I respect their feelings even when I disagree. A good example was my Uncle Szolem, who was delighted and proud that (excluding a few scary episodes that produced amusing anecdotes) his life ran according to a well-worn pattern and was not interesting. Thus, after I had taped hours of conversation with him, and eventually reduced this mass to an attractive text, he did not like it and was relieved when I offered not to publish it in his lifetime. To some extent his reaction was justified, because the "ghosted" autobiography that I had produced focused solely—as did the recent autobiography of one of my uncle's old friends—on superficial aspects of his life. I added glimpses of the daily life of Polish and French mathematicians in the 1920s and the 1930s but said nothing of the topics my uncle had worked on or of the ways his choice of topics had conformed to or contradicted his environment. The resulting contribution to history was amateurish and meager, but it has helped me to appreciate the magnitude of the task that Joseph Dauben chose to face, and the quality of his achievement.

Of course, good history had best begin with a good story, and the case of Abraham Robinson, truly singular for a mathematician, allows Joseph Dauben to alternate two narratives. The first is the story of a life, with all its richness, its many picaresque and incongruous episodes.

The second recounts the progression of Robinson's works, from airplane design to mathematical logic to nonstandard mathematical analysis.

What has emerged is a weighty book that covers many topics in great detail. Some readers will find themselves skimming over some details, but different readers will skim different parts. And I feel that in this instance there is great value in thoroughness. Let me elaborate. Many people in academia feel (as I do) that we have reached, not the end of a cycle, but the end of an era; to put it mildly, academic policy making will no longer be overwhelmed by headlong growth. If this is true, there is an urgent need, today, for documentation of the recent past. For the future historian, Robinson's tale has a marvelous virtue. As we follow in near-chronological order the life of a single individual of amazing brilliance, stamina, and versatility, we are guided back and forth without artificiality through at least three widely disparate academic cultures (pure and applied mathematics and logic-philosophy); six countries, representing six distinct flavors of Western culture; and, within the United States, two very different institutions. Indeed, Dauben depicts Robinson's long-term moves and his short-term moves during summers and sabbaticals against a rich background of general history, and he avails himself of this scope to discuss his subject's personal motivations carefully as well as delicately.

More specifically, in order to understand human creativity, I think we must investigate in detail how major creators have balanced the necessary but conflicting needs of deep-rootedness and personal boldness, how they have assessed the relative importance of their own private drumbeat, the drumbeat of the family and the professional community, and the drumbeat of society at large. For a reader craving variety in the backgrounds against which this question can be raised, Robinson's life has been a near-unique case. Thus, while it is not universally conceded that "God lives in the details," we should all be grateful to Joseph Dauben for having given future historians so much detail to think about. This tome's very size may insure its durability.

Throughout this book, Robinson is called a mathematician. I fully agree with this characterization, but I also hear the voice of the devil's advocate who would turn the same evidence around, interpret Abby's achievements backward, and assert that a person who spent much of his life outside of mathematics departments, working on airplane design and symbolic logic, was ipso facto not really a mathematician. Therefore, this book necessarily poses a larger question: What is mathematics? Opinions range the spectrum from a wide, liberal Open Mathematics to a small Fortress Mathematics. For proponents of the former school, which I favor and which I am sure Abraham Robinson also favored, mathematics is a big rambling building permanently under construction, with many doors and many windows revealing beautiful and varied landscapes. For proponents of the latter, the highest ambition is to wall off the windows and preserve only one door. Fortress Mathematics is intolerant of individuals like Robinson, who, even in his specifically mathematical work, followed not the drumbeat of others but his own.

Incidentally, the influence of Leibniz on Robinson is well documented in this book. Today, other "hard" scientists (I am one of them) turn to Leibniz, but Robinson has preceded us, and I wonder how explicitly and to what extent he should be viewed as a forerunner on this account, as he is on so many others.

Once again, this book taught me a great deal and I recommend it heartily. My deep regret at not having known Robinson better has been to some extent mitigated, and the reasons for the existence of a Yale chair bearing his name have been clearly illuminated.

Benoit B. Mandelbrot
Abraham Robinson Professor of
Mathematical Sciences at Yale University
IBM Fellow Emeritus

ACKNOWLEDGMENTS

> If the history of science is a secret history, then the history of mathematics is doubly secret, a secret within a secret.
>
> —*George Sarton*

ANY BIOGRAPHY written over the course of fifteen years will have incurred many debts. It is with pleasure and gratitude that I devote the next few pages to thanking the many individuals and institutions whose resources, assistance, and support have proven invaluable to the reconstruction of the life and work of Abraham Robinson.

The idea for this book began with a telephone call I received late in 1979. My first biography, a study of the nineteenth-century German mathematician Georg Cantor, was devoted to the origins of transfinite set theory, a revolutionary new conceptualization of the infinite. The book had just appeared, and I had given a lecture on the subject at the California Institute of Technology in Pasadena that fall. Shortly thereafter, W.A.J. Luxemburg called from Caltech to ask if I might be interested in writing a biography of the mathematical logician, Abraham Robinson.

What I knew of Robinson at that point was limited to nonstandard analysis—Robinson's own extension of the real numbers which was as controversial and compelling in its own way as Cantor's introduction of transfinite numbers. Using the power of model theory and mathematical logic to define infinitesimals (as well as infinitely large elements), Robinson's work complemented what Cantor had achieved a century earlier. In Robinson's case, however, it was the rigorous treatment of infinitesimals that was especially ingenious.

The infinite has always been regarded as troublesome in the history of mathematics, and infinitesimals especially so. Following their prominent introduction by Newton and Leibniz as key elements of the calculus in the seventeenth century, debates over their legitimacy (and apparently self-contradictory qualities) were often intense. By the nineteenth century Cauchy and, later, Weierstrass and their followers found an effective way to bar infinitesimals from mathematics (through the well-known method of deltas and epsilons), while retaining the basic idea of arbitrarily small quantities. In a similar but even more dogmatic spirit (despite his own success in creating transfinite set theory), Cantor condemned infinitesimals as "the cholera bacillus of mathematics." Thus the fact that Abraham Robinson succeeded a century later in providing a rigorous foundation for infinitesimals, although others before him had largely failed, is all the more remarkable.

What Luxemburg did not know when he first contacted me was that I was already familiar with the details of nonstandard analysis, specifically in the form Luxemburg himself had made popular and widely accessible to mathematicians. As an undergraduate mathematics major at Claremont-McKenna College (one of the Claremont Colleges in California), I had written a senior honors thesis on nonstandard analysis in 1966, taking Luxemburg's approach via ultrafilters to the subject, rather than Robinson's own path to discovery which had relied specifically on logic and model theory. With respect to my earlier work on Cantorian set theory, Luxemburg emphasized how, in a very direct way, a biography of Robinson would be a natural sequel.

Since Luxemburg was planning to visit Yale University that spring (1980), he suggested we meet there. He arranged for us to have lunch with Renée Robinson, and for me to meet members of the Yale Mathematics Department who had been Abby's colleagues before his untimely death in 1974. I wanted to be certain that Mrs. Robinson and the Department at Yale were amenable to the idea of my writing Robinson's biography, for their cooperation was clearly essential. It was decided that we would meet for lunch at the Park Plaza Hotel; Mrs. Robinson had just returned from a trip to Morocco and much of our conversation was devoted to the many trips she and Abby Robinson had taken in the course of their life together. They had been inveterate travelers, and in this we had an immediate common interest, as it turned out we did in archaeology as well.

Agreed that I could rely upon Mrs. Robinson's cooperation in writing the biographical portions of Robinson's life, I was subsequently assured of the support of the Yale Department of Mathematics. This much settled, I arranged to spend a week working in the Yale University Archives, where I made a preliminary survey of what was available for study in seventeen boxes of Robinson's collected writings, correspondence, papers and various miscellanea. Two years later, the Department of Mathematics invited me to give the Harvard Lectures at Yale (in the spring of 1982), which were devoted to "Professor Abraham Robinson: The Man and His Mathematics," with a second, more critical consideration of "Abraham Robinson: Infinitesimals, Nonstandard Analysis and the Foundations of Mathematics." I was equally pleased to have the opportunity to present a revised version of these lectures as a plenary talk for a meeting to mark the tenth anniversary of Robinson's death. This was at Yale in October 1984, for the International Mathematics Conference on Model Theory sponsored by the National Science Foundation.

By this time I had begun systematically to interview colleagues, former students and friends of Robinson, all of whom were generous with their time in responding to my questions by letter or in person. Opportunities to lecture on Robinson and his work at a wide variety of meetings, colleges and universities also proved to be immensely helpful, for inevitably, wherever I was asked to speak about Robinson and his work, former students and colleagues were present. In every case they too have helped to add further detail and color to the portrait that slowly emerged through the early drafts of this biography.

I am especially grateful for invitations, *inter alia*, from the Wolfson College Philosophy Colloquium (Oxford University), the Institute for History and Philosophy of Science (University of Hamburg), the Institutes of Mathematics and of History of Natural Science (The Chinese Academy of Sciences, Beijing), the Academia Sinica (Taipei), from Departments of Mathematics at the University of Strasbourg, the Universidad Nacional Autónoma de México (Mexico City), the University of Utrecht (Netherlands), the Castelnuovo Institute of Mathematics (University of Rome), the Departamento de Matemática Aplicada, Universidad de Zaragoza (Zaragoza, Spain), and the Departamento de Filosofía, Universidad del País Vasco (San Sebastian, Spain), as well as, in the United States, from Boston University, the California Institute of Technology, the University of Virginia, the Claremont Colleges, and from graduate and faculty seminars in philosophy and mathematics at both the Graduate Center of the City University of New York and at the Courant Institute of New York University. Each of these forums provided valuable opportunities to present and discuss material that now appears, sometimes in a greatly revised form, here.

Equally important, visits to the major institutions with which Robinson was affiliated made it possible to study local archives, to examine administrative records, and to interview colleagues who had worked or studied with Robinson. I am indebted to the Department of Mathematics at Yale University for making accommodations available to me in New Haven, especially at the beginning of this project, and to the Research Foundation of the City University of New York for contributing financial support that made it possible for me to visit the many institutions where Abraham Robinson either studied or taught. For permission to quote from departmental records at Birkbeck College (University of London), the University of Toronto, The Hebrew University in Jerusalem, the University of California at Los Angeles, and, of course, Yale University, including Robinson's departmental files at each of these universities, I am grateful.

All of the individuals I have interviewed, or with whom I have corresponded, are acknowledged either in the list of interviews at the beginning of the bibliography, or in the notes that accompany the text itself. To each of them I owe my heartfelt thanks. Additionally, the Tarski papers in the archives of the Bancroft Library of the University of California, Berkeley, as well as departmental, faculty and administrative records in the university archives on both the Los Angeles and Berkeley campuses, along with the Gödel papers at Princeton University, have proven invaluable in clarifying various details of Robinson's biography. Permission to quote from the latter, given by the Institute for Advanced Study, is gratefully acknowledged. The late I.M.H. Etherington sent me the bulk of his wartime correspondence with (and about) Robinson, and these letters have now been added to the Robinson papers at Yale. H. Donia MacLean kindly gave permission to quote from her father's correspondence. The Jewish National and University Library in Jerusalem is also gratefully acknowledged for its permission to quote from correspondence in its possession. For their help in facilitating my use of the papers in each of these archives, I appreciate the generosity of a host of librarians, including Dennis L. Bitterlich, Margot Cohn, Mark L. Darby, Ruth Gay, Abraham Neyman, William M. Roberts, Robin Ryder, Diane Wells, and David Wilson. To the librarians and archivists at the New York Public Library, the Courant Institute of Mathematics, the D. E. Smith Library (Columbia University), and the library of the Graduate Center of the City University of New York, additional thanks are due. Specifically, I want to acknowledge as well the very substantial help given me throughout the duration of this entire project by Paul J. Lukasiewicz, the librarian of the Yale Mathematics Department, and by Judith Ann Schiff, the Chief Research Archivist, Department of Manuscripts and Archives, Yale University Library.

Finally, there are a number of debts that deserve separate mention. It was only as I began to examine Robinson's contributions to aerodynamics beginning in World War II, and as I started to investigate the important first postwar International Congress of Mathematicians held in 1950, in Cambridge, Massachusetts, that I interviewed Garrett Birkhoff, professor emeritus of mathematics, Harvard University. I immediately came to appreciate the extent to which his own interests, especially as they turned to fluid dynamics as a result of the war effort, matched Robinson's. From a very early point, Garrett Birkhoff has shown his support in a variety of ways, and I am grateful to him for reading virtually every chapter in earlier drafts with care and insight, always offering helpful comments. Similarly, the biography has been read at least in part by a number of colleagues whose suggestions have been welcome. Of these thanks are due to my colleagues in New York City: to Evelyn Ackerman, Abraham Ascher, Martin D. Davis, Melvin C. Fitting, Jacob Judd, Elliott Mendelson, Rohit Parikh, Hans Trefousse, and Kurt Weil. Sheila Rabin offered substantial help in translating the diary Robinson kept during World War II, from the original Hebrew into English. So too, in various ways with particular details of Robinson's life, did Hilde Robinsohn (the wife of Robinson's brother, Saul). Others who have read substantial portions of the biography and offered useful suggestions include Chimen Abramsky, L. Lorne Campbell, A. John Coleman, H.S.M. Coxeter, John W. Dawson, George F.D. Duff, Paul A. Gilmore, Donald Kalish, David Kaplan, Hannah Katzenstein, Azriel Levy, Moshé Machover, Gregory H. Moore, Yiannis N. Moschovakis, George Daniel Mostow, Gert H. Müller, Jeanne Peiffer, Eliezer Pinczower, Paul G. (Tim) Rooney, Peter Roquette, Roderick A. Ross, Christoph J. Scriba, Hourya Sinaceur, and Angus E. Taylor. Both Ivor Grattan-Guinness and Detlef Laugwitz read the entire manuscript in both its original and revised versions, and the acumen of their criticisms, both in general and in detail, have proven especially valuable in enhancing both the accuracy and readability of this book.

Finally, a special word of appreciation to W.A.J. Luxemburg and to George B. Seligman. They have read more than several versions of this biography, from the earliest drafts to the most recent, final version of this manuscript. George Seligman's insightful account of Robinson's life, which appears as the preface to each of the three volumes of Robinson's *Selected Papers*, was the starting point from which this biography began, and his knowledge of Robinson as a mathematician, colleague and friend, has been invaluable in saving me from many errors of detail, judgment, and spelling (in multiple languages)! Likewise, to David Grose, chairman of classics at the University of Massachusetts, Amherst, who twice copyedited the entire manuscript, I also wish to express my appreciation.

The debt of gratitude that, above all, I am most pleased to acknowledge one last time here is to Mrs. Renée Robinson. Her own life, devoted as it has been to the theater, to the arts, to fashion design and the world of *haute couture*, complemented Robinson's academic life in ways that are emphasized throughout this biography. I have relied on Renée Robinson's generosity and hospitality on too many occasions over the past fifteen years even to count. She has made her home, and Robinson's study, available to me, including her vast collection of photographs that add a special dimension of personal immediacy to this book. Unless otherwise noted, all of the photographs which illustrate this book are from her personal collection and are used with her permission.

Above all, Renée Robinson has read every page of this biography, and while we may not agree on every detail of interpretation, she has provided the touchstone for the veracity of the day-to-day details of Robinson's life as a mathematician, and the general character of his life as a scholar of international renown. It is to Renée Robinson that I owe a great debt of thanks in writing this biography.

As I look back on the last fifteen years, several especially vivid memories stand out. One is the first time I met Renée Robinson at the very beginning of this project. At the time she hoped this might be finished in a matter of years, as did I. But as the complexity of Robinson's life became increasingly apparent, I was convinced that learning more about the historical details, interviewing as many of the individuals who have had a part in this story as possible, was a higher desideratum than completing this work in some predetermined period of time. I am grateful for Renée Robinson's patience, and for her unflagging encouragement to keep at it!

I also think of the meeting this past February (1994) at the Mathematisches Forschungsinstitut at Oberwolfach (Germany), a meeting organized by W.A.J. Luxemburg and Detlef Laugwitz to commemorate, in part, the twentieth anniversary of Robinson's death, and to evaluate the current state of nonstandard analysis as a viable, active part of contemporary mathematics. There I met many of the well-established mathematicians who use nonstandard analysis to surpassing effect in their own work, as well as many of the younger generation who will also, in various ways, make nonstandard analysis a part of their work in the future. This alone gives Robinson's best-known work special promise, a sense of vitality to nonstandard analysis that indeed, just twenty years after Robinson's death, makes clear that it has become an established tool useful to diverse branches of both pure and applied mathematics.

In fact, the greatest pleasure I have had in writing this biography is the chance it has provided to meet mathematicians in all parts of the world. I have interviewed more than seventy colleagues, former students and friends of Abraham Robinson, sometimes repeatedly, often at length. While I know that there are still gaps, stories, anecdotes, and details that have not found their way into this version of Robinson's life, I am confident that what it does provide is a vibrant portrayal of what it means to be a mathematician in the twentieth century. To be a mathematician, clearly, can mean a life that has its moments of adventure, as well as quiet satisfaction both personally and professionally. There is always the possibility of making substantial contributions to students, departments, universities, and above all, to mathematics itself. Robinson's story in this regard is unusually rich, for not only did he enjoy lecturing and traveling widely, he also lived through times of extraordinary historic change. And yet his life reflects in many respects the life of a typical mathematician, one working today at international levels of exchange. If I have managed to convey the spirit of what such a life can mean, for the individual who lives it, as well as for those who came to know him in the process, then this book will have achieved one of its primary goals.

For assistance with all of the technical details of turning a manuscript into a published book, I offer my editors at Princeton University Press sincere thanks. Above all, for much of the course of its writing, I have benefited from the encouragement and interest of Edward Tenner, Princeton's science editor when I began working on this biography. Trevor Lipscombe, Princeton's current science editor, has been responsible for seeing the book through press, and for his steadfast support of the book and tireless editorial efforts on its behalf, I am pleased to offer my warmest thanks. For the design, copy editing and coordination of all the many stages of getting this book into print, I would like to thank Harriet Hitch, Jan Lilly, Sandra Lynn Malyszka, Anju Makhijani, and Janet Stern of Princeton University Press as well. Preparation of the camera-ready copy for the book has been due largely to my colleague in New York, J. Cozby, whose computer wizardry has enhanced both text and illustrations, providing a publishable version that is visually a pleasure to read, and I applaud his efforts in producing the final version of this work for publication.

As I reflect on these last few words of acknowledgment to all of the individuals who have had a hand in the writing of this biography, I am sitting at Robinson's desk. His study, downstairs, in the home where he and Renée lived on Blue Trail in Sleeping Giant State Park in Hamden, Connecticut, is cool and quiet. The window to his study looks out onto a world full of serenity, contentment, peace. To whose who have given my own life exactly that over the duration of this project and more, I am truly grateful. If I have managed to return even part of what I owe them, it is still not enough to say how important and special they are to me. To Ruth Zoe Ost and David Frederick Grose, no amount of thanks can truly be enough.

Joseph W. Dauben
Oberwolfach, Schwarzwald, April 1994
Hamden, Connecticut, May 1994

ABRAHAM ROBINSON

CHAPTER ONE

Family and Childhood: Germany 1918–1933

עברי אנכי
–Jonah 1:9[1]

ABRAHAM ROBINSON was born early in October 1918 in the small Silesian mining town of Waldenburg (Prussia), now Walbrzych in Poland. By the end of the year the First World War had been won by the Allies at a cost of eight-and-a-half million soldiers killed, another twenty-one million wounded, and an estimated seven-and-one-half million taken prisoner or otherwise missing in action. In the aftermath of the war a worldwide influenza epidemic claimed an additional twenty-two million lives. Oswald Spengler's *The Decline of the West*, which first appeared in 1918, must have struck contemporary readers as a grim prophecy that, if borne out, would soon lead to a future where the masses would be easily manipulated by dictatorial governments, and human values would be measured solely on cold, rational, materialistic grounds. Prussia, given its authoritarian tradition, was Spengler's choice as the emergent power while the West continued its inevitable decline.

Clearly the world into which Abraham Robinson was born offered little to a helpless infant and his recently widowed mother with two sons—especially in a Prussia where Jews were soon to be persecuted with a vengeance unprecedented in modern history. Survival, however, was something Abraham Robinson learned at an early age. As a boy he fled with his mother and brother Saul from Germany to Palestine; shortly thereafter he joined the Haganah (the illegal Jewish civilian militia) to defend his new homeland; a decade later he narrowly escaped from Paris as the Germans invaded France; the rest of the Second World War was spent as a refugee in England where he experienced the harrowing bombing raids on London. After the war he moved first to Canada, then to Israel, and finally the United States. The only threat he was unable to survive was the fatal diagnosis of pancreatic cancer when he was fifty-five. But by then, Robinson was Sterling Professor at Yale University, and a renowned mathematician who had done much to establish mathematical logic as a significant factor in twentieth-century mathematics. Above all, he remains best known for his creation of nonstandard analysis, a rigorous theory of mathematical infinity, including infinitesimals.

[1] *'Ivri anokhi,* "I am a Hebrew."

Despite his own remarkable talents, there seems to have been no precedent among Robinson's ancestors for genius in mathematics. His forebears, however, had worked assiduously to free themselves from the Eastern European shtetels, with their uncertainties of business and the drudgery of subsistence farming. In this, they succeeded admirably. Both Robinson's father and uncle were part of the growing Jewish intelligentsia and professional class; his father, in particular, did all he could to promote the Zionist cause that eventually led to the creation of the state of Israel from the ashes of the Second World War. His father's side of the family, in fact, could trace its roots back to Poland where his great-grandfather was a moderately successful tenant farmer, estate manager, and occasional businessman.

The Robinsohn Family: Nineteenth-Century Roots

Robinson's father, Abraham Robinsohn,[2] was born in 1878 near Dabrotwa, a small town in Galicia where his father, Dawid Robinsohn, farmed a leasehold estate. Abraham was the second son of two, his older brother Isak having been born several years earlier, in Brody (near Lvov, then part of Austria-Hungary but now in the Ukraine), where their mother Sara's family lived. Sara's father, Sender Achtentuch, was a man of the "old" school. Although her marriage to Dawid Robinsohn was arranged by their parents, the evidence of a brief autobiography written by Sara Robinsohn (in German) around 1917 suggests that it was a happy one, even if their lives were not always easy.[3]

Anti-Semitic feelings forced the Robinsohns to give up their leasehold shortly after Abraham was born.[4] For a while the family lived with Dawid Robinsohn's parents in nearby Kamianka, but eventually (through his brother-in-law Feiwel Achtentuch) Dawid Robinsohn succeeded

[2] Robinson was named after his father, Abraham, an anomaly according to Jewish custom explained by the fact that his father died *before* his son was born. Robinson decided to change the spelling of his name from Robinsohn sometime between 1939 and 1941, shortly after he had become a refugee in England. His brother Saul never changed the spelling of his surname. Throughout this chapter, Abraham Robinso*h*n refers to the father, Robinsohn Senior; Abraham Robinson or Robinson always refers to his son, known affectionately to family and friends as *Abchen* or Abby, the subject of this biography.

[3] Much of what Robinson knew of his paternal ancestry came largely from a brief history his grandmother, Sara Robinsohn, wrote when she was sixty-four years old, at a time when she was living in Austria with her elder son, Isak. Her original manuscript, Robinsohn R-1917, is now part of the Robinson papers in the archives of the Yale University Library (hereafter, RP/YUL). It serves as a major source for the family history related here. An equally representative account for this period of Jewish life in Eastern Europe and Russia is given in Davidson 1946.

[4] Robinsohn R-1917, p. 24. Although the Robinsohns instituted a lawsuit in hopes of retaining their property, they lost their case and were left "without money or position."

in finding work in Odessa. Not wishing to remain alone with her in-laws, Sara decided to return to live with her own parents in Brody.

There she enrolled Isak in one of the best local cheders (ḥeder), a private Jewish elementary school. As Sara Robinsohn writes poignantly in her autobiography, "I wanted to raise my children to be educated, independent people who didn't have to battle with fate the way we did. Although I was uneducated, I came to this conclusion."[5]

Abraham (although he was only two at the time) regularly accompanied his brother to school. Both were quick learners. Three years went by, however, and Dawid was still living and working in Odessa. Meanwhile, a special teacher was hired to teach Isak Hebrew, especially reading and writing. It soon became clear, however, that he preferred German, to which his grandfather, Sender Achtentuch, strongly objected. Earlier he had forbidden any schooling for his daughter, and now he protested adamantly against Isak's reading any German books, hoping instead that his grandson would concentrate on Hebrew and become a rabbi. Sara Robinsohn, however, was committed to seeing that her elder son become an "educated person." Secretly, she hired a teacher from the local gymnasium who promised to prepare students within six months for the entry examinations.

When Isak successfully passed the examination, Sender Achtentuch reacted as if struck by lightning. He refused to eat or sleep, and wept "without pause." Nor did he come out of his room for three days, regarding it a disgrace that one of his grandchildren was about to begin a secular, German education. Isak had to be sent away from the house and stayed for a time with his Aunt Rachel.

Eventually Sender Achtentuch relented, but only because of the impending wedding of his son Moses. Under heavy pressure to reunite the family, Sender Achtentuch carefully questioned his grandson, worried that he would discover adverse effects from the boy's new schooling. To everyone's relief, Isak responded correctly to every question in Hebrew, and he convinced the grandfather that he had not forgotten any of what he had learned earlier at cheder.

Dawid, following a brief period of unemployment, eventually found work in Kopolowka, near Brody, so that he could at least be with his wife and children on Sundays. Meanwhile Abraham was also doing well at cheder and learning Hebrew at home. In fact, he was regarded as a model child, for which he was "famous in all Brody." When a rich and religious man by the name of Popper agreed to take him on as a private pupil, Grandfather Achtentuch saw this as yet another step toward his dream of having a rabbi in the family. Abraham soon proved him-

[5] Robinsohn R-1917, p. 27.

self to be an exemplary pupil, and one day Popper felt moved to tell him that "something great will come of you."[6]

Isak soon passed the *Matura* examination with distinction and decided to go on to university in Vienna. Abraham wanted to go as well, to obtain his *Matura* there rather than in Brody, where he had only studied privately. Apparently arrangements were made for both boys to go to Vienna together because their mother now decided to move to Kopolowka (where Dawid was working) rather than continue living alone in Brody. After moving several more times over the next decade in hopes of finding better positions, they settled briefly on an estate near Chernowitz, but after a time gave that up to buy a partnership in a timber business with their cousin Hersch Robinsohn.

Following university, Isak Robinsohn went on to medical school. It was not long before he had an established practice in Vienna and was prospering enough to send his parents fifty florins every month. Within a few years he was married, and bought a villa in Weidling, on the outskirts of the city, where he and his wife Josefine (a devout Christian) raised their two children, Lizzie and Alexander. Somewhat later Dawid and Sara Robinsohn moved to Weidling and spent their remaining years living with Isak and their grandchildren. While Isak was becoming an internationally recognized doctor, his younger brother Abraham had finished his education and was becoming increasingly involved with Zionist ideas and his own literary pursuits.

Abraham Robinsohn (1878–1918)

Life as a child growing up in Brody, with a frequently absent father, was not the easiest for Abraham Robinsohn. From an early age, however, he seems to have devoted himself to Hebrew studies, primarily of the scriptures and the Talmudic tradition. Brody was an advantageous city for a boy with such interests. By the end of the eighteenth century, so many Jews were living there that the Emperor Joseph II called it a "new Jerusalem" when he visited in 1790. As Abraham Robinsohn himself once described it, the city offered a "typically Jewish milieu" that was especially fruitful for his upbringing. Brody, in fact, was a prominent center of the *Haskalah*, or Jewish "Enlightenment."[7]

[6] Robinsohn R-1917, p. 38.

[7] Unfortunately, the year after Robinsohn was born, Brody lost its status as a free commercial city. This greatly depressed its formerly vital business activity. According to figures for 1880, there were about fifteen thousand Jews living in Brody, comprising more than 76 percent of the total population. The pogroms in Russia during this period brought tens of thousands of refugees to Brody, although most stayed only temporarily. The total population of the city actually declined steadily for the rest of the century. By 1900, fewer than twelve thousand Jews were left. See Balaban 1916 and Wasiutyński 1930.

Consequently, Robinsohn was able to find serious encouragement in Brody for his studies. Soon he was reading and commenting on the Talmud, which he regarded as a "Labyrinth, from which most who enter it are freed only by the liberation of death."[8] But despite his grandfather's hope that he might become a rabbi, this was not to be.

Abraham Robinsohn (1878–1918), photographed in Vienna.

Robinsohn considered it inevitable that eventually he was exposed to a few scraps of "profane knowledge," including arithmetic and a little German. More influential than such "innocent" things, as he put it, was his discovery of the new Hebrew literature, which he undertook to study intensively and which had a profound effect. Such literature, sharply critical of everything old and outmoded, inspired him because of its liberal spirit. At a time when most European students had finished their studies at the gymnasium, Robinsohn was just beginning to cover the same ground, and he did so through private study and at times even by self-study. Soon he was advanced enough, however, to leave Vienna and begin his university studies in Switzerland.

It was at the University of Bern that Robinsohn began to study philosophy and literature. He became especially interested in American pragmatism, and eventually received his doctorate in 1909, having written a dissertation on William James. At the suggestion of his professor, Ludwig Stein, he then went to Oxford to work with F.C.S. Schiller, one of the most important advocates of pragmatism outside of the United States.

It was in this period that Robinsohn became increasingly active as a writer and journalist. During the few years he lived in England, he spent a part of his time in London working for Joseph Hayim Brenner of *Hameorer*, a Hebrew magazine that was briefly published between 1906–1907.[9] Later he helped to edit the book *Der polnische Jude*, contributed to the *Bibliotheca iwrith* of Ben Avigdor, and was involved in the produc-

[8] Robinsohn R-1912, RP/YUL; many of the biographical details which follow are taken from this account.

[9] *Hameorer* was founded by Brenner, a novelist and journalist, who was born in the Ukraine in 1881. He taught Hebrew for a time, was caught up in the new Zionism at the turn of the century, and after entering the Russian army in 1902, deserted shortly thereafter, making his way to London. When *Hameorer* failed in 1907, he returned to the continent, where he worked for a Yiddish daily newspaper in Chernowitz (Lemberg),

tion of two newspapers, *Haschiloach* and *Hador* (published in Cracow between 1900–1904).[10] *Hador* appeared weekly, in Hebrew, and was devoted primarily to literary material. Robinsohn also worked for a year as editor of a weekly in Chernowitz, but this eventually folded for financial reasons. He also published several articles in *Bund* (Bern) and the *Tagwacht*.[11]

Despite his degree from the University of Bern and increasing experience in journalism and writing, Abraham Robinsohn had not as yet found a permanent position worthy of his talents. On the other hand, he was becoming increasingly involved in the movement to promote "Hebrew through Hebrew,"[12] as well as the Zionist cause in general. During his student days in Switzerland, he had been active as one of the founders of the Ivriah, an organization that was a precursor to the worldwide Hebrew Language Union. This was a reflection of Robinsohn's lifelong commitment to the Hebrew language as a means of uniting the Jewish people. His chance to do something definite along these lines came in 1912, when David Wolffsohn decided he could no longer continue his work on behalf of the World Zionist Organization without help.

Wolffsohn had succeeded Theodore Herzl in the World Zionist Organization upon Herzl's death in 1904.[13] As a successful businessman in Papenburg, East Frisia, Wolffsohn had prospered sufficiently in a timber firm to found a synagogue there. From 1887 he worked as a successful merchant in Cologne, where he had already cofounded (in 1883) the

---

Galicia. Later he emigrated to Palestine, where he was killed in Jaffa by an Arab extremist in 1921.

[10] These details of Robinsohn's early literary efforts are drawn from an obituary by Chaim Tartakower that appeared in the *Jüdische Rundschau*, 19 (May 10, 1918). A typewritten draft of this obituary is among Robinson's papers, YUL. At the bottom, handwritten in pencil, is the warning *Falsche Lebensdaten*, with the note *Hauptstudium Bern: Philosophie.* Tartakower incorrectly reported that Robinsohn had studied chemistry in Zürich and philosophy at the University of London, where he was said to have earned a Ph.D. Although Robinsohn does not mention chemistry in his own autobiographical letter of 1912 (Robinsohn R-1912), his wife later noted that *mein Mann anfangs Chemie studiert hat. Aber das ist wohl ohne Belang.* Lotte Robinsohn in a letter to Martin Buber dated May 14, 1918, Arc. Ms. Var. 350/622c, The Jewish National and University Library, Department of Manuscripts and Archives, Jerusalem, Israel.

[11] Robinsohn R-1912, p 3.

[12] The movement was known in Hebrew as *'Ivrit be-'Ivrit.* Emphasis was placed upon learning Hebrew by using it, and was an approach favored by and promoted among Zionists in Germany and elsewhere at the time. See *Haschiloach,* 4 (Berlin 1898), p. 385.

[13] A new executive was elected for the World Zionist Organization at the Seventh Zionist Congress held in Basel in July. The president of the Executive Committee, of the Inner Action Committee, and of the Zionist movement now was Wolffsohn. For details, see Laqueur 1976, p. 138.

Kölner Verein zur Förderung von Ackerbau und Handwerk in Palästina with his friend Max Bodenheimer. A few years later, upon the appearance of Herzl's *Der Judenstaat*, Herzl and Wolffsohn became good friends. Not long thereafter, in 1898, as part of Herzl's long-range plans for world Jewry, Wolffsohn spearheaded the plan to establish a bank to help finance Jewish settlement in Palestine, namely the Jewish Colonial Trust.

When Herzl died in 1904, Wolffsohn seemed to many the most appropriate choice to succeed him as head of the World Zionist Organization. For the next seven years Wolffsohn worked unceasingly for the Zionist cause. By 1911, when the Tenth Congress was held in Basel in August, he was fifty-four and in very bad health. In the face of mounting opposition, especially from Russian Zionists who favored immediate settlement activity in Palestine on a much larger scale than Wolffsohn felt prudent, he was ready to resign.[14]

Although Wolffsohn did give up the presidency, he still retained his positions within the Jewish National Fund and the Colonial Trust. Meanwhile, ill health continued to plague both Wolffsohn and his wife. She had a stroke in July and died shortly thereafter, on September 12. Throughout the ordeal, Wolffsohn sat continuously by her bedside, hardly moving for thirty-six hours. A month later Wolffsohn himself suffered a severe heart attack and collapsed in Cologne. Thereafter, he was frequently ill. He slept nearly half of every day, and found it impossible to keep up with his business, his correspondence, and above all with his efforts on behalf of the World Zionist Organization.

Although Wolffsohn had always written his own letters, this now proved physically impossible. It was at this point that he engaged a secretary, Abraham Robinsohn, to help with virtually all his work. As one of Wolffsohn's biographers put it, Robinsohn became Wolffsohn's Boswell.[15]

Because Robinsohn was an excellent Hebraist, Wolffsohn also decided to take lessons in modern Hebrew from his new secretary, and soon Wolffsohn was speaking and writing the language with considerable competence. Wolffsohn had decided that it might be best to retire to Palestine, now that his work for the World Zionist Organization was nearly at an end. He was therefore eager to prepare himself to leave Germany for the new homeland, where he planned to devote the rest of his life to some sort of work program. It was there that he hoped at last to find, as he liked to put it, peace.[16]

[14] Laqueur 1976, p. 147. [15] Cohn 1944, p. 257.

[16] Robinsohn 1921, p. 110. For Robinsohn's comments on Wolffsohn's progress in learning Hebrew, which apparently was so good that it astonished one of his friends from Palestine, see p. 112.

In May of 1914, following his doctors' orders, Wolffsohn went to Switzerland, where he intended to spend two months vacationing in Lucerne and Hertenstein. Robinsohn accompanied Wolffsohn on this trip. Frequently they would talk of Wolffsohn's increasing desire to be in Palestine, where he had already bought land near Jaffa. He also discussed with Robinsohn his plan to move Herzl's remains there as well, and in his own last will prescribed that "I wish to lie at Herzl's side."[17]

Late in June Wolffsohn decided he was ready to make the final arrangements for his trip to Palestine. On the way back to Cologne, he stopped at Homburg for a checkup with his doctors at the sanatorium, but during a routine examination he suffered another severe attack and was immediately hospitalized.[18] As Wolffsohn's life progressively deteriorated, World War I broke out, he suffered another embolism, and finally pneumonia set in on September 15. He died within a day, on his wife's birthday.[19]

Subsequently, Robinsohn assumed responsibility for both the Herzl and Wolffsohn archives in 1915. By then he had also become an important figure in the work of the Jewish National Fund. As Wolffsohn's private secretary, Robinsohn had often been concerned with its activities. In fact, he devoted much of his own time to business conducted in the Fund's central office. One colleague described him as a "selfless, silent co-worker" who often worked late into the night for the Zionist cause. Above all, his knowledge of many languages was particularly useful to the main office, which had to deal with Jews dispersed across all of Europe. This caused endless difficulties for the Jewish National Fund, and Robinsohn was instrumental in resolving many of them. In addition to translating brochures and other items used by the Fund into Hebrew, Robinsohn also wrote materials for its various publications. Among the articles he contributed to *Eretz Israel*, the newspaper of the Jewish National Fund, was a piece on David Wolffsohn's parents for its first issue.

Of particular appeal to those working for the Jewish National Fund was Robinsohn's dream of using Hebrew literature to carry out one of his major projects—namely, through the Jewish education system and especially through study of the Hebrew language, to bring all Jews together; in particular, he wanted to bridge the gap between younger people in the East and the West. Considering the extent of Robinsohn's commitment to the Fund, it is no surprise that he was considered "irreplaceable."[20]

[17] Cohn 1944, pp. 270–271.

[18] Robinsohn 1921, p. 112.

[19] Cohn 1944, p. 271.

[20] "Unersetzlich ist uns Robinsohn beim Jüdischen Nationalfond, dem er durch Jahre ein stiller selbstloser Mitarbeiter gewesen ist." From p. 1 of a typescript draft obituary

At last, in charge of the Herzl and Wolffsohn literary estates, with a reliable position working for the Jewish National Fund and a sufficient income, Robinsohn decided it was time to marry. Through his interest in the socialist movement, he had met a young woman in Cologne who shared not only his deep love of literature, but his Zionist vision as well.[21] Early in 1916, Robinsohn married Hedwig Charlotte (Lotte) Bähr. Born in Oppenheim am Rhein, Germany, in 1888, she was the daughter of Jakob Bähr, and at the time she met Abraham Robinsohn, was working as a teacher in Cologne.[22]

In 1916 the couple's first son, Saul Benjamin, was born. The Robinsohns were only waiting for the end of World War I to realize their joint dream of immigrating to Palestine, where Robinsohn had been selected to be the first director after the war of the Jewish National Library in Jerusalem. As Chaim Tartakower put it, as soon as the hard times of conflict and material deprivation were over, Robinsohn seemed destined to make great and lasting contributions to the Jewish people.

Abraham Robinsohn's sudden death in Berlin of a heart attack on May 3, 1918, at the age of forty-one, came as a cruel shock to all who knew him.[23] It also precluded his completion of a biography of Wolffsohn on which he had been working, as well as other writings he had in manuscript at the time. Many of these still remain unpublished in the archives of the Jewish National and University Library in Jerusalem.[24]

Among those who wrote about Robinsohn at the time of his death,

---

dated May 1918. It may have appeared in *Erez Israel*, the official publication of the Jewish National Fund. The typescript is among Robinson's papers, YUL.

21 Pinczower I-1982.

22 Virtually nothing is known about her mother, except that she was still alive when Abraham Robinsohn died in 1918, since she is referred to—*meine Mutter war bei mir*—in a letter Lotte Robinsohn sent to Martin Buber and his wife on May 14, 1918. See Arc. Ms. Var. 350/622c, The Jewish National and University Library, Department of Manuscripts and Archives, Jerusalem, Israel.

Little information is available for any other members of the family except for her father, Jakob. There are, however, newspaper clippings among Robinson's papers describing the achievements of Dr. Bert Bahr of Grand Island, Nebraska. He was commander of the Disabled American Veterans of the World War chapter of Grand Island, and was active in lobbying for a federal government hospital to be built in Nebraska. Unfortunately, there is no information concerning his exact relationship to Hedwig Lotte Bähr, when he left Europe for the United States, or of any family he may have had.

23 Tartakower 1918, p. 5.

24 Nevertheless, the draft manuscript of Robinsohn's biography of Wolffsohn was edited to a modest 113 pages and published in 1921 with another fifty pages or so of Wolffsohn's correspondence with Herzl included as an appendix. Apparently much of the controversial material concerning disputes within the Zionist movement over Wolffsohn's leadership was removed before publication. In a brief preface, J. H. Kann

Abraham and Hedwig Robinsohn, with their elder son, Saul, 1916.

the Zionist educator, philosopher, and writer Martin Buber was the most prominent to reflect on his character.[25] Robinsohn and Buber had corresponded as early as 1915, when Buber was trying to solicit new writers for *Der Jude*, a monthly he headed as editor from 1916 until 1924. It was doubtless in hopes of securing Robinsohn's services as a contributor that Buber wrote to him about the magazine. Robinsohn replied in November of 1915, wishing Buber every success, but declining to promise any immediate contributions to *Der Jude* on a regular basis. Although he was appreciative of the difficulties attending the creation of any new publication, Robinsohn felt that he was overcommitted and, consequently, he was not prepared to collaborate as yet.

Several months later, Robinsohn wrote to Buber again about *Der Jude*,

---

noted that although Wolffsohn left few notes behind, *sein treuer Sekretär Dr. Robinsohn* carefully wrote out many spontaneous sayings and many conversations he had had with Wolffsohn. See Robinsohn 1921. Kann also wrote a brief obituary notice of Robinsohn's death for one of the Dutch newspapers, *Joodsche Wachter*, which appeared on June 21, 1918. Both the original clipping (in Dutch) and a German translation are preserved with Robinson's papers, YUL.

[25] Martin Buber also wrote an obituary of Robinsohn for the *Jüdische Rundschau*, 19 (May 10, 1918), p. 144. See also Renée Robinson to JWD, letter of February 16–17, 1990, RP/YUL.

and even submitted a short piece he thought Buber might want to consider. Robinsohn added that he was following the successful development of Buber's *Der Jude* with great interest, although because of his many responsibilities, he still could not consider participating regularly as Buber had wished. Robinsohn did report, however, that his own idea of establishing a journal to promote Hebrew among Jewish-German youth was beginning to take on definite shape in his mind. But even that, Robinsohn said, would have to wait, for he had no time at the moment to give the project greater direction or support.[26]

Clearly Buber and Robinsohn had much in common. The closeness of their kindred spirits and aspirations was reflected in Buber's comments about his friend when Robinsohn died in 1918:

> From the time I met him, I knew that he was a soulmate. . . . He was an intellectual man, not an "intellectual," but an intellectual *man*. . . . I got to know him late, and he, who spoke so little of himself, almost never spoke to me about his life. But I know that his youth was one of strife and struggle. . . . He had a Jewish soul and a Jewish destiny. And the innermost content of his spirit was Judaism. In life his face had a tense heaviness; when you looked at him, one knew of his struggles and his hardships.[27]

In 1918 the struggles and hardships for Abraham Robinsohn were at an end, although they were just beginning for the family he left behind.

Maternal Ancestry: The Bähr Family

At the time of her husband's unexpected death in May of 1918, Lotte Robinsohn was expecting her second child. Accordingly, she decided to return to Waldenburg where her father, Jakob Bähr, was a local teacher. There she could tend to her eldest son Saul, still an infant, while awaiting the arrival of the new baby. Her parents could also help care for the children while she tried to find work and plan for the children's future in the years immediately following World War I.

Lotte Bähr's father, born in Königsberg (East Prussia), spent his youth *zu Füssen von Grössen Israels* studying the Torah and Talmud. His teacher, the Rabbi Dr. Bamberger, insisted that in addition to examining the classical Jewish texts, he should also become familiar with "profane" literature as well. Although Jakob Bähr's family was a traditional and

[26] Robinsohn in a letter to Buber dated Berlin, July 23, 1916, Arc. Ms. Var. 350/621c, The Jewish National and University Library, Department of Manuscripts and Archives, Jerusalem, Israel.

[27] Recollections quoted from Martin Buber in the introduction by Hellmut Becker to Saul Robinsohn 1973, p. 7. Also contained in a draft obituary of Saul Robinsohn by Hellmut Becker, typescript, RP/YUL.

learned one, meaning that it was natural for him to plan on becoming a rabbi, the early death of his father required a change of plans. Instead, he studied for examinations that would permit him to pursue a career in education, and in 1880 he entered the Prussian civil service as a teacher. From 1883 to 1886 he attended the Jewish Teachers Academy in Berlin, after which he returned again to teaching.

As soon as Jakob Bähr passed his examinations and began to work, he married. Doubtless it was an arranged marriage, just as Abraham Robinsohn's parents had been matched by mutual agreement of their parents. At first, Jakob Bähr taught in Oppenheim (Rheinhessen), where his daughter, Hedwig Lotte, was born sometime in 1888.[28] Later the family moved to Mainz, then Reichenbach, and finally, in 1896, to Waldenburg.[29]

Hedwig Robinsohn's father, Jakob Bähr (photographed in Frankfurt am Main).

As a man of considerable learning, Jakob Bähr was well known as both a teacher and preacher. He was also a figure greatly respected in the local community for his knowledge of Jewish culture. Some called him rabbi, although lacking university training, he was officially a prediger. One obituary later praised him in particular for his great erudition and many contributions to the local community, remembering him as one of the "old guard of German Jewry."[30] As a strong advocate of Zionism, Jakob Bähr firmly believed that religious and national Judaism were inseparable. According to one newspaper account at the time of

[28] Unfortunately, many details about the Bähr family are entirely lacking. There is no record of the exact date of birth for Hedwig Lotte, nor any information about her mother's side of the family, including even her mother's name! One detail, of which Saul Robinson was always proud to relate, was her family's participation in the Democratic Revolution of 1848. Pinczower I-1993.

[29] These details of Jakob Bähr's career are contained in a one-page, single-spaced testimonial prepared for the occasion of his sixtieth birthday. It is signed Peritz, Königsberg. It was presumably delivered before a meeting of the Jewish Teachers Union. His sixtieth birthday was on February 17, 1920.

[30] The obituary appeared in one of the local papers, and a clipping is contained among Abraham Robinson's papers, YUL. Unfortunately, neither the newspaper nor the dateline appear to identify the source more precisely.

his death, he always found an appreciative audience in the Jewish press for which he often wrote on behalf of his Zionist ideals.

In addition to his devotion to the local Jewish community, Jakob Bähr was also head of the organization of Jewish teachers in Silesia.[31] Frequently he would lecture exuberantly about bettering their position. He almost never missed meetings of the various teachers groups organized in Silesia, including the Verein israelitischer Lehrer Schlesiens und Posens. He was a member of the union's board of directors, worked on behalf of its benevolent fund, and for a time served as its vice president. He also served as president of one of the trade unions of Jewish teachers of Silesia. Consequently, due in part to his untiring efforts, the status of Jewish teachers throughout the region progressively improved.[32]

Abraham Robinson: Early Years

It was in Waldenburg, a small coal mining town in Lower Silesia where the Bähr family lived, that Abraham Robinson was born on October 6, 1918. Doubtless the small family of grandparents, mother and two sons suffered the same sort of material deprivations reported by most Germans in the first years following the Great War. Spiritually, however, the Robinsohn-Bähr household was alive with ideas and ideals that must have been working to enrich Abby Robinson's earliest years. One advantage he always enjoyed was the example of his older brother, Saul. Abby would often accompany him to school or to lessons, helping to accelerate his early education.

Lotte Robinsohn's first concern after Abby was born was to find work to support the family. As early as August of 1918, Martin Buber had offered to do what he could, and he may have played an active part in helping her to find a job.[33] Realizing that there was nothing to expect in the way of employment in Waldenburg, she hoped at first that something satisfactory for her might turn up in Berlin, but nothing did.

[31] This was the Verein israelitischer Lehrer in Schlesien. Jakob Bähr was a member of the union's board of directors for thirty years, and president for the last ten years in which he served. There is a two-page, single-spaced typewritten transcript of what appears to have been a memorial tribute to Jakob Bähr delivered six weeks after his death by a former schoolmate and colleague of forty-six years.

[32] From an unsigned, two-page obituary of Jakob Bähr, presented to the Verein over which he had presided for ten years, RP/YUL.

[33] Shortly after Abraham Robinsohn's death, Buber and his son visited Isak Robinsohn in Vienna for some professional advice about Buber's son and his heart. Robinsohn examined the boy and reassured Buber that there was nothing congenital to worry about. Apparently Buber then offered to find a position for Lotte Robinsohn [*eine offizielle Anstellung*], but at the time she was preoccupied with the last months of her pregnancy (as related in a letter Isak Robinsohn wrote to Martin Buber on August 6, 1918). Lotte

In October of 1925, after Jakob Bähr retired from teaching, the family left Waldenburg for Breslau, the capital of Silesia. There Lotte Robinsohn managed to obtain a position as secretary for the Keren Hajessod, the Zionist organization set up to aid the emigration and settlement of Jews in Palestine.

Breslau had an established and substantial Jewish population. Richard Wagner, in his autobiography, recalled a concert at which the audience was almost entirely Jewish. With more than twenty thousand members, the Breslau Jewish community was the third-largest in Germany, second only to those in Berlin and Frankfurt. This was due in part to the great influx of Jews after 1918, when parts of Upper Silesia were ceded to Poland as a result of the partition worked out by the system of spoils following World War I.

Breslau was famous as the home of Robert Koch, the bacteriologist who discovered the bacilli responsible for tuberculosis and anthrax. It was also the home of one of the most significant educational centers of Jewish religious education: the Breslau Rabbinical Seminary, which succeeded in reconciling strict orthodoxy with its reform-minded opposition. This resulted in the "Breslau tradition"—combining an historically aware Judaism with the open-mindedness of new thinking. The seminary also reflected the cooperation between the different varieties of Judaism in Breslau. As a result, the German-Jewish youth movement, with its emphasis on teaching Hebrew as a living language and its interest in Zionism, was exceptionally strong.[34]

Breslau offered the Robinsohns a stimulating cultural life, one imbued with Jewish tradition, intellectual seriousness, and excellent educational opportunities. While Lotte Robinsohn was busy working for the Keren Hajessod to support her family, Saul and Abby spent most of their time in the care of their grandfather until his death on October

---

Robinsohn wrote to Martin Buber on the same day, August 6, 1918, thanking him for his condolences occasioned by her husband's death, and for Buber's promises to help. As she said, "meine Einkünfte sind derart, dass ich nicht nur zu meiner Befriedigung arbeiten darf, sondern die Erhaltung der Familie mir zum grössten Teil obliegt," Arc. Ms. Var. 350/621d, The Jewish National and University Library, Department of Manuscripts and Archives, Jerusalem, Israel.

[34] The Breslau tradition came to an end in 1938 when the New Synagogue was burned on November 9. As one writer put it, this was a tragic symbol, opening much too late the eyes of many to the fact that "the hour of departure for the homeland had fallen irrevocably." The Jewish Theological Seminary was also closed that year. Some Jews still managed to emigrate, but of the ten thousand or so who stayed in Breslau under the leadership of the last rabbi, Dr. Bamberger, few survived. After the war, the library and manuscripts of the Jewish Theological Seminary were dispersed to a number of different centers throughout Europe, including the Zidovske Institute in Warsaw and the Wrocław (Breslau) University Archives. See Milton 1985, p. 319.

Little Abby, *Abchen*, with toy bear in the summer of 1920. This was *Opas Lieblingsbild*—his grandfather's favorite portrait.

Abraham Robinson (first row, far right) with classmates at Rabbi Simonson's school, Breslau.

Mug shots: two boys, three poses, 1927.

15, 1928. Thereafter, the formal education of the two boys was largely in the hands of Rabbi Max Simonson, who had founded his own private school in Breslau to offer a religious, Jewish education rather than a secular, German one. At the end, just before World War II, the school had nearly one thousand students and exerted an enormous influence on the local Jewish community.[35]

Simonson, following Jakob Bähr's death in 1928, became something of a second father to the two Robinsohn boys. Often the family was invited on Friday evenings for dinner and lively conversation at the Simonson home. Because they had no children of their own, the Simonsons regarded Abby and Saul as adopted sons, and the rabbi took great care and pleasure in their early training. Although he was aware that both were talented, he nevertheless observed that "the big boy [Saul] was an extremely gifted child, but the little one [Abby] was a genius."[36]

Uncle Isak: Summers in Vienna

With her father's death in 1928 and her mother's, apparently sometime in the 1920s, Lotte Robinsohn had no close relatives left in Germany.

[35] Pinczower I-1993. Pinczower was also a student of Rabbi Simonson, in Breslau, and recalls the private lessons he had, along with Saul and Abby Robinsohn, before they all left Germany for Palestine in the 1930s.

[36] This according to an account of the rabbi's widow, Simonson R-1975. In particular, Mrs. Simonson recalled that Abby Robinson "war nie ein kleiner Junge, als achtjähriger war er schon ein Mensch. Mein Mann hat sich sehr viel mit ihnen abgegeben; wir hatten keine Kinder. Er hat mit ihnen Hebräisch gesprochen, denn sie konnten Hebräisch nicht bevor. Er hat Talmud mit ihnen studiert. Er hat sie sehr geliebt und sie sehr geschätzt." See also Seligman 1979, p. xv.

The two boys, January 24, 1922.

Abby and Saul with their grandmother, Weidling, Austria.

Her husband's parents, however, were still living with Robinsohn's older brother Isak in Vienna.

By now, Isak Robinsohn had become an internationally renowned surgeon and head of the X-ray Department of the Rothschild Hospital in Vienna. Above all, he was among the first to explore the ramifications of Wilhelm Röntgen's discovery of X-rays in 1895. Together with the renowned Professor Holzknecht, with whom he had studied, he established the first X-ray unit at the Allgemeines Krankenhaus in Vienna. Robinsohn went on to build the first X-ray machines for the examination of patients in the prone position, and invented several other useful devices, including the *trochoskop*. He was especially well known for applications in urology, published widely on new techniques using X-ray technology in dentistry, and demonstrated the value of chest X-rays in pulmonary diagnosis.[37]

Uncle Isak's comfortable estate at Weidling, in the Vienna woods near Klosterneuberg, offered a special haven for his two nephews who often spent a portion of their summers there. Among the most vivid memories of those Austrian summers, in fact, were steamer trips down

[37] From an obituary by Gottwald Schwarz, chief surgeon of the Elisabethspital, Vienna, in a newspaper clipping dated Thursday, May 19, 1932, but otherwise unidentified, RP/YUL.

Summer in Kolberg on the beach, 1927.

July 1931: Mother and two boys.

the Danube from Melk to Vienna through the scenic beauty of the Wachau.

When Robinson was only twelve years old, he made a list of all the vacations and trips he had ever made—from the first he could recall (to Berlin before his second birthday) to numerous outings with his school classmates from Breslau, most often to the nearby mountains. Hiking seems to have been a favorite sport, along with excursions with his friends to the Ostsee or the Kahlengebirge near Vienna. Between 1929 and 1933 he also enjoyed several trips with the youth group Brith Hano'ar schel Ze'irei Misrachi. From an early age, Abby Robinson was clearly destined to become an inveterate traveler.

At the Jewish High School in Breslau, Robinson made a strong impression thanks to his diligence and native abilities. Later, his diploma from the Höhere Jüdische Schule für Knaben und Mädchen noted his attendance as regular, with marks of "good" or "very good" in all fields except penmanship, which was only "satisfactory," as were his marks for music and physical education.[38] In addition to his schoolwork, however, he also wrote numerous short stories, poems, and plays. He even wrote a five-act comedy entitled *Aus einer Tierchronik*, and in July 1930 composed a group of poems especially for his mother.[39] In fact, through-

[38] RP/YUL. Although he was in school until the very day he left Germany in 1933, Robinson did not stay in Breslau long enough to graduate officially. Later, Rabbi Simonson sent the diploma to Palestine so that Abby could enter school in Tel Aviv. Pinczower I-1993.

[39] These were addressed *Zum Empfang meiner lieben Mama Lotte,* and like the stories

Abby with his mother. On the back he wrote to her, "'*Bist Du mir guuuut?*' Grüße, Abby."

Abby reading, December 30, 1930.

out his childhood Abby Robinson seems to have been a voracious reader, writer, and serious student.

Fleeing Nazi Germany: April 1933

Until the fateful year 1933, Breslau was one of the largest and most intellectually active Jewish communities in all of Europe. But when Adolf Hitler came to power officially on January 30, 1933, many Jews felt the time had come to emigrate, and for a Zionist like Lotte Robinsohn, the logical goal was Palestine. Following her father's death, Lotte Robinsohn had no compelling ties to keep her and the boys in Germany–and none to keep them in Europe except for Uncle Isak in Austria, who had died in May of 1932.[40]

Lotte Robinsohn may have been reassured about moving to Palestine knowing that two of her former schoolmates had recently emigrated and were already teaching there. Moreover, Isak Robinsohn's daughter Lizzie had married a former student of her father's named Druckmann, who had become a prominent radiologist at the Hadassah Hospital in Jerusalem. Even a rather distant family tie in Palestine must have been a comfort.

By the early part of 1933, political events in Germany were accelerating at an alarming rate. The election of March 5 gave the National Socialists 44 percent of the vote, and was promptly followed on March 23 with the Enabling Act, which gave Hitler dictatorial powers. On March 29 the Nazi Party announced a boycott of Jewish businesses

---

and plays from this period, are among Robinson's papers, YUL.

[40] Robinsohn was fifty-eight at the time of his death; according to Gottwald Schwarz, he suffered an unexpected stroke. See n. 37 above.

throughout Germany, scheduled for April 1. The next day, newspapers in Breslau reported that the passports of Jews were to be invalidated for foreign travel by order of the chief of police, Edmund Heines.[41] This development, of all the ominous signs mounting against Jews in Germany, seems to have convinced Lotte Robinsohn that this was the moment to leave, even if it meant abandoning everything they had in Breslau.

Lotte Robinsohn had been thinking about a trip to Palestine for some time. She had even purchased tickets for herself and Saul for a brief visit, having originally planned to leave Abby at home with friends, perhaps feeling that he was still too young to make such a long trip. She may also have intended to see Palestine for herself before making a final decision about going there for good. The sudden threat that their passports would be canceled persuaded her to take more decisive measures.[42]

After all, she had been prepared once before to immigrate to Palestine directly after World War I, and only tragic events had intervened to preclude that. Now there was evidence everywhere that the time had come to leave Germany for a new homeland for which they had all been working, not only by virtue of Lotte Robinsohn's efforts on behalf of the Keren Hajessod, but through savings that Abby and Saul had been accumulating since babies—money for their trip to Palestine.

Lotte Robinsohn's immediate concern was to get the two boys out of Germany as quickly as possible. She thought about sending both Abby and Saul to friends in Brno, Czechoslovakia, but then decided that Abby should go to Brno while she and Saul went on to Berlin. There she

[41] Abby Robinson, in a diary he kept during this period, mentioned Heines specifically. See Robinson R-1933. As for Edmund Heines (1897–1934), he joined the Nazi Party early in his career, and soon became a member of the inner circle of the SA, the *Sturmabteilung*, or so-called Storm Troopers, which constituted the early private army of the Nazi Party. In 1929 Heines was imprisoned for political assasination, but he was released shortly thereafter in a general amnesty (Robinson described him in his diary as the *Fememörder Heines*. See Robinson R-1933, p. 2). In 1933 Heines was appointed SA-Obergruppenfuehrer assigned to Silesia, where he was made commissioner of police at Breslau. It was there that he issued the order to invalidate Jewish passports, to which Robinson refers in his diary. On June 30 (1934) Hitler and a small entourage surprised a select group of Ernst Roehm's SA officers at the Hanselbauer Sanatorium in Bad Wiessee on the shore of the Tegernsee, where Heines was found in bed with a young man, his chauffeur. According to eye witnesses, Hitler ordered the "ruthless extermination of this pestilential tumor," and Heines was immediately shot. The incident was among the first in the so-called Blood Purge through which Hitler eliminated those he believed disloyal to him or to the ideals of the Nazi Party. See Tolstoy 1972.

[42] Apparently, a friend had written at the end of 1932, suggesting that Saul should get a scholarship to finish his high school studies in Tel Aviv. Lotte Robinsohn seems to have decided to follow through on this idea. Pinczower I-1993.

could settle their affairs before leaving for Palestine. Once this decision had been made, time was critical.

Saul and his mother gathered up what belongings they could and left Breslau for Berlin a little past midnight on March 31, 1933. Abby was to have left later that same morning for Czechoslovakia, where he would be out of Germany within hours and presumably safe from danger. Because of a misunderstanding about when he should arrive at the station, he was half an hour late and missed his train.[43] There was nothing else to do, as he notes in his diary, but try catching up with his mother and Saul. Waiting for the next train to Berlin was an ordeal, but at last, at 7:32 A.M. (as he noted in his diary) he was finally on his way to the German capital.

As he sat looking out of his train window, watching the familiar avenues of Breslau, the parks, even his own street and home pass by, he wondered when, if ever, he would see them again. Soon the train was speeding through the countryside, past forests, and Breslau was but a memory. He read a newspaper. The train stopped in Liegnitz long enough for a telegram to be sent to his mother in Berlin, alerting her of his arrival at Bahnhof Zoo at 12:07 P.M. It was nearly noon when the train passed by the burned out cupola of the Reichstag, a dramatic reminder of the rising political tensions in Germany.

Robinson was immediately impressed by the modernity of the city, and by the fact that some of the suburban train stations even had escalators! Although he was now fourteen, he had only been to Berlin twice before, once as a baby, and again when he was six, but then only for an overnight stay.

He was relieved to see his mother, who was waiting for him as the train arrived at the station. From there they took the U-Bahn to Schmargendorf, where the family spent the night with a friend of Lotte Robinsohn. The next morning, the Sabbath, was also the date set for the boycott of Jewish businesses throughout Germany. Abby went for a walk that morning in one of the western suburbs of Berlin and saw the posters in bold, red lettering: *Kauft nicht bei Juden!* Shop windows and doctors' signs were covered with anti-Semitic slogans. Otherwise, there seemed to be little activity and he continued his sightseeing without event. He made a point of visiting two churches, as well as a Turkish mosque. But upon returning, he learned that his mother's friend had just lost her job, another sign that life for Jews in Germany was becoming increasingly difficult.

Meanwhile, Lotte Robinsohn had been busy trying to secure a place

[43] Robinson's diary says it was because of a *'Hörfehler'* that he arrived at the station late. He was to have met a friend of his mother's there from Czechoslovakia; posing as the man's son, he was then to have proceeded to Brno. See Robinson R-1933, p. 3.

for Abby on the "tour" to Eretz-Israel. Although she already had tickets for herself and Saul, everything was sold out by the time Abby unexpectedly appeared in Berlin. Luckily, at the last minute, someone returned a single ticket.

Because of the British mandate, immigration to Palestine was not simple. Worst of all, there was no legal way to establish residency there as a refugee. The Robinsohns, however, *were* refugees, desperate like thousands of others for a new home. Legally, however, it was not even possible to enter Palestine unless one had ongoing travel arrangements. Consequently, many wishing to leave Germany for Palestine had to do so under the guise of tourism to the Holy Land.

The primary agency in Berlin helping to arrange the so-called tours was at 100 Friedrichsstraße, just under the Hauptbahnhof. The firm was associated with the Cunard line, and seems to have developed something of a specialty in managing the *Einbahnstraße* to Israel as it was called.[44] In fact, it was the center of such activity in Berlin in the early 1930s, and eventually came to enjoy something of a monopoly in helping to arrange trips to Palestine for large numbers of German Jews.

In March of 1933 the agency was offering a Mediterranean cruise on an Italian ship, the *Volcania*, including a stop in Haifa. The ship actually sailed from New York, but because of the depression, the *Volcania* steamed into Naples with ample room available for the Cunard "tour" organized in Berlin. Nearly three hundred passengers from all parts of Germany were booked in the group prepared to sail for Palestine from Naples, including the Robinsohns. Passports and exit visas had all been arranged, but at the last minute there suddenly arose the alarming possibility that everything would collapse.

Unexpectedly, the Bavarian government decided to require exit visas for anyone traveling through its territory. And so, at virtually the eleventh hour, nearly three hundred additional visas had to be negotiated for the overland connection from Berlin across Bavaria. At the top of the passenger list, Cunard's agent Henry Wellman put the name of the only gentile in the group, a liberal Lutheran pastor by the name of Reverend Maas, who was not only pro-Jewish, but pro-Zionist as well. The ploy succeeded, and a transit permit was issued for the entire group.

The Robinsohns, along with everyone else on the Cunard "tour," were

[44] I am indebted to Mr. Henry Wellman of New York City for much of the information that follows concerning arrangements made for those like the Robinsohns who wished to leave Germany for Palestine in the early 1930s. He was the person responsible for the Cunard group leaving Berlin to join the *Volcania* in Naples. From there it sailed for Haifa via Athens. The details described here were provided by Mr. Wellman during an interview in the spring of 1982 in his office at Rockefeller Center, and I am grateful for his help and willingness to cooperate in supplying information for this biography.

scheduled to leave Berlin for Naples on a train shortly before 9 P.M. The Anhalter station was congested, teeming with confusion. At first the Robinsohns boarded the wrong train, but eventually found the right track with their train bound for Italy. More confusion followed in trying to find an empty compartment.

As soon as everything seemed settled at last, Abby realized that he had lost his cap. By quickly retracing his steps, he was able to retrieve it. Along the way, he was delighted to discover that Fritz Stein, one of the teachers from the Jewish High School in Breslau, was also on the train. Stein was immediately invited to join the Robinsohns. They all exchanged stories of the day's adventures until the train stopped in Halle an hour and a half later and Lotte Robinsohn decided it was time to sleep.

When Abby awoke the next morning, the train had reached Augsburg; soon they were in Munich. An hour's wait before the train continued on to Italy left time for a quick visit to the Palace of Justice, not far from the train station. Abby then thought about seeing the Propylaeum. He asked a local Bavarian how best to get there, and recorded verbatim the heavy local accent which fascinated him in his diary.[45]

By the time Abby returned to the station, the train was about to leave and great consternation had arisen over his absence. As soon as the family was again reassembled, they all clambered back onto the train and the journey continued, south to Rome. What impressed Abby now was the countryside. He liked the farmhouses of Bavaria (which reminded him of Swiss cuckoo clocks), and found the Alps even more impressive than he had expected from pictures. He was keen on examining them with a small telescope he had brought with him, and described them as protective guardians watching over those below.

The train stopped for some time at Innsbruck, then continued on to the Brenner Pass. At the first station in Italy, Abby was immediately aware of the Fascist soldiers assembled on the platforms. He found the Dolomites and South Tyrolean Alps more intimidating than those in Bavaria, but continued to watch the landscape until night fell and it was too dark to see. At some point an Italian professor who was trying to learn German boarded the train, and he was pleased to find that Abby and his family were happy to provide uninterrupted instruction well into the night.

In Brenner the family had to change their seats because the third-class part of the train only went as far as Bologna. Upon finding all of the second-class sections fully occupied, they were treated to "complimentary" seats in first class, which Abby seemed to like very much. He

[45] *Am best'n is' aber, wenn'S' d'Straß' da in d'r Mitt'n geh'n, da komm's gradaus hin!* See Robinson R-1933, p. 13.

had successfully negotiated this transaction with the conductor, who fortunately spoke English. Even at age fourteen, Abby already exhibited an exceptional capacity for foreign languages, a skill that was to serve him well throughout his life.

The next day, Abby was disappointed to find heavy fog and an overcast sky. Not realizing that mornings in Latium are frequently hazy, he complained that it should have been warm and sunny in the land where lemons grew. But the haze soon gave way to sun, and Rome was more than obliging; by midafternoon he found himself wishing for the cooler weather they had enjoyed in Bavaria. When they visited the baths of Diocletian, Abby could not resist joking that had it been possible, they would all have gladly taken a bath.

Abby took a special interest in the Italian language, and his diary records both his approach to learning it as well as the quick progress he made. For example, the first three words he learned were *ritirata, libero,* and *ocupato*.[46] Their pragmatic value cannot be denied, as any visitor to Italy will attest. Soon he had mastered *Si Signore, Quanto costa?*, and *Quando parto ille treno da Napoli?* He was pleased to discover that it was possible to build many Italian words simply by adding an "o" to the endings of French words he already knew. Even if he was not grammatically accurate, he could usually make his point, and that was all that mattered. His friend Fritz told him about *prego* and *gracía* and explained that *mille gracía* was especially polite. If he wished to be even more erudite, he should say *octo tscenti mille gracía*.[47]

On occasion, when his newly acquired Italian failed him, Abby quickly discovered that either French or English often worked. The only problem he noticed was a big difference between the French spoken by Germans as opposed to that of the Italians. Having asked a policeman how to get to the train station, the answer came out something like *A la Stazion? Si, Signore, voi allez cet cammino. Bueno jour*. Only after walking for ten minutes in the wrong direction did he realize that the policeman had been trying to speak French! Despite momentary difficulties of communicating, Abby found that talking to people was a better strategy than trying to use a dictionary. Dictionaries, he quipped, only contained words that "one never needed."[48]

Rome, the *ewige Stadt,* impressed Abby greatly. He thought it odd,

[46] The spellings here are transcribed *exactly* as they appear in Robinson's diary. They clearly reflect a German sense of orthography and the way in which Robinson's ear heard spoken Italian. I have not corrected mistakes, nor have I followed each misspelling with an intrusive *sic*. See Robinson R-1933, p. 11.

[47] Eighty, he claims, was the only other number he knew in Italian! See Robinson R-1933, p. 12.

[48] Robinson R-1933, p. 23.

however, that ancient ruins suddenly appeared sandwiched between or incorporated into ordinary buildings. On the other hand, he was greatly taken by the Colosseum, which he described as greater than many a mountain, yet situated in the middle of the city. The Italians themselves seemed *nett und hilfsbereit,* and local officials correct and pleasant. Whenever one was lost, there always seemed to be a policeman ready to give helpful advice. He noted that there were as many people selling postcards as there seemed to be ruins; the only difference was that one went to the ruins and was pleased to see them, while the peddlers went to the tourists and were pleased to see them—if they bought something! Abby was also fascinated to discover that among the street sellers a number were Jews. They always said "Shalom!" as soon as they met another Jew—hoping, Abby thought, that the tourist would be so pleased that he would buy something immediately.

Above all, Abby liked the pastries in Rome, which he found went well with *espresso*. And nothing seemed to cost very much. Not everything, however, caught his enthusiasm. In contrast to the cleanliness of the *latterias* were the butcher shops, he said, which produced the worst smells he could remember. Most offensive, he found, were the salamis hanging outside.

Abby regarded Rome as having something exotic, oriental about it. This was due in part to the city's many small side streets and alleyways, made all the more congested by the donkey drivers transporting their wares. Washing was hung out to dry on lines strung between buildings, a sight that especially caught Abby's imagination. Rome and its daily life seemed idyllic and of great beauty, as he wrote in his diary. The one object he did not like in Rome was the immense Victor Emmanuel Monument—*schrecklich* was the word he used, objecting to anything so ornate and overdone in the middle of the city.

Not far from the Piazza Venezia Abby and his entourage went on to visit the Jewish ghetto, the Pyramid of Cestus, and St. Paul's outside the walls—all stops on the typical tourist itinerary of the city. Abby found the simplicity of St. Paul's and its wonderful mosaics more to his liking than St. Peter's. The catacombs moved him deeply, especially the inscriptions of the early Christians, who Abby thought had been imprisoned there under the Emperor Nero's orders.[49]

From the Janiculum they could see the Colosseum, where later in the day Abby was intrigued by the tunnels through which wild animals were said to have been brought into the arena. He also wrote in his diary that the Colosseum had originally been built by twelve thousand

[49] Abby may have picked up this bit of misinformation from a guide at the catacombs, or from one of the guidebooks he may have been using.

captive Israelites as a place for nautical games. Gladiatorial combats, he added, only came later, during which criminals or prisoners were sometimes thrown to the wild animals. Abby finished his thoughts on the Colosseum by noting the Venerable Bede's prophecy that until the Colosseum collapsed, Rome would continue to exist. That anything could ever happen to the Colosseum, however, seemed inconceivable.

The next morning, there was time to visit the Vatican with Fritz. Abby was delighted by the *braccia* of the Bernini colonnade, by its ellipse of columns, and, as his Baedeker guidebook explained, by the two Bernini fountains at each of the foci of the ellipse. Upon entering St. Peter's, he exclaimed that the entry portico seemed as large as most churches. Abby was naturally impressed by the immensity of the building, the richness of its decoration, the works of art, the crypt below the church, and, of course, by the cupola above. Later that afternoon in the piazza atop the Pincio, he commented on the pagan obelisk with its crucifix on top. Despite his being in the most Christian of Christian cities, "Who ever thought of that?" he wondered. He could not understand what the one had to do with the other.

At 5:00 P.M. the Robinsohns were back on the train headed for Naples. Along the way they passed the newly finished town of Littoria, which the Fascists had built on an earlier swampland. Abby likened this feat to what the Jews had accomplished in transforming Eretz-Israel from a desert wasteland into a land of promise.

It was well after nightfall when their train finally reached Naples. Deciding upon a hotel was not easy. The first they came to was too elegant, but as they paused in front of it, someone rushed up to recommend a different hotel. Before they knew it, they were being hustled to a seedy, run-down establishment, the Livorno. The Robinsohns were shown a large room with three beds, not freshly made, and without a washbasin. Everything, to Abby's eye, seemed to be falling apart. Lotte Robinsohn, completely dissatisfied, went out to find another *albergo*. Meanwhile, the porter from the train arrived with their bags, and a great hullabaloo ensued with the manager of the hotel, who threatened to call the police if they left. After bribing him with a few lire, however, the Robinsohns eventually moved to the hotel Cavour, which apparently satisfied everyone—it even had an elevator and a restaurant.

Once the question of hotel had been settled, Fritz disappeared with a friend and did not return until after 1:00 a.m. Since they were staying in the same room with Abby and Saul, they had to wake the boys up in order to get in. To their chagrin, both of the boys were so soundly asleep that even loud banging on the door and repeated telephone calls from the concierge did nothing to rouse them! Finally, after much commotion, Saul heard the noise and got up to open the door. Abby, dead to the world, slept through it all, and learned what had happened only the next morning. Why the two men had been out so late, Abby was not sure, but he was unconvinced by their explanation that they had simply gone out to a café and forgotten about the time. Later, Abby jokingly threatened to write home to all the relatives about the goings-on that night.

If one could measure Rome by its postcard peddlers, it was the street urchins that gave Abby a yardstick for Naples. Fritz had decided to take a photograph of the dirtiest and most bedraggled example he could find, but every time he took a picture, another child would turn up looking even worse. On the morning they were to leave Naples, the Robinsohns were having their breakfast in the hotel restaurant when an old woman hoping to sell them something came up to the window. When they expressed no interest, she eventually left. Then one of Naples' famous street urchins appeared, begging for some bread which they gave him. To their amazement, he was soon back with the same old woman who had approached them earlier. Now she wanted something too, and so they felt obliged to give her a piece of bread as well. By then, word was out. Suddenly two old women with two dirty children appeared, also asking for something to eat. This doubling process, Robinson noted, continued as they left their hotel, where a crowd of eight children and three adults had now gathered and began to follow them. Only four blocks later were the Robinsohns able to escape them.

They walked to the waterfront, where Abby was immediately struck by the deep, shimmering blue of the Bay of Naples. Vesuvius rose majestically in the distance, a little cloud of white steam rising from its crater. Their next stop was the Naples Aquarium, which provided an hour of pleasant diversion. The starfish and sea horses, as well as the sharks, drew Abby's attention. He was also fascinated by the electric rays. These were in open tanks and could therefore be touched to the delight—and shock—of visitors who wanted to put their hands into the water.

The *Volcania*

At 4:00 p.m. the *Volcania,* already in port, awaited the embarkation of her passengers. To Abby she was a majestic sight. No sooner were they aboard and shown to their rooms than Abby was up on C-Deck. As a third-class passenger, he was restricted to lower decks, but soon he was well acquainted with every passageway and promenade of the ship. While orienting himself, Abby met two boys and a girl about his own age who turned out to be the children of Dr. Hermann Badt. Badt was a former Social Democrat who, as Abby knew, had lost his position in Prussia due to the rise of National Socialism and the ensuing political changes in Germany.[50]

As the *Volcania* departed at 6:00 P.M. it was already dark, and the lights of Naples receded along with the shadowy outline of Vesuvius, barely visible in the distance as night fell. The next morning Abby was on deck with his telescope, the Straits of Messina already past and land close at hand—Calabria.

For four days the Robinsohns enjoyed the lazy life aboard the *Volcania*. Abby was proud to be on a ship of such size, 24,000 tons, no "tin can" or "leaky crate," as he said, but a thoroughly seagoing vessel. To be seasick on the *Volcania* would have been a real feat, he noted, because it moved so smoothly in the water. It was easy to forget that one was not on dry land.

On Thursday, the Peloponnesus loomed off the port side of the ship, but then disappeared within a few hours. Work on the Corinth Canal prevented the *Volcania* from passing through it, and so the ship took the longer route around the Peloponnesus toward the Piraeus and Athens. That evening Abby wandered above decks and enjoyed the breeze and sound of the sea. He sat in a deck chair and watched the stars overhead as he listened to the renowned cantor Yossele Rosenblatt, who was singing in the second class salon for the benefit of the Jewish National Fund. Rosenblatt was a world-class tenor from the United States who was on his way to make a Yiddish film, *The Dream of My People*, in Palestine. A heart attack unfortunately cut this venture short; Rosenblatt died three months later and was buried in Jerusalem on the Mount of Olives.[51]

[50] Dr. Hermann Badt (1887–1946), born in Breslau, was the son of a classical scholar, Benno Badt. As a young man he joined the Mizrachi Party, and in 1919 was the first Jew in Prussia to be admitted to the civil service after the Revolution of 1918. From 1922–1926 he was a Social Democratic member of the Prussian Diet, and then became the highest ranking Jewish civil servant in Germany. As Ministerialdirektor in the Prussian Ministry of the Interior, he was in charge of constitutional affairs. In 1932 he represented Prussia before the German Reich in an unsuccessful legal action against Chancellor von Papen, who had deposed the legal government and instituted himself as a dictatorial Reichskommissar in Prussia.

Badt was active in the Zionist movement, and visited Palestine several times before deciding to settle there. Due to the mounting attacks to which he was exposed in Prussia as a Jew and a Zionist, he finally decided to leave with his family in 1933, and thus was on the *Volcania* for the same reasons as were the Robinsohns. Badt founded the Kinneret Company to promote middle-class settlement in Palestine on the land where kibbutz Ein Gev was later founded. See Brecht 1967 and Braun 1940.

[51] Rosenblatt was one of the most popular and internationally renowned *ḥazzanim* (the cantors who officiate in a synagogue) of his day. Born in Belaya Tserkov, Russia, in 1882, he toured Eastern Europe as a child prodigy, conducting synagogue services with his father. In 1912 he was hired by the Hungarian congregation, Ohab Zedek, in New

On the Acropolis, Athens, Greece, April 1933.

As the *Volcania* approached Athens, most passengers were on deck to watch the island of Salamis on the left. It was there, Abby mused, that the king of the Persians must have sat as he watched the advance and eventual defeat of his naval flotilla. Because of its size, the *Volcania* could not actually dock at Piraeus, but had to drop anchor in the harbor. Launches ferried passengers to shore, where the Robinsohns spent the day seeing the sights and sites of Athens.

Any faith they might have had in the honesty of the Greeks, Abby reported, was soon lost when someone offered them postcards at a very low price. The deal included stamps, which they also bought. Only later did they learn that Fritz had purchased the same stamps at a much lower price. At the post office they were dismayed to discover that they had actually paid twice as much as they should have.

The Robinsohns took a trolley car to the foot of the Acropolis. The temple of Theseus (Hephaestus) did not impress Abby; he described it as looking like a junk room on the inside. The Acropolis itself was a different matter. Abby was the first to reach the top, and while he was waiting for the others, one of the guides shouted, "Shalom!" This immediately sparked a conversation in English. Soon the Robinsohns were proceeding through the Propylaeum to the Parthenon and then the Erechtheum. Abby was amazed at how well preserved the caryatids were, and was dazzled by the magical combination of blue sky against the white marble of the monuments. There is a particularly poignant photograph of the family taken in front of the Erechtheum, its caryatids standing expressionless in the background. As Abby said, it was hard to tear himself away from the magnificence of the build-

York City to serve as its *ḥazzan.* With a tenor voice of two-and-one-half octaves and an agile falsetto, he ranked among the most brilliant of coloraturas. He was equated with the great operatic tenors of his day, composed hundreds of liturgical melodies, commanded enormous salaries, and received high fees for High Holy Day services. In 1928 his voice was heard in the first full sound film, Al Jolson's *The Jazz Singer.* Through extensive concert tours and numerous recordings, he became nearly as famous as Caruso. For details consult Rosenblatt 1954 and Goldstein 1940, p. 335.

ings, and he never forgot the impression that they made on him as an ensemble on the Acropolis.

Back on board ship, the last Sabbath before reaching Palestine was a particularly moving one. Aware of the significance of this trip for many of the passengers, their cabin steward surprised the Robinsohns with a garbled Hebrew-German greeting: *Halleluya, Palestina–frei!* [52]

Abby spent the afternoon above decks, as far front as he could, looking out to the horizon. There lay Palestine, the land of Israel, the land for which he had worked and saved, for which he had longed and dreamed ever since he could remember having heard of it as a small child. He stood there by himself, alone, staring ahead, trying to imagine what his new home, the promised land, would be like.

The next morning Abby was back at this same post, but now hundreds of fellow passengers were swarming on the decks of the *Volcania.* The sun had just risen over Palestine, and he could see the Lebanon and Galilee.

It was Sunday morning, April 9, 1933. As the *Volcania* finally steamed into Haifa, virtually everyone aboard had but one thought. As they looked out toward Palestine and their new home, they were all united, Abby believed, "through one great love."[53] With a strange feeling in his throat, he went down to find his mother and Saul. As the three of them made their way to the upper decks for disembarkation, the ship was already in the harbor. Mount Carmel could be seen, rising above the small port town of Haifa. The blue-and-white Zionist flag fluttered from the ship's mast. The anchor chains, Abby recalled, slipped slowly from their windlasses, as if not wanting to disturb the peace that everyone felt.

Behind him, in front of him, all around him people were singing the *Ha–tikvah.*[54] As Abby closed his diary, a new chapter in his life was about to open. The last words in his account of the voyage, in fact, were from the anthem that everyone was singing: "*Od lo avedah tikvatenu, Ha-tikvah ha-noshanah, La shuv le-erez avoteinu . . .*"[55]

After discharging most of its passengers, the *Volcania* sailed on to Port Said with only a handful of Americans. They probably had little idea of what was about to happen to their recent copassengers, as many of them sought to begin new lives in a strange yet familiar land—Palestine.

[52] Robinson R-1933, p. 47.

[53] *Vereint waren wir alle in einer großen Liebe,* Robinson R-1933, p. 48.

[54] "The Hope," the national anthem of the State of Israel.

[55] "Our hope is not yet lost, the age–old hope, to return to the land of our fathers. . ." The *Ha-tikvah* was based upon a Polish patriot's song which became the national anthem of the Polish Republic: "Poland is not lost yet, while we still live." Although it was sung at the early Zionist Congresses, the *Ha-tikvah* was only made the official Zionist anthem by the Eighteenth Zionist Congress meeting in Prague in 1933.

CHAPTER TWO

Life in Palestine: 1933–1939

> Jeruschalajim, Stadt aus Gold,
> Aus Kupfer und aus lichtem Schein,
> Oh, könnt' für all' Deine Lieder
> Ich Harfe sein.
>
> —*Das Goldene Jerusalem*[1]

DESPITE THE euphoria the Robinsohns must have felt at having reached the promised land, it was clear from the beginning that life in Palestine would not be easy. From Haifa, where Lotte Robinsohn and her two sons had disembarked in early April 1933, the family moved to Tel Aviv, where they settled shortly after their arrival in Palestine.[2]

TEL AVIV

Tel Aviv, in the 1930s, never failed to make an impression. Arthur Koestler, correspondent and author living in Palestine at the time, viewed it with a rather jaundiced eye:

> There was a main street named after Dr. Herzl with two rows of exquisitely ugly houses, each of which gave the impression of an orphanage or police barracks. There was also a multitude of dingy shops, most of which sold lemonade, buttons and flypaper.[3]

The Robinsohns, however, saw Tel Aviv—and indeed all of Palestine—quite differently. Like their fellow passenger on the *Volcania*, Yossele Rosenblatt, the Robinsohns felt a tangible euphoria upon reaching Palestine. Rosenblatt, for example, had no sooner landed in Haifa than he

[1] Translated by Saul Robinsohn from the Hebrew, song by Naomi Shemer, with philological and historical commentary. Robinsohn 1969. Quoted in Becker 1972, p. 4, and Becker 1973, p. 435.

[2] Lotte Robinsohn already knew two women in Jerusalem, teachers she had met in Köln before she was married. Because of her husband's prominence and her own contacts in Palestine, especially her connections with the Keren Hajessod (United Israel Appeal) and other Jewish agencies, the Robinsohns could count on considerable help, especially from prominent people, as they began the task of resettling in Palestine. Renée Robinson to JWD, letter of November 10, 1987, RP/YUL. Details presented here of what the Robinsohns faced are also based to a large extent upon the recollections of Miriam Hermann and Eliezer Pinczower. See transcripts of Hermann R-1975 and Pinczower R-1977.

[3] Koestler 1980, p. 45.

felt as if an oppressive weight had been lifted from him. Tel Aviv seemed a dream come true:

> Can you believe it? It is really Tel Aviv that we are in, our own Jewish Tel Aviv, built by Jewish hands, run by a Jewish mayor, guarded by Jewish policemen, everything as it was foreseen by our prophets.[4]

Rosenblatt was even more emphatic about the beauty, vitality, and promise of the land itself. "You wouldn't recognize it, Shmuli," he wrote to his son, adding that Palestine was "veritably a 'land flowing with milk and honey,' possessing all of the virtues of the modern countries of the world with very few of their defects."[5]

The Robinsohns, however, were not in Palestine as celebrities. Despite their commitment to the idea of Eretz-Israel and the promise of a new life, they initially had little to rely on in the way of material assets. To begin their new life in the midst of an economic depression and the escalating political disorder sweeping across Europe was a formidable task on any terms.

Nor were the Robinsohns on favorable grounds legally. Having entered Palestine as "tourists" with ongoing tickets to Trieste that were never used, they remained in Tel Aviv without legitimate immigration papers. Despite British efforts to deport illegal aliens, the local Jewish community had developed means to help those who wished to stay. Lotte Robinsohn, well connected as she was, fortunately succeeded some years later in obtaining the documents she and her children needed to maintain permanent residency as citizens of Palestine.[6]

Meanwhile, to support the family, Lotte Robinsohn found rooms in Tel Aviv on Rothschild Boulevard and opened a pension, which she operated with the help of another woman, Grete Ascher.[7] During these years life was financially difficult, but Lotte Robinsohn succeeded in carefully managing what resources she had. Despite what must have been

[4] Yossele Rosenblatt, quoted in the biography by his son, Samuel Rosenblatt. See Rosenblatt 1954, p. 342.

[5] Rosenblatt 1954, pp. 342–343.

[6] On August 29, 1935, Lotte Robinsohn was officially granted a certificate of naturalization, for which she had applied on March 3. Accompanying the application is a wistful photograph of Hedwig Robinson; the certificate, no. 25136, is among Robinson's papers, YUL. For a description of what immigrants to Palestine like the Robinsohns faced in Palestine, see Bentwich 1936.

[7] Eliezer Pinczower surmises that Lotte Robinsohn must have sent some money out of Germany via her relatives in Vienna in order to have enough capital on hand to open the pension. Despite the fact that exporting any funds from Germany was illegal, Pinczower recalls that Lotte Robinsohn had good connections through German Zionists in Palestine. The pension was quite successful because many prospective settlers first went as tourists to "look around," and then decided whether or not to emigrate. Pinczower R-1977, I-1982, and I-1993. See also Seligman 1979, p. xii.

Mother and sons.

a meager income, the two boys never lacked for anything and were always dressed as "little princes."[8] The impression, even then, was that Lotte Robinsohn believed each of her boys was destined for something special.

Soon after the family was settled in Tel Aviv their old friend and the boys' former teacher, Rabbi Simonson, came from Breslau with his wife to see Palestine for themselves. Lotte Robinsohn had reserved the best room for them in her pension, which Mrs. Simonson found comfortable and very pleasant. Palestine in general must have impressed them favorably, for after her husband's death in 1936, Mrs. Simonson decided that she too should emigrate. Once settled, she frequently saw the Robinsohns. Just as in Breslau, Abby continued to impress her—particularly because of his seriousness. "He was never a child," she thought, "at the age of eight he was already an adult."[9]

Several years later, after Mrs. Simonson had begun to work at the Hadassah Hospital in Jerusalem, her original impression of Abby remained unchanged. Upon meeting him again with his mother, Saul on one arm, Abby on the other, Mrs. Simonson automatically said to Abby, "Shalom, Herr Professor."[10] So taken was she with Abby's mind, she

[8] Pinczower I-1982. The word he actually used was *Fürsten*. Frau Robinsohn, he noted, instilled a feeling in the boys that *wir sind für etwas Besonderes bestimmt*. One thing Robinson later appreciated as an adult, in fact, was tailored clothing. Although he usually dressed casually, he also had a fine wardrobe of handmade suits and shoes ready for special occasions. In Palestine, however, dressing two children well was a great effort, but one Lotte Robinsohn was determined to make.

[9] Simonson R-1975. See also Pinczower I-1993.

[10] Simonson R-1975. Robinson was sixteen or seventeen at the time. Mrs. Simonson also described Robinson's appearance, remarking that "Er hatte einen besonderen Kopf, schon als Junge. Das war ein besonderes Gesicht. Es kann sein, dass Shaul der Schönere war, aber Abby's geprägtes Gesicht, wer das einmal gesehen hat, das war was Ungewöhnliches."

recalled, that the greeting seemed entirely natural. It was absolutely instinctive—and unintentionally prophetic.[11]

Jerusalem

Jerusalem has always been a city of history, heterogeneous in its populations, a crucible of religions, races, customs, and languages. Saul moved there in 1934, as soon as he received his *Abitur* (or high school diploma), to continue his education at the Hebrew University. There he studied both general and Jewish history with Richard Köbner and Jizchak Baer, as well as philosophy, sociology, and pedagogy.[12]

Within the year Abby and his mother moved to Jerusalem too. Not only did Lotte Robinsohn have more friends there, but it was possible to live less expensively than in Tel Aviv. More important, she and Abby could also be closer to Saul as he began his university studies. In addition, there was also the matter of Lotte Robinsohn's health. By 1934 she had developed a disabling thyroid condition, and had to give up hard work.[13]

At first she and Abby stayed with Dr. Abraham Bartana and his wife. Bartana was vice director and later director of the Rehavia Secondary School where Abby enrolled to finish his last year of high school in Jerusalem. The Bartanas were good friends of the Robinsohns; they were able to live harmoniously together in the same apartment as a close-knit family, celebrating holidays together and inviting mutual friends, like Eliezer Pinczower, to join them for special occasions such as Purim and the Seder at Passover. Later, however, Abby and Lotte Robinsohn moved to a house of their own on Gaza Road, and thereafter to a neighboring building on what is now ha-Palmaḥ Street.[14] Although for a time she

[11] At the time, Abby had no idea that scholarship—let alone mathematics—was destined to be his career. His gift for mathematics only became clear during his last year at the local gymnasium in Jerusalem. His surpassing talent, however, was discovered by Abraham Fraenkel soon after Abby had enrolled at the Hebrew University. Fraenkel had a "sixth sense" for mathematical ability and recognized a similar gift for mathematics in Michael Rabin, who also immigrated as a child to Palestine from Breslau. For details, see Rabin R-1976.

[12] Saul passed his M.A. examinations in 1940, qualifying thereby for a teacher's certificate. See Goldschmidt 1972, p. 82. Goldschmidt reports that Robinsohn's M.A. thesis, "The Israeli State as Model in the Writings of Political Thinkers of the 16th, 17th and 18th Centuries" [Biblicism in Political Theory from Machiavelli to Adam Müller, in Hebrew] is to appear in a memorial volume to be published by the Hebrew University.

[13] Lotte Robinsohn apparently sold her interest in the pension in Tel Aviv to Grete Ascher, who continued to run it with two of her friends until 1960. Presumably, sale of her share in the pension gave Lotte Robinsohn sufficient funds on which to live in Jerusalem. Pinczower I-1982.

[14] Later Bartana became director of the Ministry of Education. Wherever the

took on a position as director of the dormitory at the Hadassah Nursing School, Lotte Robinsohn had to give this up after only half a year for reasons of poor health. To help make ends meet, Abby, like his brother Saul, began to give private lessons on any subject he could.[15]

Rehavia Secondary School

Soon after their arrival in Jerusalem, Abby enrolled in the Rehavia Secondary School, the second oldest Jewish high school in Palestine. There he found himself in a lively class of students about to finish their last year of high school, and the marks of his strict, thorough, German education were clear to everyone. As soon as they got to know him, his classmates immediately recognized Abby as an outstanding student, but also one of principle and remarkable maturity. As one of his classmates admitted:

> We were a gang of young rebels and the whole atmosphere at that time was against study. One considered it unimportant. Learning was nothing. One just did it so as not to make our parents too angry. The country was in ferment—it was in the thirties—there was constant fighting with the Arabs even then. It was the great Arabic rebellion against the mandate, against Zionism.[16]

Abby was a prominent newcomer in part because he spoke Hebrew better than most of his classmates. Although he was apparently well liked, he often went his own way, unlike most of the others who ran in groups and followed the views of the majority.

Peer pressure ran high. For example, when the time came for final examinations, it was decided that they should be stolen and copies dis-

---

Robinsohns stayed, in Tel Aviv or Jerusalem, Pinczower recalls that they always lived in the best neighborhoods, and in Jerusalem this meant either Rehavia or Kiryat Shmuël. See Pinczower R-1977.

[15] According to Pinczower, both of the boys gave Hebrew lessons, a measure of how gifted they were in languages since neither was a native speaker. Hebrew was regarded by young people at the time as an extremely important part of their life in Palestine. It was not only a symbol of their culture, but a clear sign that they were pioneers, even "creators and parents" of the language, as Pinczower put it. For some the attention they paid to Hebrew was a conscious and deep experience.

As for Saul's life at the university, Pinczower recalls that he lived very modestly, on extremely small means. Typically, they never spoke of their financial problems. Both of the boys knew that it was up to them to find the necessary resources to survive and to support their mother at the same time as best they could. Although Pinczower recalls that Lotte Robinsohn had a brother who owned a farm in Pardess Hanna, he apparently had no money to spare for his sister. Pinczower I-1982.

[16] Hermann R-1975.

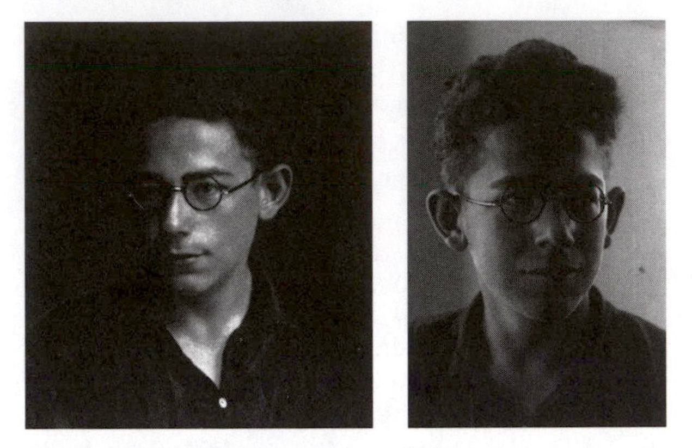

Abraham Robinson, Jerusalem, 1935.

tributed. Anyone with special abilities was expected to contribute answers, and it was assumed that Abby would help with mathematics. When he refused, as Miriam Hermann recalls:

> We were absolutely flabbergasted. It did not occur to us that anybody in his right mind would refuse to do what the whole class did. Abby was very pleasant and very firm and very quiet about it, but he said that he wouldn't—it was against his principles.... He was ostracized and we were willing to say the most awful things about him and it didn't seem to sway him in the least.... I remember for weeks on end when we passed the exams successfully with the aid of this fraud, we couldn't get over the fact that this boy would not do as we did.... I knew that deep in our hearts we all admired him terribly for it—for his courage, his integrity, for his consistency.[17]

In 1936, as Robinson continued his studies at the Hebrew University, he was a modest and reserved young man, but with a quiet sense of humor that was one of his most charming qualities.[18] Saul was the extrovert, Abby the introvert; everyone knew Saul, but fewer knew Abby. Thus his impressive development as a student and his later recognition as a mathematician of international standing were all the more remarkable and unanticipated. Nevertheless, the first evidence of his mathematical interests dates from this time, namely a set of notes in German that he kept on properties of conics.[19]

[17] Hermann R-1975.

[18] At home, his brother and mother usually referred to him by the affectionate diminutive *Abchen*. The words Eliezer Pinczower used to describe Robinson at that time were *bescheiden und zurückhaltend*. Pinczower R-1977 and I-1982.

[19] George Seligman describes these as "carefully written"; there are also notes among Robinson's papers in Hebrew, dated 1936, on the geometrical optics of lenses with surfaces generated by conic sections. See Seligman 1979, p. xii.

The Haganah

Tensions between Arabs and Jews mounted in the 1930s as more and more Jews arrived, primarily from Europe, to settle in Eretz-Israel. On April 19, 1936, Arab rioting was widespread and severe throughout Palestine, but especially in Jerusalem. In response, Saul joined the Haganah, the illegal organization for the Jewish defense of Palestine. From 1936 onward there was continuous trouble, and the Haganah developed more and more into a military force. This was felt to be necessary since the English were not prepared (or willing) to intervene on behalf of the growing Jewish population.[20]

Following Saul's lead, Abby soon joined the Haganah and was often involved on a daily basis, usually assuming a night watch. Not only was this dangerous, but sometimes there were heavy exercises for weeks at a time during which it was impossible to study or attend classes at the university.[21]

What was happening across Palestine in the late 1930s was extremely serious, and the Hebrew University was no ivory tower immune from the dangers of escalating violence. Between May and August of 1936, six students were killed and the university library on Mount Scopus was attacked, although this was repulsed by a team of special constables and watchmen. Nevertheless, the danger persisted, and many students were wounded while on military duty.

The first fatality among the faculty made an especially dramatic impression. Levi Billig—ironically a lecturer in Arabic literature—was shot

[20] For a detailed history of the cycles of terrorism and counterterrorism that plagued Palestine from about 1900 onward, erupting into full-scale civil war between Arabs and Jews in 1936, readers should consult the comprehensive *A History of the Israeli Army*, Schiff 1974 (Schiff was military correspondent for the newspaper *Ha'aretz*). As he writes, "from the moment Jews decided to return to Palestine and cultivate their own lands the need to protect life and property became a major concern." His coverage of the various Jewish underground groups, including Etzel (also known as Irgun) before and during World War II, is incisive. Robinson himself would no doubt have agreed with much that Schiff has to say, especially about the futility of violence as a solution to the Arab problem in Israel today. See Schiff 1974, p. 19.

[21] Pinczower R-1977 observes that "Abby participated in especially strenuous exercises in the mountains near Jerusalem. This means that he had made it clear he was ready and willing to undertake the most difficult and dangerous missions. . . . Abby was a through and through Zionist, who was always ready—he avoided no danger and never complained that he was overworked. The construction of a free Jewish state was the duty of every person, especially the young and able."

According to W.A.J. Luxemburg, although Robinson rarely spoke about his student days, he was very proud of the work he did with the Jewish underground as part of the Haganah, especially the fact that from time to time he was called upon to help defend Jerusalem. During this period, in fact, his mathematics was done on the side, whenever he could find time. Luxemburg I-1981.

Robinson on duty with the Haganah, Palestine.

on the night of August 20–21, 1936. He had come to Palestine from Jews' College, London, and was not only the founding member of the university's School of Oriental Studies, but was one of the very first appointments made at the Hebrew University.[22]

Because the situation throughout Palestine was becoming increasingly serious, the Haganah selected a number of students to be trained as junior officers, and Robinson was among those called for special duty. It was during his service in the Haganah that Abby first met Chimen Abramsky.[23] Both of them were at Kiryat Anavim, just outside Jerusalem, where an officers' training camp had been set up in 1937. At the time, Abramsky was studying philosophy and history at the Hebrew University, and the two young men soon found themselves discussing philosophy rather than defense. As Abramsky put it, "He was a few years younger than I with an enormous wide erudition in every branch of German philosophy. I sat almost spellbound at his immense knowledge of Kant and Hegel, and we discussed various problems until far in the night."[24]

By 1937, because their studies were not so closely related, Robinson

[22] The Hebrew University 1936a, pp. 1–2, 31–32.

[23] Abramsky was born in Minsk, Russia, on September 12, 1916. He received his B.A. with distinction from the Hebrew University of Jerusalem, and his M.A. from Oxford University. From 1940 to 1965, he worked in the family business (of his wife), the famous firm of Shapiro, Valentine and Co., publishers and booksellers of Jewish books. In 1965, he joined the Department of Hebrew and Jewish Studies of University College, London, later becoming professor and head of the department. Since 1967 he has also been a senior reader and associate fellow of St. Anthony's College, Oxford.

[24] Abramsky R-1976.

and Abramsky began to drift apart. Not only were their academic interests taking them in very different directions, but they were engaged in different branches of the Haganah. In 1939 Abramsky left for England (as he often did on holiday), but this time war broke out and he did not return to Palestine.

Abraham Robinson in the 1930s

Despite the demands of his studies, in addition to his tutoring and his commitment to the Haganah which often required late and unpredictable hours, Robinson still managed to find time to indulge his aesthetic interests, especially art and music. Pinczower recalls going with Abby, sometime towards the end of the 1930s, to an exhibit of works by Israeli-German artists. They also heard a humorous chamberpiece by Mozart, additional evidence that Abby was not just a "narrow-minded mathematician," but a young man of wide cultural interests. And although Abby usually struck his peers as quiet and sensitive, Pinczower remembers that he could also laugh "full out."[25]

From an early age Robinson had always enjoyed writing and tried his hand at both poetry and plays. Among his papers in the Yale University Archives are poems he wrote as a child in Breslau, including "Der Frosch" and "Die Sperlinge." Both are dated January 27, 1928. Nor did his interest in writing poetry diminish as he grew older. Sometime during his first year as a student at the Hebrew University he wrote "Der Attentäter" and "Mathematischer Morgenstern." The latter, a dialogue between an angle and its two defining sides, was doubtless a parody modeled on the German poet Christian Morgenstern (1871–1914). It reflects nicely the lighter side of Robinson's poetic flair as a first-year student majoring in mathematics at the Hebrew University.[26]

The Hebrew University of Jerusalem

The Hebrew University was built on the highest ridge of Mount Scopus, just on the outskirts of Jerusalem, overlooking the city. Off in the dis-

[25] Pinczower I-1982. Pinczower also recalls that Abby once visited him in Tel Aviv at Purim, and they went to an exhibition in which Abby (who was nineteen at the time) apparently expressed considerable interest in drawings by the artist Nachum Guttmann. In addition to art, music and literature, Pinczower adds that Abby was also deeply interested in Hebrew culture. See also Pinczower R-1977.

[26] Drafts of a number of Robinson's poems are to be found among his papers, YUL. Morgenstern not only wrote a poem of his own, "Die Sperlinge," but also used mathematical metaphors in his poems; see, for example, "The Two Parallels," in Morgenstern 1965, pp. 298–299.

tance, the hills of Moab in Transjordan provided a silent, forbidding background.[27]

When Chaim Weizmann sank twelve foundation stones—one for each of the twelve tribes of Israel—into the summit of Mount Scopus for the new university on July 24, 1918, Palestine was still at war. As Lord Allenby recalled, "Mount Ephraim, Samaria, Carmel were all in the hands of the enemy . . . and within the hearing of gunshot [Weizmann] laid the foundations of the university."[28] Weizmann himself acknowledged the symbolic importance of the university as a special haven in a world full of chaos and danger in his dedicatory address: "In the darkest ages of our existence we found shelter within the walls of our schools."[29] So it was to be again, although no one then foresaw World War II, nor the holocaust to come.

The university was officially opened on April 1, 1925, with a ceremony held in a natural amphitheater built into the slope of Mount Scopus. Lord Balfour, then seventy-six years old, was present, along with representatives of forty-one universities and twenty learned academies. Two departments had already begun operation in 1924—Jewish Studies and Chemistry. Mathematics was added in 1927.

From the beginning, the administrative philosophy of the university was to avoid becoming a "degree mill." Because of the country's limited resources, it was understood that the Hebrew University "would have to begin as a research institution adequate for its small faculty and qualified student body."[30]

The nucleus of operations, which had begun on Mount Scopus as a limited venture, soon grew to distinguished proportions. Largely because of the Nazi persecution of Jewish intellectuals, the "ingathering of exiles" throughout the 1930s added dramatically to the size and prestige of the university.[31] Although this indeed furnished a great opportunity, there were serious academic difficulties associated with the sudden arrival in Jerusalem of so many scholars and scientists from abroad. As Norman Bentwich described it:

> The university was an infant. In 1933 it consisted of a small institute of Jewish Studies, a small Institute of Arabic and Oriental Studies, a few

[27] For a general history of the Hebrew University, see Bentwich 1961. Selected aspects of the university are also covered in Bentwich 1960, esp. p. 161. Among publications of the University itself, see The Hebrew University 1936b and 1939.

[28] Levensohn 1950, p. 29. Lord Allenby entered Jerusalem on December 9, 1917, having liberated the southern part of Palestine from the Turks. Within weeks, Weizmann was urging that the foundations for the new university be laid, although the north was still in Turkish hands.

[29] Weizmann, quoted in Levensohn 1950, pp. 30–31.

[30] Levensohn 1950, p. 37.

[31] Levensohn 1950, p. 65.

departments of the general Humanities–which together comprised the faculty of Humanities–and a few small institutes of the Biological Sciences. Its academic staff did not exceed 50, its equipment was very limited, its library, though growing rapidly, included less than a quarter of a million books. Its students numbered less than 200. It was, too, a handicap for possible recruits from Germany that the language of the university, used for teaching and for all meetings of the academic bodies, was Hebrew, which few of them mastered.[32]

The Einstein Institute of Mathematics

Although it was not qualified at first to offer either minors or majors, and despite its being lumped administratively with the Humanities, the Mathematics Department was especially strong. Edmund Landau (1877–1938), internationally prominent for his important contributions to analytic number theory, was called from Göttingen where he had been extremely successful as a teacher, drawing doctoral students from all over the world.[33] When the Hebrew University's Board of Trustees met in Munich following the Fourteenth International Zionist Congress in Vienna in 1925, Landau was the preferred candidate to occupy the first chair in its newly created Einstein Mathematics Institute.[34] He was pleased to accept, and agreed to begin his new position two years later.

Meanwhile, Landau immediately set about to learn Hebrew with a tenacity that surprised no one who knew him well. He quickly made great strides, and his lifelong friend Abraham Fraenkel recalls that soon Landau was even sending telegrams to him in Hebrew.[35]

Because Landau was not scheduled to leave Göttingen immediately, it was left to Benjamin Amirà to teach the first mathematics courses as

[32] Bentwich 1953, p. 57.

[33] Fraenkel 1967, p. 163.

[34] Einstein himself apparently took part in this decision, made at one of the last meetings he ever attended of the university's board of trustees. This according to recollections in Fraenkel 1967, p. 164. For biographies of Landau, see Hardy and Heilbronn 1938, Knopp 1951, and Mirsky 1985. None of the "official" biographies of Landau even mentions the fact that he was called to the Hebrew University as head of the Einstein Institute of Mathematics, nor that he spent any time at all in Palestine. Fraenkel fortunately provides considerable detail about Landau and the Hebrew University. See Fraenkel 1967, pp. 162–167.

[35] Fraenkel also describes a visit he made to Landau in St. Moritz following the International Congress of Mathematicians in Bologna in 1928. Landau and Fraenkel were conversing in Hebrew when the telephone rang, and Landau was momentarily called away. Fraenkel was briefly interrupted by a hotel porter, and following their exchange of words, without thinking, he returned to his conversation with Landau, but in German, whereupon Landau exclaimed, "באיזה שפה אתה מדבר אלי" (In what language are you speaking to me?), Fraenkel 1967, p. 165.

senior assistant. He was also charged with supervising the actual construction of the Mathematics Institute on Mount Scopus and with organizing its first library during the academic year 1925–1926. The library, which amounted to some four thousand volumes, was built around the collection of Felix Klein, whose library was purchased after his death with funds donated by Philip Wattenberg of New York City. Wattenberg also paid for the construction of the institute's first building, erected in 1927–28 and appropriately named the Philip Wattenberg Building. It housed the Einstein Institute of Mathematics, and included two lecture halls.[36]

Systematic instruction in mathematics only began in the winter semester of 1927, when Landau finally arrived in Jerusalem to give courses. Unfortunately, his first semester was not a happy one, and proved to be his last. Due to differences with the university administration and its board of trustees, he left Palestine at the end of the term, never to return. Back in Göttingen, he was dismissed five years later by the Nazi authorities:

> The political situation had forced him to give up his professor's chair and to continue research privately during the last years of his life. His genius and work were deliberately not honored; no collection of his writings, no retrospect, no formal praise followed.[37]

A few years later, Landau died from complications following an unexpected heart attack in 1938.

Following Landau's resignation at the end of the winter term, 1927, the university's board of trustees again found itself having to find a suitable candidate for the suddenly vacant chair of mathematics. When the board met in London late the following spring, the choice fell this time upon Abraham Fraenkel, who had just left the University of Marburg for a position at Kiel.[38]

By the time Robinson enrolled as a student in 1936, Fraenkel was undoubtedly the best-known member of the faculty, having already earned an international reputation. Others like Michael Fekete and Jacob Levitzki, however, were also excellent mathematicians, good teachers, and well known within more specialized circles of research. Among the

[36] The Hebrew University 1939, p. 64.

[37] Editors' preface in Landau 1985, 1, p. 11.

[38] In 1927, according to Fraenkel, Landau had already expressed his intention of recommending Fraenkel for a position in the Einstein Institute. When Landau's position had to be filled, Hadamard nominated the Polish-French analyst Szolem Mandelbrojt as Landau's replacement. Mandelbrojt, however, did not want to leave Paris, and so declined. Fraenkel had only just begun his new positon at Kiel as a *professor ordinarius* when he received a telegram from the university on June 5, 1928, inviting him to assume the professorship in Jerusalem. See Fraenkel 1967, pp. 164–165.

most influential members of the faculty for Robinson's future course of study were the following.

Abraham Adolf Fraenkel (1891–1965)

Fraenkel was born in Munich on February 17, 1891.[39] After studying in Munich, Berlin, Breslau, and Marburg, he received his Ph.D. in 1914 (Marburg) and then went on to serve in World War I. Thereafter he returned to Marburg where he advanced from *Privatdozent* to *ausserordentlicher Professor*.

Fraenkel had only just accepted a position at the University of Kiel when he received the telegram from the trustees of the Hebrew University (June 5, 1928), inviting him to join the faculty of the Einstein Institute. The offer, he said, came as a complete surprise. Although he lost no time in accepting, after three years in Jerusalem he went back to Kiel as head of the seminar for mathematics. But when the Nazis came to power in 1933, Fraenkel and his wife returned to Palestine, where he was again offered his old position in the Einstein Institute.[40]

At the time, the president of the Hebrew University was Judah Leon Magnes, an American whose ideas of running the university were usually at odds with Fraenkel's. Fraenkel, in fact, was not alone in his opposition to Magnes, and was but one of a formidable group comprised, according to one report:

> chiefly of the British members of the several governing bodies led by Prof. Einstein with substantial support from Dr. Weizmann. . . . [They] were dissatisfied with the academic administration and progress of the University, especially in the scientific departments.[41]

Fraenkel, of course, familiar with the tradition of European universities, may well have resented Magnes' American assertiveness, and he clearly preferred the model of European universities where the faculty Senate made most of the decisions concerning overall issues of policy and development. In his confrontations with Magnes, Fraenkel could count on strong support, especially from Chaim Weizmann.[42] Ultimately,

[39] His "official" German name was Adolf, but his Jewish name was Abraham Halevi. In the 1930s, for obvious reasons as Fraenkel himself says, he dropped the name Adolf and adopted Abraham at just the time he made the decision to settle permanently in Palestine. See Fraenkel 1967, p. 56.

[40] Mrs. Fraenkel recalls immigrating to Palestine by ship from Trieste. Although at first conditions in Jerusalem seemed primitive—in particular, the roads were difficult to walk on, and the children were always losing their shoes—the Fraenkels were delighted to be in Palestine, Fraenkel I-1982.

[41] Parzen 1974, p. 33.

[42] Pinczower I-1982. Fraenkel was a controversial personality at the Hebrew Univer-

a revolution sympathetic to Fraenkel's views occurred which resulted in the naming of an academic rector, a position Fraenkel held shortly thereafter, from 1939–1940.

Because the Einstein Institute of Mathematics, like the university as a whole, was small, Fraenkel had few students, and so especially bright ones like Abraham Robinson quickly came to his attention. When Robinson entered the department in 1936, it counted eighty students who were either majoring or minoring in mathematics.[43] But of this group, Robinson was clearly the one who stood out. Fraenkel seemed to have a knack for spotting mathematical talent; he soon realized that there was great potential in Abby's abilities, and was quick to take him under his wing. By 1938 Fraenkel believed that Abby had advanced so far that he had nothing more to teach his brightest pupil.[44]

Of special importance for Robinson's future career as a mathematician were Fraenkel's interests in mathematical logic and axiomatic foundations. At first Fraenkel worked in algebra, providing axiomatizations for Hensel's p-adic numbers and ring theory, areas that would also play a role in Robinson's later mathematics. In trying to prove the independence of Zermelo's axioms for set theory put forward in 1908, Fraenkel saw that the system failed to provide a satisfactory foundation for set theory. Not only did he realize that a stronger axiom of infinity was needed, but he also saw how to avoid Zermelo's imprecise notion of "definite property."[45] Robinson's first mathematical publication, in fact, which concerned the independence of the axiom of definiteness, was set firmly in this context of Zermelo-Fraenkel set theory (see below,

---

sity from the start. He also clashed with Dr. Franz Bodenheimer, director of the Zoology Department, who accused Fraenkel of intervening in the department's affairs on behalf of Fraenkel's nephew, Gottfried Fraenkel, then a junior assistant who had been denied a promotion (allegedly for financial reasons). Fraenkel bluntly threatened Bodenheimer, "especially exposure of his professional shortcomings at the university." Full details are provided in Parzen 1974, p. 36.

[43] The Hebrew University 1936b.

[44] Seligman 1979, p. xv. At the same time, Robinson greatly appreciated what Fraenkel had done for him. When the earliest appropriate opportunity came, he expressed himself in the preface to his first book, *On the Metamathematics of Algebra*, dated March 1950, as follows: "I should like to acknowledge on this occasion the debt of gratitude which I owe to my teacher and friend, Professor A. Fraenkel of the Hebrew University of Jerusalem under whose guidance I made my first original contributions to mathematics and to symbolic logic," Robinson 1951, p. vi.

[45] This has nothing to do with later axioms of infinity related to assumptions about the existence of large cardinal numbers. As Martin Davis stressed in a note to me (September 7, 1994), for Fraenkel, the axiom of replacement "permits one to form the set of elements satisfying some condition if this set is the image of an already existing set (and hence no larger)."

pp. 53–54). Later Fraenkel became increasingly well known as an advocate of transfinite set theory, for his many contributions to the foundations of mathematics, and for the many textbooks he wrote on these subjects.[46]

Benjamin Amirà (1896–1968)

Amirà immigrated to Palestine in 1910 from Mohileff, Russia, where he was born in 1896. After attending the University of Geneva from 1915–1919, he received the D.Sc. in Mathematics (with a thesis on entire functions, written under Landau's guidance), and in 1921 went to Göttingen, where he taught for three years as a privatdozent (1921–1924). After moving briefly to Geneva, he agreed to return to Palestine as the founding member of the Mathematical Institute at the Hebrew University in 1925. In 1934 he became a senior assistant, and then lecturer in mathematics. Later he was the founding editor of the *Journal d'analyse mathématique*, and from the appearance of its first volume in 1963, was a member of the editorial board of the *Israel Journal of Mathematics*.[47]

Michael Fekete (1886–1957)

Fekete was born in Zenta, Hungary.[48] After receiving his Ph.D. from the University of Budapest in 1909, he spent a postgraduate year at Göttingen where he too studied with Edmund Landau. He then returned to Budapest where he served as an instructor in municipal and Jewish secondary schools until 1928. Simultaneously, from 1912–1919 he was an assistant in the Mathematics Seminar at the University of Budapest, where he also taught from time to time in various other capacities as both a *Privatdozent* and lecturer in an advanced seminar for teachers in commerical schools. After 1920, however, his contacts with the university were apparently severed, and from then on his teaching seems to have been confined to secondary schools.

[46] See among others Fraenkel 1928, 1953, and 1973.

[47] Amirà devoted years to the idea of establishing an international mathematics journal emphasizing analysis and to be published in Jerusalem. This he succeeded in launching in 1950, with support from the Rockefeller Institute. See Bentwich 1978, p. 78. Volume 23 of the *Journal d'analyse mathématique* was dedicated to Amirà, and noted that "In the life of science we need catalysts, organizers and selfless men of high vision perhaps more than narrow specialists. Amirà was the pioneer of mathematical organization, par excellence." See Amirà 1970, pp. xii–xvi, esp. p. xvi.

[48] For biographical details, see Balázs 1958, pp. 197–224. A bibliography of Fekete's works appears in *Matematikai lapok*, 9 (1958), pp. 1–5. See also Rogosinski 1958, pp. 496–500.

Following Edmund Landau's unexpected departure from the Hebrew University in the fall of 1927, the Department of Mathematics was desperately in need of a new administrative leader. Fortunately, Fekete accepted the university's offer of an appointment, and in 1928 he arrived in Jerusalem to take up his position as lecturer and director of the Mathematical Institute (he was promoted to the rank of professor in 1935). Later he went on to other important administrative positions as well, and served as dean of science and rector of the university from 1945–1948. In 1955, upon his retirement, he was awarded the Israel Prize for Exact Sciences.[49]

Fekete's major research interests centered on the theory of functions of real and complex variables. Among the courses he taught at the Hebrew University were number theory, theory of functions, and Fourier and Laplace series. He also worked on polynomials, both algebraic and trigonometric, and contributed to the theory of Fourier series and analytic functions.

Fekete was universally regarded as an excellent teacher, largely owing to his enthusiasm for the subject and his interest in helping younger mathematicians.[50] Physically, he was a small, energetic man with fiery eyes and an unruly mane of white hair, on a "fine large head," reminiscent of Einstein. Even as he grew older, Fekete retained a youthful delight in his work, and actually died at his desk "doing mathematics."[51]

Jacob Levitzki (1904–1956)

Levitzki was born in Cherson, a city in the Ukraine, but his family left Russia for Palestine when he was still a child. After attending the Herzlia Gymnasium in Tel Aviv, he went to Germany in order to study mathematics. In 1929 he received his Ph.D. from the University of Göttingen, and then taught as an assistant instructor of mathematics at Kiel. In 1930 he was awarded a special stipend and spent the next year in the United States as a Sterling Research Fellow at Yale University. Thereafter, he returned to Palestine where he joined the Mathematics Depart-

[49] Rogosinski 1958, p. 496.

[50] Agmon 1956, pp. 1–8. This article is in Hebrew, and pp. 2–8 constitute Fekete's bibliography. Agmon emphasizes that mathematical geniuses are not just born but require "a *living* mathematical atmosphere" for inspiration in order to thrive. Consequently Agmon suggests that the decision of someone like Fekete to settle in Palestine in the 1930s was not easy. He adds that Fekete's main influence was in seminars. There he dealt with proofs just beginning to develop in his head, and this meant that students could observe a mathematician at work as he approached a problem in different ways in search of a solution.

[51] Rogosinski 1958, pp. 496–497.

ment of the Hebrew University first as a junior assistant, and then, in 1934, as a senior assistant.[52]

Levitzki was highly regarded by his colleagues as an excellent teacher, and was appreciated above all for having a special feeling for getting the important recent advances of mathematics across to his students. He was well known for believing that the most modern ideas and notations should "prevail in modern mathematics," and consequently, that "the young generation of mathematicians must encounter them as early as possible."[53] This approach to teaching mathematics at the most advanced and creative levels was surely of special benefit to students as bright and receptive as Robinson.

Levitzki taught basic courses in algebra and geometry, including analytic geometry and abstract algebra. His major research interests in the 1930s when Robinson was a student were rings (especially nilpotent elements), subrings, matrix rings, and chain conditions.[54] In 1954 he was awarded the Israel Prize for Exact Sciences.

Theodore Motzkin (1908–1973)

Motzkin was born in Berlin on March 26, 1908. Later his mathematical training took him to Switzerland, where he received his Ph.D. from the University of Basel in 1934. As a member of the faculty at the Einstein Institute of Mathematics, he taught geometry, differential geometry, and topology, with strong research interests in algebra and convex point sets.[55] From 1936–1949 Motzkin served as president of the Mathematical Society, a responsibility he relinquished when he left Israel in 1950 for a position at the University of California at Los Angeles. In 1947 Motzkin and Robinson published a jointly authored paper, "The Characterization of Algebraic Plane Curves," in the *Duke Mathematics Journal*.[56]

Otto Toeplitz (1881–1940)

Otto Toeplitz came to Palestine from Breslau in 1938, another casuality of Nazi anti-Semitism and relentless persecution. Toeplitz was born in Breslau, studied at Göttingen, and later taught at Kiel and then Bonn until he was dismissed by the Nazis from teaching in 1933. For the next

[52] The Hebrew University 1939. Levitzki, at the time he submitted an article to *Mathematische Annalen* for publication, listed himself in 1930 as Sterling Research Fellow at Yale University. See Levitzki 1931, p. 620.

[53] Amitsur 1974, p. 2.

[54] Amitsur 1957.

[55] The Hebrew University 1939, p. 64.

[56] See Robinson and Motzkin 1947. For discussion of this work, see below, chapter 4.

five years he worked with the declining Jewish community in Germany until he could bear it no more and immigrated in 1938 to Jerusalem, where he became an administrative advisor at the Hebrew University. He even taught a private seminar in which he reported results of the work he had been doing with Gottfried Köthe prior to his departure from Germany.[57]

As Robinson later wrote:

> Toeplitz was a typical German-Jewish intellectual who, while retaining an interest in Jewish mathematics, felt himself to be a part of his country of birth *to an extent which is no longer remembered even in the more well-meaning Germany of today. The events of 1933 and after came as a great shock to him and he never recovered from their impact.*[58]

Courses at the Hebrew University

One of the first courses Robinson took as a new student at the Hebrew University in the spring term, 1936, was Fraenkel's seminar on the foundations of mathematics. During the academic year 1936–1937 he also enrolled in Fraenkel's lecture-seminar course on set theory and a colloquium offered by Benjamin Amirà on number theory.[59] In all, during

[57] Robinson wrote the article on Toeplitz for the *Dictionary of Scientific Biography* (hereafter *DSB*), Robinson 1975. Although he cites the most detailed obituary notice on Toeplitz following his death, namely Behnke and Köthe 1963, Robinson's article contains details that he could only have included based upon his own personal knowledge of Toeplitz. For example, neither Behnke 1949 nor the only other notices of Toeplitz to appear before Robinson's *DSB* article–i.e., Born 1940; an anonymous and brief obituary in *Nachrichten von der Gesellschaft der Wissenschaften zu Göttingen* (1937–38), p. 10; a short biography with portrait that appeared in *Reichshandbuch der deutschen Gesellschaft*, 2 (Berlin 1931), p. 1060; or Behnke and Köthe 1963–makes any mention of the fact that Toeplitz actually taught a private seminar in Jerusalem. For more on Toeplitz, see Pinl 1969, pp. 201–202.

Robinson himself asked the Editor of the *DSB*, Charles Gillispie, if he might write the article on Toeplitz. It is more than likely that he did so because he had known Toeplitz and had a high regard for his work, having been exposed to it at the Hebrew University. Toeplitz, who was deeply interested in history of mathematics, held that "only a mathematician of stature is qualified to be a historian of mathematics." Robinson's own later interest in history of mathematics certainly fit Toeplitz's expectations. See Robinson 1975, p. 428.

[58] From Robinson's own typescript draft of his article on Toeplitz written for the *DSB*. The italicized portion of the quotation was deleted from the published version of the article. The draft is among Robinson's papers, YUL. For the published article, see Robinson 1975.

[59] From records in Robinson's student file in the Einstein Institute of Mathematics, the Hebrew University. Fraenkel notes that Robinson took "active part" in his seminar (dated June 16, 1936). Amirà passed Robinson with "excellence" in his colloquium (July 5, 1937).

his three years as a student majoring in mathematics, Robinson completed more than forty different courses. The bulk of these were lecture courses or seminars offered by Fraenkel, Levitzki, Fekete, Motzkin, and Amirà.[60]

In addition to the standard introductory courses in mathematics, Robinson also did advanced work in courses devoted to the theory of functions, algebra, and geometry. He also took a number of courses in theoretical physics (as well as a private seminar in experimental physics) with Shmuel (Samuel) Sambursky. Other courses in natural science included the theory of electricity and a course on inorganic chemistry.

Nor did Robinson neglect the humanities as a student at the Hebrew University. "Greek for beginners, part 1," must have complemented nicely the reading he did in ancient philosophy from Protagoras and Plato. He also studied philosophy in general, including logic, ethics and politics, with one course devoted specifically to Leibniz. He took history of modern philosophy and an introduction to logic and theory of knowledge, as well as a number of education courses, including one on educational psychology with Moses Brill.[61]

Among his fellow students, Robinson was known as the person to ask for clear, understandable explanations. As Ernst Straus (a somewhat younger student at the time) recalls, "When we did not understand something we would ask him to explain it to us later and I do not recall a single case when this did not work."[62] In addition to tutoring other students with their mathematics, Robinson was an active participant in the Hebrew University mathematics club, which he had helped to organize. On one occasion he gave a lecture on the zeta function, which Straus always remembered as having made a lasting impression on him because it was his first introduction to the subject.[63]

At the beginning of 1938 the secretary of the university wrote to Robinson, announcing that he had won a prize from B'nai Brith of

[60] There is a receipt from the secretary of the university showing Robinson's payment for part of his fees for 1937, the rest to be paid in 1938. All records are part of Robinson's student file, Department of Mathematics, at the Hebrew University, Jerusalem.

[61] Brill joined the faculty in 1938 as external teacher of educational psychology. Formerly he had been assistant clinician at the Psycho-Educational Clinic at New York University from 1933–1935. See the Hebrew University 1938, p. 42.

[62] Ernst Straus of the University of California, Los Angeles, and a fellow student of Abby's at the Hebrew University in the late 1930s, quoted from Straus R-1976. See also Seligman 1979, p. xv.

[63] Later, Robinson insisted he had never given such a lecture, and was certain that he had only learned analytic number theory much later. But as Straus recalls, "I could never convince him later that this ever happened. He said he never knew what analytic number theory was until much later, but I remembered this lecture very vividly and apparently it made a much bigger impression on me than it did on him," Straus R-1976.

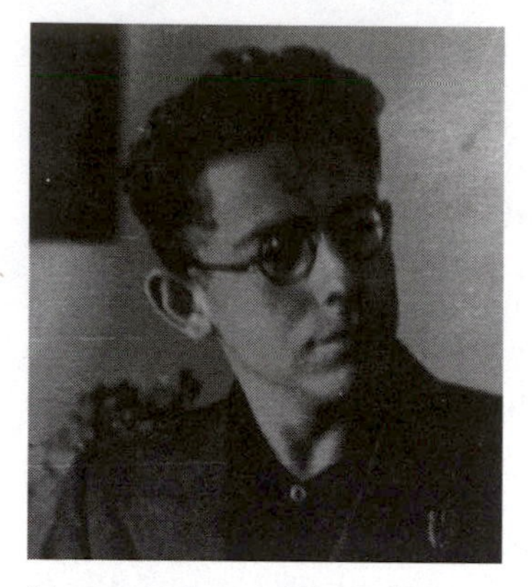

“A good photo of Abby”—May 1938.

Canada amounting to twenty-five dollars, which he shared with another student.[64] The money was extremely welcome, for Abby was among the poorest of students at the university. In 1937 he was only able to make partial payment of his tuition. At first he petitioned for permission to pay in installments, which was approved. Later, after 1937, he asked to be released from paying fees altogether. In one of his letters appealing to the secretary for cancellation of his tuition, he noted that the only money he had came from whatever private lessons he could give to students needing help with their studies.[65]

The university, in agreeing to waive Robinson’s tuition, recognized his financial dilemma, but his intellectual merits were equally acknowledged, not only through the B’nai Brith awards (he also received another small sum from B’nai Brith of Portland, Oregon, in 1938), but through a major foreign scholarship. Thanks to negotiations that had only recently been concluded in Paris by President Magnes, the French government agreed to establish a special scholarship fund for deserving students enabling them to study at the Sorbonne in Paris.[66] Robinson, who was clearly regarded by his professors in mathematics (and by the

[64] The letter of award is dated January 12, 1938, and is part of Robinson’s student file at the Hebrew University. B’nai Brith (Sons of the Covenant) is the oldest and largest Jewish service organization in North America.

[65] For example, Robinson gave private lessons in Hebrew to one of his fellow classmates, Kattie Hayden. About his financial problems, Robinson wrote to the secretary of the university on October 31, 1937, asking to be allowed to pay his tuition in installments. The first letter asking that he be excused altogether from paying tuition is dated July 1938. Other letters making similar appeals are also to be found in Robinson’s student files at the Hebrew University. Letters from the administration of December 6, 1938, and another for 1939, officially released Robinson from payment of all fees.

Abby’s brother Saul also asked to be allowed to pay tuition in installments in 1937. By this time, however, Saul was married and had found a position as librarian to the great private collection of Salmann Schocken which today forms the core of the Schocken Institute. As Pinczower notes, this greatly eased Saul’s financial concerns, and made it possible to earn enough to help support himself and his wife (who also worked at the Schocken Library). All along, Saul always managed to contribute something to the support of his mother as well. Pinczower R-1977. The letters Robinson exchanged with the university are among the papers contained in his student file, Department of Mathematics, at the Hebrew University, Jerusalem.

[66] Funds were also contributed for the library. See The Hebrew University 1938, pp. 51–52.

university administration) as one of their most talented students, was among the first selected to participate.

In addition to his strong record as a student, Robinson had already begun to publish, and this must have impressed the committee charged with reviewing applications for the French government scholarship. By the end of 1939, Robinson's future career as a mathematician was already well on course, and extremely promising. Not only did he have one article in print, but another had just been accepted for publication and was being readied for press.

Robinson's First Publication

Six months before he left Palestine for France, Robinson's first publication appeared (under the name Robinsohn) in the *Journal of Symbolic Logic*.[67] It was dated August 1939, and drew on previous work by Vieler, Zermelo, Ackermann—and of course, by his mentor Fraenkel. In his paper Robinson was interested in showing that the axiom of definiteness (i.e., the axiom of extensionality) was independent of the other axioms of Zermelo-Fraenkel set theory. This axiom holds a special place among the Zermelo-Fraenkel axioms because it is a purely relational one; it is the axiom which establishes the character of equality within the system.

Although Robinson cited Vieler's dissertation, "Untersuchungen über Unabhängigkeit und Tragweite der Axiome der Mengenlehre" (Marburg 1926), he did so to provide a contrast with his own approach to the problem. Vieler had already proven the independence of the axiom, but Robinson criticized the proof for being overly complicated because of its appeal to ordered sets and application of the well-ordering theorem. Instead, Robinson took a simpler approach in order to produce a much more natural proof showing that the axiom could not be derived from the other axioms of Zermelo-Fraenkel set theory under several definitions of equality.

Zermelo had originally introduced equality intentionally, writing = (x, y) if two symbols x and y represented the same object. Robinson took two alternative approaches. In the first he assumed equality as a primitive concept required for defining an equivalence relation and satisfying the axioms; the second introduced equality as a derived concept by means of the following definitions:

Definition I: $= (\ ,\ ) =_{\mathrm{Df}} \hat{x}\hat{y}\{(z) \cdot e(z, x) \equiv e(z, y)\}.$

Alternatively, he defined equality by

Definition II: $= (\ ,\ ) =_{\mathrm{Df}} \hat{x}\hat{y}\{(z) \cdot e(x, z) \equiv e(y, z)\}.$

[67] Robinson 1939.

In either case, Robinson was able to show that the defined equality satisfied the axioms but that it was then necessary to postulate satisfaction of the axiom of definiteness/extensionality—establishing its independence.[68]

Fraenkel had sent an earlier version of Robinson's paper to Paul Bernays, who replied with a number of suggestions for revision. Once a final draft was ready, Fraenkel wrote again, forwarding Abby's paper and doubtless hoping this time that Bernays would find it acceptable for publication. Bernays was indeed pleased with the alterations Robinson had made, although he still had a few more suggestions for revisions. Nevertheless, he promised to recommend the final, revised version of the paper to Alonzo Church for publication in *The Journal of Symbolic Logic*, and within the year, Robinson's first publication had appeared.[69]

Robinson's Second Publication—Almost

At about the same time that his first paper was published in the *Journal of Symbolic Logic*, Robinson was hard at work on his second article. He was only twenty years old and still an undergraduate when this latest effort was also accepted for publication in August of 1939 by *Compositio Mathematica*, a journal "intended to further the development of mathematics and at the same time of international cooperation."[70]

What Robinson had devised was a very simple proof of the theorem that for rings with the minimal condition for right ideals, every right nilideal is nilpotent. Doubtless the inspiration for the paper—and the suggestion that Robinson submit it to *Compositio Mathematica* for publication—came from Jacob Levitzki, with whom Robinson had studied advanced algebra in 1936–37, and more advanced topics on algebra in the following two academic years.

The theorem itself was not a new discovery. The American mathematician Charles Hopkins had first proved the result in a paper on "Nilrings with Minimal Condition for Admissible Left Ideals," published in the *Duke Mathematics Journal* in 1938. Levitzki, independent of Hopkins, had also proved a similar theorem in his article "On Rings which Satisfy

[68] According to Definition I, two sets are taken to be equivalent if they have the same members; Definition II regards sets as equal if they belong to the same sets. The first is better known as the principle of extensionality, and, as Martin Davis suggested to me (in an E-mail note of September 7, 1994), "the second may be thought of as formalizing Leibniz's principle of the identity of indiscernibles." See Fraenkel 1926 and 1928, §16.

[69] P. Bernays to A. Fraenkel, August 28, 1938, RP/YUL.

[70] This is how *Compositio Mathematica* was described on the inside front cover of its first issue; the description remains the same to this day.

the Minimum Condition for the Right Hand Ideals," which appeared in 1939 in *Compositio Mathematica*.[71] It is more than likely that Levitzki also presented his own version of the theorem on nilideals in one of his advanced algebra seminars, which in turn may have directly inspired Robinson's interest, prompting his discovery of a simpler proof. This was the major virtue of his version of the theorem, which he was able to prove using much simpler ideas than either Hopkins or Levitzki had employed. It thus portended the future because Robinson, as a model theorist, was to become well known as a mathematician skilled at reducing mathematical problems to their simplest, most elegant terms.

Levitzki and Robinson: the Algebra of Ideals

Levitzki's interest in algebra and especially ring theory had been inspired by Emmy Noether. In fact, Levitzki turned to mathematics as a student at Göttingen thanks to one of her lectures. She so inspired him that he gave up chemistry in order to devote his life primarily to algebra, and chiefly to ring theory. At the time, attempts to extend Wedderburn structure theorems of associative algebras had met serious difficulties posed by nilpotent elements, and these, along with nilsubrings, became an early focus of Levitzki's research.

Artin's subsequent extension of Wedderburn structure theorems to rings with ascending and descending chain conditions on one-sided ideals in 1927 still left open the question of whether both were necessary conditions.[72] Here the importance of nilpotent elements had been emphasized by Köthe in an article which appeared in the *Mathematische Zeitschrift* for 1930, and these were pursued by Hopkins and Levitzki simultaneously but independently. Hopkins showed that in a ring satisfying a minimal condition for left ideals, any subring of only nilpotent elements was itself nilpotent. Levitzki (using right ideals) provided an alternative and somewhat longer proof of this same result, along with a number of related theorems.[73]

Robinson's reworking of the Hopkins-Levitzki result is remarkable on several counts. Not only is it evidence of his keen insights at a very early age, especially in abstract algebra, but its style was already indicative of certain features of his later, more mature mathematical personality. It was short, simple, direct, and provided a much clearer statement of the essence of the theorem than had been given in either of the earlier versions of the theorem due to Hopkins or Levitzki.

Unfortunately, despite the interest of Robinson's result, it was never

[71] See Hopkins 1938 and Levitzki 1939.
[72] See Van der Waerden 1985, pp. 210–211.
[73] Levitzki 1939.

published in his lifetime. Even though it was set in page proof in 1939 and announced as "soon to appear" in *Compositio Mathematica*, it never did owing to World War II. Although the journal managed to produce one issue in 1940, it was forced to suspend operations entirely before Robinson's paper could be printed. Even so, the paper that never appeared in *Compositio Mathematica* was a clear sign that Robinson already had the soul of a mathematician.[74] As Fraenkel already knew, there was not much more that Robinson could hope to learn by staying in Palestine.

Scholarship to the Sorbonne

When Abby heard about the availability of scholarships to support study in Paris, he applied immediately. By the summer of 1939 his publications in print (and promised) would have assured that his credentials would stand out, wholly apart from the honors he had already received for the quality of his work as a student.

His application included a brief statement of purpose which outlined what living abroad would mean to him. He explained that he had already made a systematic study of the limited mathematical literature available to him in Jerusalem as an undergraduate. The aim of further study in Paris was to widen his mathematical horizons and to improve his "mathematical methodology."[75] The possibility of studying at the Sorbonne would enable him to pursue his research interests, especially by making use of literature not available in Palestine.

Official word of Abby's scholarship to study in France was not received until late in 1939. Amédée Outrey, the Consul Général de France for Palestine and Transjordan, wrote to Ibn Zahav, secretary of the Hebrew University, on December 9 with the news that two awards would be made, one to Abraham Robinson and the other to Jacob Fleischer (later Jacob L. Talmon).[76]

It may seem unthinkable that as war threatened Europe in the late 1930s, Robinson would seriously have considered going to Paris at all. In October, at the opening address of the academic year for 1939, Presi-

[74] The paper has recently been printed in Vol. 1 of his *Selected Papers*, from a copy of the original proof sheet from *Compositio Mathematica*. See Robinson 1939a.

[75] Included with the application was a copy, in English, of Robinson's *Curriculum Vitae*, RP/YUL.

[76] Letter from Outrey to Ibn Zahav, December 9, 1939, Jerusalem; in Robinson's student file, Department of Mathematics, the Hebrew University, Jerusalem. Fleischer is better known by the name he adopted in England, Jacob L. Talmon. He was Robinson's senior by a year and a friend of his brother Saul. For an appreciation of Talmon's life and work, see Arieli 1982.

dent Judah L. Magnes reminded the students of the Hebrew University that many of their classmates, and in fact Jews everywhere, were facing imminent danger:

> As we open this term, students are still in Europe, caught by the war while visiting their parents.... Many of the great centers of Jewish learning in Central and Eastern Europe have been destroyed within the past few years, and as we meet here the ruins of some but recently destroyed may still be smouldering. Moreover, thousands of Jews have been driven from the halls of various universities and colleges, and thousands from professions which require university training.[77]

As Magnes went on to say, European Judaism was being destroyed before their very eyes, especially in Poland. Magnes was speaking within days of November 9, when only a year earlier hundreds of synagogues had been set ablaze in Germany. At the end of his address, Magnes called on the assembled students to stand in silence for a moment out of respect for "the killed, the wounded, the imprisoned, and for the torn and burnt and bespattered scrolls of our imperishable Torah."[78]

What was already happening in Germany and Poland, however, despite the attention Magnes drew to the progressive destruction of European Judaism, failed to impress most of his listeners at the time. Later, Talmon (Fleischer) recalled that as he and Abby were making their final preparations to leave for France, no one warned them of the serious danger they faced in travelling to Europe in 1939:

> In retrospect, it may look as an act of madness: two boys in their early twenties (I was two years older than Abby) go in the middle of an international war to study in a country at war. To both of us it was the first voyage into the world, neither of us had any means except the promise of the French Government scholarship, which was to be paid out to us in monthly installments. The truth is that in the period of the *drôle de guerre*, between the destruction of Poland in September 1939 and the invasion of Norway in March-April, nothing was happening on the Franco-German border, and we in Israel, terribly ill informed as we were at the time, were mistakenly sure that Nazi Germany would be crushed in the first real encounter with the Anglo-French armies or would be brought to her knees by the blockade. Moreover, both the French authorities (more directly the French Consulate General in Jerusalem) and the Hebrew University took it as self-evident, natural and wholly unproblematical that we should be going. The University in Jerusalem was still a very small institution, with just a few hundred students, and it was only as far as we knew, the second time

[77] Magnes 1939, in Goren 1982, p. 358.
[78] Magnes 1939, in Goren 1982, p. 365.

> that a foreign government had placed at the disposal of the Hebrew University scholarships to study abroad. It was considered almost a matter of national importance, and a signal distinction. Our young imaginations and ambitions were greatly affected by the honour as well as by the visions of Paris, the capital of the world, not to speak of the chances and vistas opened for the more distant future. In a way, Abby took greater risks than I, because I had already received in 1939 my M.A. degree, whereas Abby was still an undergraduate.[79]

Robinson and Fleischer lost no time in booking passage on a ship leaving Beirut for Marseilles. "The moment of that journey remained with us for all the years after," said Fleischer, "and we would often remember it wistfully in our conversations."[80] From Jerusalem they went by *sherut* (a public taxi) to the Lebanese border at Russ-al-Nakine. There a sharp turn in the coastline afforded one last look at Palestine. Having made their way to the highest vantage point within reach, they looked down on Eretz-Israel, which stretched away as far as the eye could see to the south and east:

> A vast panorama, as of the whole of the Holy Land was spread out before our eyes to be seen before departing from it. When indeed we reentered the car and crossed the border into the Lebanon, there was no looking back upon the Land of Israel. Little did we know that we had plunged into a world that was to suffer a deluge of blood and fire, and that for many years we [would] not be able to return to our homes, as far as Abby was concerned, in fact, never, except for two or three years some twenty years later.[81]

[79] From a manuscript in Talmon's own hand, "Our Escape from France in 1940, and Early Days in England," Talmon R-1940. Talmon died June 16, 1980, during heart surgery at the Hadassah Medical Center in Jerusalem, at age sixty-four. He was born in Poland in 1916 and immigrated to Palestine in 1934. After studying at the Hebrew University (and briefly at the Sorbonne), he studied for his doctorate at the London School of Economics. Towards the end of World War II he worked with the Board of Deputies of British Jews, and after the war returned to Jerusalem where he accepted a position in the Department of History at the Hebrew University. Talmon is best known for his studies of the evolution of totalitarian ideologies and aberrations of democracy from the time of the French Revolution onwards. Obituary notice in the *New York Times*, Wednesday, June 18, 1980.

[80] Talmon R-1940.

[81] Talmon R-1940.

CHAPTER THREE

Robinson in Paris: January–June 1940

תקל תקלתא במאזניא והשתבחת חסיר
–Daniel 5:27[1]

HITLER'S ARMIES invaded Poland on September 1, 1939, having annexed Austria and taken Slovakia the year before.[2] On September 3, to make good their treaty obligations with the Poles, England and France simultaneously declared war on Germany. But instead of war, only diplomatic skirmishes followed, and neither the English nor the French offered more than token opposition, thereby allowing the German army to occupy and subdue Poland. For the next six months, Europeans tried to persuade themselves that the world could live with totalitarian, expansionist Germany and still return to peace.

Meanwhile the *drôle de guerre* (or "phony war" as U.S. Senator William E. Borah dubbed the pause in aggression) created a bizarre mixture of apprehension and ennui. As Simone de Beauvoir wrote in her journal, "this war seems a phantom war—not the least hint of the shadow of a German is perceptible."[3] By the end of 1939, many Europeans and most French had been lulled, if only temporarily, into a state of nervous indifference about Germany and its true intentions. Claude Jamet, an army lieutenant, wrote disarmingly: "*Et la guerre?* Frankly, one isn't interested in it. One does not think of it. Does it really exist?"[4]

[1] "T'kel, you have been weighed in the balances and found wanting." Robinson quoted this passage from the Book of Daniel in an account he wrote (in Hebrew) of the few brief months he spent in Paris during the spring of 1940, including a vivid description of his escape as France fell to the Germans that June. Although at least part of this was written several years later, it is listed in the bibliography as Robinson R-1940, the year in which the events it describes occurred. The English translations throughout are by Sheila Rabin.

[2] For a detailed study of the initial months of World War II, especially the events leading up to the fall of France in June of 1940, see Chapman 1968. Other general accounts include Horne 1969 and Werth 1940.

[3] De Beauvoir 1960, p. 444. Throughout the war Simone de Beauvoir kept a record of her daily life, one that seems to have continued without much change in 1939–1940, despite the dramatic events of those years. Other personal accounts of Paris before and during the German occupation that have been used here include Amouroux 1961, Bardoux 1957, Baudouin 1948, Boothe 1940, Gide 1946, Langeron 1946, Murphy 1964, Porter 1942, Sartre 1984, and Spears 1954. These serve to corroborate the general descriptions found in Robinson R-1940, or add details to the account he provides.

[4] Horne 1969, p. 102. Robinson also logged his impressions of the *drôle de guerre* during the spring of 1940, and was particularly struck by the French refusal to consider

In February Jean-Paul Sartre echoed these sentiments in one of his diaries: "Everything has become simplified, everything grown relaxed."[5] But at the highest levels, this lack of urgency would soon prove disastrous. Typical of the shortsightedness and poor judgment of those in command was a remark made all too casually by General Billotte to some of his corps commanders. In response to complaints about the lack of arms for their men, the general showed no concern at all: "Why bother yourselves? Nothing will happen before 1941."[6]

From Palestine to Paris: January 1940

In hindsight, only a startling naiveté can explain why Abraham Robinson and Jacob Fleischer, in January of 1940, could have thought it prudent to leave Palestine for Paris. No one, it seems, seriously questioned whether or not the French capital was a safe place for two Jewish students from Jerusalem once war had been declared in September. On the other hand, considering the escalating violence in Palestine, Paris may actually have seemed a better, perhaps even safer alternative, and without question it must have seemed an exciting prospect.[7]

In January, Robinson and Fleischer sailed for Marseilles, with a brief stop in Alexandria. Any misgivings Robinson may have had about the war would have been in sharp contrast to the total lack of concern he observed among French soldiers returning home on furlough from duty in Syria. He was amazed at the extent to which the French simply regarded the war as "a source of some serious disturbances to normal life, but as a passing difficulty." The army, he observed, seemed more concerned with furloughs than with fighting:

> Coming to France, you might think that the "furlough organisation" was the focal point of the whole war. The boat *La Providence* on which I sailed from Beirut to Marseilles was a vacation boat bringing a unit of East-army soldiers back home from Syria to spend a month with their families. When

---

the possibility of any serious consequences to France as a result of Germany's aggressive actions in other parts of Europe only months earlier.

[5] Sartre 1984, p. 354. Paul Baudouin, secretary of the War Committee in Reynaud's government and later foreign minister, also noted (as did many others) the general apathy among the French, and how the German threat seemed to fade from memory as the *drôle de guerre* drew out month after month in early 1940. Baudouin 1948, p. 87.

[6] Quoted in Horne 1969, p. 201.

[7] As a member of the Haganah, constant vigilance was a daily part of Robinson's life. What he faced on active duty in Jerusalem must have made the rising tensions in Europe seem remote. In the first half of 1939 alone, 643 victims had been killed in Palestine as a result of the ongoing demonstrations, riots, shootings, bombings, and persistent terrorism practiced by both Jews and Arabs. *New York Times*, July 2, 1939, p. 6, col. 7.

> we reached Alexandria we saw another ship at the pier, *Le Champollion*, on its way to Syria taking back soldiers who had finished their vacations.[8]

Marseilles, as Robinson and Fleischer first saw it, was also teeming with military personnel on leave. Everywhere the city was postered with instructions for soldiers in transit. Because all of the night trains were packed, with no extra space to Paris, Robinson and Fleischer were forced to wait an extra day before they too were able to find a day train going north.

Mathematics in Prewar Paris

Once in Paris, Robinson and Fleischer found rooms in a small, family-run hotel in the Latin Quarter, not far from the Sorbonne. Abby enrolled in the Faculté des sciences, for which he was issued an official student card by the Ministère des affaires etrangères.[9]

In January, just before Robinson left Palestine, Fraenkel had written a spirited letter of introduction on Abby's behalf to the French philosopher Léon Brunschvicg (1869–1944). Fraenkel recommended him highly, and hoped that Brunschvicg might agree to take Robinson under his wing, advise him about his studies, and introduce him to local mathematicians and philosophers.

Brunschvicg, born in 1869, was over seventy years old when Robinson arrived in Paris. A graduate of the École Normale Supérieure and the Sorbonne, he quickly became one of the most prominent French philosophers of his generation. In 1909 he was named professor of general philosophy at the Sorbonne, where he continued to teach until the German occupation. After 1940 he moved to Aix-en-Provence, and somewhat later to Aix-les-Bains where he died in 1944.[10]

Brunschvicg was not only a founder of the *Revue de métaphysique et de morale*, but was well known as an editor of Pascal and an authority on the works of Descartes and Spinoza. In his philosophical work, often characterized as "critical idealism," he attempted to understand the nature of human thought as revealed in the history of science, mathematics, and philosophy. Brunschvicg was especially interested in trying to account for the reasons why mathematics was so successful in applications, especially in physics. Basically, he believed that mathematics was so well suited for science because it was a free creation, independent of any physical interpretation. At the same time, mathematics was

[8] Robinson R-1940, p. 1.

[9] Official document signed and dated February 19, 1940, RP/YUL.

[10] For details of the life and philosophy of Brunschvicg, consult Deschoux 1949 and Messaut 1938.

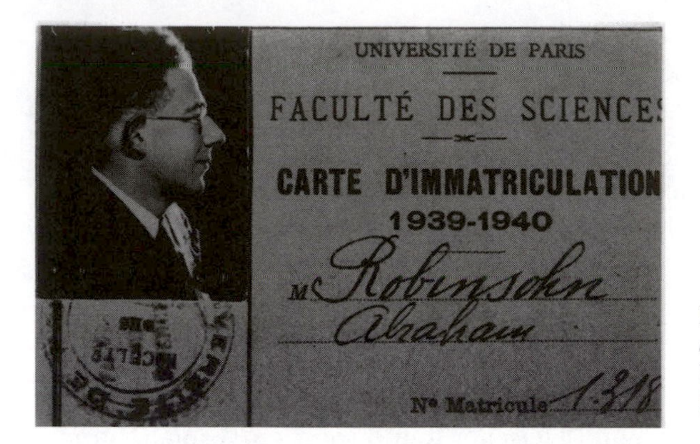

Robinson's *carte d'immatriculation*, University of Paris, 1940, RP/YUL.

inseparable from experience in both its origins and in its "collaborative task of assimilating being to the understanding."[11] One of his most influential books was *Les étapes de la philosophie mathématique*, first published in 1912.

Clearly such ideas would have given Robinson a great many things to discuss with Brunschvicg, if in fact they ever met under conditions suitable for serious exchanges. In one of Fraenkel's letters to Robinson written that spring (April 12, 1940), he acknowledged having heard from Robinson, and specifically noted how pleased he was to learn that both Robinson and Fleischer had been "well received by Professor Brunschvicg."[12] Fraenkel was especially happy to know that the Hebrew University was being appreciated abroad, thanks to the favorable impression Robinson (and Fleischer) were making.

Unfortunately, there is nothing to indicate how often Robinson may have been able to see Brunschvicg, and no record of what they talked about when they did meet. Since works of a religious nature seem to have dominated the last decade of his life (prior to his death in 1944), Robinson may have found little beyond academic courtesy to motivate further contacts. Some idea of Brunschvicg's interests when Robinson was in Paris may be gleaned from the title of a book by M.A. Cochet that appeared in 1937: *Commentaire sur la conversion spirituelle dans la philosophie de Léon Brunschvicg.*

Fraenkel's letter to Robinson, however, does suggest several other

[11] For assessments of the significance of Brunschvicg's philosophy of mathematics, especially as it was developed in *Les étapes de la philosophie mathématique*, consult the essays by Louis de Broglie, "Léon Brunschvicg et l'évolution des sciences," and "La philosophie scientifique de Léon Brunschvicg" by Gaston Bachelard, pp. 73–74, 77–84, respectively, in Brunschvicg 1945.

[12] A copy of Fraenkel's letter to Robinson in Robinson's files, Department of Mathematics, the Hebrew University, Jerusalem.

mathematical contacts Robinson may have had during the few months he was in Paris: "Please give my greetings to Professors Hadamard and Cartan." If Robinson indeed had occasion to do so, he would have met two of the world's greatest then-living mathematicians. But like Brunschvicg, Jacques Hadamard, born in 1865, would have been in his mid-seventies, and Elie Cartan, born in 1869, was as old as Brunschvicg. In fact, of all the celebrated elder statesmen of mathematics in France at the time, Henri Lebesgue was the youngest, born in 1875; Emile Picard was the oldest, born in 1856. In between were Ernest Vessiot, longtime director of the École Normale Supérieure, who was born in 1865, and Emile Borel, born in 1871.

Of the younger mathematicians in Paris with whom Robinson might have had some rewarding contacts, Paul Montel (born 1897) and Gaston Julia (born 1893) were the best known. Robinson may also have sought out younger colleagues with interests similar to his own, like the young Jean Cavaillès, who was one of the brightest talents in France before the war. Cavaillès had written a good deal about set theory and foundations of mathematics, and it is easy to imagine Robinson enjoying lively conversations with him on these and related subjects. Tragically, Cavaillès was arrested during the Nazi occupation of France, and did not survive.[13]

The Diversions of Paris

Although Robinson's own account of the months he spent in Paris in 1940 gives no indication of what he accomplished mathematically, his description of Parisian life in general was vivid. To Robinson Paris was, in Montaigne's words, the "crown of France and, moreover, one of the most beautiful monuments in the world."[14] Compared with the austerity of Tel Aviv and Jerusalem, the vitality of the city's social activity and the extent to which its cultural, scientific, and educational institutions still flourished, despite the onset of the war, was amazing. Audiences continued to crowd the operas, theaters, nightclubs, and dance halls, so much so that Robinson thought the best slogan for France at the time was: "We shall sing and dance in spite of it all."

Mixed with this blind refusal to face the reality of the war was a staunch sense of optimism; posters and sidewalk billboards proclaimed: "With

[13] Cavaillès was arrested by the Gestapo in Paris on August 28, 1943. From a prison camp in Compiègnes, he was taken to Arras where he was executed. Later discovered in an unmarked grave with twelve other bodies, Cavaillès is now buried not far from Descartes in the Sorbonne Chapel. Santiago Ramírez in a letter to JWD, September 8, 1994. For accounts of Cavaillès's life and work, see Ferrières 1950 and Sinaceur 1994.

[14] Robinson R-1940, p. 2.

patience will come victory." This same motto was also used by the French post office on stamps sold throughout the country. On the lighter side, street vendors were selling pictures with Hitler's head on the bodies of gorillas, pigs, and elephants. The Germans, however, were so bold as to warn the world of what was to come. On April 15 Hitler announced that he would be in Paris by the middle of June, but as Simone de Beauvoir remarked, "no one took these impudent boasts seriously."[15] Instead, as Robinson observed, all the French could think to do was sit and wait:

> And so the days passed, and so millions of soldiers along the Maginot line and in French barracks, as well as hundreds of thousands of soldiers of the Eastern army and in the French colonies, waited to see what would happen. So did the civilians, and as it turned out later, so did the cabinet ministers and the generals.[16]

Spring came late. Despite the sandbags and the annoying air raid sirens (to which Parisians rarely responded), life carried on. Although France had instituted the blackout, Robinson was amazed that no one took such emergency measures seriously. Most ironic of all, he thought, was the fact that each night as he walked up the rue Soufflot towards the Panthéon, "On the right was a window where a bright light would shine, and that was a window in the police station of the fifth *arrondissement*." This nonchalance also struck the British correspondent Anthony Gibbs, assigned to Paris, as most peculiar. Despite the blackout, "the glare and dazzle of head-lights was like returning from the Aldershot Tattoo," he wrote. "Every shop in St. Germain and in the outskirts of Paris blazed to high heaven."[17]

The traffic and exuberance of the streets was matched that spring by the conviviality of the sidewalk cafés. Not far from Robinson's hotel in the Latin Quarter, two of the city's best-known establishments served the local literati, including artists and writers like André Gide, Simone de Beauvoir, Pablo Picasso, and Simone Signoret. Local personalities and international celebrities alike could be found at the Café de Flore (one of Sartre's favorites) or the Aux Deux Magots, both of which vied for patrons on the Boulevard Saint-Germain.[18] Across the street, the Brasserie Lipp offered good meals at inexpensive prices. As one Parisian *habitué* described them just before the war:

> At these two cafés, the chicory-roasted coffee was the most emphatic, their croissants and brioches the freshest, their light meals the tastiest and the house wine the most respectable, of all Paris. This was in the very heart of

[15] De Beauvoir 1960, p. 445.
[16] Robinson R-1940, p. 3.
[17] Quoted in Horne 1969, p. 102.
[18] De Beauvoir 1960, p. 435.

> creative Paris, where writers and publishers, artists and gallery directors lived within a block or two of each other.[19]

How much of Parisian café society Robinson saw can be guessed by the extent of his budget, limited as it was to his monthly stipend provided by the French government. But there was plenty to do in Paris that could accommodate a student's resources, and both Robinson and Fleischer seemed to enjoy the active social and cultural life in the French capital. There was the city itself, of course, free for the walking, as well as the parks, public concerts, street plays, and exhibitions. Robinson even found time for a late winter holiday, the only time he ever went skiing.[20] On one occasion, he was especially taken aback when one of the two major orchestras in Paris, to the consternation of public opinion, offered an entire program devoted to Wagner![21] He was also an avid theater-goer and always enjoyed visiting art galleries and museums, interests that he nurtured first in Paris and, later, in London.

At the Théâtre Montparnasse Robinson saw Margit Jouvet in *Maya*, set in the red-light district of Marseilles but "full of vision" in Robinson's opinion. At the Pierre Blanchard another play dealt with premarital cohabitation, while at the Athénée, Louis Jouvet was starring with Madeleine Ozaret in Jean Giraudoux's mythical play *Ondine*.[22] Robinson regarded Giraudoux as an excellent playwright whose success was all the more remarkable because he was also a politician. In fact, he was the Minister of Information in Daladier's government at the beginning of the war. His play, adapted from de la Motte-Fouqué, expressed with typical "French *esprit* and *clarté* " everything that was only conveyed indirectly between the lines of the German novella. Robinson regarded the original as one of the great pieces of German literature of the romantic period, but he was clearly impressed by its French incarnation as well.

Most of the entertainment in Paris, Robinson was surprised to find, had nothing to do with the war, and rarely dealt with contemporary issues. Of all the plays he saw that spring, only one referred specifically to the war: "a rather banal spy play in which the well-known actress Gaby Morely participated."

Nor did the major theaters show any interest in the current political situation. Only standard productions of the classics were presented at the Opéra, the Comédie-Française or the Opéra Comique. One excep-

[19] Harris 1987, p. 5.

[20] Renée Robinson I-1982.

[21] Robinson R-1940.

[22] For details of the Parisian theater and cultural life in general at this time, see Knapp 1958.

tion was *1939*, which ran at the Odéon for several weeks before "the end," as Robinson put it. On the other hand:

> Attending the Opera in March, only three things made you aware of the war: the large number of officers uniforms in evidence, particularly Polish, the tapes which were stretched across the palatial mirrors in order to prevent their breaking from unexpected tremors, and the sandbags surrounding the sculptures inside the building.[23]

Cinemas, however, were more sensitive to timely topics. The war, Robinson found, as well as war-related themes, were given more prominent (if not more insightful) treatment at the movies than they were on the stage.

As for art, the galleries in Paris, like the theaters, were mounting new shows as usual. Sometimes, next to a painter's name in a catalogue or accompanying a picture on exhibit, it would be noted that the artist had enlisted. Although most painting seemed detached from recent events, Robinson observed that "the greatest influence of the war was noticeable among the independent, young painters, those who voluntarily or involuntarily did not find themselves among the well-known painters of the day."

Abby with his French béret, Paris, 1940.

Robinson was impressed, for example, that the April exhibition of modern painters included works by Jewish refugees from Germany. But he was disappointed when an exhibition of conventional academic painters (entitled simply French Art) did nothing to reflect on the major concerns or events of the late 1930s, although one of the galleries did mount an exhibit of toys, including war toys, and these made quite an impression. But none of the galleries in Robinson's opinion mounted anything of serious importance considering what was happening in the world at the time.

The only exception was an exhibition in the rue de La Boétie of recent work by the Belgian artist Frans Masereel, who Robinson felt was appreciated more in Israel (and outside of France in general) than he was in Paris. Masereel was a member of the Association of Revolutionary Artists and Writers, an antifascist group. He had been sympathetic to the Republican cause in Spain, supported Léon Blum's Popular Front government in France, and stressed socially progressive, even radical

[23] Robinson R-1940, p. 6.

themes in much of his artistic work.[24] Robinson was sensitive to the importance of Masereel's ideas, and lamented that "the French people never quite understood the deep meaning of his work."

One exhibit was noteworthy for its direct relation to the war. *Keep your mouth shut*, prophetically enough, opened just before the Germans attacked, and was devoted to the need to guard military secrets. Robinson's eye could tell that it had been thrown together by amateurs whose intent was to display works done by French soldiers, including examples from all of the battalions along the Maginot Line. As Robinson commented dryly, this all "went back to the days when the soldiers spent their time playing, because there was no more serious business."[25]

The Invasion of Scandinavia: April 1940

On April 9 the Germans marched into Denmark and seized the country with hardly a shot being fired. At the same time they launched their offensive against Norway.[26] As news of these events grabbed headlines in France, the faces of people on the streets in Paris at last began to turn serious. Within minutes *Après-midi* disappeared from all the newsstands. The evening edition of *Paris Soir* came out in one, then a second and finally a third edition, each one supposedly a "final" but with different headlines.

Tensions eased momentarily when the English navy seemed poised to intervene, although the hopes this raised were clearly premature. Robinson was startled when usually so careful a reporter as Henri de Kérillis entitled one of his editorials in *L'Époque* over-optimistically as "Victory." Some actually took Germany's invasion to the north as a blessing, hoping this would mean that France might still be spared the destruction of battle on its own territory.[27] At least it was hoped that Germany, now venting Nazi aggressive energies against the Norwegians in particular, would buy France some time. Indeed, life in Paris returned all too soon to its previous casualness, and only those "with information" had a deep fear of what might come next.

The Battle of Europe: May 1940

Shortly before dawn on May 10, Paris was jolted from sleep by the scream of air raid sirens. Most Parisians did not bother to respond, but Robinson and Fleischer took refuge in the Métro Station Monge. It was only a

[24] For more on Frans Masereel see Avermaete 1975, Egbert 1970 and Herman 1980.
[25] Robinson R-1940, p. 7.
[26] For details of the German advance into Scandinavia, see Horne 1969, pp. 170–173.
[27] Murphy 1964, p. 38.

matter of hours before the all-clear was sounded. Again the Germans were on the move, and this time their sights were aimed directly at France. Except for the sound of a few shells, however, nothing dramatic was heard in the city until later the following morning when the dreaded news arrived: the Germans had simultaneously invaded Holland, Belgium, and Luxembourg. As the Allied forces scrambled to counter the enemy, a steady stream of reports from the front began to flow into the capital.

Almost immediately, fear and suspicion of a "fifth column" began to grow. As Alistair Horne described it: "Like wildfire the rumors of ubiquitous fifth columnists—the nuns in hobnailed boots, priests with machine pistols under their soutanes—spread."[28] Police started to carry weapons and identity papers were frequently checked. By an order of Georges Mandel, in his capacity as minister of Interior Affairs, the cafés along the Boulevards Saint-Michel, Saint-Germain, and the Champs-Elysées were routinely searched.[29]

Within days King Leopold of Belgium capitulated to the Germans. In Paris, an atmosphere of panic set in. As Fleischer recalls, increasingly he and Abby began to feel that any day they would have "to run for their lives." But where? Robinson wanted to set out for Nice in hopes of returning to Palestine. Fleischer preferred going in the direction of Bordeaux, where Spain offered a possible escape. The chances, he felt, were also better on the channel coast for embarking either to England or perhaps to the United States.

On June 10 Mussolini put an end to their indecision by proclaiming war against France and England. By then France was about to fall. Robinson and Fleischer, still in Paris, suddenly found their options severely limited. Any hopes they may have had to return directly to Palestine through the Mediterranean were now effectively blocked.[30]

As the Germans invaded Benelux, rumors about their progress began to spread. Robinson noticed small groups of women standing in the streets, whispering. One day it was Saint-Quentin, completely destroyed. Another day it was Reims, where a German division had annihilated the entire city! As the rumors became increasingly extreme, even ridiculous, Paul Reynaud found it necessary to give an encouraging public speech extolling the heroic deeds of French forces and promising severe punishment for anyone found guilty of spreading rumors. "They say that the enemy has got to Reims—at the time when our armies

[28] Horne 1969, p. 223. The role that journalists played in popularizing the idea of a "fifth column" in France is discussed in Murphy 1964, p. 34.

[29] Simone de Beauvoir comments on the increased police checks at the sidewalk cafés in Paris. See de Beauvoir 1960, p. 436.

[30] Knapp 1972, p. 7.

are gaining fame. They say that our front was broken through at a time when there is no front, only a 'pocket.' But there were other pockets that we straightened out in 1918!"[31]

This of course was all vain hope, and far from the reality of the situation. The very next day, following Reynaud's brave reassurances, Robinson met a Red Cross nurse who told him she had just passed through a camp for French refugees retreating from the north, and that indeed the Germans were directly in front of Reims. Soon Paul Reynaud would be forced to appear before the Senate and admit the worst.

Meanwhile, the boulevard theaters, dance halls, and night clubs were closed. The government theaters, oddly enough, remained open but with diminished audiences. As late as June 9, a stormy Sunday evening just before the final phase of the Battle of France, Simone de Beauvoir was at the opera with a friend to see *Ariane et Barbe-Bleue*, but the theater was empty. One had the impression that it was "a final swaggering, symbolic demonstration against the enemy."[32]

Prior to that, the only concession to reality came at the beginning of every performance, when instructions were given to facilitate an orderly exit to nearby shelters should the air raid alarms go off. And, in a much-needed gesture to patriotism, at the end of every performance the "Marseillaise" was played with greater force and enthusiasm than ever.

Suddenly one morning, all of the buses disappeared from the city. Later, Robinson learned they had been commandeered to save refugees from the north where the battle was intensifying. For the first time, he observed, the inhabitants of Paris actually felt the battle "on their own flesh," whereas the bombing raids, limited primarily to military targets on the fringes of the capital, never seemed to make much of an impression on public opinion. He thought it ironic that the only real disaster to befall Paris prior to May 10 was a stray shot from a French cannon. But all too swiftly, within days, General von Rundstedt's panzer divisions broke through the French positions near Sedan and, after May 15, disaster followed disaster.[33]

The seriousness of the German advance was reflected in a solemn mass held in the Cathedral of Notre Dame on Sunday, May 19. The entire government with representatives from all sectors of public life in France were present, including many members of the foreign diplomatic community. A large crowd assembled on the square in front of the cathedral, and as Senator Jacques Bardoux described it in his journal, there was "a splendid evocation of the Saints of France; while the relics of Ste. Geneviève, of St. Louis were paraded, . . . each of the saints

[31] Robinson R-1940. [32] De Beauvoir 1960, p. 451.
[33] Warner 1968, p. 159.

was invoked and the crowd responded in chorus. Bullitt [the American Ambassador], who was in the front pew, was unable to hide his tears."[34]

That Sunday was also the day that General Gamelin was relieved of his post, to be replaced by General Weygand who arrived at Vincennes the next day to take command of the Allied armies in hopes of reversing the calamitous setbacks of the past week.[35] But May 20 only brought further news of triumphs for the Germans. Amiens was virtually empty when the advancing Nazi army arrived. By midday the Germans had taken the city with little resistance, and then pushed southwards to secure the bridgeheads over the Somme in preparation for phase two of the Battle of France. As the German General Guderian noted in his orders that night for his forces the following morning: "Today's battles have brought us complete success. Along the whole front the enemy is in retreat in a manner that at times approaches rout."[36]

Finally, the moment for honesty had to be faced. The next day, May 21, Reynaud appeared before the Senate to admit publicly for the first time how bad the situation really was: "The country is in danger!" he reported. "It is my first duty to tell the Senate and the nation the truth." One observer, who was in the gallery at the time, described how "a gasp of bewilderment rose from the senators' benches."[37] After Reynaud's speech, Senator Bardoux described the atmosphere in the corridors as "terrifying." People appeared in the streets of Paris with tears in their eyes.

General Weygand's appearance at the head of the army, however, inspired new hope among the French. By May 23 the booksellers had reopened their stalls along the quais of the Seine, and at a cocktail party, Alexander Werth found people expressing confidence that things were "going far better" especially now that Weygand "had organized his Somme front." Everyone agreed that "Hitler made a mistake in not attacking Paris on May 16."[38]

[34] Bardoux 1957, quoted in Horne 1969, p. 476–477. See also Pryce-Jones 1981, p. 6. A week later more solemn rites were performed throughout Paris. This time the relics of St. Geneviève were displayed in front of the Panthéon, where Senator Bardoux described the "stricken and silent crowd, which has lost its voice so that it can no longer even sing the *Marseillaise* and recites the litanies mechanically," quoted in Horne 1969, p. 540.

[35] After a long military career, especially distinguished during World War I, Maxime Weygand was called back to Paris from Lebanon where he was commander in chief of the Eastern Mediterranean theater headquartered in Beirut. Born in 1867, he was over seventy at the time, but it was hoped that his experience and the high esteem in which he was held by all would be sufficient to reverse the fortunes of France in the war with Germany.

[36] Guderian, quoted in Horne 1969, p. 488.

[37] Werth 1940, and Horne 1969, p. 515.

[38] Werth, quoted in Horne 1969, p. 539.

Champagne and martinis may have fueled optimism among the well-to-do in Paris, but the Germans continued to pile success upon success. Enemy forces soon threatened to encircle the French and British at Arras, leaving only two roads open for retreat. On May 23 the decision was finally made to abandon the city and begin a massive, unprecedented evacuation of Allied forces from France. Between May 27 and June 4, 337,000 troops were ferried across the English Channel from Dunkirk. Of these, nearly a third were French.[39]

Meanwhile, the Belgian capitulation to the Germans on May 28 left the French aghast—and wholly vulnerable.[40] General Weygand, whose troops suddenly had to fend for themselves without British support as the Expeditionary Force was being evacuated from the beaches of Dunkirk, described what was left of his defenses extending from the Somme to the Maginot Line as a "wall of sand."[41] It was now clear, said an editorial in *L'Oeuvre*, that the French had been hopelessly unprepared. "One can only think with a shudder of our unawareness during the past eight months, in which our noblest duty seemed to consist of making the leisure hours of the soldiers at the front pleasant."[42]

Days of apprehension during the extraordinary evacuation of troops from Dunkirk ended, oddly enough, with a sense of success greater than anyone had anticipated or thought possible. As Robinson described it in his diary, "the press celebrated [Dunkirk] as a tremendous victory, and perhaps even exaggerated a little the part the French had played in the crossing."[43]

After Dunkirk, Paris simply held its breath. Everyone waited for the last "Battle of France," as it was called even before it had begun. There was not long to wait. Hitler, drawing upon German experience from World War I, did not allow his generals to repeat an earlier generation's mistake by hesitating a second time.

Paris, now virtually without defenses, was bombed for the first time on June 3. Casualties, it was reported, amounted to 250 dead.[44] Less than a week later, the sounds of distant cannon and the staccato detonation of bombs were almost continuous.

[39] Horne 1969, p. 551. The loss of Arras and its consequences are discussed in detail in Chapman 1968, pp. 192–194.

[40] The French, it seems, were completely unprepared for such quick resignation on the part of the Belgians. Not only did they regard this as betrayal, but Reynaud called it "a deed unprecedented in history." Parisians were so enraged that they threw Belgians out onto the streets and burned their belongings. For details, see Horne 1969, p. 541.

[41] Horne 1969, p. 542.

[42] *L'Oeuvre*, May 30, 1940, quoted in Horne 1969, p. 554.

[43] Robinson R-1940, p. 11.

[44] Horne 1969, p. 561. The major targets of the German Luftwaffe on June 3 were the Paris airports. See Pryce-Jones 1981, p. 6.

The Battle of France: June 1940

The Germans began their final assault early on June 10, and yet the next day Robinson heard Reynaud on the radio broadcasting the unbelievable news that all indications were encouraging for France. The army was retreating "in good order," he said, "according to plan." Reynaud went on to praise the acts of heroism at the Seine as being even greater than those remarkable feats the Allied troops had endured just days earlier at Dunkirk. Despite the government's efforts at optimistic propaganda, the Germans were fast approaching Rouen, and it was not long before they reached the northeast outskirts of Paris.

Already by June 8, a Saturday, it was clear that the "die had been cast" as Robinson put it. The next morning long caravans of automobiles began to make their way down the Boulevard Saint-Michel, slowly, heading south. As Robinson walked through the streets of the Latin Quarter to the seventh *arrondissement* that Sunday, it seemed that all of Paris was fleeing. He went into shops. He took the Métro. Everywhere he went he heard the same thing: "the verb *partir* in all its conjugations"; people had left, were leaving, would leave soon, will leave tomorrow, next week, soon, never.[45]

By now it was virtually impossible to get a taxi. Any cab not going to one of the train stations was hired to evacuate people to the south. Late that afternoon Robinson passed by the Gare d'Orléans-Austerlitz, not far from Notre Dame on the left bank. Thousands of people were jammed together, crowding the station and the streets around it. Robinson heard that many of them had spent the last few nights sleeping nearby in the Jardin des Plantes.

All of the Métro trains going to the *grandes lignes* were overcrowded. Those desperate to leave were no longer the "well-mannered French," Robinson recalled, but "frightened people who only sought to run for their lives. They pushed and they yelled and they had absolutely no consideration anymore for each other." Day by day there were fewer people on the Métro. Towards the end, in fact, all was quiet, "the ladies elegantly combed and dressed in high style, just like any other day."[46]

When Robinson stopped at the British embassy on the rue Saint-Honoré Monday morning, only the doorman was left. In front of the office of the military attaché (on the rue D'Aguesseau), a group of English soldiers was simply waiting, gathered around a parked truck. They were going on leave, they said, at the "last moment." Meanwhile, anyone arriving at the British consulate for advice was referred to the embassy, and from the embassy they were sent to the apartment of the military attaché.

[45] Robinson R-1940, p. 12.

[46] Robinson R-1940, p. 13.

Robinson did not know as yet that the entire British embassy staff, like most of the other foreign diplomatic corps (except for a few, like the Americans), had already abandoned Paris early on Monday morning, *en masse,* along with the French ministries. Later, foreign correspondents who were under the nominal protection of the Ministry of Information complained bitterly that despite solemn promises to the contrary, they had been given no warning to leave as well.

The government, in fact, had been living from day to day, with no real foresight or planning. As Robinson observed:

> When, in the course of a private conversation, Paul Reynaud's secretary was asked whether the government had prepared a future plan of action in the event of an invasion of Paris, he replied that it had not, and sadly added: "If only we had a few more 75mm guns (the type used for anti-tank warfare), the situation would have been different." Talk of the chicken and the egg! [47]

On Monday the Sorbonne was officially closed. Only the custodians remained, and they clustered about in small groups between the university buildings. Questions about final examinations (which were supposed to have been given just a few days later) were answered with the shrug of a shoulder.[48]

By evening the city was covered with a heavy "fog." This Robinson attributed to fires and the bombardment of industrial plants on the outskirts of Paris. Rumors, running rampant as usual, held that the "fog" had actually been produced by the Germans trying to conceal their activities. Others claimed that it was caused by the French to prevent air raids. Robinson described it as "black, dry and warm, consisting of a thin film of soot which penetrated the nostrils and soiled the handkerchiefs on which people blew their noses."[49]

[47] Robinson R-1940, p. 14. This same bad news stunned the prime minister as the French army was losing its hold along the Meuse in mid-May of 1940. Reynaud "rang up Daladier to ask him what were Gamelin's counter-measures, to which Daladier replied, 'He has none.'" Clearly lack of planning and implementation plagued the French at all levels. See Horne 1969, p. 30.

[48] It was only on June 10, when Simone de Beauvoir learned that examinations had been cancelled at the University of Paris, and that professors were free to leave, that she made up her mind to go as well. De Beauvoir 1960, p. 451. Since Robinson does not seem to have made up his mind to leave Paris until June 10–11, perhaps it was also the closing of the university that effectively forced him to decide, finally, along with Fleischer, to abandon Paris at the eleventh hour.

[49] Robinson R-1940, p. 14. In fact, the dense, acrid smoke was due to the burning of petroleum depots. Robert Murphy, a foreign service officer with the U.S. State Department, was counsellor of the American embassy in Paris during the early years of World War II. His memoirs recall being telephoned one morning at 4:00 A.M. by the manager for Standard Oil of France, William Crampton, who had been given "irrevocable" orders

Next morning the haze was even thicker, and Paris so dark that the sun could not be seen. "This is how free Paris sank," observed Robinson, "in the midst of the storm which, one might say, was developing more rapidly than events at first suggested." One day, anyone daring to mention plans to leave was considered a defeatist, but by the next day, the stream of cars and bicycles moving out of the city was underway in earnest.

Although Paris was declared an open city in hopes of saving it from as much senseless bombing and mutilation as possible, the police force stayed, disarmed. Rather than leave the cultural monuments and the city itself open to destruction, the government had wisely decided to abandon the city rather than try to defend it. As the French army retreated, the Germans marched into Paris on Friday, June 14, without opposition.

Fleeing Paris: June 11, 1940

The inner quarters of Paris provided a dramatic contrast to the mayhem elsewhere. As Robinson walked about on Tuesday morning, all seemed quiet. The streets were empty, patrolled only by national guardsmen who were busy examining papers. The Boulevard Haussmann was deserted. Near the Arc de Triomphe, at the top of the Champs-Elysées which was as silent and abandoned as the rest of the city, Robinson watched two elderly ladies trying in vain to push a little cart. It was little more than a wheelbarrow into which they had piled all of their most precious belongings.[50]

---

from the French general staff to destroy enormous stocks of petroleum that had been built up in the Paris area. "He said he wanted to touch base with the embassy before blowing up all of the oil accumulated at great expense over many months. . . . For days thereafter the city was covered with heavy black smoke, which provided a kind of Dante's Inferno background for the pitiful refugees from several countries who flowed in and through and out of Paris," Murphy 1964, p. 40. Pryce-Jones 1981, p. 3, also reports the burning of the fuel storage depots.

[50] There are many vivid firsthand accounts of the fall of Paris in June of 1940. Very similar to what Robinson saw, for example, is a description by Ilya Ehrenburg, a Russian correspondent in Paris who watched while "An old man laboriously pushed a handcart loaded with pillows on which huddled a small girl and a little dog that howled piteously," Horne 1969, p. 562. Similarly, Paul Léautaud in one of his diaries published in the *Mercure de France* wrote that "Along the boulevard Saint Michel, an uninterrupted flow of people is leaving Paris with the most varried means of transport: cars crammed with luggage, heavy trucks loaded with people and suitcases, people on bicycles, others pushing a small handcart, with a dog tied underneath on a lead, huge country carts going as fast as two or three slow horses can pull them. . . . At midday, Boulevard Saint-Germain, same spectacle. Rue Dauphine, luggage all over the pavements, trucks loading up. Shops shut. Some are leaving with nothing more elaborate than a pram," Pryce-Jones 1981, p. 9. See also Murphy 1964, p. 42.

By 11:00 A.M. Robinson was back in the Latin Quarter. The sun barely managed to break through the gloom, and the city must have seemed dreary and desperate. He had just heard that the Germans were as close as Pontoise, only some 30 kilometers northwest of Paris. It was clearly time to leave, dangerously close to being too late, but with his friend Fleischer, they decided to make their way towards Bordeaux.

In trying to find a map before they left, Robinson discovered that every regional guide to the south of France from Paris to Orléans, along with compasses, backpacks, bicycles, and any other basic travel items, had long been sold out. It was equally impossible, Robinson noted wistfully, to get a decent lunch on the Boulevard Saint-Michel, since most of the restaurants were closed. But he did manage at last to lay his hands on a map, and a fairly good one at that.

Returning to their hotel, Robinson and Fleischer found the place in total chaos. Residents were abandoning their rooms, running in and out with suitcases and parcels, leaving whatever instructions they could think of about what they had left or where they were going, if they knew at all. The owner, who was loading his own car to leave with his family by way of Fontainebleau, was amazed when the two foreign students paid him the balance of their rent. "Apparently," Robinson surmised, "he had given up on it."

The immediate problem was finding their way out of Paris itself, since no long-distance trains were running from the city center. Robinson also foresaw that the stations would be chaotic. Fortunately they heard of a suburban train to the Chevreuse valley that was still operating as usual, and this got them as far as Orsay—about 20 kilometers south of Paris. From there they continued on foot.

Orsay, a small, normally sedate town with only occasional visitors, was brimming, alive with activity. As in Paris, there was a constant stream of vehicles of every variety and description imaginable—small cars loaded to capacity, carriages, bicycles, motorcycles, even trucks, some jointly hired by entire groups. Everything was moving south; nothing was headed toward Paris.

> The caravan moved slowly. People were not following any plan or purpose, but were just drifting along. When we arrived at the road junction, we realized that almost all of the vehicles had taken the main road to Orléans, even though there was a perfectly good side road by which it was possible to get there faster. Moreover, cars traveling on these roads could cover no more than 40 km a day. Often we saw people stopping for a picnic, or camping towards the end of the day at the side of the road or sleeping in their cars. There was no one on foot, and people regarded us with amazement. I already knew from before that the French are not fond of hiking. Other than the bicycle, they can conceive of no other way of

> traveling on their own. Apparently, a day or two later when no other choice was left, some people did leave Paris on foot, but by that time it is unlikely that they managed to escape.[51]

Robinson was an honest observer, not only of what was happening to France, but of his own feelings about the hundreds of thousands of refugees from Paris fleeing the advance of German troops. He viewed the French with as much detachment as they did him:

> I bear no resentment to those miserable people who were traveling along that road, at the time more anxious and worried than either of us were. But it is a fact, and perhaps not an accidental one, that although several half-empty cars have passed us, not one of them ever offered us a lift, not even to the nearest town from which the trains were still operating regularly. On the other hand, several drivers actually stopped to consult my south-of-Paris road map, or to ask me for directions when they had lost their way.[52]

Robinson, as he and Fleischer passed through the Île-de-France, mused that this was the "golden setting of the gem of France, the enchanted region where Pascal had philosophized, Racine had written his poetry and Corot had painted." But now, all of that was forgotten. Instead of poets or painters, tanks—French tanks—were passing on the country roads, edging through the masses also heading south. Occasionally, explosions thundered in the distance, bombs and antiaircraft guns could be heard from afar, and every so often a plane would buzz overhead. Soldiers were setting up beacons to help spot aircraft. Others were digging trenches at road junctions, while elsewhere young villagers armed with rifles and members of the National Guard, formed largely as a defense against foreign agents and presumed fifth columnists, filed along.

Robinson and Fleischer planned to sleep in the forest of the Chevreuse that night, but the buzzing of mosquitoes and the rustle of leaves during a constant drizzle conspired against them. Even without such distractions, the noise of constant traffic on the roads made sleep impossible.

Shortly before midnight Robinson suddenly began to hear a different kind of explosion, one that came in pulses, with bright flashes of light brimming up on the horizon. Worried that this might be the result of cannons signaling the advance of the German lines, they thought better of sleeping and decided to get back to the road and hurry on.

Soon after leaving the soggy forest, Robinson and Fleischer were overtaken by a convoy of trucks. They were promptly stopped by soldiers

[51] Robinson R-1940, p. 16.
[52] Robinson R-1940, p. 17.

who found their looks suspicious. After brief questioning they were ordered onto one of the trucks and told to sit in full view of the headlights on the truck behind them. Periodically the caravan would stop and one of the officers would come around to borrow Robinson's map, which was more detailed than any the soldiers themselves had!

Slowly this caravan of trucks moved westward, away from the French lines stretched along the Seine. When they finally reached the town of Dourdan, the papers Robinson and Fleischer carried were carefully scrutinized by a policeman. A tense moment followed when it was discovered from Robinson's passport that he had been born in Waldenburg, Silesia. Fortunately, since both he and Fleischer were traveling on British passports—with special status from the French government because of their scholarships—it was decided that nothing was really amiss, and they were allowed to continue.

Dourdan was overflowing with refugees. Thousands of the displaced and homeless were stretched out with blankets on wooden stalls and planks strewn about the local market place, where Robinson and Fleischer tried to find a free spot. Throughout the night a constant commotion of horses, engines, screeching wheels, cars, trucks, and tanks made sleep impossible. Adding to the noise, orders were constantly being shouted back and forth between army officers headquartered in the square, without interruption. And always, in the background, there was the relentless pounding of artillery and antiaircraft guns.

After only a few hours rest, fitful at best, Robinson was once more on the alert. Again he recognized the same kind of explosions that he had heard earlier, prompting their sudden flight from the forest. Now fully awake, he overheard two Frenchmen next to him:

"What time is it?"
"Half past three. There's an air raid alarm on just now."
"Quite so. And what are these explosions?"
"Artillery (and he mentioned the size of the guns)."
"So they are not far away?"
"No."[53]

How far away the Germans really were, Robinson never knew. Similarly, he was not certain if the explosions were actually the result of German or French artillery as the battle seemed to move ever closer. Rather than invade Paris hastily, the Germans first surrounded the city on all sides before actually moving in to claim the capital of France several days later. They did so, as everyone who witnessed the final moment observed, with great order and "control."[54]

[53] Robinson R-1940, p. 18.

[54] For descriptions, see Horne 1969, p. 563, and Murphy 1964, p. 42.

By 4 A.M. (June 12), Robinson and Fleischer were again on the road. Cars were parked everywhere, packed with sleeping passengers. As the sun slowly rose and the day progressed, so did the traffic, which soon grew as heavy and lethargic as before.

After several hours Robinson and Fleischer stopped at a small village for breakfast, hoping for some news. All they could determine, however, was that all road connections with Paris had been severed. Neither mail nor newspapers had arrived. Otherwise, absolutely nothing was known. At one shop they were able to buy strawberries very cheaply (having been intended for sale in Paris, which was now out of the question).

By lunchtime Robinson and Fleischer passed a farmer plowing his field, a sign that they had now reached calmer surroundings. An hour later they arrived at a large village where they tried to find something to eat in one of the local restaurants, but to no avail since all were full. This made little difference, they were told, because all of the food was gone anyway. As Robinson and Fleischer continued on, they eventually chanced upon a small bistro where they hoped at least to get something to drink. But before they could ask for anything, the young proprietor declared that he was a member of the National Guard, along with his friends. As Robinson observed, "with the courtesy and shyness of a new trainee—this apparently was his first chance to carry out his duty—he requested to see our papers."[55]

As before, their passports were duly examined and returned without further difficulty. This time, however, there was a chance for some polite conversation. They asked about the latest news on the radio, and were told that the French were retreating in "orderly fashion." They also learned that the train in the neighboring town of Angerville had resumed normal service. Immediately they decided to make it their next destination. Upon reaching the Angerville train station at about 3 P.M., Fleischer sat down on a bench while Abby went into the station master's office for information. He reappeared with two tickets.

According to the ticket seller, there was to be a train leaving Angerville at 6 P.M. The crowd of waiting passengers grew steadily as numerous trains passed by, one of them full of soldiers who laughed and waved their hands. The trains going north were empty; those going south sped by without stopping. Finally the station master took it upon himself to halt an approaching freight train, onto which he put everyone waiting in the station. It was about midnight when Robinson and Fleischer finally reached Orléans.

There were no lights in the station. Thousands of people were

[55] Robinson R-1940, p. 19.

milling about in total darkness, trying to find trains as directions were periodically shouted by a soldier over a loud speaker. Fortunately, from Orléans Robinson and Fleischer were eventually able to continue their journey by train, haltingly but without too much delay, first along the Loire and then across the plain of southwest France. Their plan was to reach Bordeaux within the next twenty-four hours.

At times the train would stop between stations or before entering a new region. Between Tours and Angoulême it crawled at what Robinson described as "an almost democratic pace. Timetables no longer existed, and trains moved seemingly by their own wills, carrying their passengers, all of them refugees, anywhere to the south." While on the train, with nothing else to do, there was ample time for conversation:

> One of the passengers in our compartment, a woman whose nerves must have been affected by all these upheavals, made a spontaneous speech against all aliens, pointing out the extreme generosity of the French people. Her neighbor (by his accent from Alsace) contributed an explicitly anti-Semitic remark. The rest of the passengers commented indifferently from time to time.
>
> I reacted with a few words, and this was enough to change the atmosphere. Yes, French anti-Semitism, which came to the surface once again during the period preceding the armistice and, more intensely under German influence afterwards, is entirely artificial. While there are circles in which it is deeply rooted, the concept of race is quite foreign to the majority of the French people and they lack the prerequisite of "true anti-Semitism": the ability to recognize a Jew.
>
> One of the women asked our pardon. The Alsatian worker also sought to explain his behavior and the instigator of this incident chose to move to a nearby compartment where she let her nervousness expose itself in some other way. Afterwards, the man from Alsace informed us (without being asked) that the factory where he had worked—some unnamed factory for airplanes—had been transferred from Paris to a place on the Spanish border which he mentioned, in flagrant violation of the order "Quiet—the enemy is listening."[56]

Further south, when the trains stopped, French Red Cross volunteers came up to the cars offering refreshing drinks, and the military opened its own canteens to refugees wanting something to eat. By the time they

[56] Robinson R-1940, pp. 20–21. On anti-Semitism in pre-Vichy France, see Murphy 1964, p. 37. Greater detail on the subject of Jews and anti-Semitism in Vichy France may be found in the section on "France and the French," in Paxton 1972, pp. 168–185. See also his book *Vichy France and the Jews,* Marrus and Paxton 1981.

crossed the Garonne Bridge to Bordeaux it was well past nightfall, and soon, as they approached the Saint-Jean station in the center of the city, Robinson saw a remarkable sight. For the first time since he had been in France, the railway station into which they disembarked was fully illuminated.

Bordeaux

Robinson was surprised to find that Bordeaux, one of the most important provincial towns in France, was to his eye so drab and lackluster. He compared it with Haifa and Tel Aviv, which he thought were much livelier despite far fewer inhabitants. Haifa and Tel Aviv, of course, did not have to compete with Paris.

Robinson did not suppress his disappointment at finding so little of interest in Bordeaux: "three or four squares with monuments such as the Girondins statue, or buildings such as a church or a theater—and the boulevards connecting them—this is all there is to the 'city' of Bordeaux."[57]

The city, however, was certainly not at its best, crowded as it was with hundreds of thousands of people. Every room in Bordeaux was strained to capacity and beyond. Many, recognized as Parisians by their dress and their speech, crowded into the cafés like Le Régent in the Place Gambetta. It was there that Georges Mandel was arrested the day after his resignation, only to be released a few hours later with an apology from the new prime minister.[58]

Some, such as students and factory workers, were lucky to sleep in schools either in Bordeaux or the suburbs. Others less fortunate slept under the gates and archways at the train station. One typical conversation that Robinson overheard was between a British vice consul and an English woman:

> SHE: Can you tell me where I can rent a room while I am in Bordeaux?
> HE: I am sorry to tell you that there are no more empty rooms in the city. Where did you sleep last night?
> SHE (excited): In the gutter! After all, it is your duty to get me a room. You too have a room where you can sleep.
> HE: I attribute your excitement to your nervous state. I got here two weeks

[57] Robinson R-1940, p. 21.

[58] "Pétain, acting on rumor, ordered the arrest of Georges Mandel, who had been Minister of the Interior [and] who believed—as Reynaud—that the war should be pursued at any cost. Mandel was Jewish. Some sort of denunciation had been made that he was planning an armed coup against the new government. So absurd, Mandel was let go almost at once with a written apology from Pétain," Pryce-Jones 1981, p. 11.

> ago from Dunkirk and today I sleep in a kitchen with my wife and two children.[59]

Robinson and Fleischer whiled away their time by walking around the city. The old town with its narrow streets dated from the Middle Ages, although much of it had been destroyed during the French Revolution. Robinson was particularly interested in observing the city's inhabitants and their adjustment (or maladjustment) to the war. Posters lavished the usual propaganda on walls and bulletin boards, among them: "Buy war bonds." Long lines in front of the District Savings Fund, however, primarily served clients making withdrawals.

By Friday everyone knew that the French government itself had come to Bordeaux, seeking refuge. To Robinson this seemed especially ironic. The place so intimately linked with the birth of the Third Republic was now destined to become, as he observed, "its grave."

Occasionally a crowd would assemble to watch the president of the Republic alight from his car, or to cheer Paul Reynaud as he drove to one of his last meetings. Throughout the weekend soldiers, black and white, stood in front of the municipality buildings in Bordeaux where the government was deciding the fate of France, to the extent (as Robinson commented) that it was at all within its power to do so.

As the government met, trying frantically to decide what to do next and whether the time had indeed come to surrender, the Germans continued to advance and refugees continued to pour into the city:

> Bordeaux, normally a quiet, provincial town, buzzed with millions of human beings. At the same time German tanks and armoured cars roared on the highways of central France, just as German and Italian planes flew over Bordeaux and the Gironde valley. That night Bordeaux was bombed for the first time—the city which believed itself so safe that there were no shelters. We, the refugees from Paris, who only a week ago thought we would be spared having to see the capture of the capital, were now witnessing the end of the French government.[60]

On Friday morning, June 14, Robinson took a taxi from the train station to the British consulate. Driving along the Garonne River, the driver said dramatically, while pointing at the immense river: "Here they will not pass." But Robinson knew better, and laconically observed that less than two weeks later, pass they did.[61]

At the British consulate that morning, anyone seeking advice was told that, in an emergency, an English boat anchored at the mouth of the Garonne would be confiscated to evacuate refugees. No one, of

[59] Robinson R-1940, p. 22.

[60] Robinson R-1940, p. 23.

[61] Robinson R-1940, p. 23.

course, was certain what would constitute "an emergency." Under the circumstances, since Bordeaux was only a few hundred miles from the approaching German front, didn't that constitute emergency enough? A note, posted outside on the consulate gate, ended tersely: "There are no more rooms available in Bordeaux."

Once again, events began to develop with terrifying speed. Radios that evening broadcast the news that the enemy had again broken through the French lines. At the final hour, Paul Reynaud called upon the Allied forces to come to the rescue of France. He also made one last desperate appeal to President Roosevelt for the United States to enter the war, but to no avail. Throughout the weekend, the government in exile continued to meet, until Reynaud resigned and Marshal Pétain was appointed in his place. It was Pétain who insisted that the fighting must be stopped.[62]

On Sunday, the British consulate announced that there would soon be a boat ready to leave. British citizens "desiring to reach England" were instructed to assemble with enough food for themselves at the consulate. Anyone with a British passport, from tourists to those who had resided in France for years, including people with relatives in England or having any other reason to hope they might be included, queued up. Everyone waiting had stories to exchange about the last few weeks, in particular how they had managed to get to Bordeaux from all parts of France.

Eventually tickets were distributed by a consulate official. In the afternoon a special train took everyone to La Verdun, at the mouth of the Gironde. There, English and French sailors helped to transfer passengers via a small motorboat onto the rescue ship anchored only a short distance away.

Once aboard, however, the ship—actually a coal tender—did not sail immediately. In the evening, as Bordeaux was being shelled by the Germans, La Verdun did not escape either. Several enemy bombs dropped very near the ship, whereupon the boat's cannon responded with fire of its own.

The following morning the ship was still at anchor. More refugees arrived, packing the ship to the brim. Some were English and French officers, others journalists and politicians. All the while they were boarding, planes and antiaircraft guns were fighting a battle above the port. And in one unforgettable moment, as those on board watched, a plane was shot down, its tail afire.

Finally, towards evening, the boat sailed. It was only one of many leaving France for England (or for North Africa) in those days. Many

[62] The story of Reynaud's resignation is told succinctly in Chapman 1968, pp. 311–313.

were fishing boats from the shores of the Bretagne. The ship Robinson and Fleischer were on, meant to ferry them across the channel to England, was now full with refugees of every nationality and religion. It was also the last ship to escape Bordeaux.

Reflections on the Fall of France

No sooner were they on board, the ship not as yet underway, when the French refugees began to ask themselves, "How did this all happen?" Others, of course, wanted to know who was to blame? Robinson had already begun to formulate some answers of his own:

> The French themselves have a strong tendency to answer here, as they have on similar occasions in the past: "We were betrayed, we have been sold out." The idea that Gamelin was not only unfit for his position of great responsibility, but that he did in fact betray France, was very widespread after the catastrophe of the Meuse.
>
> When I got to England, where I had the opportunity to talk to French refugees, I heard, "What can we do? We were sold out everywhere." And the same tendency, to throw the blame on the bad intentions of a few, turns up again at the Riom trials, at the court of law established by the Vichy government. But considering the list of lawsuits executed by the courts up to now, one can see immediately that there is nothing in them but a search for scapegoats.
>
> In the English literature on the subject you will often find, as the explanation of these events, the Fascism prevalent among statesmen and high-ranking officers, as well as their defeatism. Others say that the workers and peasants did not want the war, primarily under the influence of the communists. This opinion is also supported by people who were involved with French workers. There are two versions of this interpretation. One claims that the German-Russian pact had a direct influence on the stance of these classes; the other, expressed also by Lloyd George, claims that their opposition to the war originated from the persecution of the communists after the signing of this pact.
>
> "We were not prepared, yet we entered this war," said Marshal Pétain on the radio. Paul Reynaud's secretary, as mentioned before, also delivered a speech in this same spirit. Their meaning: what we mostly lacked was equipment. If only we could have had it in time, even at the beginning of June, aeroplanes, tanks and anti-tank artillery, then we could have held out. . . .
>
> In a conversation during the fighting in Belgium, I also heard: "France has no ideals"—I was told this by young people, high-school graduates who came to join General de Gaulle's movement to free France.
>
> What degree of truth is there in any of these claims? Apparently open

> treachery, which played such an important role in Norway and Holland, was not so important in France where the Germans won by sheer military strength. On the other hand, it is true that the French were not prepared, but this fact became apparent to them only as the catastrophe was already taking place.
>
> At the beginning of April I heard these words from Mr. Herriot's own mouth, spoken with that special charm so peculiarly his: ". . . and so, ladies and gentlemen, our victory is completely assured." This was said in the middle of a lecture on the differences between the French and German cultures, not after long deliberation of the chances and possibilities, but completely casually, as if it were a self-evident thing.
>
> Big propaganda posters proclaimed: "We will win because we are the stronger." And in every bookstall in the Latin Quarter, a booklet on the French army was on sale which offered proof that the French armies were ready for every eventuality. This booklet was written by the same man who today is one of the main pillars of Pétain's government, none other than the Commander-in-Chief of the French army before the armistice, General Weygand. This same man also carries some of the responsibility for the Riom trials, in which they are trying to find reasons for the failure of the national defenses.[63]

Robinson also pondered what he called the "Fascism of defeatism" that had developed in France, something he regarded in itself as a strange phenomenon. It had, he said:

> all of the negative elements of Fascism, all the hatred of equality and freedom, and none of the driving force of the concept of national greatness. And though it was true that there were Nazi sympathizers in France before the war (among them some government members like Bonnet), and even though some like Baudouin tried to encourage a spirit of defeatism among his colleagues in the cabinet in the last few weeks, the vast majority wanted victory, including the military and naval commanders from Gamelin to Weygand, from Auriol to Darlan, and most importantly statesmen from Daladier to Reynaud, from Herriot to Lebrun.[64]

On the other hand, Robinson believed that the great majority of the workers did not realize the true significance of the war. Like the peasants who were crucial to the spirit of the army (in which the percentage of workers was comparatively smaller), French workers were largely sym-

[63] Robinson R-1940, pp. 25–26. This last remark suggests that Robinson must have been working on his reminiscences of his student days in Paris, and his subsequent escape from France, as the Riom trials were taking place, i.e., between February and April of 1942. See Aron 1958, pp. 298–300, and Hytier 1958, pp. 329–330.

[64] Robinson R-1940, pp. 26–27.

pathetic to the communist slogan, "We have no interest in the war of our capitalist government."[65]

Robinson was also extremely critical of newspaper censorship. From time to time, blank spaces appeared in newspapers of all persuasions. Sometimes, Henri de Kérillis' articles in *L'Époque* were completely struck out. One of the editors of *Le Temps* told Robinson that the Ministry of Information used to send instructions about what should be emphasized and what was to be suppressed.[66] For example, two days before Italy declared war on France, all of the anti-Mussolini propaganda (which had only begun to appear a few days earlier) disappeared entirely from all the newspapers. Apparently, the overriding opinion in the government was that it would be best to avoid any trouble with Mussolini which might unnecessarily precipitate Italy's declaration of war.

Robinson found it remarkable that the French press itself seemed so indifferent to the issue of censorship. Unlike Britain, where stormy discussions in Parliament prior to the formation of Churchill's government exerted crucial influence on subsequent events, Robinson found that there was no such debate in France. André Chamm, author of some of the most important editorials in *Paris Soir*, called the British debates nothing more than "chatter." Only *Le Temps* was willing to stress the importance of the freedom to criticize, even in times of war. Robinson believed it was significant that *Paris Soir*, which he felt was "characterless and capricious," nevertheless was a better reflection of public opinion in France than was *Le Temps*, although he regarded the latter as "an excellent paper in itself."[67]

L'Information, Robinson recalled, was the one paper to react strongly against censorship. From time to time, at least, it filled the empty spaces gouged out by the censors with political caricatures as a gesture of protest. Meanwhile, the major debate in the government over censorship and the activities of the Ministry of Information in France met with little interest. Robinson followed carefully the "brilliant speech" by Léon Blum before the French Senate on the subject, but was disappointed that the debate "continued a few days later in front of a half empty house."[68]

Worse than its passive acceptance of censorship, however, was the way the French press abused what little freedom it did have to delude and mislead the public to an even greater extent than official government announcements did. When the Germans invaded Belgium, for example, Robinson scoffed at the newspaper accounts which reported that "Our generals foresaw these developments." And when General

[65] Robinson R-1940, p. 27.
[66] Robinson R-1940, p. 27.
[67] Robinson R-1940, p. 28.
[68] Robinson R-1940, p. 28.

Corap's army was annihilated on the Meuse, the newspapers spoke preposterously of a dangerous "lengthening of transportation lines that would be harmful to the enemy." [69]

Similarly, the gap on the front created by the Germans was stubbornly referred to as a "pocket." Robinson could scarcely contain his contempt for this kind of reporting:

> When German armored cars and tanks started through [the "gap"] towards Calais and Boulogne, [the press] decided that now a guerrilla war had begun, and after all, in guerrilla warfare inventiveness, quick thinking, instant decisions and immediate actions were required—all of which are great characteristics of the French. Therefore, do not be afraid, we shall win! [70]

But as Robinson noted, the hopes raised by such hyperbole in the press were never to be realized. Despite the clear fate of France sealed by the evacuation from Dunkirk, the French seemed determined even then to ignore Churchill's warning that one does not win by retreats! Even at the beginning of June the French press quoted a Swedish newspaper as saying that the latest events in Western Europe were proving once again that the French soldier was still the best in the world. Robinson, however, knew how self-deluding the French had always been in this regard:

> Following in Paul Reynaud's steps, who was ready to believe in miracles if a miracle were necessary to save France, Louis Madelin of the Academy maintained in *L'Époque* that in the darkest days of the previous war, he used to find comfort in saying to himself from time to time: "France cannot be lost, France cannot be lost. . . ." And again they mentioned Marshal Joffre, who said that he would retreat until the moment when he had created enough distance between himself and the enemy, and then he would begin the battle.[71]

In those days, Robinson found that the only accurate information about what was happening on the front was to be found in the *Petit Parisien*. Even there, unfounded optimism could get the upper hand, as in the case of the paper's military correspondent Charles Maurice just before he left Paris. Having spoken to General Weygand, Maurice wrote: "Now, when all seems lost, he [General Weygand] is still hoping. Let us believe in him! He reminded me of Marshal Foch's words when he said

[69] General André Corap, whose career had been made largely in North Africa, was named to command the Ninth Army, which was responsible for defense of the Ardennes, on October 15, 1939. For details of the battle on the Meuse, until Corap was relieved of his position as head of the Ninth Army on June 15, see Chapman 1968, pp. 106–142.

[70] Robinson R-1940, p. 29.

[71] Robinson R-1940, p. 29.

that he was waiting for one mistake on the enemy's side, and then he would show them." But as Robinson dryly added, the opportunity which presented itself to Marshal Foch in 1918 did not, unfortunately, repeat itself in 1940.

Robinson attributed French complacency to the country's full recovery following victory in World War I. France, he believed, was one of the happiest countries in Europe, despite all of its strikes and scandals. This was so, he thought, because the French were actually not very interested in the cultures and fates of other countries. As long as they were left alone, they had little concern for the rest of the world. Although many in France mixed hatred with fear of the Germans—*Les Boches*—they also understood that any new war in Europe would only bring disaster to France. Even so, they seemed to care little for what happened beyond their own borders.

This was especially true in the Midi where, according to French in the north, Robinson understood that any national feeling whatsoever was entirely alien. One story he thought typical concerned a government minister who reached Bordeaux in 1914, when Paris was partially evacuated during the First World War. Upon his arrival the official was greeted with the exclamation "Oh, but you have a war going on up there!" As Robinson added with his own emphasis: "*You*—but not they."[72]

This explained for Robinson why most Frenchmen did not understand the need to fight, and moreover, why they did not want to fight, even after the government stepped up its efforts to justify the necessity of mobilizing France for war. As he saw things: "It was difficult for [the French] to grasp that sometimes there is not only a need for a war 'against'—but also a need for some future vision, an 'ideal' to fight for, but this was not offered to them by their leaders." To a young man who had joined the Haganah to defend *his* homeland and protect *his* ideals, this response of the French was difficult to comprehend.

Robinson also thought carefully about the social and political reasons why France had failed to mobilize effectively against Germany, and even more tragically, had failed in its own defense:

> "Soldier, you are fighting for your own home!" the government said. But how could a peasant, living in the center of France, far from that part of the country where the war began, believe this when only a year earlier the Prime Minister of that same government was celebrated as a great hero for refusing to enter the war to save a similar country?
>
> The French soldiers who I met on the boat that took me to Marseilles did not understand, and all those who I met later, who were and stayed good-humoured, did not understand. The need for this war eluded them.

[72] Robinson R-1940, p. 30.

I was in the company of young Frenchmen who came from civilian neighborhoods, and they too did not bother much about international affairs. They talked about their approaching enlistment into the army without fear, but also without understanding.

At Bordeaux I heard from the lips of a soldier who claimed to have fought at Dunkirk: "One: they have broken through the Belgian front; Two: they have broken through the continuation of the Maginot Line; Three: they have broken through the Seine front. And now it's all over!" All this was said with a smile, and it seemed not to worry the soldier at all. At this same time, as the Reynaud Government was sitting at its crucial meeting only a few steps away, I saw a group of students outside (not far from the chain of soldiers), who were laughing and making fun, as if it were a holiday.

This fact is more staggering and frightening than any calamity caused by the few (or many) who had sacrificed their own country for the sake of a party's ideal, or for the sake of a foreign country. They did not understand . . . and all their beliefs and opinions were nothing but attributes, or manifestations of this lack of understanding that was nourished by an unwillingness to give up a life of peace and tranquillity. One of the most serious hours for France found the French people groping in the dark. And their leaders, except for two or three of them, were people of small and limited vision.[73]

The Writing on the Wall

As if to summarize his feelings about the French and what had happened to produce such a swift and staggering defeat, Robinson ended his account of the six months he had spent in France by quoting a line from the Book of Daniel, Chapter 5:27: "T'kel, you have been weighed in the balances and found wanting." This passage held a special significance for Robinson as he reflected on the past few months. Its reference to T'kel was, in fact, to none other than the famous "writing on the wall" which appeared ominously during Belshazzar's infamous feast, and foretold his imminent death.

Now the fate of Europe hung in the balance. It was up to the British not to fall short. Robinson was prepared to contribute in whatever way he could to the Allied war effort, hoping to insure that Hitler and fascism would eventually be defeated.

As the tender carrying Robinson and Fleischer to England left on June 18, France was in a state of collapse. The last words Robinson saw in one of Bordeaux's evening newspapers were etched in his mind—the

[73] Robinson R-1940, pp. 30–31.

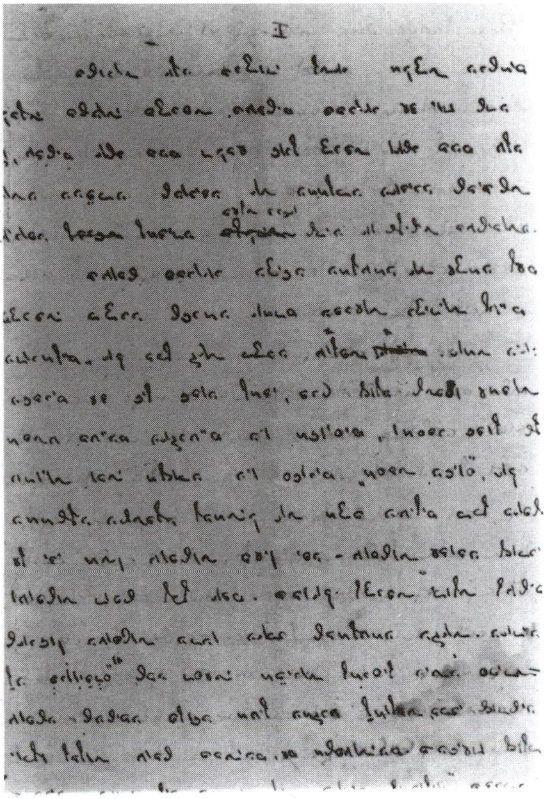

The first page of Robinson's record, in Hebrew, of the six months he spent in Paris and his escape to England with Jacob Fleischer (Talmon) in June 1940, RP/YUL.

words were those of Marshal Pétain: "It is time to stop the fighting."[74]

Once at sea, no one, Robinson recalled, could accept what was happening as real. One Polish diplomat who had managed to scramble aboard kept muttering to himself: "I don't understand, I don't understand." The incredulity and shock of it all seemed comparable to the Fall of Rome. Although, as Robinson commented, there was no real comparison, the analogy was one that many people seemed to draw. It was as if they "wished to secure their equilibrium by finding a precedent in history to these events."[75]

Robinson and Fleischer now found themselves along with hundreds

[74] An account of Pétain's speech and the immediate reaction to it, stirred largely by the ambiguity of what the phrase "it is time to stop the fighting" actually meant (especially because it misled many in France to believe that the war was over), is to be found in Freeman and Cooper 1940, pp. 291–292. At Pétain's trial it was later said that his words were responsible for "thousands of prisoners and even dead, for in the villages, people stopped taking shelter from aircraft, believing the war to be over," Léon Noel, quoted in Chapman 1968, p. 313.

[75] Robinson R-1940, p. 4. Upon his arrival in England, Robinson found exactly this

of others, exhausted and apprehensive, not knowing what the future would bring. Crossing the channel itself proved a harrowing experience. The coal tender, overladen as it was, made slow progress. Their route was not a direct one, but seemed to meander. In all, the trip took *four* days—made all the more frightening by bombardments from a German ship and strafing from enemy planes overhead. Facilities on board were minimal. Everyone slept on the decks, crowded together, making do as best they could with what little space there was. But finally the ship reached Falmouth intact. Upon its arrival, a beautiful moon was shining in a cloudless sky over Cornwall.

Robinson's first taste of England was tea and biscuits. These were distributed by local English ladies on hand to ease the plight of the tired and hungry passengers disembarking after their long ordeal. Later the refugees (including Robinson and Fleischer) were taken by train to London. There they were billeted in the Anerley School for the Deaf and Dumb, along with thousands of others, for processing. For Robinson it was the beginning of an entirely new life.

---

comparison drawn in a column by the correspondent George Slocomb in the *Daily Express*. Witnessing the fall of France to the Germans, Slocomb wrote, "Now I know how Rome fell."

CHAPTER FOUR

Robinson and the War: London 1940–1946

> I was glad that, if any of our cities were to be attacked, the brunt should fall on London. London was like some huge prehistoric animal, capable of enduring terrible injuries, mangled and bleeding from many wounds, and yet preserving its life and movement.
>
> —*Winston S. Churchill*[1]

In June of 1940, as France fell prey to the Germans and their Vichy collaborators, England braced herself for an invasion from across the channel. Britain, as Churchill said, was now alone. After the desperate yet heroic evacuation of troops from Dunkirk, the military at home was in disarray, "almost unarmed except for rifles."[2] On July 19, bold from his victories in Western Europe, Hitler delivered a triumphant speech in the Reichstag. He was looking forward to the speedy collapse of Britain and the capitulation of the British. Throughout July, however, a steady stream of American armaments poured into England. Thousands of loyal citizens spent sleepless nights unloading the precious military supplies, making certain they were widely distributed and ready for use. By the end of the month, Churchill was confident, despite the heavy losses of materiel and weapons at Dunkirk, that Britain was again an armed nation.

Interned and Released

Meanwhile, Robinson and Fleischer had witnessed firsthand the shocking defeat of France. Luckily they had made good their escape, and were now safely in England. Having disembarked in Falmouth, their first night was spent sleeping on the floor of a local school. The next morning, a Sunday, was bright and clear, which seemed an auspicious sign. Suddenly they were in a different world, far indeed (at least in spirit) from the beleaguered chaos and despair of now-occupied France. To Fleischer:

> The sight of orderly life, people going to Church, flower pots in nicely curtained windows, the courtesy of bypassers—all that was in such stark

[1] Churchill 1959, p. 377.

[2] Churchill 1959, p. 334.

> contrast to the total disintegration of French society, the panic and exasperation of its people. From sworn Francophiles we turned ardent Anglophiles.[3]

As subjects of the British Crown with passports from Palestine, Robinson and Fleischer were entitled to board a special train to London reserved for British subjects. That Sunday morning, while waiting for their train at the station in Falmouth, Fleischer happened to overhear a conversation between two fellow refugees that sparked his anger. Two Polish politicians were making a bet. Both were well known: one a prominent Zionist leader, the other a Polish Socialist. The Zionist was prepared to wager that within a week the British Empire would fall to Hitler, and this so enraged Fleischer that he could barely contain himself. Just as he and Robinson were about to board the train, he faced the "prophet of doom." All he did was quote a famous Hebrew saying: "The eternity of Israel will not fail," whereupon the Polish pessimist burst into tears. The indignation in Fleischer's voice must have been plain, and his words had their effect.

Upon their arrival in London, Robinson and Fleischer, like thousands of other refugees, were initially interned at the Anerley School for the Deaf and Dumb. There they were kept for preliminary surveillance, to be examined, sorted out, and eventually released (or so they hoped). The school itself was completely isolated, its gates always locked, with sentries posted night and day. Worse than the discomfort and isolation, however, was the boredom.

There was no way of knowing how long it might be before they would be allowed to leave. After talking it over, Robinson and Fleischer decided to contact the only person they knew in London, their former classmate from the Hebrew University, Chimen Abramsky. Finding him, however, was not a simple matter, since neither Robinson nor Fleischer knew his address. All they knew was that Abramsky's father was an eminent rabbi, and a member of London's Rabbinical Court, so the message they sent was mailed in his care, bearing only the designation "Whitechapel," the middle-class Jewish quarter in London's East End. All they could do now was bide their time.

Meanwhile, Chimen Abramsky had just been married—on Thursday, June 20, 1940. When the card arrived from his former schoolmates, Jacob Fleischer and Abby Robinson, he was completely taken by surprise. The card was brief but explicit, saying only that they had managed to escape on the last boat from Bordeaux, and that they were now

[3] Recollections of Jacob Talmon, hereafter cited as Talmon R-1940. Fleischer changed his name to Talmon in 1950, according to Chimen Abramsky in a letter to JWD of June 13, 1994.

interned in a camp at the Anerley School near London. Without delay Abramsky and his wife set off to find them, on Sunday, June 23.

When they got to the Anerley School, Robinson and Fleischer were sitting together, talking on the lawn. Suddenly their conversation was interrupted by a guard shouting someone's name, a name that neither of them could make out until Fleischer realized that it was his name, terribly mispronounced. In a flash both he and Robinson were up, running to see who was calling for them at the school gate. Despite their shouts of welcome, Abramsky did not return their greetings right away. Instead, with a ceremonial flourish:

> [Abramsky] solemnly stretched out his hand in the direction of his lady companion.
> "Please meet my wife."
> "Since when?"
> "Since yesterday," came his reply.[4]

Although this initial visit from Abramsky and his wife—who had brought a basket of food with them—must have been reassuring, Robinson and Fleischer were not immediately free to go, but remained at the Anerley School for another week or so. Meanwhile, Abramsky appealed to Norman Bentwich, who was very active in England at this time on behalf of the Hebrew University. Abramsky hoped Bentwich might intervene on behalf of Robinson and Fleischer.[5] After assurances that they were neither German spies nor foreign agents, but simply two students who had been living on government scholarships at the Sorbonne when the war broke out, Bentwich managed to secure their release.[6]

Bentwich proved to be one of the angels of mercy watching over Robinson and Fleischer in London. In fact, he had already received an enquiry from the Hebrew University about them, and he was anxious to ascertain their whereabouts and circumstances. Since normal postal communications had now been interrupted for weeks, no one in Jerusalem had heard anything from either of the young men. Bentwich, a conduit for much of the business concerning the Hebrew University in Great Britain, was the logical contact for the university to use in hopes of getting some word of its scholarship students.

[4] Talmon R-1940. Obviously there is some discrepancy here, since Abramsky was married on June 20, but did not see Robinson and Fleischer (Talmon) until the following Sunday, June 23.

[5] Bentwich did much to aid refugees, especially scholars, and after the war wrote a book on the subject, Bentwich 1953. He was also a great supporter of the Hebrew University, and wrote his own account of its history: Bentwich 1961.

[6] Abramsky R-1976.

Although Bentwich was pleased to meet with Robinson and Fleischer, he was not very encouraging. Doubtless he had no wish to raise false hopes, but he did give them letters of introduction to the International Student Service. For Fleischer, at least, this proved to be of inestimable value. With his M.A. degree in hand, and having already begun work on a doctorate at the Sorbonne, he was an easy candidate for graduate work. When asked to admit him as a research student, the London School of Economics not only gave him a free place, but later awarded him a scholarship. Robinson, with no such formal credentials, was at a considerable disadvantage.

Help finally did come from the Jewish National League in London. Thanks to its efforts, at the end of July both Robinson and Fleischer were sent to live with a working-class family in Brixton, "a quarter of ill repute."[7] Since neither of them had any money, they were housed with a British family at government expense. They were also given some pocket change by the Jewish Agency for Palestine.

At their new home, Robinson and Fleischer were the youngest of a small group of foreign boarders that included three medical doctors, two businessmen, and a Polish diplomat. This motley crew was squeezed under the same roof and, for many months, all were unemployed. As Fleischer described it, the prevailing atmosphere in Brixton was far from edifying:

> The main, if not sole topic of conversation of our cohabitants was sex, and their occupation was hunting for women. And they were, especially the doctors (one a gynecologist and abnormal philanderer and the other a specialist in venereal diseases) mighty hunters. Abby and myself were terribly inhibited and shy boys, and the discovery of all that "cynicism" and promiscuity was a shock to us.[8]

London During the Blitz

Life in London during the "phoney war" was as seemingly carefree as it had been in Paris prior to the German invasion. Just as Simone de

[7] Talmon R-1940.

[8] Talmon R-1940. The Poles in England were notorious, it seems, as ladies' men. Large numbers were in England, both as refugees and as members of the Allied forces. Even so, they were not universally loved by the British. "The lukewarm attitude towards the Poles may be explained by the fact that they were a dashing body of men and, so far as many an English male was concerned, far too charming and gallant. Their success with women had been considerable," Mosely 1971, p. 302. In fairness it should be added that of all the foreign Allies in London, Americans were said to have been held in the lowest esteem of all. In part this reflected the fact that there were far more of them than any other foreign group in England. For details, see Mosley 1971, pp. 304–307.

Beauvoir was attending the Opera in June only days before the Nazis arrived, the English too were enjoying the theater as usual. Prior to the fall of France, the hit of London was the romantic revue *Me and My Gal*, in which the show-stopping number involved the whole company "Doing the Lambeth Walk," described by one critic as a sort of cockney conga.[9]

Just as Robinson and Talmon had been struck by the seeming normalcy of life in England despite the war, others in the know were similarly impressed:

> The tranquility of England seemed unreal and not altogether reassuring. In the capital traffic circulated normally, people went unhurried about their business, theatres and restaurants were open, taxis plied for hire; it was like Paris before May 10, almost dangerously nonchalant. . . . There were some coils of barbed wire in the parks and in the sidestreets near Whitehall; and on the lawns men in denims went through the motions of rifle drill with makeshift weapons. . . . The ubiquitous gasmasks—in Paris too they had been dutifully carried—showed that most of the inhabitants were aware that a war was in progress somewhere.[10]

But by the end of June, with the evacuation of Dunkirk as fresh in their memory as the recent defeat of France, the English braced themselves for a German invasion, which seemed inevitable and imminent. Sholto Douglas (Lord Douglas of Kirtleside), marshal of the Royal Air Force, anticipated the worst for London: it was "heavy bombing that we expected from the Germans."[11] As Virginia Woolf, who like many others feared the apocalypse, wrote in her diary: "I have my morphia in my pocket."[12]

Douglas was right about the bombing. The infamous air raids over London began in earnest on July 10, 1940, and the Battle of Britain was underway. On August 6, with her novelist's eye for detail, Woolf observed:

> men excavating gun emplacements in the bank. . . cementing floors; sand bagging walls; . . . no one pays any attention—so blasé are we. Guns along

[9] Mosley 1971, pp. 29–30.

[10] Like Robinson and Fleischer, the writer Duncan Grinnell-Milne had also been in Paris just before it fell to the Nazis in June of 1940. See Grinnell-Milne 1962, p. 109.

[11] Douglas 1966, p. 47. Virginia Woolf recorded the "black news" about the fall of France in her diary for June 12, 1940, noting that André Maurois had been on the BBC "begging for help last night." Maurois was an observer to the British high command in France, and was evacuated to London on June 11, where he made his BBC broadcast.

[12] On June 20, Woolf wrote in her diary: "The French stopped fighting. . . . I have my morphia in pocket," in Woolf 1984, pp. 294–297.

> the river, boughs for camouflage, excite no one. It is like the raising of the gallows tree, for an execution now expected in a week or fortnight.[13]

Among the first architectural casualties was St. Giles' Church in Cripplegate. The photographer Cecil Beaton described the picture in his diary for August 24:

> Scattered cherub's wings and stone roses were strewn about—whole memorial plaques of carved marble had been blown across the width of the church and lay undamaged. The entire frontage of the deserted business premises opposite was wrecked. And Milton's statue had been flung from its plinth. Yet the lamp post was standing erect with no pane of its lantern broken.[14]

Two weeks later, after a performance of Gounoud's *Faust* at the Sadler's Wells, having just seen Faust and Mephistopheles descend into the fiery depths of hell, the audience "came out to a night sky glowing fiercely from the fires in the docks." The Germans were preparing to open the way for a full-scale invasion from across the channel.

Hermann Goering, in command of Luftwaffe strategy, decided in early September to switch from daylight raids to night attacks. In doing so, central London was now the primary target. For fifty-seven nights the bombing was unceasing.[15] The German Blitzkrieg, intended to break the English spirit, was designed to bring London along with the rest of England to her knees.

The weekend of September 6–7 was brutal; hundreds of enemy aircraft began a large-scale assault which persisted until the 15th.[16] Escorted by swarms of fighter aircraft, 625 bombers concentrated on London's East End and the shipping terminals along the Thames. On one night alone, 330 tons of high explosives and thousands of incendiary devices wreaked havoc upon the city. This was only the beginning, however, of what in many ways was the worst part of the war for Britain, and certainly the worst for London:

> The air raids are now at their prelude. Invasion, if it comes, must come within three weeks. The harrying of the public is now in full swing. The

[13] Woolf 1984, p. 310.

[14] Hewison 1977, p. 28.

[15] Churchill 1959, p. 367. See also the account of the Blitz in Johnson 1978, p. 39.

[16] See "The Battle of Britain," chapter 12 in Churchill 1959, esp. pp. 357–365. Churchill regarded September 15 as one of the decisive battles of the war. Two days later, Hitler decided to postpone indefinitely Operation Sea Lion, the German code name for the invasion of England. A month later, on October 12, Operation Sea Lion was officially put off until the following spring. It was in reference to the fighting of mid-September that Churchill made his famous remark before the House of Commons: "Never in the field of human conflict was so much owed by so many to so few," Churchill 1959, p. 366.

> air saws; the wasps drone; the siren—it's now Weeping Willie in the papers—is as punctual as the vespers.[17]

Despite the threat of impending disaster, the English maintained a bold front. As the Blitz was in full progress, the newspapers often seemed to cover the war as if reporting the results of a test match or soccer cup final. Headlines often carried figures which read like cricket scores: "It's 65 for 12" was typical, meaning sixty-five German planes had been shot down against twelve losses for the home team.[18]

During the air raids, Robinson and Fleischer spent long periods, night after night, in London tube stations. Although greatly improved as the war dragged on, at first these were makeshift and frankly squalid quarters. Henry Moore, who took to sketching the scenes he found in the underground at the beginning of the war, described the huge East End Tillbury Shelter in one of his diaries:

> Perambulators with bundles. Dramatic, dismal light, masses of reclining figures fading to perspective point—scribbles and scratches, chaotic foreground. Chains hanging from old crane. Sick woman in bath chair. Bearded Jews blanketed sleeping in deck chairs. . . . Dark wet settings (entrance to Tillbury). Men with shawls to keep off draughts, women wearing handkerchiefs on heads. Muck and rubbish and chaotic untidiness around.[19]

Actually, at the beginning of the war the government, fearing that once down in the tubes the population might never resurface, tried to forbid their use as shelters during air raids. This of course was to no avail. Anyone with the fare to enter simply did so. Eventually, with the addition of cots, portable lavatories, and modest medical facilites, not only was livable space eventually provided, but culture as well. In Bermondsey, amateur actors performed Chekov's *The Bear*, and the Council for the Encouragement of Music and Arts gave over 150 concerts in tube stations during the Blitz. Some of the larger shelters had film shows and libraries, and the London City Council soon arranged for evening classes. The Swiss Cottage Station even had its own magazine, published sporadically as *The Swiss Cottage*.[20]

[17] Virginia Woolf, her diary account for August 28, 1940. In a postscript she noted that while in the garden that afternoon, a German plane was downed before her very eyes. See Woolf 1984, p. 313. Several weeks later, on Wednesday, September 11, she heard Churchill on the radio, delivering "a clear, measured, robust speech. Says the invasion is being prepared. It's for the next 2 weeks apparently if at all. Ships and barges massing at French ports . . . ," Wolff quoted by Brewster 1960, p. 317.

[18] Mosley 1971, p. 75

[19] Moore, quoted in Hewison 1977, p. 40.

[20] Hewison 1977, p. 34. Hewison also adds however a corrective of sorts, in a section of his book captioned "The Myth of the Blitz," in which he describes it as "part false-

Chimen and Miriam Abramsky

At the beginning of the London Blitz, Chimen Abramsky and his wife saw quite a lot of Robinson:

> In those days every day counted an era. To be alive was something of a miracle. Abby and I used to converse very far into the night, and very often right through the night on problems of mutual interest, particularly philosophy, the question of peace and the eternal peace as formulated by Kant. He always came back with Kant against my Hegellian views. He countered my beliefs in Hegel and Marx with the absolute value of truth instead of the relative value as put forward by Hegel.[21]

Clearly Robinson and Abramsky enjoyed a very close friendship. Abramsky was impressed by Robinson's wide range of interests, which included not only philsophy but a very broad spectrum of German culture. Abby's substantial knowledge of Goethe was, he thought, remarkable. The two of them liked to talk widely on topics ranging from medieval philosophy to modern Jewish literature. Not surprisingly, Abramsky recalls that Robinson's Hebrew was "chiseled," meaning that he spoke it carefully, with great accuracy and erudition. In fact, they normally spoke to each other in Hebrew, although whenever others like Abramsky's wife were present, they switched easily to English.

Abramsky was also quite taken by Robinson's penchant for quoting large chunks of modern poetry, especially that of Chaim Nachman Bialik, from memory.[22] But Robinson was just as happy discussing Maimonides, which Abramsky admitted, "again surprised me because Maimonides, I thought, was a bit remote from modern aspects of mathematics."[23]

Homeless in London

Meanwhile, the London Blitz continued, and life settled into a routine of sorts for Robinson and Fleischer:

---

hood of glossified and sanctified 'finest hours,' part truthful courage and endurance, part nostalgia for a moment of genuine common exaltation," Hewison 1977, p. 37.

21 Abramsky R-1976.

22 Chaim Nachman Bialik (1873–1934) is one of the most significant figures of modern Hebrew literature. His essays, stories, and poetry reflect a simple nostalgia for the countryside of his Russian childhood, as well as broader themes associated with Jewish culture generally and Palestine in particular, where he settled as an immigrant in 1924.

23 Abramsky R-1976. Maimonides was not so far removed from mathematics as Abramsky may have thought. In one of his most influential works, *The Code of Maimonides: Sanctification of the New Moon* (1956), Maimonides demonstrated considerable mathematical talent. This work provides rules for determining the Jewish calendar, which in turn raises one of the most difficult theoretical and practical problems of medieval astronomy—

Night after night (and the days were getting shorter and shorter)—we would repair to the shelter in the nearest subway station, and stay there for the whole night in the midst of a large and very mixed crowd. Bombs were falling outside, but inside there was much song, hilarity.

In day-time between one alert and another we would wander about in the drizzling, chilly weather, drop in on the Abramsky's over the week-ends or on a Jewish Hungarian family with whom we had become friendly, and which had two pleasant and nice looking daughters of our age, who quite clearly liked us, but we were too shy and too poor to date with them.[24]

Brixton, where Robinson and Fleischer were living in South London, was a high-risk area, and finally one morning the worst actually happened:

The nadir in our fortunes was one early morning when upon our return from the shelter we found our house wrecked by a bomb. For 8–10 days we were quite homeless, wandering about in the wet, freezing cold, camping in railway or subway stations.[25]

This was a scene repeated daily throughout the city. The conflagrations of a single night would leave thousands of Londoners with nothing. Fortunate to have escaped injury, and buoyed by a courage and tenacity of their own, Robinson and Fleischer survived as best they could, scarcely more than hand to mouth. From one day to the next there was little to do, as Fleischer said, but wander the streets in the cold. As soon as the fall term began, however, Fleischer was off to Cambridge to take the place Norman Bentwich had helped to arrange for him at the London School of Economics. Robinson was not so fortunate. Alone in London, he decided to enlist with the Free French Air Force as soon as it became clear that he would not receive financial help either from Bentwich or the Friends of the Hebrew University. By November of 1940 he was wearing the French uniform in exile bearing the cross of Lorraine.[26] With no other means of support in London, Robinson may simply have taken this route as the most expedient for his own survival.

---

determination of the exact moment of first visibility of the new crescent moon. Futher discussion is provided in a Hebrew commentary (with English summary) in Kalikstein 1978.

[24] Talmon R-1940, p. 12.

[25] Talmon R-1940. Leonard Mosley described 1940 as "a noisy winter for London, and a killing and miserable winter too. The capital was ripped and scarred everywhere now and no community had been spared. Every night the bombers came over and then the capital went dead," Mosley 1971, p. 185.

[26] Seligman 1979, p. xiv. Seligman notes that "Although Abby tried to join the British armed forces, they were not yet ready to accept so new an arrival, especially a born German. In November 1940 he finally found a way to contribute to the war effort by

Robinson and the Free French

Like Robinson, Charles de Gaulle arrived in London just as France was collapsing in June of 1940. In de Gaulle's mind, the refugee French reached Britain as if "washed up from a vast shipwreck upon the shores of England."[27] But de Gaulle was not one to be easily daunted. No sooner had the Pétain government begun negotiating for an armistice with the Germans, than de Gaulle took to the airwaves. At 6:00 P.M. on the evening of June 18, the BBC gave him the microphone. The "Appeal by General de Gaulle to the French" was an emotional broadcast urging his countrymen not to lose heart, but to believe, as de Gaulle said, that "France is not lost." His brief address ended with a flourish, a rallying call for support:

> I, General de Gaulle, now in London, call on all French officers and men who are at present on British soil, or may be in the future, with or without their arms; I call on all engineers and skilled workmen from the armaments factories who are at present on British soil, or may be in the future, to get in touch with me. Whatever happens, the flame of French resistance must not and shall not die.[28]

Less than a month after this stirring appeal, the general was reviewing his first troops in Whitehall. The occasion was Bastille Day, July 14. First, de Gaulle inspected his men and then, befitting the French holiday, he placed an official tricolor wreath at the foot of the statue of Marshal Foch in Grosvenor Gardens, surrounded by the French Renaissance-style buildings of Grosvenor Square. It was, as he said himself, "in the midst of a deeply moved crowd."

Despite the drama of the Bastille Day ceremony, by the end of July de Gaulle commanded scarcely seven thousand Free French troops. Nevertheless, he was optimistic: "Little by little, in spite of everything, our first units took shape, equipped with an odd assortment of weapons,

---

joining one of the regiments being organized by the Free French under de Gaulle. . . . It is not clear why Abby was assigned to the air force, whether by choice or by chance. In either case, there could hardly have been much forethought. The affiliation shaped his career for many years to come."

[27] De Gaulle 1959, p. 83.

[28] De Gaulle 1959, p. 84. The complete text of de Gaulle's broadcast of June 18 is transcribed in Mosley 1971, p. 48. Mosley, in describing de Gaulle, characterized him as the complete antithesis of Churchill. De Gaulle, he said, was "totally lacking in humour, never for a moment relaxed or human. Never once did a smile crack that grave face on top of the tall eucalyptus tree of a body. But he exuded sincerity and patriotism." As someone else put it, "the destiny of France oozes from his pores." Churchill, who approved the broadcasts from London, told de Gaulle: "You have my backing. I will let you have the BBC as your platform," Mosley 1971, p. 47.

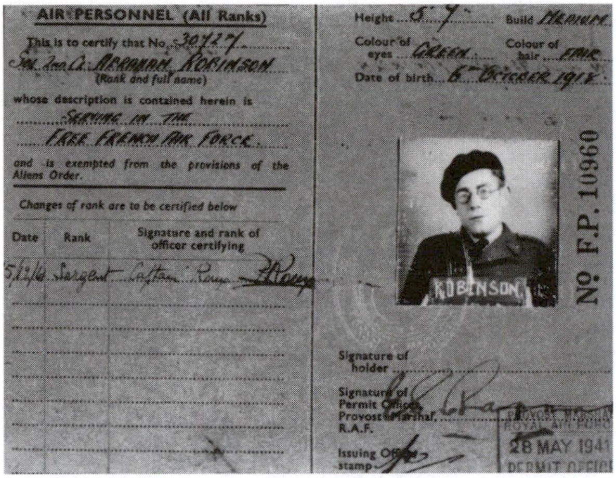
AIR PERSONNEL (All Ranks)
This is to certify that No. 30727
(Rank and full name) ABRAHAM ROBINSON
whose description is contained herein is
SERVING IN THE
FREE FRENCH AIR FORCE.
and is exempted from the provisions of the Aliens Order.
Changes of rank are to be certified below

Date	Rank	Signature and rank of officer certifying
5/12/41	Sergent	

Height 5' 7" Build MEDIUM
Colour of eyes GREEN Colour of hair FAIR
Date of birth 6th OCTOBER 1918
ROBINSON
No. F.P. 10960
Signature of holder
Signature of Permit Officer, Provost Marshal, R.A.F.
Issuing Officer stamp
28 MAY 1941

Robinson's identity card, listing him as a sergeant in the Free French Air Force, December 14, 1941, RP/YUL.

but formed of resolute men." To a man like Robinson, familiar with service in the Haganah, this would have struck a very familiar chord.

> De Gaulle's major goal in 1940 was to establish a Free French army in Africa: in the vast spaces of Africa, France could in fact re-create for herself an army and a sovereignty, while waiting for the entry of fresh allies at the side of the old ones to reverse the balance of forces. When that happened Africa, being within reach of the peninsulas of Italy, the Balkans, and Spain, would offer an excellent base for the return of Europe, and it would be French.[29]

Unfortunately, most of northwest Africa seemed securely in the hands of the Vichy French. Although the puppet government's hold over Algeria, Morocco, and Tunisia seemed for the moment impregnable, demonstrations on behalf of the Free French in Dakar, Saint-Louis, Ouagadougou, Abidjan, Konakry, Lomé, Duala, Brazzaville, and Tananarive were greatly encouraging. De Gaulle was especially hopeful for quick support in Chad and the Cameroons.

In fact, on August 26 the local authority, Félix Eboué, proclaimed that Chad was joining de Gaulle. Similar support came from the Cameroons, and in fact de Gaulle was pleased and relieved that "the greater part of the Equatorial Africa-Cameroons block was attached to Free France without a drop of blood having been shed."

The crucial target on which both the French and English were agreed, however, was Dakar. Churchill confided as much to de Gaulle:

[29] De Gaulle 1959, p. 105. De Gaulle's concern was not simply that France should eventually be restored to the French, but that while she was at her lowest ebb, the English and Americans should not take an undue interest in any of the French colonies. De Gaulle 1959, p. 106.

> We must together gain control of Dakar. For you it is capital. For if the business goes well, it means that large French forces are brought back into the war. It is very important for us. For to be able to use Dakar as a base would make a great many things easier in the hard Battle of the Atlantic.[30]

Unfortunately, de Gaulle's plans went badly from the beginning. In fact, the attempt to take Dakar from the Vichy government was a fiasco. De Gaulle's failure was especially bitter because it left a negative impression on the British and Americans. Afterwards, the Americans in particular refused to include de Gaulle in any major offensive actions of the war.[31]

With the collapse of the Dakar operation, de Gaulle next set his sights on Eritrea, the region of northern Ethiopia fronting the Red Sea. There he hoped to engage the Italians, setting the stage for French intervention in the Middle East. For this action to succeed it was all the more necessary to recruit forces to serve French interests in equitorial Africa, and it was as part of this effort that Robinson soon expected to be sent abroad.

The War Progresses

Meanwhile, the Axis powers were making grim progress on all fronts. In November of 1940 Italy attacked Greece. On March 1, 1941, Bulgaria was forced to join the Axis. Alarmed by Nazi advances in Eastern Europe, the U.S. Congress was finally persuaded to vote the Lend-Lease Act. America, by becoming (in Roosevelt's phrase) "the arsenal of democracy," was at last taking "a gigantic step towards war."[32]

In April, German troops overwhelmed both Greece and Yugoslavia. In the Far East, too, the Japanese menace (in de Gaulle's understated words) "was becoming definite." Meanwhile, the British and their allies were equally preoccupied with matters in North Africa. At home, the English were also having problems of their own, not the least of which were the continuing air raids on the capital.

On April 16, 1941, London suffered its worst bombing yet, advertised by the Germans as "the greatest raid of all time." Londoners simply called it "the Wednesday."[33] The attack itself was devastating, lasting eight hours during which 450 planes dropped 100,000 tons of bombs. Incredibly, this was followed by an even bigger raid on May 10 which set fire to the House of Commons and part of Westminster Abbey. The fire fighting went on for eleven days.

And then, as if miraculously, the Blitz came to an end. There was

[30] Details provided in de Gaulle 1959, p. 115.

[31] See de Gaulle 1959, pp. 123–130.

[32] De Gaulle 1959, p. 165.

[33] Hewison 1977, p. 33.

only one more sizable raid in 1941, and this was aimed not at London but at Birmingham.[34] Perhaps it was thanks to the relative lessening of tension in London with the slackening of German air attacks that Robinson found time to return to mathematics and finish another of his early research papers.

Robinson's Continuing Work in Algebra: Quasi-Fields (1940)

Despite the demands that training with the Free French must have made, Robinson still found time to pursue his love of mathematics. He had always been a diligent and consistent worker, and by the late spring of 1941 he had managed to finish writing a short paper "On a Certain Variation of the Postulates of a Commutative Algebraic Field."[35]

This had its origins in material Robinson had studied at the Hebrew University in the late 1930s. It drew upon ideas contained in two papers by Wilhelm Dörnte and Heinz Prüfer.[36] Prüfer had considered special cases of multiple groups, generalizations of groups where an n-ary rather than a binary operation is considered. This approach was taken up in greater generality by Dörnte, who called the structures he considered "n-groups." In going over these papers, Robinson found that the study of multiple fields could be reduced to what he called "quasi-fields"—commutative fields for which the usual distributive law $a(b + c) = ab + ac$ is replaced by a general n-ary law:

$$a(b_1 + b_2 + \ldots + b_n) = ab_1 + ab_2 + \ldots + ab_n \text{ (for } n > 2).$$

Robinson's paper of 1941 was devoted to exploring the implications of such quasi-fields, whose most interesting property is the fact that the zero element for addition possesses an inverse.

[34] Douglas 1966, pp. 132–133.

[35] A copy of Robinson's typescript draft of this article was kindly made available to me by Professor I.M.H. Etherington, to whom I am pleased to express my gratitude for his help in answering a number of questions related to his contact with Robinson in the 1940s. Copies of the material in his "Robinson file," including correspondence, notes, Robinson's original paper on quasi-fields, and the galley proofs of the article, have subsequently been given to the archives of Yale University, where they are now part of the Robinson papers there.

How much of the work Robinson did on this paper in England is not clear. According to W.A.J. Luxemburg I-1981, Robinson had brought a lot of mathematical material with him from France, which presumably included ideas on which he had already been working at the Hebrew University in Jerusalem prior to his leaving for France. Luxemburg recalls that later, Arthur Erdélyi also remarked that when Robinson got to England "he had stuff with him from France." Luxemburg surmises that Robinson must have done a considerable amount of work in Paris. The fact that he was able to send his paper on quasi-fields to Selig Brodetsky soon after his arrival in London (see below) lends additional support to this view.

[36] Dörnte 1929 and Prüffer 1924.

After introducing ideals in such fields, with a suitable definition of "derived" multiplication, Robinson showed how to reduce the quasi-field to a ring in such a way that the ideals in the quasi-field become ideals in the ring. He also found that by taking a similar approach to defining "quasi-rings" and "quasi-domains-of-integrity," he could obtain analogous results. Specifically, he showed that for quasi-domains-of-integrity, these could always be embedded in a quasi-field in the way an ordinary domain of integrity can be embedded in an ordinary field.

Unfortunately, having joined de Gaulle's Free French forces, Robinson was afraid he would not be in England long enough to arrange for the publication of his paper. Anticipating de Gaulle's plans—especially for a trained soldier like Robinson whose experience in the Haganah in Palestine must have made him an especially valuable recruit—Robinson had received all of the inoculations necessary for service abroad. He fully expected that any day he might be ordered to leave England for active duty in Africa.[37]

In hopes of finding some means of getting his paper into print, Robinson first sent a copy to Selig Brodetsky at Leeds, to whom he had been highly recommended by Fraenkel. Brodetsky was an eminent mathematician, a senior wrangler at Cambridge, whose interests had turned increasingly to applied mathematics. With no special expertise in algebra, he forwarded Robinson's paper to E. T. Whittaker at the University of Edinburgh. Whittaker in turn asked his younger colleague Ivor M. H. Etherington for his opinion—an ideal chain of events since Etherington was himself a specialist on just the sort of abstract algebra to which Robinson's paper was devoted.

Etherington was highly impressed, and immediately recommended Robinson's article for publication in the *Proceedings* of the Royal Society of Edinburgh. He changed the title to reflect more accurately its contents: "On a Certain Variation of the Distributive Law for a Commutative Algebraic Field." Etherington also made minor modifications at several points to avoid unnecessary misunderstandings or to spare readers confusion due to what he felt were vague explanations or sometimes inconsistent use of notation. Minor problems of English usage were corrected and several bibliographic references that Etherington felt should be mentioned were added.[38] Having made what changes he felt necessary, Etherington communicated Robinson's paper directly to the Royal Society of Edinburgh, fully in support of its publication. In turn, he was asked to send brief, nontechnical descriptions of the ar-

[37] Seligman 1979, p. xiv.

[38] In addition to providing full references for the papers Robinson cited in his paper, Etherington also added a recent title that Robinson had not himself had a chance to consult, namely, Richardson 1940. See Robinson 1941.

ticle to local newspapers, including the *Scotsman* and the *Glasgow Herald!*

Meanwhile, Etherington had written back to Brodetsky with a number of questions about Robinson and his paper. Brodetsky replied in a letter of June 30, 1941:

> Abraham Robinsohn is not a student of mine,[39] but a student of the Hebrew University of Jerusalem, highly recommended to me by his professors of the Mathematics Department there. I am very grateful to you for the trouble that you have taken about his paper, and I am sure that he will be at least as grateful to you as I am.
>
> I cannot say whether Mr. Robinsohn is now in England, because he was expecting to be sent out to Africa; but his last address is as follows: 37 Barrington Road, London SW 9.[40]

On July 3 Etherington received a formal letter from the Royal Society of Edinburgh, asking him to review Robinson's paper officially. The society was especially concerned for his opinion on two matters:

> Does the paper contain any information which might be of value to the enemy?
>
> In view of the high cost of publication the attention of Referees is especially directed to the possibility of curtailment of the paper . . . .[41]

Etherington replied favorably, pointing out that Robinson's article complemented work then being done in algebra by a number of noted mathematicians in England and the United States, but was "entirely new," with "interesting results." Etherington also noted that he had revised the paper before communicating it to the society, and that he thought it was now ready for immediate printing. He was specific in adding: "It could not be further abbreviated without sacrifice of its unity. It could not be of any assistance to the enemy."[42]

When Etherington received galley proofs in early August, he wrote to Robinson, hoping that he might still be in England.[43] Above all, he wanted to explain the alterations he had made, and give Robinson a

[39] Robinson had not as yet it seems consistently begun to spell his name without the *h*.

[40] Brodetsky to Etherington, June 30, 1941, in the possession of I.M.H. Etherington. A copy of the letter is now on file in the collection of Robinson papers, YUL.

[41] Letter from the general secretary of the Royal Society of Edinburgh, dated July 3, 1941, to I.M.H. Etherington, RP/YUL.

[42] Etherington to the general secretary of the Royal Society of Edinburgh, July 13, 1941, RP/YUL.

[43] The proofs were dated August 2, 1941. Etherington must have received them within a few days. The title page indicated that the paper had been received June 26, 1941; read July 7, 1941. Robinson's affiliation was listed as the Hebrew University of Jerusalem, RP/YUL.

chance to make any further changes he might feel were warranted before the paper was actually printed.

Etherington's letter included nearly two pages of notes detailing what he had done to Robinson's article prior to its acceptance by the Royal Society of Edinburgh for publication. Etherington never received a reply to this letter, and assumed Robinson must indeed have left London for Africa. Robinson *had* left Brixton, but not for the Allied front and de Gaulle's command in Africa. Instead, he had been reassigned to service with the British Air Force, and had subsequently been transferred to the Royal Aircraft Establishment in Farnborough, southwest of London in Hampshire.[44]

The Royal Aircraft Establishment, Farnborough

Precisely how Robinson managed to impress his mathematical talents on the Free French—and the British—came about quite by accident. According to Chimen Abramsky:

> [Robinson] joined the Free French Forces when he realized that he would not receive financial help from Professor Norman Bentwich and the Friends of the Hebrew University. At first, he had hoped to be sent as a meteorologist to French Equitorial Africa, but because the Free French could not find a boat, they made him clean the latrines. By chance, he met a French Captain [whom Robinson knew from the Sorbonne], and the Captain asked him for help in preparing a memorandum on aircraft wings for the Ministry of Aircraft Production. Robinson wrote the report, which was subsequently submitted and made a strong impression. When the French Captain was praised for the effort, he told the Ministry that it was actually Abraham Robinson who had written it. As a result, the British Government asked the Free French to have Robinson transferred to the Ministry of Aircraft Production.[45]

By the end of 1941, Robinson had been advanced to the rank of "sergeant" in the Free French Air Force, but was already (as of December 20, 1941) an assistant (grade 3) in the British Ministry of Aircraft Production.[46] Within a month, when he officially left the Free French in

[44] Robinson eventually reestablished contact with Etherington after the war (see chapter 5). Meanwhile, he had no idea that his paper had actually been published in the *Proceedings* of the Royal Society of Edinburgh for 1941.

[45] Chimen Abramsky to JWD in a letter of June 13, 1994.

[46] Robinson's official "Identity Card for Foreign Air Personnel" (issued by the Royal Air Force) is dated May 22, 1941. The identity card describes him as 5' 9", green eyes, fair hair, and lists him as a "Soldier Second Class" serving in the Free French Air Force. His promotion to the rank of sergeant is given as December 15, 1941. See Seligman

January 1942, he was given several photographs as souvenirs from officers with whom he had been working, and these remembered him as "friend, collaborator, professor."[47]

Abraham Robinson in his Royal Air Force uniform, January 27, 1942.

Robinson first went to work at the Ministry of Aircraft Production for a probationary period of three months, with a salary of £250 a year and a war bonus of £26.2.0 annually. This appointment to a technical post of such responsibility was, in every respect, a remarkable achievement, especially because Robinson still did not have a university degree.[48] Nor was he as yet a specialist in aerodynamics. His best published work in pure mathematics had been devoted to problems of abstract algebra and axiomatics, hardly the sort of applied mathematics needed at Farnborough.

At first Robinson was assigned to the Department of Structures. In keeping with his new position, he began to read assiduously about aerodynamics, trying to learn as much applied mathematics as possible. He gave himself a crash course on wing theory, and in a short time had proven himself a specialist on the subject of high-speed aerodynamics. His friend, Hermann Jahn, who shared a room with Robinson in Farnborough, recalls that he used to read under the blankets at night:

> He was working under great difficulty because we had a mean landlord who used to switch off the electric light at 10:00 o'clock. And Abby used to ask me if he woke me up or not when he used a torch to read in bed. But somewhat later he told me that he had passed an examination. I do not know if it was for an associate or membership, but he felt he needed professional qualification to work as an engineer. Because he was a pure mathematician he [thought he] needed qualification as an aeronautical engineer.[49]

What Robinson had done was to study on his own time for an exami-

---

1979, p. xiv, and official certificates documenting Robinson's war service in box 4 of Robinson's papers, YUL.

47 Seligman 1979, p. xiv.

48 Abramsky R-1976.

49 Jahn R-1976.

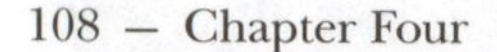

Contact sheet: the many faces of Abraham Robinson, 1942.

nation administered by the Royal Aeronautical Society. Not only did he pass the examination in three different areas: aerodynamics, pure mathematics, and meteorology and its applications to aeronautics, but later he discovered that in pure mathematics he had received first place. By June 22, 1942, Robinson was officially an "Associate Fellow" of the Royal Aeronautical Society.

This had valuable repercussions, for by the end of the year he was promoted to junior scientific officer, retroactively effective as of October 1, 1942. This new appointment brought with it a salary increase to £325 annually, but with a war bonus of only £19.12.0 and a stipulation that he take no more overtime.[50]

[50] All of the relevant documents concerning Robinson's "Associate Fellowship" examination for the Royal Aeronautical Society, as well as his letters of appointment to the Ministry of Aircraft Production, are among Robinson's papers, YUL.

At Farnborough many heard of Robinson for the first time when the head of his department, P. E. Montagnon, on one rare occasion exclaimed: "Ah, Robinson has made his first mistake!" Actually, having galvanized everyone's attention, almost immediately Montagnon was heard to say, "No, he hasn't!" Within minutes Abby appeared with a surprised look on his face: "What's this about a mistake?"[51]

After writing several reports that proved to be of considerable theoretical interest, Robinson was shifted to the aerodynamics department, and before long it was apparent that he was one of the group's best applied mathematicians. Moreover, his theoretical skills were enhanced by an outstanding gift for understanding physical problems. Almost always Robinson was able to apply mathematics in ways that were as practical in their results as they were elegant in their solutions. It was no surprise to Abramsky that:

> He was immediately given high rank because the superior officers or the research scientists who worked at the Royal Aircraft Establishment . . . recognized instantly that they had in front of them a young man of exceptionally wide knowledge both in mathematics and in the applied fields relating to aeronautics.[52]

Daily Life: 1941–1942

Once he began working at Farnborough, Robinson had more time to call his own. One of the first priorities was to renew his earlier contact with the Abramskys. Abramsky too was pleased because Robinson now came again to visit:

> . . . once or twice a month for long discussions on the problems of war and peace. At that time we were still not fully aware of what was facing the Jews and we were talking about how long the war would last, how long it would take us to go back to Israel, as he wanted very much to go back, and what Europe would look like after the Nazi's were defeated. All of us had full confidence that Britain, though standing alone at the time, would somehow emerge victorious.[53]

As usual, Robinson was ever the optimist:

> He was the source of immense encouragement and hope in the future. I found him a constant companion in his quiet, slow voice. He used to cheer me up enormously with regard to the prospects there would be after the war. Then when the news later on in 1943 began to emerge of the disasters facing the Jews in Europe, a certain pessimism penetrated my mind,

[51] Jahn R-1976. [52] Abramsky R-1976. [53] Abramsky R-1976.

> and he again began his philosophical arguments, always reverting to Kant somehow to encourage me—and very often with our friend Talmon—not to lose hope, that somehow Europe would emerge victorious and defeat the Nazis.[54]

To get away from the pressures of work at Farnborough, Robinson would sometimes bicycle into London for a lunchtime concert at the Church of St. Martin in the Fields, or to see a play in the West End. Often he would just stop by to talk with the Abramskys, and it was on one such visit—January 30, 1943—that he met Renée Kopel.

Renée Kopel

Renée Rebecca Kopel was the elder (by a year and a half) of two daughters of a prosperous Viennese family, and grew up with a special taste for the arts. Her father directed his own successful chemical firm, and was careful to see that both Renée and her younger sister Sulamith (Sully) received an excellent education, beginning with Hebrew which they studied at home even before they were sent to primary school. Encouraged by their parents, both daughters were introduced to music at an early age. Even while they were still quite young, they were taken regularly to the lively Viennese theater. As Renée recalls:

> We had piano lessons at home, went to dance classes, performed on stage, danced and sang together, even entertained when we were invited to the homes of parents' friends. But we also went with the school to performances at the *Burgtheater* and *Opera*. I never wanted to be anything but an actress.[55]

At home, Renée often played with the one doll she had; it was never a *baby*, she insisted, but an actress or a dancer for whom she designed special costumes. She remembers herself as a shy child, often hiding behind the curtains or under the tablecloth whenever strangers would visit the house.[56] Yet she also loved to dance, sing, and entertain these same visitors.

When she first began secondary school, her father encouraged her to follow in his footsteps, studying science and chemistry, but Renée was determined to find her own way as an artist. After five years, she was anxious to leave school for a career in the theater, but she knew that her parents would never approve:

> I had to find a way by myself. At school we had a non-obligatory class on

[54] By this time Fleischer had changed his name to Talmon. Abramsky R-1976.
[55] Renée Robinson to JWD, letter of March 15, 1989.
[56] Renée Robinson I-1980; to JWD, letter of March 5, 1994.

Renée Kopel, fashion photographs, London, July 5, 1938.

> Tuesday afternoons. The teacher was a well-known radio actress, and this was an acting class. I told her what my intentions were, but she said that I should not leave school until I had matriculated. Instead, she offered to give me private lessons, so that by the time I finished school, I would also be a trained actress. My parents would not have to know anything.[57]

Renée agreed to this arrangement, and it worked well for a time. By the end of the year, when the class performed a play by Schiller, Renée had a large part, and everybody commented: "You are an actress, do you know?"

She did, but she was becoming increasingly dissatisfied with the arrangement she had made with her teacher at the gymnasium. "I did not like her ways. It was not good for me, so I looked for something else." What she found was a place at the Vienna Conservatory, which was not too far a walk from her school:

> It was in the *Musikvereinsgebäude*, and there I spoke with the head of the acting class, a well-known, veteran actor from the *Burgtheater* who I had seen on the stage. I explained my situation to him, and that I would have to have a free place. He auditioned me and then agreed to take me on. He also gave me private lessons in his home.[58]

57 Renée Robinson to JWD, letters of March 15, 1989, and March 5, 1994.
58 Renée Robinson to JWD, letter of March 15, 1989.

Unfortunately, her teacher died about a year later. In his stead a younger actor, very well known, took over the directorship of the conservatory. Renée's parents still knew nothing about her acting classes, until one day a notice arrived at home announcing a student performance of *Romeo and Juliet* in which Renée was to play Juliet. "They came to the performance and realized that if I could get so far on my own, there must be something in me. And so I became an actress and not a chemist!"

Renée Kopel, photographed by Hedda Medina in Vienna, 1936.

Thereafter, Renée's contacts and experience in the Viennese theater world increased dramatically. She met the well-known actress Margarete Schell-Noé, who became something of a patron for her. When Schell played Iphigenia, Renée was the leader of the chorus. Once she even stood in for Schell, replacing her for an evening when Schell was scheduled to give a poetry reading but was unable to do so.

Soon Renée was a member of Austria's Actors' Equity, earning money dubbing films (including a French film on the life of Jesus in which she took the part of Claudia Procula, the wife of Pontius Pilate). She played in the famous *Cabaret am Nachtmarkt*, and was well on her way to a promising career in Vienna when fascism began to change everything:

> Then came Hitler. I left for London, left everything behind. My sister joined the "Hellerau Dancing Group" and went to Switzerland. In London I stayed with a family in Golders Green. I knew that I now had to earn my living, and all I could think of to do was sketching.[59]

Renée prepared some samples of her work so that she would have a

[59] Renée Robinson to JWD, letter of March 15, 1989. Renée was staying with Mrs. Nirenstein (who had a Jewish bookstore in the East End) and her three daughters. It was Mrs. Nirenstein who signed the guarantee for Renée's stay in England. One of the daughters, Miriam, became one of Renée's closest friends; later she married Chimen Abramsky, and it was through the Abramskys that Renée Kopel met Abraham Robinson; Renée Robinson to JWD, letter of March 5, 1994; and Chimen Abramsky to JWD, letter of June 13, 1994.

Renée Kopel, two publicity photographs, 1938.

portfolio to show, illustrating her abilities. Armed with about twenty-five drawings of models in a variety of fashionable clothes, she made an appointment with the director of Rita and Company, a leading fashion house in London:

> I told the director of this firm, which was in a building on Great Marlborough Street next to Liberty's, that I had worked for a firm in Vienna (which was of course not true), and gave him the only name I knew of such a firm. He asked me if I could sketch something when I saw it. I said "Yes," and he said "We shall see."
>
> Then I said "Why not now?" So he told his driver to take us to Mayfair. There we got out and he showed me some clothes in a window. It was actually a firm for which he was a supplier. It was a big firm selling clothes all over the country.
>
> When we returned, I sketched what I had seen, and although he tried to confuse me, I was firm about what I saw. That's how I became his sketcher and soon a designer. I stayed there for years.[60]

Eventually, Renée was put in charge of the gown department, and as her responsibilities increased, her salary increased as well. She felt sufficiently established in London that she had time to think about the stage again. First she got in touch with some actors she knew from Vienna. Not all were Jewish or Austrian, and some in fact were from Berlin and other parts of Germany. But all were anti-Nazi, and within this group Renée made some very good friends. They began rehearsing in the evenings and performed (only in German) in Swiss Cottage. Shortly

[60] Renée Robinson to JWD, letter of March 15, 1989.

thereafter, Renée met a number of aspiring and soon-to-be-famous actors, including Herbert Lomm, Arnold Marlé, and Frederik Valk. She also did some work in German for foreign broadcasts at the BBC, but quickly realized that she would have to quit her job as a designer if she wanted to work regularly on the radio:

> Marlé, with whom I played in *Nathan der Weise* (I was his daughter), became a great friend of mine. His wife was the niece of Sigmund Freud. Valk also became quite well-known. Marlé once asked me "Why do you do this [referring to her acting]? After a long day's work at your job, then rehearsing at night, if any of us had a job like the one you have, we would never do it."[61]

"Love at First Sight"

Renée Kopel never forgot the day she met Abby Robinson. It was a Sunday, and she happened to be visiting her friend Miriam Abramsky. Robinson, too, happened to be there, having ridden his bicycle in from Farnborough:

> We all went out to hear a concert on records at a friend's. Afterwards, Abby brought me from there to Golders Green, where I was living. He was quite different from other boyfriends I had known.[62]

Initially, Renée was impressed by Robinson's fair complexion and curly hair, which she thought could benefit from a hair net to flatten it down. After mentioning this to him, he got a crew cut! But it was the refined character in his face that was especially appealing. Abby, in turn, was fascinated by Renée's earlier experiences in Vienna, and by her work as a designer and actress.[63]

Increasingly, they came to spend what time they could spare from their jobs together, often with trips to the nearby countryside, or in the city where they would meet at lunchtime for concerts or to visit museums.[64] One weekend, Abby invited Renée to Farnborough, where she

[61] Renée Robinson to JWD, letter of March 15, 1989. In Vienna, one of her treachers had given her the stage name "Renée Worth," and this was the name she continued to use as an actress in London.

[62] Renée Robinson to JWD, letter of March 15, 1989.

[63] Renée Robinson I-1983.

[64] Robinson seems to have had a natural interest in the arts and theater from the time he was a small boy, doubtless thanks to his mother's own cultural interests. But during the war there was a general revival of interest in the arts, one that seems to have been universal, at least in England. Stephen Spender attributed this directly to the war itself: "This arose," he wrote in his autobiography, "spontaneously and simply, because people felt that music, the ballet, poetry and painting were concerned with a seriousness of

Renée and Abby, their wedding photograph.

remembers having a "lovely time." It was immediately clear that the pair had a lot in common, and almost as quickly, Robinson made up his mind that Renée was the woman he intended to marry:

> Once after he had come to see me in a modern play in which I played a very coquettish girl who smokes, etc., he escorted me to the underground station where, as we were waiting, he gave me a little calendar which he asked me to look at. There I suddenly saw that, on the 30th of January, he had written: "we shall get married." I did not want to get married at all! We had never spoken of an engagement. I did not even want a wedding ring.[65]

Despite her reservations about marriage, Renée knew that Robinson was someone special. "He gave you something," and she felt this immediately. "We used to walk in London, the palace on Hampstead Heath, Farnborough, Guilford (where they carved their initials into a railing)."[66] They both enjoyed nature, walking, and above all, music and theater. As she quickly came to realize, they complemented each other nicely. He

---

living and dying with which they themselves had suddenly been confronted," Spender quoted in Hewison 1977, p. 50.

[65] Renée Robinson to JWD, letter of March 15, 1989. Although she agreed to the marriage, Renée made it clear that she was against most conventions. For example, she rarely wore jewelry; even so, she eventually relented about a wedding ring and agreed to have a German friend design a single engraved gold band. Inside, the date of their wedding was inscribed, and from the day they were married, she never took it off. Renée Robinson I-1983.

[66] Renée Robinson I-1983.

was sometimes too modest for her taste, but she was outspoken enough for both of them, and would usually tell him exactly what she thought! Once, years later, Robinson confided to Renée's sister that she was a very wise woman, and that in her way, she gave him confidence.[67]

Renée and Abby Get Married

From time to time Robinson's old friend Jacob Fleischer would also visit the Abramskys. This usually happened on weekends whenever he could get back to London from Cambridge. On one of these trips, he heard the latest news about Abby and Renée from Miriam Abramsky:

> A young woman was staying with them and we were introduced to each other by Mrs. Abramsky. The girl left the room very soon and I think went to town. As soon as she was out Miriam turned to me and asked what impression the young lady had made upon me?
>
> "Why are you asking me this question? We hardly exchanged more than a few sentences."
>
> "Well," Miriam replied, "She made a terrific impression on Abby."
>
> "On Abby?!"
>
> "Yes, it is love at first sight."
>
> Then I remembered that a few weeks earlier (on Abby's last visit to Cambridge), my friend and companion confided to me that he was feeling very lonely at the station at Farnborough, and would very much like to find a wife. He was ripe for Cupid's arrow.[68]

Cupid's arrow, when it came, was also fairly swift. Exactly one year to the day after they had met, just as he had hoped and predicted, Abraham Robinson and Renée Kopel were married in a small, modern synagogue in Golders Green, Temple Fortune, close to where Renée was living. It was a small, private wedding, limited to only three of their closest friends.

At first, Renée had wanted a civil ceremony, to which Abby agreed if she would make the necessary arrangements. But when she attempted to work out the preliminary details, the magistrate told her that he would find someone off the street to serve as a witness. Renée felt that this was simply too impersonal, and so agreed to a religious service.[69] This was also politically expedient, for as Abby explained to her, if they ever wanted to live in Palestine it would be much easier, especially for Renée, if they were married in a synagogue.

January 30, 1944, was a beautiful day, but almost too warm for the navy-blue suit Renée had planned to wear. Talmon was best man. In fact, he was so nervous that the rabbi assumed he must have been the

[67] Renée Robinson I-1983.
[68] Talmon R-1940.
[69] Renée Robinson I-1980.

The Robinsons at home, 23 Chumleigh Road, Surbiton, 1945.

bridegroom instead of Abby. Miriam and Chimen Abramsky completed the party. Renée did not invite any of her colleagues from work, but they gave the newlyweds a present anyway.[70] After the ceremony, the wedding party enjoyed a simple lunch together. Then the newlyweds took the train to Devon, where they spent a week. Renée remembers that while they were there a priest they happened to meet gave her a bouquet of violets when he heard they had just been married.[71]

Life Together

Finding a place to live after they were married, one that was both affordable and conveniently located for Renée's job in London and Abby's work at Farnborough, was not easy. The first house they saw was, as Renée recalls, "something depressing." But as luck would have it, they found another place, a "dream cottage" in West Byfleet, Surrey. It was nearly equidistant between London and Farnborough. Even better, the local woodlands were perfect for long hikes, and since Abby had taught Renée how to bicycle, they could also enjoy more ambitious trips into the surrounding countryside. As for their "dream cottage":

> A woman kept it for a friend who owned it. There was a bedroom and a living room. The woman cooked—she was from Yorkshire—and she made a different pudding every day. We lived there for a year, but we might both have died of pneumonia, it was so cold! But a year later her friend decided to sell, which was a great blow because we had to move. It was such a beautiful cottage. The food, full board, it would be impossible to match.[72]

70 Abby wrote to his mother in Jerusalem about their wedding, and sent photographs. Renée never met her mother-in-law, who died in Jerusalem in 1949. Renée Robinson I-1980.

71 Renée Robinson I-1980.

72 Renée Robinson I-1980.

The Robinsons, however, managed to find what Renée describes as a "nice house" in Surbiton, where they were to live for the rest of the war. It was a somewhat longer commute to Farnborough for Abby, but only twenty minutes into London for Renée. The house in Surbiton also came with a housekeeper, "but this woman was the worst cook, everything looked like a gray stew."[73]

Even so, they found it an enjoyable walk from Surbiton to Richmond, famous for its park. At times, Renée was left alone on weekends when Abby had to be away serving with the Home Guard, on the coast. Renée remembers that occasionally he would return muddy from long details in the rain. Apparently, in addition to his war service at Farnborough, Robinson also wanted a more active, physical level of involvement with the war effort. Above all, he worried about a possible invasion of Britain, and thus contributed much of his off-duty time from the Royal Aircraft Establishment to serve with the Home Guard, which had a local unit at Farnborough. Robinson joined with the rank of private. After passing the required field tests and training in the use of weapons and tactics, he was awarded a proficiency badge which he first wore on his uniform as of February 28, 1944.[74]

Robinson's Aeronautical Research

Upon his arrival at the Royal Aircraft Establishment in early 1942, Robinson's first assignments were related to aircraft design. As a member of the Structures and Mechanical Engineering Department headed by Alfred Pugsley, he was put on a project with R. H. Whitby to analyze the relative merits of single-engine versus double-engine planes for deployment on aircraft carriers. The question arose due to changes in admiralty regulations which had recently permitted a greater folded width for wings on planes used on carriers. Was there now any advantage to introducing twin-engine planes?

Analysis of the most significant factors included such parameters as wing span, aircraft weight, available engine power, and wing loading. In light of the increased overall weight and fuel requirements that twin-engine aircraft would entail, Robinson and Whitby found that the ratio

[73] Renée Robinson I-1980.

[74] Seligman 1979, p. xv, and Renée Robinson I-1980. Robinson had joined the Twenty-seventh Hampshire Regiment Home Guard in 1943, which was affiliated with the Royal Aircraft Establishment. Among his official Home Guard papers is one issued to "Private Robinson" which details a number of "Proficiency Badge tests" he had taken at various times in 1943 and 1944, including ones for general knowledge, rifle, 36-M grenades, battlecraft, and map reading. Robinson continued his Home Guard service until the guard was disbanded in December of 1944.

of range to weight was still greater for single engines. Even though single-engine aircraft were generally limited to about 3000 BHP or less, this was not so severe a limitation as to cancel the overall advantage of single- over twin-engines.[75]

Another of the early questions Robinson worked on concerned compound lifting units. Compound units involve any combination of air foils, for example, a tail unit with one or more fins, or the two wings of a biplane. The unit might be connected or disconnected in separate parts. In either case, the problem is to determine the total distribution of lift over the entire structure.

For the simpler case of single wingspans, this problem had been dealt with by Felix Ziller, who considered both rigid and torsionally elastic wingspans. His approach succeeded by considering total lift as a variational problem using lifting-line approximations introduced by the father of German aerodynamics, Ludwig Prandtl. Robinson wanted to find a similar variational principle for compound lifting units, including wing and tail units in various combinations, both separate and connected. Not only did he succeed in doing so, but Robinson's method held several advantages over Ziller's; primarily, Robinson took into account the characteristics of the wing along its entire span and not just at a relatively few points (as was the case in Ziller's analysis).[76]

One of the most interesting problems presented to Robinson, however, concerned problems of structural fatigue or outright collapse:

> A large flying boat had suffered a main spar failure during a heavy alighting and Robinson was asked to investigate this accident. This led him to study how stress waves are propagated, reflected, and dispersed in structures in response to imposed impulses. At that time this was a novel approach to dynamic problems but its power and value were readily recognized, particularly for the insight it gave into possible causes of structural failure due to impulsive loads.[77]

Robinson's study of the flying boat failure led to an important paper

[75] It was noted, however, that any reduction in the maximum single power unit available would tilt the balance in favor of twin-engine models. Other practical limits on the Robinson-Whitby analysis were also noted, including the suggestion that should very fast torpedo bombers or reconnaissance aircraft be needed on short notice, it might be best to modify an existing twin-engine aircraft, like the Mosquito, using the most recent low-altitude engines available and enlarging the body. Otherwise, it would be necessary to wait for larger horsepower engines (which were not anticipated for about four years).

For details see Whitby and Robinson 1943. Although this paper is not cited in the bibliography of Robinson's works in Robinson 1979, a copy is preserved among Robinson's files in the archives of Yale University Library; it may only have been allowed to circulate within the research staff at Farnborough.

[76] See Robinson 1945a.

[77] Young 1979, p. xxx.

on "Shock Transmission in Beams of Variable Characteristics." Here the question of shock transmission was directly relevant to his analysis of the flying boat accident.[78] After calculating the stresses an aircraft must endure in landing, he also interpreted the experimental data available and then formulated practical strength requirements. Like his earlier work on single-engine planes aboard aircraft carriers, his shock transmission analysis also looked farther, beyond the specifics of the concrete problem he had been given, to consider larger questions of theoretical and practical interest. His analysis of the flying boat accident served to clarify factors that contribute to physical stresses that lead potentially to structural failures in general.[79]

Thanks to the success of his work in the Structures and Engineering Department, which clearly demonstrated the penetrating mathematical insights Robinson typically brought to his research, he was soon transferred to the Aerodynamics Department headed by Brian Squire.[80] It was in the new department that Robinson met Alec Young, who was impressed by the fact that "in a very short time [Robinson] had mastered the subject of wing theory and in a number of extraordinarily brilliant papers expanded its frontiers, particularly in the field of supersonic aerodynamics."[81]

Serious consideration of supersonic aircraft came relatively late, just as rocketry was also in its infancy during the war.[82] In fact, one of the last projects on which Robinson worked at the Royal Aircraft Establishment was the remarkable attempt to reconstruct an entire German V-2 rocket from bits and pieces of debris that the Royal Air Force had succeeded in recovering from Sweden and Poland.

Robinson and the Farnborough V-2

Following the Japanese attack on Pearl Harbor on December 7, 1941, the United States finally declared war on the Axis powers. The Americans spent most of 1942 in skirmishes in North Africa before Allied

[78] Robinson 1945b.

[79] Robinson returned to this problem after the war, when he was teaching aerodynamics at the University of Toronto. See Robinson 1957.

[80] In addition to the three papers already discussed, Robinson wrote two others while working in the Structures and Engineering Department, both of which were issued as internal publications of the Royal Aircraft Establishment in 1945. Robinson 1945a, and Robinson, Fagg, and Montagnon 1945.

[81] Young R-1976. Robinson's contributions to supersonic aerodynamics are considered in greater detail in the next chapter.

[82] There is extensive discussion of both German and British efforts to produce jet aircraft during the war, neither of which was very successful until the spring of 1944, in chapter 5, "Misfortunes of War," in Johnson 1978, esp. pp. 275–292. After the war, what

troops were ready to begin their move up the Italian Peninsula under Alexander's and Montgomery's commands. Meanwhile, the Russians succeeded in pushing further west towards Germany in the fall of 1943 and early in 1944.

In England, when the dense population of American soldiers suddenly disappeared from Mayfair in late spring of 1944—and Londoners could again get a taxi—everyone realized that something momentous was about to happen. Long in the planning, Eisenhower's strategy for D-Day succeeded in surprising the Germans in France. Within the first week of Operation Overlord, 326,000 troops landed on the beaches of Normandy.[83] As D-Day commenced on June 6, "London was silent all day long, waiting for the news."[84]

The bad weather across the English Channel that initially threatened Overlord actually served the invasion by catching the Germans off guard. Nevertheless, despite the initial success of the Allied armies on the beaches of France, the war was by no means over. A week after Overlord had been successfully launched, the first of Germany's new flying bombs hit London.

Early on the morning of June 13, a number of low-flying aircraft hurled across the English Channel. One of them, upon reaching London, slammed into Grove Road, Bethnal Green. A block of homes was suddenly in ruins, and six people were dead—the first casualties in a new era of warfare made possible by a new weapon, the long-range guided missile. Within the next twenty-four hours, 393 more were fired, killing 6000 people, mostly in London, with another 18,000 seriously wounded.[85]

At first the newspapers called them "Robot Planes," but popularly they were known as "Buzz Bombs" or "Doodlebugs." The Germans called them *Vergeltungswaffe Eins*, "Revenge Weapon Number One," the "V-1."

German jets survived (like the Blitzes, Messerschmitts, Junkers, Arados, and Komets) were taken to Farnborough where many were extensively flown. Only one of the Blitzes still exists today, as part of the Smithsonian Institution's National Air and Space Museum in Washington, D.C.

[83] In the first two days, 176,000 men and 20,000 vehicles were successfully landed—in itself a remarkable achievement of logistics and planning. As D-Day was underway, members of the Home Guard, like Robinson, were on duty manning antiaircraft stations and coastal defenses. They also took over many routine and security duties as all of southern England became one massive military camp, housing and provisioning the nearly one million Allied troops that were waiting to invade France. Churchill 1959, p. 820.

[84] Hewison 1977, p. 168.

[85] Johnson 1978, p. 123. Ten "flying bombs" were launched against England on the night of June 12. Four crashed shortly after launching, and one went astray. One fell in Sussex, and the others came down in Gravesend and Sevenoaks, in addition to the one mentioned above at Bethnal Green. For details, see Jones 1978, esp. chapter 44, "V-1," pp. 413–429.

These caused tremendous destruction and loss of life, especially in densely populated areas. The attacks were mostly remembered as ones heard but not seen. This was due to the fact that V-1s flew at relatively low altitudes, approximately 3000 feet:

> The buildings in the city areas prevented one from seeing any distance; the first warning would be the deep-throated throb of the pulse jet, growing rapidly louder. . . . If the engine suddenly cut, then everyone would dash for the cover of doorways or even lie flat on the pavements: there would be a wait that could seem an eternity, then a shattering explosion followed by a deep roar as some building or other dissolved into rubble; then silence. Smoke would billow up behind the skyline and then the bells of the fire engines and rescue services would be heard and probably the distant deep throb of another bomb.[86]

Or as George Stonier put it in his *Shaving Through the Blitz*:

> Never can a secret weapon have become so immediately public; English conversation, switched from the weather, developed a sudden high-pitched buzz. On roof-tops and in pubs one heard of nothing else; it called for drinks, it put off dinner hours, it brought out the theoretician and the bogey man. And while the knots of talkers were still speculating, another of the assailants would come bumbling over . . . the guns would crack; the drone would suddenly cease, and for a few instants sickening uncertainty took the place of gossip.[87]

George Orwell described his impressions of the V-1s in the *Partisan Review:*

> After the wail of the siren comes the zoom-zoom-zoom of the bomb, and as it draws nearer you get up from your table and squeeze yourself into some corner that flying glass is not likely to reach. Then BOOM! The windows rattle in their sockets, and you go back to work.[88]

Renée remembers the V-1s just as vividly:

> The "Flying Bombs" also came in the mornings and we went under our tables at work. Once I arrived at Victoria Station when the warning came and we all lay down on the platform until it was over. Abby wanted me to evacuate and go into the country as many did, but I never considered this and commuted to London every day by train and underground.[89]

The V-1, however, proved to be a minor-league cousin to an even

[86] Johnson 1978, p. 165.
[87] Stonier as quoted in Hewison 1977, p. 168.
[88] Orwell as quoted in Hewison 1977, p. 169.
[89] Renée Robinson to JWD, letter of March 15, 1989.

more destructive weapon that the Germans were desperately trying to perfect in 1944 in hopes of still winning the war: the V-2. This was a genuine guided missile, one that travelled at very high altitudes and at supersonic speeds.[90]

What made the V-2s potentially more dangerous than the V-1s was not only their much greater range and speed, but the fact that there was virtually no defense against them. Unlike the "flying bombs" (which were little more than pilotless airplanes moving at relatively low speeds and altitudes—which consequently meant they could be downed by conventional antiaircraft weapons), the V-2s could not be shot down. Guided by gyroscopes, their range was set by cutting off the rocket's fuel when a predetermined velocity had been reached.[91]

Worse, the V-2s were all the more elusive because they did not require permanent launching facilities, but could instead be fired from mobile units. Under normal circumstances it was possible to set up, fuel, arm, and launch a V-2 in about four hours. "To be able to launch a sophisticated missile in this way was an extraordinary achievement and it explains why the rockets were next to impossible to find and attack."[92]

On the same day that the first V-1s had been used successfully against London (June 13, 1944), one of the new V-2 rockets was fired in an unsuccessful trial from Peenemünde (the German Army Experiment Station on the island of Usedom, just off the Baltic coast). The device veered wildly off course and exploded at low altitude over Gräsdals Gäro in Sweden.[93] The main debris of the rocket was subsequently lodged in a crater some 200 meters from a local farm. The wreckage was quickly removed by the Swedish military, and shortly thereafter was acquired by the Royal Air Force in exchange for some British radar equipment (which the Swedes did not have at the time).[94]

The Germans were also testing V-2 rockets at Blizna, an experimental site in Poland. Realizing that important clues might be obtained from the wreckage of test firings, an underground organization of Poles had been formed to retrieve the debris found at nearby impact sites before the Germans could do so. Over seven months they collected what they

[90] Johnson 1978, p. 165.

[91] Johnson 1978, p. 176.

[92] Johnson 1978, p. 177. If there were any consolation to be had, the V-2s in one respect proved to be less destructive than the V-1s. This was because of the much higher speed at which they traveled; as a result, V-2s upon impact tended to bury themselves into the ground before exploding, often resulting in comparatively less damage. See Johnson 1978, p. 164.

[93] Jones 1978, p. 431. Chapter 45 is devoted to the "V-2," pp. 430–461. It is worth noting that among the German rocket experts working at Peenemünde was Wernher von Braun, who later contributed substantially to the development of rocketry and manned space flights in the United States.

[94] Johnson 1978, p. 167.

could, including remains from one rocket that had fallen without exploding on May 20, 1944, near Sarnaki (about eighty miles east of Warsaw, near the Russian border). In this case the V-2 was mired in marshland on the banks of the Bug River. The Poles simply pushed the rocket ever deeper into the water until a German search party sent to retrieve it gave up. Later, some of the recovered parts were then taken by bicycle some two hundred miles southeast to Tarnow.

Since it was impossible to avoid the Germans retreating before the advancing Russian army, this bicycle trip was a dangerous one. The parts had to reach Tarnow because arrangements had been made for a British Dakota to fly in and retrieve all of the rocket parts collected by the Polish underground. In late July a New Zealander, Guy Culliford, was sent with his team to rendezvous with the Poles.

Unexpectedly, a small Luftwaffe unit with two aircraft temporarily set up camp near Tarnow where Culliford was to land. Fortunately, the Germans left at sunset. As soon as the Dakota arrived, the assorted rocket parts (nearly 50 kilos worth) were quickly packed on board, along with Jerzy Chmielewski, the young Polish engineer who had biked the two hundred miles to make sure the parts got safely to Tarnow. Due to the added weight, however, the Dakota became mired in the mud and could not take off. Fearing that its bright landing lights and the noise from its engines might attract the enemy's attention, a small group of farmers hurried to scrape the mud away from the plane with their shovels and bare hands, while local Poles tore up nearby fences for planks which they forced under the wheels. At last freed, Culliford got his plane into the air just as the Poles on the ground had to contend with a German patrol attracted by all the commotion.[95]

On July 28 the rocket parts arrived in England, but Chmielewski, who spoke no English, refused to let anyone near them until a Polish officer he knew gave official approval. Meanwhile:

> [He] drew a knife whenever one of our Intelligence Officers made any move. . . . This scene continued for some hours, reaching peaks of embarrassment whenever he wished to fulfill one of his natural functions, while we searched for the [Polish] General and the Colonel.[96]

Within days of the Polish rocket reaching London (and eventually, Farnborough), a Halifax bomber landed at the Royal Aircraft Establishment with twelve large packing cases containing two tons of wreckage from the V-2 that had crashed in Sweden the month before. These were unpacked under conditions of great secrecy in a hangar of the

[95] Jones 1978, pp. 443–444.
[96] Jones 1978, pp. 444–445.

Aircraft Crash Investigation Unit to which Robinson had been assigned:[97]

> Gradually the rocket took shape as the twisted metal parts were laid out in their correct order. . . . As the pieces of this giant jigsaw were fitted together the picture became clear. A 12-ton missile had crashed at around 2500 feet/sec (1700 mph). . . . It was now clear that the V-2 rocket was a most brilliant scientific achievement. The Germans had solved a number of very complex problems and were years ahead of any other nation in long-range rockets. The value of such a missile as a military weapon with a conventional warhead was another matter.[98]

A formal report of the findings made by the Farnborough team with which Robinson was working was presented to the government on August 31, 1944, exactly one month after the V-2 remains had arrived at the Royal Aircraft Establishment.[99] Included in the report was a detailed (and essentially accurate) drawing of a V-2 rocket.[100]

In the meantime, the Allied armies had advanced from their beachhead at Normandy, and by August 19 had reached the Seine. Paris was liberated by the 25th, and soon the Allies were pushing on through Belgium.[101] The war now seemed as good as won. With the flying-bomb sites overrun and the suspected launching sites of the V-2s captured, Duncan Sandys (chairman of the Committee of Service Chiefs and Scientists responsible for coordinating the defense of London against the flying bombs) and Herbert Morrison (minister for Home Security) held a press conference to announce that "Except possibly for a few last shots, the Battle of London is over. . . ."[102] On the next day, Friday, September 8, the newspapers carried banner headlines with pictures of Sandys proclaiming the end of the fighting (at least by air over London).[103]

97 Luxemburg I-1981. See also Johnson 1978, p. 172.

98 Johnson 1978, pp. 172–173.

99 Johnson 1978, p. 174. The results were presented officially at a meeting of the War Cabinet's Crossbow Committee, which had been formed by Churchill in June to report on the V-1s and V-2s, and the effectiveness of countermeasures being taken. Sandys became chairman of the committee on June 20. See Jones 1978, p. 425.

100 The drawing was by Ronald Lampitt, and was made before any of the V-2s were fired at London. The drawing was based on papers captured in France, decoded Enigma messages, and the detailed information obtained from the Swedish rocket studied at Farnborough in June of 1944. The only major inaccuracy later found in this drawing was that the hydrogen peroxide container should have been below the pump. See Jones 1978, p. 454.

101 As one observer put it, the liberation of Paris was indeed "occasion for relief and small celebration, but it was galling to know that Paris was liberated while London was still under aerial siege," Hewison 1977, p. 170.

102 Johnson 1978, p. 176.

103 Jones 1978, pp. 458–459.

This news came as a great relief—or would have had there not been a tremendous explosion early the next evening at Chiswick. The first German long-range rocket, the dreaded V-2, had just fallen on Britain, and it had been fired only five minutes earlier from a site just outside The Hague, in Holland.[104]

Over the next few months many more V-2s were launched from the Netherlands. The last of these arrived as late as March 27, 1945, when one fell on a block of flats in Stepney killing 134, and another on Orpington that same day.[105]

The war in Europe, however, persisted stubbornly for another month. Then, tragically, on April 12, 1945, President Roosevelt died. The British were as devastated by the news as were the Americans: "His death, before he could see the victory he had helped to organise, hurt them deeply," wrote Leonard Mosley. "The Underground was full of tearful faces—far more than if Winston had died, I'm sure."[106]

VE-Day: May 8, 1945

Meanwhile, events of the war were proceeding at breakneck speed:

> In only a few days at the end of April, and the beginning of May, 1945, a sudden and momentous rush of events brought the war in Europe to an end. . . . Within forty-eight hours there occurred the savage butchering of Benito Mussolini in northern Italy, and the suicide of Adolf Hitler in a hole in the ground in Berlin. That the deaths of the two dictators should have happened within hours of each other was additionally telling in the impact that they made on one's mind because these violent deaths were in no way related, other than in marking the macabre ends of two tyrants.[107]

The war in Europe was not officially over, however, until the Germans finally agreed to unconditional surrender early in May: "The instrument of surrender was signed by the Germans in the small hours of the morning of the 7th of May, with all operations to cease by midnight

[104] Another V-2 had already been launched against newly liberated Paris, falling just outside the capital. Johnson 1978, p. 176.

[105] Of the V-2s launched against England, 1115 actually arrived, 517 of them in London. In all, 2754 people were killed and 6523 seriously injured. Conventional bombing had killed 51,500, with 61,400 injured. From a strategic, military point of view, the V-2s fortunately accomplished far less than intended.

[106] C. P. Snow, quoted in Mosley 1971, p. 374. Mosely also noted that "The House of Commons adjourned for the day on 13 April, the first time it had ever adjourned over the death of a foreign statesman. All over London the people walked around as if they had suffered a death in the family. As, they felt, they had," Mosely 1971, p. 374.

[107] Douglas 1966, p. 280. Hitler committed suicide with Eva Braun in his bunker beneath the ruins of the Berlin chancellery on April 30, 1945.

of the next day, the 8th of May, which became known as VE-Day."[108] The timing, however, created considerable confusion and uncertainty in England: "So far as Londoners were concerned, 7 May 1945 was the messiest day of the Second World War. They knew that the war was over but nobody would make it official, and no one quite knew what to do."[109]

"VE Day May be Tomorrow," said newspaper headlines, but the pubs were already full of expectant revelers. Finally, Churchill officially proclaimed May 8 as VE-Day with a massive celebration in London. Robinson, wearing his R.A.F. uniform, made the trip in from Farnborough to be with Renée, and together they joined the crowds pouring into central London to hear Churchill address the nation from Whitehall:

> That afternoon a vast river of Londoners flowed down the Strand and the Mall and through Trafalgar Square into Whitehall, where it had been announced that the Prime Minister would speak. At 3 P.M. the great bell of Big Ben struck and over loudspeakers came a voice to tell them that Churchill was coming.[110]

Winston Churchill came onto one of the balconies at the Ministry of Works, and looked down on the cheering crowds amassed below. The scene was awash with "paper hats and favours, and rattles and whistles of a kind which hadn't been seen since 1939."[111] And then, suddenly, there was a heartrending silence, a solemnity and gravity that everyone could feel as Churchill began to speak:

> People hung onto every word he said. When he told them that as of midnight hostilities would cease, there were loud cheers and a waving of hats and flags; and then a louder cheer when he said: "The German war is therefore at an end." People began to cry and laugh and cheer at the same time.[112]

Churchill ended his speech with "Advance Britannia." The Scots Guards, brass bugles gleaming, sounded the traditional, ceremonial cease-fire. "Then the band struck up the National Anthem and looking round I saw everyone, young and old, civilians and soldiers, singing with such reverence that the anthem sounded like a sacred hymn."[113]

To Renée, VE-Day was the most impressive event she remembered in London: "Abby came to the city and we went to see Churchill when he passed outside St. James Park."[114] All afternoon, people were dancing

[108] Douglas 1966, p. 277.
[109] Mosley 1971, pp. 374–375.
[110] Mosley 1971, p. 376.
[111] Mosley 1971, p. 375.
[112] Vera Hodgson, quoted in Mosley 1971, p. 376.
[113] Vera Hodgson, quoted in Mosley 1971, p. 376.
[114] Renée Robinson to JWD, letter of March 15, 1989.

in the streets, and "The Lambeth Walk" from *Me and My Gal* was a popular favorite. At the Ritz Hotel, the celebrating was just as genuine if more reserved. Henry Channon, MP for Southend-on-Sea (a cockney seaside resort), remembers the Ritz: "beflagged and decorated: everyone kissed me, Mrs. Keppel, the Duchess of Rutland and Violet Trefusis all seized me alternately." As Leonard Mosley put it, "Nobody said no to anybody in London on 8 May 1945."[115]

Robinson's Trip to Germany for British Intelligence

With the war officially over, one of Robinson's last official duties at the Royal Aircraft Establishment was a matter of reconnaissance. Shortly after the armistice with Germany was signed, both Robinson and his friend Hermann Jahn were sent (together with a contingent of fellow officers from Farnborough) on a scientific intelligence-gathering mission to Germany. Before he left, Abby went into London to meet Renée on her lunch break, and together they walked down Regent Street, talking about how much better things would be now that the war was over. Robinson could begin to think again about his scientific career as a scholar and mathematician.[116]

The BIOS mission to which Robinson was assigned had orders to investigate and report on what the Germans had accomplished in aerodynamic research during the war.[117] The head of the team was A. R. Collar. Robinson and Jahn were included doubtless because of their scientific expertise and fluency in German. According to Jahn:

> We flew over to Germany, first to Frankfurt, where I remember one incident. Abby must have been prepared beforehand, because he told me: "I want to see the village where Riemann was born [Breselenz], which is not far from where we are at the moment." I do not know, nor do I remember, how he knew the way, nor how we got there, but suddenly we were standing in the village and he was asking people, "Do you know the house where Riemann was born?" In some way, I think, there must be a strong similarity between Riemann and Abby. Riemann, the famous pure mathematician, also made his mark in fluid mechanics, as did Abby.[118]

[115] Mosley 1971, p. 377.

[116] Renée Robinson to JWD, letter of March 15, 1989.

[117] British Intelligence Objectives Subcommittee. Many of England's research establishments sent out reconnaissance parties to Germany after the war under the aegis of BIOS.

[118] Jahn R-1976 and Seligman 1979, p. xvi. Given that Breselenz is not far from Hamburg, the BIOS team must have gone north from Frankfurt, when a visit to Riemann's birthplace would have been more likely, before the team flew south. Renée remembers that Abby indeed once mentioned Hamburg, and told her how much of the city had

Robinson and Jahn (both dressed officially in their R.A.F. uniforms), then flew from Frankfurt to Chiemsee in Upper Bavaria:

> I remember we were in a van, there was a German driver, and Collar was in the front. Abby was saying to Collar: "Can't we arrange it so that some of us can also sit in the front seat, so that everyone has a chance to see some of the landscape?" So it was done, and nobody missed anything.
>
> I also remember another occasion. [Robinson] told me after we had started on a trip, "You know, I had a high temperature this morning, but I took some aspirin because I did not want to stay behind." He wanted to see everything, and did not want to miss anything.[119]

Beginning Anew

Abby and Renée, London, Easter, 1946.

Within months of VE-Day, life in London returned to normalcy, or at least to as much of the old and familiar daily routines as could have been expected. By midsummer of 1945, even before the war in the Pacific was over, the Sadler's Wells reopened with Benjamin Britten's opera, *Peter Grimes,* and the Comédie-Française completed its first official postwar visit to Great Britain. A year later, in April of 1946, Covent Garden was prepared to reopen, one correspondent noting that "American soldiers will remember it as a dance hall where jitterbugging was not allowed."[120]

A few months later, Robinson officially tendered his resignation to the Royal Aircraft Establishment, which was accepted on August 27 with a termination date of September 7, 1946. It was time, at last, to get on with other things. Robinson had been offered a position at the newly established College of Aeronautics in Cranfield, about fifty miles northwest of London, where he was to teach mathematics in the Department of Aerodynamics under A.D.

---

been destroyed, "flattened" by the time he saw it just after the end of the war. Renée Robinson to JWD, May 23, 1989.

[119] Jahn R-1976. The specific sites that Robinson and the team from Farnborough were authorized to visit included installations at Oberursel and Darmstadt.

[120] Littlefield 1946, p. 24.

Young as head. Renée would continue to work in London, but she and Abby soon moved from Surbiton to a house provided by the College of Aeronautics, which celebrated its inaugural term on October 19, 1946.[121] Cranfield was to mark a transition not only in their lives, but above all, in Robinson's career as a mathematician.

[121] *Potential,* 1 (July 1947), p. 10.

CHAPTER FIVE

Robinson after the War: London 1946–1951

> I see that Parliament's first act after the end of the war in Europe has been to rescind the Bill making it a punishable offence to spread gloom and despondency. So great was our danger in certain years that we were forbidden to look miserable. Now we can be as unhappy as we please. Freedom is returning.
>
> —*Vere Hodgson*[1]

ALTHOUGH the end of World War II did not immediately mean freedom for Abraham Robinson—at least not from his military service at the Royal Aircraft Establishment in Farnborough—he and Renée had survived the war happier than many. They were married, Renée had a well-paying, interesting position as a fashion designer, and Abby could finally think seriously again about the future. He looked forward to his release from government service, and longed to return to pure mathematics. But first there was a period of transition.

ENGLAND AFTER THE WAR

When the facts and figures—and especially the photographs—of the devastation wrought upon the rest of Europe as a result of the war became known, it was clear that Britain had not fared as badly as her neighbors across the channel. England had at least been spared actual combat on her own soil, but the German Blitz and prolonged air attacks had done substantial damage. Recovery was slow from the destruction caused by German bombs and rockets, and the frequently devastating fires that, in their wake, had destroyed considerable parts of London and such memorable targets as Coventry.

Potentially more problematic was the immediate economic disaster facing the country. As soon as the two American atomic bombs had been dropped on Hiroshima and Nagasaki, abruptly ending the war in the Pacific, World War II was officially over. So too was the American lend-lease program. Throughout the war the United States had provided billions of dollars in contributions and credits to England and

[1] Mosley 1971, p. 378.

the rest of Allied Europe, but overnight these credits turned into debts. As Churchill realized, this could well mean financial ruin.

Churchill's conservative government called for elections on May 23, 1945, only two weeks after VE-Day. Scarcely a month later the electorate went to the polls, and the Labour Party's resounding victory on July 25 came as a shock to many, especially because the mandate was so clear—in the House of Commons, Labour won by a margin of 183 over all of the other parties combined. That same evening Churchill rode to Buckingham Palace in his chauffeur-driven Rolls-Royce and tendered his resignation to George VI. It was a sign of the profound changes to come that the newly elected prime minister, Clement Atlee, arrived moments later in an ordinary Standard Ten, his wife at the wheel. The king invited Atlee to form a new government, and for six years Labour set about to create a new era in England.[2]

The Labour government was faced with the immediate task of rebuilding the country—economically as well as socially. As the world looked tentatively ahead to the atomic age, England soon found herself not at peace but mired in the Cold War. The brilliant success of the Royal Air Force during World War II, however, would mean that considerable emphasis would now be placed upon developing air power. Robinson was well placed to benefit from this shift in strategic emphasis, for his successful work at the Royal Aircraft Establishment put him in an excellent position to trade on that expertise. Even before he was released from government service, Robinson had also begun to reestablish contacts with mathematicians throughout Britain and abroad, reaffirming in the process the ultimate importance of pure mathematics in his life as a theoretician.

Back to Mathematics

In February of 1946 Robinson wrote to Ivor Etherington, thanking him warmly for all of his previously unacknowledged help:

> Dear Dr. Etherington,
>
> I believe that in 1941 you were kind enough to present a paper of mine ("On a variation of the distributive law")[3] to the Royal Society of Edinburgh. At that time I was in the Forces, and I did not find out that the paper had been published until much later, when I had joined the Scientific Civil Service. However, now that the war is over, I am again taking a more active interest in pure Maths., in addition to the applied Maths. which I

[2] Addison 1951, p. 19.

[3] Robinson misquoted his own title because Etherington, in editing the paper for Robinson, had changed it. For details, see chapter 4, pp. 103–106.

have been doing during the last few years. In this connection, I wonder if you might let me have a few reprints of my paper—at the moment I haven't got a single one. I am thinking of trying to find a University job, but although I can bring good personal references of mathematicians who know me and my work in recent years, I have not really got an efficient line of attack for this problem. Thus, one reason why I am bothering you with these reprints is that they might assist me in my attempt for an Academic career.[4]

Etherington sent off a package of reprints directly, and ten days later wrote to explain what had happened earlier:

Your MS was given to me by Professor Whittaker, who had it from Professor Brodetsky in 1940 or 1941. I was asked for my opinion on its merits [~~as to whether it was suitable to publish or not~~].[5] I replied that it was a very interesting paper and well worth publishing, but that the MS was not in a suitable form for [~~sending to~~] immediate submission to the printers, and that apart from that some revision was desirable. I suggested that the MS should be returned to you for revision. But I learnt that you were expecting to leave the country very shortly for service in the forces, and Professor Whittaker asked me to undertake the revision myself and to see the paper through the press.

I made a number of alterations, and hope very much that there is nothing of which you disapprove, and that I have not introduced any errors. I would actually have preferred to have you approve of the changes, before the paper was printed, and I wrote to an address in London which Professor Brodetsky gave me, hoping that it would be forwarded to you, but I got no reply. . . .

Robinson answered promptly, describing briefly the circumstances which prompted him to write the paper in the first place:

Dear Dr. Etherington,

Thank you very much for your letter and for the reprints of my paper. I only regret that owing to the circumstances I was unable to give you credit in my paper for all the trouble you went to in connection with its edition. Having just re-read my original version of paragraph 3.2, I quite agree that it is "a little confused." My failure to quote the reference to Dörnte's

[4] Robinson signed his letter, adding parenthetically, "formerly of Jerusalem University, at present a Scientific Officer." See Robinson to Etherington, February 26, 1946, RP/YUL.

[5] Overstricken, bracketed portions of the text [———] indicate cancellations in Etherington's draft manuscript, which is reproduced here from the originals kindly supplied to JWD by I.H.M. Etherington. These originals have been deposited with Robinson's papers in the archives of the Yale University Library.

work accurately was due to the fact that I was unable to get hold of the *Math. Zeits.* and thus had to quote from memory.

As a matter of "historical" interest I might mention that I first considered "multiple fields," but seeing that their consideration could be quickly reduced to the consideration of ordinary fields with only a modified distributive law, I decided to concentrate on the latter. I hope that my work will be of some use to somebody some day. Just the connection with "multiple fields" seems to indicate that quasi-fields are more [than] mere pathological cases.

I have seen some of your work on train algebras, but I shall be very grateful if you can spare me a few reprints.[6]

Several months later Robinson wrote again, this time with some observations on Etherington's arithmetic approach to trees:

Dear Dr. Etherington,

I believe I still have not thanked you for sending me the reprints of your papers, and I apologise for it. I remember, I thought I would first read through them but somehow was unable to cope with them all at once.[7]

Now, apart from thanking you for sending me those papers, I should like to say how interesting I found their contents (from the algebraic point of view—I know little of genetics). In particular, the arithmetisation of trees is, I believe, an important step, in spite of the unwieldiness of shapes compared with ordinary numbers.

If I may add a few remarks: "shapes" appear to me to be special cases of Russell's "relation numbers" (*Principia Mathematica*, vol. II), since trees can be conceived as "relations." However, the operations (addition and multiplication) defined by you differ from Russell-Whitehead's who were primarily interested in a generalisation of Cantor's arithmetic of transfinite ordinals. Nevertheless I found it quite interesting (as an exercise for myself) to express your ideas in a more "formal" way. Another interesting exercise is to define a logarithmetical system axiomatically, and to consider subsystems of a given system (as you would in field theory, although

[6] Etherington began to apply algebra to genetics following some very simple ideas first published by Glivenko in 1936. Etherington dealt with a wide range of genetic problems, for all of which, the "genetic" algebras in question were "special train algebras." See Etherington 1939. Etherington 1940 also applied nonassociative linear algebras (namely baric and train algebras) to genetics. This approach was further developed in more detail in Etherington 1941. Many problems of genetics, for example, the frequency distributions of genes and zygotes in the offspring of successive generations, can be treated by special algebras for which the multiplication sign is interpreted literally as meaning genetic multiplication. Etherington 1941a is also illustrated with examples from genetics.

[7] Etherington noted that he sent nine of his offprints to Robinson on March 16, 1946.

the results are more trivial). However, it is probably not worth it to waste your time by going into details.

Robinson as a civilian, London, April 19, 1946.

Robinson, in fact, had gone into considerable detail on the subject of "logarithmetic" (or the arithmetic of shapes of nonassociative combinations as Etherington had defined them in a prewar paper, "On Non-Associative Combinations").[8] The basic idea can be illustrated with a binary noncommutative operation for which the products $((AB)\,C)\,D)$, $((BA)\,C)\,D)$, and $((AA)\,A)\,A)$ all have the same "shape," whereas $D((AB)\,C)$ has a different "shape." Here Robinson found that an axiomatic approach was of interest. After first defining the arithmetic system of shapes by a set of axioms, he showed that any realization of the axioms is isomorphic with the synthetic system of shapes. He preferred to call realizations of the axioms "simple forests," since "shapes" had also been called "trees" by Arthur Cayley almost a century earlier (although, as he realized, from a different point of view).[9]

From the very beginning of his career, as his letter to Etherington of May 26 (1946) makes apparent, Robinson was persistently drawn to the "clarity" that he found in writing down the axioms of a given system and then exploring their consequences. For Robinson, the precise if austere focus that axiomatizations furnished was both an aesthetic and a practical matter. Etherington understood this as well, but developed his own ideas about axiomatizations in very different directions, especially in applying ideas from algebra to genetics.

Over the next few months Robinson and Etherington continued their correspondence on the subject of trees, forests, and nonassociative systems in general. Robinson even sent Etherington notes on a paper he was writing on the "Arithmetic of Binary Root-Trees (Axiomatic Approach)." In his covering letter, he outlined briefly the kind of applied mathematics he was doing at Farnborough, in addition to explaining his current work in algebra (including some results on infinite matrices)[10] and saying a bit more about relation-arithmetics. He also told

[8] Etherington 1939a.

[9] Cayley 1857. Robinson 1949 was an early foray into model theory. See also the "Record of Meetings of the Edinburgh Mathematical Society, 65th Session, 1946–47," in the *Edinburgh Mathematical Notes,* 37 (1949), p. 29.

[10] This remark is helpful in documenting the fact that Robinson must already have

Etherington that he was about to leave for a brief visit to Palestine:

> Dear Dr. Etherington,
>
> Thank you very much for your letter. I intend to go abroad for about a month towards the end of June (to Palestine) so I think you had better not send me your manuscript now.
>
> There are two sorts of relation-arithmetics, one which does in fact depend on the individual relations involved, and one which depends only on the type (rather, relation-number) of these relations. In my last letter I was referring to this second arithmetic which however is still . . . different from yours. I enclose a few notes on the possibilities of attack which I mentioned in my letter. I am glad to learn that similar ideas have already occurred to you.[11]
>
> I have not found an academic post so far, although I have one fairly vague chance. Thanks for mentioning that there might be a vacancy in Edinburgh. I shall be on the lookout for advertisements in "*Nature*."
>
> In my professional work at the moment, I am concerned with high speed aerofoil theory, i.e. mathematically speaking, with certain problems appertaining to the theory of hyperbolic partial differential equations. Before that I was working on shock transmission in beams.[12] I hope to be able to publish my results in due course—the trouble is that in our government research work results are secret for a year or two, after which they may be published if the people on the committees which decide these matters approve of it.
>
> In my own time I have in recent years done some work on infinite matrices, as applied to general summability. I have now summed up my results, and my paper is just being typed. I haven't decided yet what to do with it, but at any rate I shall send you a copy when it is ready.
>
> Apart from the above two subjects there are of course a number of

---

been working with Richard Cooke at Birkbeck College, University of London, even while he was still on government service at the Royal Aircraft Establishment in Farnborough. Robinson to Etherington, June 8, 1946, RP/YUL.

[11] In a penciled note (undated) to himself covering the main points of his reply, Etherington offered some suggestions for improving Robinson's approach to trees, urging him to publish the results in a more general form and saying that he felt Robinson's approach was both different—and better—than his own. "My approach is more general," he explained, "because I do not limit myself to binary trees." Etherington suggested Robinson think about generalizing his treatment by studying addition as a "free" operation and generalizing his approach to noncommutative forests (Etherington had not considered quasi-forests at all), RP/YUL.

[12] This work later led to Robinson's paper on the integration of hyperbolic differential equations, which he had first considered in connection with his work on stress propagation while working at Farnborough. For details, see Robinson 1945. For details of the integration method Robinson developed for hyperbolic differential equations, see below, p. 159, n. 79.

> others in which I am interested, and which I would like to work on if I only had enough time to crystallise my ideas on paper. This, I suppose, is a feeling familiar to many other mathematicians.

Robinson's Return to the Hebrew Universtiy: Summer 1946

Throughout the war, Robinson had remained in sporadic contact with his family and mentors in Jerusalem. On December 15, 1942, he broke a long silence by writing a joint letter to Fraenkel, Fekete, Amirà, Levitzky, and Motzkin:

> You have not heard from me since I left Israel. You understand that it was not an easy road and frequently my future was dark and I did not want to write as long as I could not justify the hopes that you placed in me when I left Palestine. . . . I decided that I should take an active part in the general struggle so I went through different situations and decided that in the course of time I had an opportunity to apply my knowledge in applied mathematics. And recently I was appointed assistant in a government scientific research institute. Two years ago I would not have believed that my mathematical knowledge would help me solve technical problems, and as I did not visualise this it has happened. Please believe me that the memory of the Hebrew University and the memory of all of you has not left me for a moment. And I hope that the time will come when I shall be able to return and study and research under your supervision. Yours in great respect. AR [13]

Two years later, by the end of 1944 when the war was finally beginning to reach its end, at least in Europe, Robinson began to enquire about finishing his degree. On November 28, 1944, he received the following reply from A. Ibn Zahav, the academic secretary of the Hebrew University:

> Dear Mr. Robinson,
>
> We were pleased to receive your greetings to us from Dr. Senator. . . . We understand from him that your intentions are to come here to be examined and finish your studies. Is this so? It is possible that the university could help you to finish your studies by finding some sources of money for the short term that you might need while you stay here for the examinations. Yours sincerely, A. Ibn Zahav

Abby responded late in January the following year:

[13] This and the following two letters exchanged between Robinson and the Hebrew University are translated from originals in Robinson's administrative file, Department of Mathematics, the Hebrew University, Jerusalem.

Dear Mr. Zahav,

I was glad to receive your letter. It is my wish to get to the University to finish and submit my thesis to obtain my degree, but I cannot yet fix the date when I could carry out my plan. I serve in a job which is connected with the war effort and therefore cannot demand leave of absence till the end of the War. We can hope that even in the present conditions I could come to Jerusalem during 1946.

As is known to you I serve as a scientific officer in an important government research institute in a fairly high position which gives me status as well as decent conditions, but not sufficient to provide enough for living in Jerusalem. I want to emphasize that during the last years I dealt with applied, technical mathematics, on which I could lecture in the Hebrew University as I have already lectured on them in London University. . . . The main thing for me is that the University has not forgotten me, as I have always tried to be a worthy student of the H.U.

My kind regards to Dr. Senator and all of my teachers.
Yours sincerely, AR

Indeed, plans were laid for Robinson's return to Jerusalem to finish the formalities necessary for awarding his degree. In May of 1946 he wrote to explain that he wanted to be examined in mathematics as his main subject and physics as his minor subject, and that he expected the examination would cover algebra, analytical geometry, differential and integral calculus, the foundations of analysis, and complex function, as well as mathematical logic and foundations of mathematics.

Robinson left for Palestine early in July. Renée, who was now a manager for Rita and Company, was unfortunately obligated by her responsibilities to stay in London.[14] Abby therefore had to make the trip to Jerusalem on his own. He went by sea, which he loved to do, and spent nearly a month in Palestine. This not only gave him enough time to take his examinations at the Hebrew University, but he was also able to see his mother, brother, and friends there for the first time in six years.

Back in Palestine, Robinson formally applied on July 20 to take his examinations for the M.Sc. degree. He had already filed preliminary documents on July 10, indicating that his main study at the university earlier had been mathematics, but he listed as "secondary" fields both physics and philosophy. Upon successfully passing the examinations, he was duly awarded the Master of Science Degree from the Hebrew University.[15]

[14] Rita and Company was not far from Oxford Circus, only a minute's walk, next to Liberty's of London. Renée Robinson I-1980.

[15] Abramsky recalls that the university also offered Robinson an immediate job in Jerusalem equivalent to a senior lecturer. See Abramsky R-1976.

Abraham Robinson, with his mother in Jerusalem, July 1946.

Six months later Robinson received the official results of his examinations, which were sent to him by letter on December 15, 1946. The results: "Physics good, mathematics excellent." This, however, did not satisfy Robinson, and apparently he was very unhappy with how the examination had been conducted. He took the trouble to complain about it in a lengthy, six-page letter to Fraenkel (the length of the letter was as unusual as Robinson taking exception to what had happened), and later rectification was made admitting the justice of Robinson's complaint. As a result, Fraenkel wrote a confidential letter to the university administration "correcting" results of the examination.[16]

While in Palestine, Robinson not only used the opportunity to finish his degree, but was happy to work with his former instructor, Theodore Motzkin.[17] Together they wrote a paper on "The Characterization of Algebraic Plane Curves" which they submitted to the *Duke Mathematical Journal* in August of 1946. Basically, their paper showed that algebraic curves are completely described as sets for which the multiplicity of every point of the set on every line through it can be defined to satisfy certain preconddititons. The editors asked for revisions, and once these had been made, a new version of the paper appeared the following year. Taking work that was well known concerning properties of algebraic curves over a closed field, namely:

1) that the set has a constant finite number of intersections with every straight line not contained in it; and
2) the cyclic ratio of the set is 1 for every triangle with no vertex on the set or at infinity (Carnot's theorem),

[16] Robinson's administrative file, Department of Mathematics, the Hebrew University, Jerusalem.

[17] Motzkin had already published work on plane curves and Carnot's theorem, and as early as 1938. See Motzkin 1938 and 1945. He was about to publish another paper on "Monotone Trajectories of Unisolvent Families of Curves" when Robinson appeared in Jerusalem and they began work on their joint paper on algebraic plane curves. This was completed just before Robinson left Jerusalem for his return to London in August of 1946.

Motzkin and Robinson established that in a plane over an arbitrary infinite field, a set with properties 1) and 2) is also, necessarily, an algebraic curve. In investigating various possibilities, they considered cases in which 1) is sufficient to characterize algebraic curves, but also constructed other plane sets that are not algebraic curves and yet possess either the property 1) or 2) alone. In particular, in the case of the affine plane over an enumerable closed field, they showed the existence of a very large class of sets satisfying 2) that are not algebraic curves. After discussing Carnot sets, the paper focused on algebraic curves characterized as Carnot sets with a constant intersection number, others characterized only by a constant intersection number, and then general sets with a constant intersection number or with given intersection numbers.

The Royal College of Aeronautics, Cranfield

Following the war, which had been waged as much in the air as on land and sea, England realized that her future, both commercial and military, would be closely tied to the success of British aviation. Above all, "the establishment of an aeronautical college with wide facilities for theoretical and experimental study was thought by many to be a vital necessity in a country acutely concerned with aeronautical development for defense and commerce."[18] The idea of creating a Royal College of Aeronautics originated with Sir Stafford Cripps, formerly minister of aircraft production during the war. In 1943 he presented a confidential report to the government's Committee on Aeronatucial Research outlining details for such a college. By July 1944 a special committee headed by Sir Roy Fedden urged acquisition of a suitable airfield where the college could be established, and soon the Royal Air Force Station at Cranfield, Bedfordshire (fifty miles northwest of London), was undergoing modifications in time to begin teaching in 1946. The college's expressed aims were to provide "a comprehensive education designed to fit its students for good positions in the aircraft industry, civil aviation, aeronautical research, the Services, and in the educational field, with the hope that its best students will ultimately become leaders of aeronautical thought and practice."[19]

Having mastered the subject of wing theory during the war, in addition to writing (or coauthoring) seven research papers related to various aspects of aerodynamics, Robinson was an ideal candidate for a position at Cranfield. Both his interests and abilities were well-suited to the needs of the new college, and he was urged to apply for a post in

[18] From the College's *Bulletin*, no pagination.

[19] From the College's *Bulletin*, no pagination.

mathematics by W. J. Duncan, the first professor of aeronautics to be appointed.[20] Robinson applied by letter early in May, and following an interview, E. F. Relf, the principal of the college, wrote with good news: "I have the pleasure in offering you an appointment as Senior Lecturer." The salary was set at seven hundred pounds a year.[21]

The opening day assembly of the governors, staff and students took place at Cranfield on October 15, 1946.[22] It was a mild Tuesday, and Sir Roy Fedden's speech that day "bristled with brisk hopes" in an atmosphere, said one student, "of awed expectancy."[23] The college offered a two-year program at the postgraduate level leading to a D.C.A. certificate. After a year of introductory studies (intended to furnish a broad knowledge of aeronautics), the second year required students to specialize in a particular department. Mornings were devoted to lectures and theory, while the afternoons concentrated on practical work in the laboratories, drawing offices, or actual in-flight training. In the second year students wrote a thesis, usually involving experimental research or a design project.

Heading the Department of Aerodynamics was A. D. Young, under whom Robinson served in 1950–1951 as deputy head of the department. In addition to providing courses on basic mechanics, kinematics of fluids, dynamics of frictionless fluids, compressible flow, airfoil theory, general aerodynamics, and heat transfer, the Department of Aerodynamics was also responsible for all of the college's basic mathematics courses, including differential calculus, vector analysis, complex variables, differential equations, Fourier series, calculus of variations, determinants, statistics, and numerical methods.

Laboratory work included the aerodynamic characteristics of lifting surfaces, propeller characteristics, and experimental dynamics. The college was equipped with an impressive assortment of wind tunnels, including a low-turbulence tunnel, smoke tunnels, and high-speed wind tunnels. Renée recalls that whenever anyone wanted to find Abby, he was almost always in one of the wind tunnels, where he liked to study the practical, concrete effects of various theoretical models in experi-

[20] Young R-1976.

[21] With 5 percent deducted for the Federated Superannuation System for Universities. To this the college added a contribution of 10 percent of the annual salary. Relf to Robinson, July 13, 1946, RP/YUL. According to Olga Taussky Todd, "several distinguished researchers had applied for this position," Taussky R-1976.

[22] A group photograph, dated "Cranfield, October 15, 1946," shows the opening day assembly of governors, staff, and students. Included are W. J. Duncan (aerodynamics), A. D. Young (aerodynamics), W.S.S. Marshall (aerodynamics), S. Kirby (aerodynamics), G. M. Lilley (aerodynamics), and A. Robinson (mathematics). See *Potential*, 1 (1947).

[23] *Potential* 1, p. 10. *Potential*, a magazine of the College of Aeronautics, was published occasionally by the students' society. Its first number appeared in July of 1947.

Faculty and staff, Royal College of Aeronautics, Cranfield.

mental contexts. Cranfield also provided a hydraulics laboratory and a computing laboratory. The latter included a variety of electrical calculating machines, from hand-operated and fully automatic models, to a small differential analyzer and a harmonic analyzer. Robinson showed considerable interest in the computing laboratory, and in one paper written at Cranfield, he used methods that were only feasible in the age of high-speed computers.

Robinson's teaching responsibilities were twofold. Courses were expected not only to develop facility in day-to-day practical applications of mathematics, but to enable students to read and appreciate scientific and technical literature on aeronautics. Consequently, special attention was given to numerical and statistical mathematics, and to the practical as well as theoretical difficulties that arise whenever aircraft approach or exceed the speed of sound. In addition to significant changes in the air surrounding the aircraft, distortion of the aircraft's structure and consequent modification of related aerodynamic forces become critical. The entire subject not only received special attention in the department, but was a particular interest of Robinson's as well.

College Social Life—and Learning to Fly

The College of Aeronautics provided faculty with housing, so Abby and Renée moved to Cranfield, where they usually spent their weekends together. Because of her job, during the week Renée stayed in London with a friend, Sophie Smith, but she would return to Cranfield on Friday evenings, staying until the following Tuesday morning. Every week, on Wednesdays, Abby was in London for classes at Birkbeck College, and they would see each other then as well.[24]

For the small house they had in Cranfield, they bought some new furniture, and for the first time Renée had to learn how to manage in the kitchen. Abby could cook an egg, she remembers, but not much more.[25] Renée learned to prepare simple soups, and later progressed to more difficult recipes as time went by.

Realizing "that uninterrupted specialised study is not education," the development of cultural and social activities at the college was encouraged "as an essential complement to the academic work." There was a "flourishing" amateur dramatic society and a music circle. Informal social evenings (affectionately known as "honks") provided light entertainment during the term, including on one occasion a Hindu mind-reading act and a sketch presented at one function that produced, as the student magazine reported, "a phrase that quickly passed into the language, 'integrating all over the bed.'"[26] At the last of these evenings, one member of the staff won the rear-fuselage from a Tempest fighter aircraft, "ostensibly in a raffle, but more, we suspect (according to the student paper), as a reward for his Set Design Problems."[27]

As the year drew to an end, the staff held a fancy dress dance, for which Renée wore a splendid evening gown; Abby appeared in black tie. Indeed, when it came to clothes, according to Renée, Abby could at times "walk around in rags, but there was no limit to his taste for dressing well, to go first class. . . . Sometimes he could look like a tramp, but he felt just as comfortable dressed informally as he did in a tuxedo." Thanks to Renée's efforts, Abby was the best-dressed man at the College of Aeronautics; she had all of his suits tailor-made in Mayfair.[28]

The social life at Cranfield was lively, and when it came to social drinking, Robinson could outlast anyone. When he went with Jahn to Germany, he even won a beer-drinking competition! But Renée recalls seeing Abby drunk only once, "at the College of Aeronautics, where there were three bars and everyone drank like mad." One night Abby came home after midnight, and Renée heard him say to one of his colleagues, "Can you find the keyhole?" Robinson was jolly and in high spirits as he came up the stairs.[29]

[24] Seligman 1979, p. xx.

[25] Renée Robinson I-1989.

[26] *Potential*, p. 12.

[27] *Potential*, 1 (July 1947), p. 11.

[28] This was something that in later years he missed, especially in the United States, where it was not so easy to find tailors like the ones that they had both been used to in London. Fortunately, for years Robinson was a perfect thirty-nine, and so could easily wear clothes off the rack. Renée adds that his shoes were always impeccable, and he had two pairs that were quite exotic, of crocodile and python! Renée Robinson I-1980.

[29] Once, after flying to Paris, they went straight to a party where there was nothing

Abby's Tiger Moth.

Training at the London Gliding Club, Dunstable.

One of the highlights of Robinson's first year at Cranfield was learning to fly. At first, the long-awaited Tiger Moths–the college's training planes–arrived, only to be retired for modifications. Lack of decent weather also caused delays in his flight training schedule.[30] Undaunted, Robinson put in as many hours as were required. For fear that Renée might worry, he never told her about his lessons, although eventually she learned about them anyway. Renée remembers that he was the only one of the staff members at Cranfield who learned to fly. She also remembers one incident, his first solo flight in fact–when he returned home, "green" from having gone into a tailspin.[31]

That spring (1947) Robinson was asked to give an informal presentation on aeronautics for a general audience. It was one in a series of meetings which the student magazine described as "a fascinating and useful collection of odd lectures." It was one of the few of his lectures that Renée ever attended, and she found it an informative glimpse into the kind of work Abby was doing.

In addition to everything else, Robinson also decided to learn Russian:

---

but champagne and canapés. Renée soon felt like she was flying, dancing. As for Abby, "he could drink lots. He loved to eat, drink, and above all," according to Renée, "he especially liked chocolate. He was cross without chocolate in the house." But he usually had his cocktails by himself, because Renée did not like to drink. Renée Robinson I-1989.

[30] *Potential*, p. 12. The magazine included a photograph of the college's Tiger Moth G-AIXG used for *ab initio* flight training.

[31] Renée Robinson I-1980.

> He acquired languages almost through his paws. The number of languages which he had a very considerable command of was extraordinary. I remember at Cranfield when he decided to learn Russian. He very soon left those of us who tried to join him far behind with speed and rapidity with which he acquired some command of the language.[32]

Delta Wings and Supersonic Flight

The postwar years in Britain and the United States witnessed extensive advances in aviation and aircraft design. By 1944 it was evident that the piston engine had been pushed to its limits; more power could only be expected from new technologies. Fortunately, the gas turbine held out great promise for increased power, both in conventional propeller and in jet propulsion systems.[33] It was the jet engine, however, and the design of aircraft capable of supersonic speeds that captured the imagination–Robinson's as well.

During the war (and increasingly thereafter), research in Britain and the United States experimented with low-drag wing sections by extending the area of the wing over which the boundary layer flow is laminar. As the Royal Air Force had learned, as speeds increased in aerial combat, aircraft ran into "compressibility" problems whenever they approached the so-called sound barrier. Considerable attention was therefore given to this problem.[34]

Delta wings (in the shape of isosceles triangles with lines of symmetry parallel to the direction of flight) were of special importance. If, for example, the apex angle is small, the pressure distribution and corresponding lift may be estimated from the theory of wings of small aspect ratio. For large apex angles it is necessary, however, to employ a more general lifting surface theory. The simplest form of three dimensional wing theory is lifting line theory, in which the wing is represented as a bound vortex lying along a line normal to the stream. Lifting line theory, however, is not satisfactory for wings which are swept back, or which have a small aspect ratio.[35]

In a paper written with F. T. Davies, "The Effect of the Sweepback of Delta Wings on the Performance of an Aircraft at Supersonic Speeds," this was exactly the problem Robinson addressed. As a result of their analysis of delta wings, Robinson and Davies concluded:

> It appears that a large angle of sweepback is not uniformly beneficial for the performance of an aircraft at supersonic speeds. While it is likely that

[32] Young I-1976.
[33] Greene 1970, p. 114.
[34] Greene 1970, p. 115.
[35] For a wing of span s and surface area A, the "aspect ratio" is equal to s^2/A.

> Delta wings or wings of similar shape will be adopted in any case for supersonic aircraft for reasons of stability, the actual optimum amount of sweepback can be determined only as a function of the height and speed to which the aircraft is designed.[36]

Making Do in the 1940s

Although 1946 was a significant year for Robinson, full of turning points— his return to Palestine, his degree from the Hebrew University, his new position at the College of Aeronautics—it was still one that found Britain struggling to recover from the war. Clothes, gasoline, and many basic foods (including meat, sugar, and chocolate) remained on the ration. The first peacetime Christmas, 1946, "was a festival of scarcity":

> In Trafalgar Square the children could be taken to admire a new feature of the London Christmas, a magnificent spruce newly arrived by ship from Norway, the gift of the Norwegian people. But filling the stockings was difficult.[37]

In August of 1947, the financial crisis many had feared due to the end of the lend-lease program with the United States meant continued austerity, resulting in sharply reduced imports and the continuation of butter and meat rationing. Shopping was still a nightmare, often requiring hours of waiting in lines. As one housewife remembered, "It was queues for everything, you know, even if you didn't know what you were queuing for . . . you joined it because you knew there was something at the end of it."[38]

Victor Ceserani, chef at Boodle's, remembered the first banana he and his wife saw after the war:

> We looked at it with reverence, it was the first time we'd seen one for years. So we peeled it very carefully and ceremoniously. We then got two small plates and knives and forks and we each then carried on and ate our banana, cutting it as thinly sliced as possible so that we could really enjoy it to the full.[39]

[36] Davies and Robinson 1947. This paper is not reproduced in Robinson 1979.

[37] Hopkins 1963, p. 42. See also Addison 1951, p. 26.

[38] Mrs. Vera Mather, quoted in Addison 1951, p. 31. Rationing for bread ended in July of 1948; chocolates were taken off the ration briefly in April of 1949, but instant shortages put them back on. Harold Wilson announced a "bonfire of controls" including the end to clothes rationing, in November 1949, but nevertheless gasoline rationing continued until 1950, tea until 1952, and food rationing until 1954; meat was the last to be returned to a free market. See Addison 1951, pp. 53–54.

[39] This was just after the first bananas arrived at Avonmouth in December of 1945, Addison 1951, p. 34.

Not everyone could look forward to bananas on their plate. Due to persistent shortages and the continuation of rationing, many English families found whalemeat on their tables. Renée remembers having seen it, but the smell was so bad and the thought so repulsive that she doubts she and Abby ever ate any. As one who tried it remarked, "The only thing I can say about whalemeat is that there was a lot of it and it was very smelly. . . . I think we had it once in our house but you could smell it right through the house for the whole week afterwards."[40]

Nor did Renée ever face another meat substitute—an imported delicacy from South America called "snoek." This first appeared in May of 1948, and was described by one poor English housewife in unforgettable terms:

> I never tried snoek, it was the label on the tin that did it, with this horrible blue fish—it was a most hideous fish and I wouldn't eat it, none of my family would.[41]

Even so, the Ministry of Foods did everything it could to promote it, including distribution of a recipe for "snoek piquante" (little more than a combination of spring onions and a dash of syrup).[42] Cooking had become "the art of turning tasteless and inferior food into something palatable for the family. No wonder that the most articulate popular protest against austerity was organised by housewives."[43]

Women's fashions also mirrored the shortages of the war years. Styles tended to be plain with a military look, skirts usually only an inch or two below the knee to conserve cloth. But in February of 1947 "a revolt began when the Parisian designer Christian Dior launched his first-ever collection." The "New Look" he introduced that year lengthened skirts to the ankle and restored pleats, bows, and leg-of-mutton shoulders, recreating a society image of *la belle époque*. English women followed the "New Look" in *Vogue* and the *Daily Herald*, and Princess Elizabeth gave a royal imprimatur to the new trend (but only after having amassed enough ration coupons of her own) to put together a calf-length trousseau for her marriage to Prince Philip.[44]

To many women, however, the "New Look" was a mixed blessing:

> We were very cross about it, because there was Christian Dior, lowering the hemlines almost to the ankle with enormously full skirts and little slim

[40] Addison 1951, p. 39. [41] Addison 1951, p. 39.

[42] One critic called the two teaspoons of syrup a "master stroke"! The recipe and article by Susan Wilson, "Snoek Piquante," is reproduced in Sissons and French 1964, pp. 35–57. See also Scott-James 1953.

[43] Addison 1951, p. 29. [44] Addison 1951, pp. 51–52.

> tops like this and nothing that we had in the cupboard was any longer in fashion and an awful lot of us said, "Oh, we can't afford it, and we can't buy it," but of course we did.[45]

Dior, in fact, was entirely conscious of what he was doing: "We were just emerging from a poverty-stricken, parsimonious era, obsessed with ration books and clothes-coupons: it was only natural that my creations should take the form of a reaction against this dearth of imagination."[46] Renée, of course, followed these developments avidly and embraced the "New Look" in her own designs at Rita and Company. Soon she was back in Paris to see for herself.

Paris after the War and Christian Dior

As Harry Baehr, an editorial writer for the *International Herald Tribune*, recalls:

> Paris had not been badly battered in the war, but in 1947 food, power and francs were still in short supply, and there were black (or at least grey) markets for all of them. We wore overcoats at meetings, had small stoves under tables at meals and burned candles at restaurants at lunch. A transit strike . . . sent swarms of bicycles flowing over the city, as they had done during the war.[47]

As France began its recovery, one of the first real symbols that it was ready to regain some of what it had lost during the war was embodied in the growing interest in high fashion and the work of such designers as Christian Dior. It was with something revolutionary in mind that he opened his new house for *haute couture* in Paris on December 15, 1946. The address, at 30 Avenue Montaigne, was prized for its subdued, Louis Seize interior. The white enamelled furniture, grey hangings, glass doors, and bronze light fixtures with their small lampshades had all been crafted to Dior's specifications, and were calculated to reflect (in his words) an "unobtrusive elegance" reminiscent of the Ritz and Plaza hotels.

In the spring of 1948, Renée and Abby were among the first to be included at one of Dior's early presentations for the wealthy and well-to-do. Renée was on assignment to observe the latest fashions, and Abby wanted to see what the world of high fashion was all about.[48] Renée sat

[45] Addison 1951, p. 52.

[46] Dior 1957, p. 25.

[47] Baehr 1987, p. 5.

[48] Renée went regularly to Paris in February and August with the head of Rita and Company to visit two or three of the famous fashion houses. "In 1948 we intended to go to Dior's opening. As it happened, Abby had to be in Paris for a meeting at the same

downstairs with the buyers, while Abby found a place on one of the staircases overlooking the entire scene. What they saw is best described by Dior himself:

> It is a quarter to ten. In the hall, someone is spraying scent in the path of the guests, and on the first floor the *Service de presse,* posted at the bottom of the stairs, is distributing the programmes of the show. . . . Laden with flowers, illuminated by chandeliers and floodlights, the *salons* begin to fill up. The whole scene has an aspect, at once light-hearted and worldly, quite different from that found in the theatre. Here, there is no red impressive curtain, no armchairs drawn up in a neat line. . . . The Louis Seize chairs, in spite of their numbered cards, are obviously drawn up to watch some drawing-room comedy.[49]

And then, finally, it began:

> 10:25. . . . There sits the public, watchful, curious, equally likely to be carried away by enthusiasm as by disappointment. People are standing up and waving at friends at the other end of the *salon,* and late comers are looking for programmes. Sweets are being handed round. A girl makes her way among the close-packed chairs and distributes fans; cigarettes light up.[50]

Meanwhile, Renée was watching and remembering:

> The mannequin walks across the *salon*, turns, threads her way through the narrow space between the chairs and leaves for the *petit salon*. There at the entrance, a second announcer repeats the name and number. The announcement echos a third time on the landing.[51]

The collection that the Robinsons saw in 1948 was named that year after famous composers.[52] As soon as the last mannequin had left the salon, Renée rushed back to the hotel and began sketching, as best she could recall, each of the designs she had just seen. This was no mean

---

time, and he wanted to go to Dior's as well. It was not easy to get tickets; there were two for the Company, but these were very expensive. But Abby did manage to get in by showing his passport and explaining that he had nothing to do professionally with fashion, and they let him sit on the staircase." Renée Robinson to JWD, March 5, 1994.

[49] Dior 1957, pp. 116–117. At the time, Dior employed in all a staff of eighty-five, and he described it as "a little beehive, that is what my house was when I presented my first collection. . . . I stressed two principal silhouettes—the corolla and the figure S. These long skirts, emphasized waists, and tremendously feminine fashions were instantly baptized the New Look. In the autumn, this trend was emphasized still more. The corollas curved outwards and the skirts, becoming longer still, restored all its former mystery to the leg," p. 159.

[50] Dior 1957, p. 119.

[51] Dior 1957, p. 120.

[52] Renée recalls that Abby liked "Chopin" best. Renée Robinson I-1980.

feat, because it required her remembering scores of individual designs, in as much detail as possible, mimicking shape, contour, fabric, the colors, lines, cuts, and accessories.[53]

Dior regarded the business of presenting his creations as "an extraordinary show, in the course of which my dresses are treated like girls in a slave market."[54] Moreover, Dior was fully aware that his designs were being copied—as he put it, "during all this time the drama—or should I call it the tragi-comedy?—of the copyists is being enacted in the various bars and hotels in the district round the Avenue Montaigne."[55]

For Renée, fashion was business, admittedly a sometimes heady mixture of personal interest and artistic livelihood, but why should a mathematician like Abby have found anything of interest at Dior's, in the salons of high fashion in Paris? If it was more than a reflection of his keen interest in everything, or respect for Renée's work, it perhaps is best explained by a comment of the famous French novelist Colette:

> Quite willingly these spectators admit their enjoyment of those formal presentations of wearing apparel which every well-known dressmaker organizes with all the pomp and circumstance of a religious ceremony. Monsieur accompanies his wife to the various "showings," and Madame shrugs her shoulders and declares: "Of course he only comes to look at the *mannequins*." But she is mistaken. Men are, at times, capable of several quite disinterested emotions, among which may figure a liking for colours, line, and especially for what is new. No longer does a man when he crosses a dressmaker's threshold assume the sheepish air of a big boy who has been caught playing marbles, or the awkward manner of a shipwrecked sailor cast upon on the shore of *L'Île des Femmes*. Only a man can enjoy the spec-

[53] In his first collection, presented in 1947, Dior designed ninety models. Although he was designing, as he put it, for "an established *clientèle* of experienced buyers and habitually elegant women, I now had the pleasant surprise of finding that young women wanted to adopt the new fashion as well." See Dior 1957, p. 31. Renée sketched as many as seventy of a hundred or so designs she would see on a typical visit to Paris. Renée Robinson to JWD, March 5, 1994.

[54] Dior 1957, p. 132.

[55] Dior 1957, p. 132. In all, Dior counted five "classical methods" by which dresses were copied, "of which the most distasteful is naturally that which originates with the treachery of a member of the staff," Dior 1957, p. 133. Other forms included disclosures by the press, which might provide too many details; copiers on the spot, required to "hand over" their sketches; microscopic photography; copies made deliberately by clients; "model renters," and other ingenious subterfuges. Dior credited the forgers with being "exceptionally gifted. I have stated that 57 out of 142 of the models of my winter collection were reproduced exactly, although the copiers must have relied chiefly upon memory. . . . Of course the programme with which all the guests are provided jogs their failing memories, but the annoying standard of accuracy points to exceptional powers of observation," p. 137.

tacle these occasions offer with a spirit devoid of the least shade of envy or covetousness. While his companion, secretly frantic with hopeless longing, abandons despairingly that little creation costing only six thousand francs, the man, thoroughly at his ease, asks questions, notes that X's gowns have a lower waist-line and that Z must have hired a new cutter, just as he would follow the evolution of a painter or any other artist. . . .[56]

Birkbeck College: For the Working Student

Birkbeck College was designed to accommodate the mature, working student by providing instruction for the most part in the evening and on weekends. Shortly after World War I, under R.B. Haldane's leadership as president, the University of London resolved that "Birkbeck College be admitted as a School of the University in the Faculties of Arts and Science for evening and part-time students." Whereas most branches of the University of London were evacuated to other parts of England at the outbreak of the Second World War, due to the nature of its working students, Birkbeck was in no position to leave, and so (along with the Imperial College of Science and Technology) Birkbeck College remained in London—a matter of great pride later among those who held on no matter what.[57]

Thus the "traditional Birkbeck student" was, typically, a mature adult in full-time employment who needed evening courses to earn a university degree.[58] Abraham Robinson was certainly no exception in this regard. But he was exceptional in other respects, as those who met him at Birkbeck soon realized.

Olga Taussky Todd, for example, found that "he was a very friendly person, exceedingly so, but not a charmer. Nothing ever seemed a grievance to him; he just accepted things." Among all of her colleagues, Taussky recalled that there was no one with whom she got along better. She was especially impressed by the fact that his friendliness was always so natural, their relationship always "frictionless." Although Taussky did not see him often, she heard about him from others and knew he was "great."[59]

[56] Colette 1932, pp. 32–33.

[57] Birkbeck was almost unique among London cultural institutions, in fact, because it did not evacuate to the countryside, or close down entirely due to the war. Two days after World War II was officially over, the governors of Birkbeck College sent a message to the king (on May 10, 1945), noting that "it is a special honour for the college to have remained in London with your Majesties throughout these recent years." See Warmington 1954, p. 175.

[58] See the preface by E. H. Warmington to the *Report of the Academic Advisory Committee on Birkbeck College,* Ashby 1967.

[59] Taussky I-1981. Taussky was very early interested in logic, and remembers meeting

She also knew about Robinson's position at Farnborough, which she could not help but compare with her own (similar) position. At the time she was working at the National Physical Laboratory at Teddington, where everyone, she recalled, "looked up to Farnborough." Although Robinson did not talk much about his work, Taussky had a book on aerodynamics that he needed, and when he phoned her about it, she offered to loan it to him. She was unprepared—and somewhat taken aback—by his reply: "Why should you want to help me?"[60]

Both Taussky and Robinson were working with Richard Cooke at the time, who along with Paul Dienes was an important figure in Robinson's mathematical development during the war, and after. Taussky was assisting Cooke with a book on *Infinite Matrices and Sequence Spaces*, and Robinson had done some work that was also included.

Richard G. Cooke (1895-1965)

Richard Cooke began his academic career as an analytic chemist, but took his D.S. degree in 1928 (at age thirty-three) under E. C. Titchmarsh. He then joined the faculty at Birkbeck as a lecturer in mathematics in 1926, and was promoted to reader in pure mathematics at the University of London in 1947. His mathematical interests fell into three major areas: theory of functions of a complex variable (where his work was especially influenced by Watson, Hardy, and Titchmarsh), infinite matrices, and summability methods.

Of his two books, *Linear Operators* and *Infinite Matrices and Sequence Spaces*, the latter included a considerable amount of work by Cooke's graduate students, to whom he was always pleased to give encouragement and recognition.[61] Cooke was also interested in quantum theory and the role infinite matrices could play in modern physics (especially in terms of work like that of Heisenberg and Schrödinger).

Robinson's Work with Cooke

Among the Robinson papers at Yale University is one that was never

---

Fraenkel at the Oslo Congress in 1937. She first met Robinson at Birkbeck College, when she went to hear a lecture by A. M. Ostrowski.

[60] Although Taussky found this a rather odd and surprising remark at the time, she later surmised that perhaps it reflected the fact that often people didn't help him. Considering Robinson's career as a whole, and the fact that he rarely if ever needed help, it may have reflected the independence he had come to expect as a Jewish refugee in Palestine, and now England, where spontaneously offered help was the exception rather than the rule.

[61] Cooke 1950 and 1953. Cooke acknowledged his indebtedness to Dienes, who he

published, doubtless because the substance of the paper appeared in Cooke's book on infinite matrices. The draft of Robinson's "Linear Transformations of Infinite Sequences and the Theory of the Core" was written in February 1946, and revised the following October. In it, he drew upon the findings of Knopp, Agnew, Allen, and Winn.

When Cooke came to publishing his book, *Infinite Matrices and Sequence Spaces*, it contained material on the properties and applications of infinite matrices, above all applications to summability of divergent sequences and series. Here the links between infinite matrices and the theory of functions, modern algebra, topology, and even mathematical physics (including quantum mechanics and spectral theory) were especially important. Not only was Cooke pleased to acknowledge that it was Paul Dienes who had first aroused his interest originally on the subject, but he also thanked Olga Taussky for an original theorem (§169), and Abraham Robinson, who had contributed "original proofs concerning the core of a sequence" (§6.4, II), (§6.5, I), and (§6.5, II).[62]

But Robinson's paper as drafted offered considerably more detail than the three briefly stated theorems and proofs that were published in Cooke's book. As he elaborated:

> In some cases it is found possible to gain some further information on the asymptotic behavior of a divergent sequence by the introduction of a concept, similar to that of the core, *viz.* the concept of the parallax of a sequence. Its properties are developed here in parallel with the discussion of the properties of the core.
>
> Cancelled: [~~Somewhat different in character from the remainder of this work, are some applications to the theory of power series given at the end. It is shown that analytic continuation inside the Mittag-Leffler star is practically identical with a certain procedure of conventional summation (summability). This, in conjunction with results on the core obtained earlier on, is used to establish various simple theorems of the type of Vivanti-Pringsheim's. However, the fact that analytic continuation inside the Mittag-Leffler star amounts to conventional summation would appear to be of some interest, independently of applications. It is not to be mixed up with the fact that analytic continuation inside the Mittag-Leffler star can be replaced by conventional summation, e.g. by the Lindelöf matrix~~].
>
> I wish to express my thanks to Dr. R. G. Cooke, whose encouragement has helped me to maintain my active interest in the subject [~~under unfavourable circumstances~~].

---

says persuaded him to begin writing his a book on infinite matrices in 1944. For details, see Cooke 1960, p. 251.

[62] Cooke 1950.

Clearly Robinson felt greatly indebted to the support Cooke had given him in his mathematics at Birkbeck. But of even greater importance for Robinson's development (especially as a logician) was another member of the faculty at Birkbeck, the Hungarian mathematician Paul Dienes.

Paul Dienes (1882–1952)

Dienes was born in Tokaj, Hungary, in 1882. After studying in Budapest and at the Sorbonne (where he received his doctorate in mathematics), he taught in Budapest until the Béla Kun government fell in 1919. "Dienes had to leave Hungary in haste, with his life in danger."[63] Indeed, he escaped in an empty beer cask on a cargo barge going from Budapest to Vienna on the Danube! Dienes left Europe for Great Britain, and initially taught at the University of Wales at Aberystwyth before moving to Swansea, where he wrote a book on Taylor series. Dienes' early work on functions of a complex variable was perhaps his most important, and was largely concerned with the behavior of Taylor series and their circle of convergence. In 1929 he was appointed to a readership at Birkbeck College, where his interests switched to infinite matrices. Thanks to a seminar he took with Dienes, Robinson became interested in functional transformations and summability.[64] Drawing on Dienes' development of the Toeplitz-Silverman theorem, which gives the necessary and sufficient conditions for a matrix A to transform any convergent sequence into a convergent sequence with the same limit, Robinson used Borel's integral procedure of summation and was able to generalize his results further to cases involving unbounded linear matrices and semicontinuous matrices.[65] Among applications, he discussed a method for producing semicontinuous G-matrices of bounded linear operators in a complete metric linear space, Riemann's method for summation of trigonometric series, or the case where S is the space of all real functions on $[0,1]$ satisfying a Lipschitz condition with a metric defined such that S becomes a complete metric linear space.

Dienes, who always had a serious interest in philosophical questions, was also fascinated by symbolic logic and axiomatics as early as 1914. Between 1942 and 1948 he returned to this interest, when Robinson and R. L. Goodstein were among his research students in this area.[66] Dienes was especially important for Robinson's thesis because of all the

[63] Cooke 1960, p. 251.

[64] Although Robinson 1950b, "On Functional Transformations and Summability," was submitted to the London Mathematical Society in July of 1947 (it was officially read October 16, 1947), it was not published until 1950.

[65] For details of the Toeplitz-Silverman theorem, see Dienes 1931, p. 289.

[66] Cooke 1960, p. 252.

mathematicians at Birkbeck, he was the one most interested in mathematical logic. Just before the war, in 1938, he had published a book that was directly relevant to Robinson's own later mathematical interests, namely Dienes' *Logic of Algebra.*[67]

In addressing the "crisis" in mathematics prompted by the antinomies of set theory and logic discovered at the turn of the century, Dienes adopted a "realist" attitude. His approach was to analyze the "crisis" through an examination of the logical structures of arithmetic and algebra. "Our experience gathered in human practice are the ultimate source of knowledge and there is no higher authority for laying down the laws for science."

Working with Dienes during the war sometimes meant seminar meetings at his home in north London.[68] According to George Barnard, who was part of the group studying with Dienes at the time, as was Robinson:

> My clearest recollection of the "seminars" we used to have with Paul Dienes is of leaving his flat in Cranley Gardens, North London, a bit later than usual, so that the shrapnel from anti-aircraft fire was already falling rather thickly while we were walking rapidly (one didn't run) towards the underground station at High Gate. I think Abraham Robinson had been at that seminar, but I could not be sure. The lead in our discussions was always taken by Paul Dienes who had been a direct pupil of Borel and who shared Borel's "semi-intuitionist" views on the foundations of mathematics.[69]

Ambrose Rogers, formerly head of mathematics, University College, London, also went to some of Dienes' seminars, and knew Robinson when he was a student at Birkbeck:

> Towards the end of the war I went to R. G. Cooke's M.Sc. lectures (given at week-ends) and to P. Dienes's seminar. I continued to attend Dienes's seminar for a year or two after the end of the war. At that time R. G. Cooke seemed to be the most active of the Birkbeck mathematicians. His boundless enthusiasm for the Theory of Infinite Matrices and for Matrix meth-

[67] Dienes 1938.

[68] The description of Dienes' seminars during the war was provided to me by George A. Barnard. Barnard spent most of his career as professor of statistics at Imperial College, London, and later at the University of Essex, although he was at Birkbeck for a time during the war. In addition to a letter from Barnard of November 17, 1987 (from Mill House, Brightlingsea, Colchester, Essex), I am also grateful for collateral information provided by Philip Holgate in a letter dated October 23, 1987. Both of these are now included among the files of Robinson's papers, YUL.

[69] Barnard to JWD, November 17, 1987, RP/YUL. According to his colleague Richard Cooke, Dienes was essentially of the Paris school, and counted many French mathematicians among his friends. Immediately after the liberation of France in 1945, he arranged seminars at Birkbeck College for his French colleagues, and was among the first to invite Mandelbrojt, Dieudonné, Laurent Schwarz, Cartan, and others to London.

> ods for assigning limits to divergent series had a great influence on me and on many others. Other active members of the seminar were J.L.B. Cooper, Abraham Robinson, and P. Vermes. The whole atmosphere was very friendly and helpful and I certainly found Robinson very friendly and helpful. He was, at that stage, recognizable as a very fine mathematician with very wide interests.[70]

Robinson's "Amazing" Return to Mathematical Logic

In 1946, at the same time he accepted his appointment at the College of Aeronautics, Robinson also began "his amazing journey back to mathematical logic."[71] Robinson's interest, originally inspired at the Hebrew University by Fraenkel's early work on the subject, was still unusual in the 1940s. According to Garrett Birkhoff:

> Bourbaki, of course, emphasized informal logic, whereas most mathematicians at the time did not take symbolic logic seriously. Like mathematical physics, they knew when it would work and when it wouldn't. But Fraenkel, Tarski and Robinson were among those early on with a faith in mathematical logic, taking it literally and then seeing what logically followed.[72]

Mathematical logic early in the twentieth century was interested primarily in saving mathematics from the logical and set-theoretic paradoxes that caused such a stir following Georg Cantor's creation of transfinite set theory.[73] For those concerned with the foundations of mathematics, metamathematics offered a means of using axiomatic methods to investigate questions of what constitutes a valid proof, or the conditions under which a given theory could be considered consistent or complete. Russell and Whitehead's monumental *Principia Mathematica* was one attempt in this direction, as was Hilbert's program, developed by many of his circle, and culminating in the profound if somewhat negative results of Kurt Gödel in the 1930s.[74]

What sets Robinson's metamathematical interests apart from this earlier use of mathematical logic is the extent to which he went beyond questions of validity, consistency, and completeness to consider ways in which logic could be applied directly to specific mathematical problems. In the case of Robinson's thesis and first book, this meant the metamathematics of algebra.

[70] C. Ambrose Rogers to JWD, letter of December 1, 1987. Philip Holgate explains that Rogers "joined in the 'seminars' as a very precocious young man." Holgate to JWD, October 23, 1987, RP/YUL.

[71] Macintyre 1977, p. 649, and I-1982.

[72] Birkhoff I-1989.

[73] For details, see Dauben 1979.

[74] Mostowski 1966, p. 8.

The Progress of Robinson's Thesis

By August of 1947 Robinson had decided on the subject of his thesis. As he wrote down the first tentative title, "The Syntax of Algebra," he added that the introduction would contain "A History of Mathematical Argument." Six months later he had drawn up a "Plan for Thesis" (dated February 1948), at which point he was already far enough along to present a lecture "On the Metamathematics of Algebraic Systems" at one of the mathematics colloquia at Birkbeck College. The talk itself, given on Wednesday, February 4, was basically an expository one.[75]

A month later, on March 5, Robinson scheduled a meeting with his advisor Paul Dienes to discuss his "programme." Over that spring and summer he continued to work on his ideas, and on October 18, 1948, sent a letter to Fraenkel, with a six-page "introduction" to what was now titled "On the Metamathematics of Algebra." Clearly Robinson waited until he was well along with his thesis, at last sure of the title and the major content of his work, before communicating anything to Fraenkel.

Seventh International Congress of Applied Mechanics: London 1948

In September of 1948, the Seventh International Congress for Applied Mechanics was held at the Imperial College of Science and Technology in London. In all, nearly nine hundred participants from all over the world met to hear plenary sessions and papers on a wide variety of subjects, including aerodynamics. It was the first time the Congress was held in London, and receptions were arranged at Senate House of the University of London and at the Guild Hall, where the lord mayor presided over a welcoming gathering of delegates. Tours were arranged to various scientific establishments during the week after the congress, including the National Physical Laboratory at Teddington, the Royal Aircraft Establishment at Farnborough, and the College of Aeronautics at Cranfield. Robinson read a paper in section two for aeronautics and hydrodynamics, as did Garrett Birkhoff and Clifford Truesdell. Robinson's was "On Some Problems of Unsteady Supersonic Aerofoil Theory."[76]

Considering unsteady supersonic flow around an airfoil of infinite span, Robinson showed that under forward acceleration the pressure at

[75] One of the posters advertising this lecture is among Robinson's papers, YUL. It took place in Senate House, University of London, room 353, where the mathematics seminar met from 7:00–9:00 P.M. The poster listed him as Dr. Robinson of the College of Aeronautics.

[76] Robinson 1948a.

Formal dinner, College of Aeronautics.

any given point of an airfoil may be analyzed into three components.[77] The paper then went on to consider the oscillatory supersonic flow around a delta wing inside the mach cone emanating from its apex from the standpoint of linearized theory. Robinson showed that the velocity potentials corresponding to vertical and pitching oscillations of the wing can be represented by a series of normal solutions using a special system of curvilinear coordinates. Although two-dimensional oscillatory airfoil theory had been dealt with in detail by various authors from the point of view of linearized theory, the three-dimensional case required certain modifications.[78] Different cases had to be distinguished, the simplest being the "definitely supersonic case" (in which pressures on the upper and lower surfaces are independent of the geometry of the surfaces). "Mixed supersonic" cases, which do not satisfy this independence, are more complicated. In the case that interested Robinson, the flow around a delta wing is mixed supersonic (sometimes called "quasi-sub-

[77] These were the steady Ackeret pressure due to the instantaneous velocity, acceleration, and the square of the velocity in a limited time interval preceeding the instant under consideration. Robinson noted that, in practice, the difference between the total pressure and the "steady pressure component" could be neglected under all of the supersonic conditions that were likely to occur.

[78] Among those who had also worked on this problem were Robinson's former colleagues at Farnborough, G. Temple, and H. A. Jahn. See their *Royal Aircraft Establishment Report*, no. S.M.E. 3314 (1945).

sonic") if the leading edges of the airfoil lie inside the mach cone emanating from the apex. Applying a method of pseudo-orthogonal coordinates that he had used earlier in solving the corresponding steady flow problem, he was able to determine first the pressure distribution and then the forces on a delta wing oscillating in a given mode.[79]

In assessing the significance of Robinson's aerodynamic research during the war and later, at the Royal College of Aeronautics, no one is in a better position to do so than his colleague and lifelong friend both at Farnborough and Cranfield, A. D. Young:

> Seemingly without effort [Robinson] acquired the art of abstracting the essentials of physical problems so as to be able to model them in relatively simple but valid mathematical terms. It was soon evident to all his colleagues that he was a first-rate applied mathematician, in the sense that he had not only a wealth of powerful mathematical techniques at his fingertips but also the acute insight into the physical relations involved in his problems that is needed to separate the essential from the inessential.[80]

Robinson and the Association for Symbolic Logic

As a student at Birkbeck College, Robinson joined the Association for Symbolic Logic (ASL), and it was to the association that he first communicated some of the new results he was achieving in the course of

[79] Full details appear in Robinson 1948a. Related ideas were developed in another of his papers also published that same year in the first volume of the newly founded *Quarterly Journal of Mechanics and Applied Mathematics.* This was a much more technical paper mathematically, in which Robinson considered the hyperbolic character of the differential equation satisfied by the velocity potential in linearized supersonic flow, which entailed fractional infinities in the fundamental solutions of the equation. This introduced difficulties which he resolved using Hadamard's finite part of an infinite integral. Exploiting a natural analogy between incompressible flow and linearized supersonic flow, Robinson derived formulae for the field of flow due to an arbitrary distribution of supersonic sources and vortices that were comparable to methods used by Stokes and Helmholtz in classical hydrodynamics. See Robinson 1948. A. D. Young describes this paper as "another major contribution of outstanding importance," Young 1979, 3, p. xxxi. Note that applications of these results to airfoil theory had already been published in Robinson and Hunter-Tod 1947.

The following year Robinson published a theoretical examination of the integration of hyperbolic differential equations, Robinson 1948b, which reached a much wider audience as Robinson 1950a. Although Robinson's approach utilized a method of successive approximation, in practice he noted that "the method is not as yet very suitable for numerical work." This was so in part because a large number of integrations were involved, but as he realized, "The methods outlined in the present paper appear to have distinct prospects for numerical application. Their main advantage is that since integration is carried out along the characteristic curves, there is no possibility of a failure, such as may occur in the application of some lattice methods." See Robinson 1950a, p. 216.

[80] Young 1979, p. xxix.

finishing his thesis. Late in 1948 he was ready to announce several of his most important theorems, and for the association's eleventh annual meeting, which convened that year at Ohio State University in Columbus, Ohio, he submitted a paper that was read *in absentia*, by title.[81]

This came none too soon, for both Alfred Tarski and Leon Henkin were simultaneously pursuing avenues connecting algebra and mathematical logic that were very similar in spirit, and to some extent even in detail, to what Robinson had accomplished in his thesis. Despite the brevity of the note, as Garrett Birkhoff realized when he read it in the association's *Journal*, "here's a fresh wind." Even without the details, Birkhoff recognized that Robinson's "On the Metamathematics of Algebra" was full of fundamental new ideas.[82]

In only a few brief paragraphs, Robinson explained the two different approaches he took using logic in applications to algebra. The first, instead of proving theorems one at a time, was to consider statements about theorems in general. Robinson described how a theorem true for one type of mathematical structure could also be true for another type of structure. He gave as an example any theorem proven in the restricted predicate calculus (with addition, multiplication, and equality) for the field of all algebraic numbers, which would also be true for any other algebraically closed field of characteristic 0. Similarly, any theorem proven in the restricted predicate calculus (with addition, multiplication, equality and order) that was true for all nonarchimedean ordered fields would also be true for all ordered fields. These sorts of results, Robinson indicated, could then be used to prove actual theorems, and he offered the following case as an illustration:

> *Let $q(x_1, \ldots, x_n)$ be a polynomial with integral coefficients which is irreducible in all extensions of the field of rational numbers. Then $q(x_1, \ldots, x_n)$, taken modulo p, is irreducible in all fields of characteristic p greater than some constant depending on q.*[83]

Robinson's "second line of attack," as he put it, considered the joint properties of structures satisfying some specific system of axioms (again, formulated in the restricted calculus of predicates, including a relation of equality involving substitutivity). He then introduced "gen-

[81] The ASL met jointly that year with the American Mathematical Society and the Mathematical Association of America, December 30–31, 1948. See the *Journal of Symbolic Logic*, 14 (1949), p. 24. Seven of the sixteen papers described in the published report were presented by title only.

[82] Birkhoff regarded Robinson as very ambitious, trying "to make a killing in fluid dynamics." But he also remembers Robinson as "very scholarly, always interested in foundations," Birkhoff I-1989.

[83] Robinson 1949a, p. 74.

eral concepts" analogous to familiar algebraic ones like algebraic numbers, polynomial rings over a given ring, and ideals. The beauty of this approach lies in the fact that the general concepts Robinson introduced were not mere analogs to the algebraic concepts in question, but they actually reduced to their familiar counterparts in the algebraic systems from which they were taken. As Robinson concluded, "this shows that they can be abstracted from the specific arithmetical operations with which they are normally associated."[84] Moreover, many of their familiar algebraic properties could also be transferred to the more general cases Robinson had already explored in greater detail in his dissertation.

Even from the very brief description that appeared in the *Journal of Symbolic Logic* early in 1949, it was clear to anyone reading Robinson's half-page abstract about his "analysis and development of algebra by the methods of symbolic logic" that his concerns were much more explicitly directed towards mathematical issues than were the more logically directed interests of either Henkin or Tarski. Robinson's major interest was algebra, and logic was a means of achieving new results that could not be obtained so easily, if at all, otherwise.[85] Here again the strength of Robinson's knowledge of pure mathematics gave his work a sense of solidity that would long be the hallmark of his own approach to metamathematics.

Bourbaki and Symbolic Logic

Robinson was not the only logician represented *in absentia* at the eleventh annual meeting of the Association for Symbolic Logic in 1948. The "polycephalic" mathematical collective, Nicholas Bourbaki (the group of French mathematicians originally founded by Chevalley, Delsarte, Dieudonné, and Weil), had prepared an invited address that was delivered by André Weil on December 31:

> So far as we mathematicians are concerned, [logic] is no more and no less than the grammar of the language which we use, a language which had to exist before the grammar could be constructed. . . . Logical, or (what I

[84] Robinson 1949a, p. 74.

[85] Although, as Leon Henkin later read Robinson's thesis, he was not convinced that the generalization of mathematical concepts Robinson presented had succeeded. On the question of whether they could indeed lead to new and deeper results, Henkin was negative: "it does not appear that an affirmative response . . . can be made for Robinson's theory as contained in the later sections of the book." For additional comments, consult Henkin's review of North-Holland's published version of Robinson's thesis in the *Journal of Symbolic Logic*, 17 (1952), pp. 205–207. See also the detailed discussion of Robinson's book, Robinson 1951, and in chapter 6.

> believe to be the same) mathematical reasoning is therefore only possible through a process of abstraction, by the construction of a mathematical model.[86]

Bourbaki, however, saw logic as a means of making mathematical results precise and systematic. Robinson's interest (like those of Tarski and Henkin as well) went much further, embracing the subject of logic itself, and using logic not simply as an informal framework for presenting mathematical results, but as a powerful *means* of investigating the essence of mathematics, especially algebra. In this respect, Alfred Tarski and Robinson's nearer contemporary, Leon Henkin, were also headed in fresher and more original directions, at least so far as logic was concerned, than was Bourbaki.

Alfred Tarski (1901–1983)

Alfred Tarski's book *On Mathematical Logic and Deductive Method* first appeared in Polish in 1936, with a German translation the following year.[87] It was intended from the beginning as a popular scientific book, inspired by problems on the foundations of mathematics but meant to convey a sense of the importance and utility of modern logic. Over the years Tarski had increasingly sought to provide a "unified conceptual apparatus which would supply a common basis for the whole of human knowledge."[88]

By the time an English edition was prepared, Tarski had fled his native Poland—and forsaken Europe—for a temporary position at Harvard University during the academic year 1939–1940. Tarski, of course, was but one of many refugee mathematicians who had made their way to the United States in response to the upheavals of World War II. In his preface to the English edition of 1941, Tarski was eloquent in expressing the political dimensions he now found in his work as a logician:

> The course of historical events has assembled in this country [the United States] the most eminent representatives of contemporary logic, and has

[86] Bourbaki 1949, pp. 1–2. It is interesting to note that two years later, Bourbaki was not allowed to register for the International Congress at Harvard, the Congress secretary, J. R. Kline, being "stuffy" as Birkhoff recalls (Birkhoff I-1989). For more on Bourbaki, see Dieudonné 1970. In Dieudonné's paper it is explained that "Bourbaki" was a nineteenth-century French general. "It was part of an 'initiation' of first year mathematics students that an upper classman, pretending to be a famous foreign mathematician, would deliver a lecture to them in which the theorems bore the names of famous generals and were all wrong in a nontrivial way."

[87] The first English translation was made by Olaf Helmer in 1941; citations here are from Tarski 1965.

[88] Tarski 1965, p. xi.

> thus created here especially favorable conditions for the development of logical thought. These favorable conditions can, of course, be easily overbalanced by other and more powerful factors. It is obvious that the future of logic, as well as of all theoretical science, depends essentially upon normalizing the political and social relations of mankind, and thus upon a factor which is beyond the control of professional scholars. I have no illusions that the development of logical thought, in particular, will have a very essential effect upon the process of the normalization of human relationships; but I do believe that the wider diffusion of the knowledge of logic may contribute positively to the acceleration of this process. For, on the one hand, by making the meaning of concepts precise and uniform in its own field and by stressing the necessity of such precision and uniformization in any other domain, logic leads to the possibility of better understanding among those who have the will for it. And, on the other hand, by perfecting and sharpening the tools of thought, it makes men more critical—and thus makes less likely their being misled by all the pseudo-reasonings to which they are incessantly exposed in various parts of the world today.[89]

Whether Tarski was really such an idealist—or had actually set himself such an explicit agenda to promote mathematical logic to achieve social or political goals— it was clear to all who knew him that he was determined to do everything possible to establish and promote the subject in the United States. From Harvard, Tarski went on briefly to New York, where he was appointed a visiting professor at City College. A year later he was the recipient of a Guggenheim Fellowship which he spent as a member of the Institute for Advanced Study in Princeton. This was all prior to his appointment as a lecturer at the University of California, Berkeley, where in 1942 he began almost immediately to lay the foundations for a new school of mathematical logic, eventually making Berkeley the central focus for such activity on the West Coast, much as Princeton (with Gödel and Church) was its eastern counterpart. As Birkhoff put it, "the Eastern European idea of a school was very strong with Tarski; having established one in Poland, he did one in Berkeley."[90]

Leon Henkin

Meanwhile, at Princeton, Leon Henkin was an entering graduate student in mathematics in September 1941. Three years younger than

[89] Tarski dated his new introduction "Harvard University, September 1940." See Tarski 1965, p. xv.

[90] Birkhoff I-1989. Tarski was lecturer at Berkeley from 1942–1945, associate professor from 1945–1968, and professor thereafter. In 1944 he was elected to serve a two-year

Robinson (Henkin was born in New York on April 19, 1921), the two were virtual contemporaries. Considering how closely, yet independently, their interests in metamathematics and model theory developed in the late 1940s, it is especially instructive to compare Henkin's early development as a mathematician/logician at Princeton with Robinson's own experience at the University of London.

Henkin had begun his study of mathematics at Columbia University, where the crucial influence on his intellectual future was Ernest Nagel in the Philosophy Department.[91] But thanks to Frank J. Murray (who worked on operator algebras and had collaborated with John von Neumann), Henkin was led to a careful study of Gödel's revolutionary results on the consistency of the axiom of choice, largely on his own:

> Princeton University Press had just brought out Gödel's monograph on the consistency of the axiom of choice and the generalized continuum hypothesis. Although I had never been a student of Murray's, he knew that I was about the only one among mathematics students or faculty with an interest in logic, so he sought me out and proposed that the two of us work through the Gödel monograph together. I readily agreed and ordered a copy of the monograph. As far as I can recall, Murray and I had one or two meetings to discuss the scope and the beginning of the work, and then he found himself too busy with other projects and I was left to work through Gödel's monograph on my own. This event was probably my most important learning experience as an undergraduate.[92]

At Princeton, Henkin worked with Alonzo Church until the bombing of Pearl Harbor and the American entry into the war. Six months later he was in government service, and until 1946 he served as a mathematician on the Manhattan Project. Part of this time was spent at the Kellex Corporation in New York City (1943–1945), where Henkin's research concentrated on isotope diffusion, a matter of primary concern for the separation plant being built at Oak Ridge, Tennessee. It was at Oak Ridge where the enriched uranium and plutonium necessary for the first U.S. atomic bombs were produced. Henkin spent the next year (1945–1946) in Tennessee working with the Carbide and Carbon Chemical Company. When he finally returned to Princeton in March of 1946

---

term as president of the Association of Symbolic Logic. For recent biographies, see Addison 1984, Hodges 1986, and Givant 1986.

[91] Henkin I-1989. Brief biographical information is also provided by Alder and Niven 1990. Henkin also won the Chauvenet Prize in 1964 for outstanding expository writing. See Henkin 1962.

[92] From a draft of a paper by Leon Henkin, "The Discovery of My Completeness Proofs," presented at the Nineteenth International Congress of History of Science, Zaragoza, Spain, August 24, 1992. Quoted here by permission of Leon Henkin.

to resume his graduate work, he had already fulfilled all of the formal requirements for his Ph.D. All that remained was his dissertation.[93]

Back in Princeton, Henkin suddenly had an "astonishing experience." In 1941 he had taken a course with Church, who spent the semester going over a paper he had written on type theory using lambda notation. Six years later, in 1947, Church was using the lambda notation for another purpose—as a natural foundation in the context of Frege's use of denotation. Henkin "fell in love with it as a kind of formalization," and found that he could use it for an especially short formulation of the axiom of choice.[94]

Basically, Henkin showed how the lambda notation could be adapted to provide a semantic interpretation of a given system, and how it could be used, in particular, to name choice functions. In fact, Henkin focused on the problem of picking out particular choice functions. Here the difficulty (in making the axiom of choice precise) was in specifying the meaning of the choice functions one might pick.

Henkin began to wonder if it was in fact possible to describe a particular choice function at all. He worked intensively on this problem, with nothing to show for it after months of effort. Since he was being supported by a National Research Council Fellowship, he needed a concrete result, something to justify the continuation of his grant. Finally, in the spring of 1949, with little progress to report, the tension became too much. Henkin decided that he could not continue, and made up his mind to leave graduate school. Instead, he thought semiseriously about "becoming a farmer out west."

That night, lying in bed, Henkin ran over the letters he intended to write the next morning, one to Solomon Lefschetz (then chairman of the department) about his decision to leave Princeton. The more difficult letter was the one to his father explaining why he was abandoning his graduate work without having received the Ph.D. The next morning, however, Henkin suddenly had a new idea about his functions. Newly inspired, he got back to work at once.[95]

What Henkin had been trying to do was to develop a visualizable

[93] Having begun in 1941, Henkin realized that the war might soon intervene, and he would then have to start all over again after the war. Instead, he managed to take the qualifying examination for the Ph.D. at the end of one very intensive year of course work.

[94] Henkin I-1989.

[95] Later, when Paul Cohen introduced forcing to show the independence theorem, Solomon Feferman used Cohen's technique to show that there is no formula or formalism that names a particular choice function (with the Zermelo-Fraenkel axioms). Thus only well after the fact did Henkin realize that he had been trying to solve a problem that had no solution, at least not one of the sort on which he had expended so much effort—to his great frustration—in the late 1940s.

picture of functions that could be named in Church's lambda notation (for which, unlike Russell and Whitehead's notation in *Principia Mathematica*, the lambda notation was not well suited). What Henkin wanted was a clear picture of all nameable functions. But as the months passed, he was able to get little more than a "fuzzy picture" of the problem. And then, unexpectedly, just when he had decided to abandon his career as a mathematician:

> In the corner of my eye, I saw the completeness proof—I saw the fact of completeness and how it emerged from the formalism. This was *not* what I was looking for, but there it was, and it was fascinating, exciting.[96]

Subsequently, Henkin wondered: if his approach worked for the theory of types, why not for first order logic? But here again there was a difficulty connected with the existence of symbols that name choice functions in Church's notation. Eventually Henkin saw a key aspect of the completeness theorem for types, but only after he had tried it for first order logic where he resolved the difficulty of naming elements by bringing them in as extra individual constants.

This much seems to have come to Henkin by sheer inspiration. The second part of the proof, bringing in constants, proceeded basically by copying type theory in first order logic. As he got more results related to completeness, he then wrote all of this up carrying completeness much further, from type theory to first order logic to sentential logic, where the method was little more than a "snap of the fingers."[97]

Henkin soon found himself doing things in a more constructive way. Other completeness theorems had been established by E. L. Post and László Kalmár, but Henkin added more.[98] He was particularly pleased by the applications he was able to make of completeness results, and especially of his method of adding individual constants.

Among the graduate students then at Berkeley, Gilbert Hunt, working on analysis and later probability, was a colleague with whom Henkin would often discuss his latest ideas. One day, as they were discussing Boolean algebra, Stone's results on representations of Boolean algebra came up, something Henkin had never seen in any of his courses. Later he realized that his completeness result, or the "compactness theorem," could also be used to obtain Stone's result. Similarly, in a course with Emil Artin on field theory, he again saw another appplication of the "compactness theorem" that he also incorporated into his thesis.[99]

[96] Henkin I-1989.

[97] Henkin I-1989.

[98] Kalmár 1935.

[99] The "compactness theorem" is put in quotation marks here because the reference in this context is somewhat anachronistic. Although everyone was using versions of the

Between the spring and summer of 1947, Henkin wrote his doctoral dissertation, "The Completeness of Formal Systems," which was accepted at Princeton in October.[100] At the time, Henkin had not yet heard of Abraham Robinson. Nor did he really know Tarski, either. The first time he ever saw Tarski was in a seminar at Columbia University that Ernest Nagel had arranged in the fall of 1939. But Henkin did not speak to Tarski then, and it was not until well after the war that he first had a chance to talk to Tarski directly about mathematical logic.

The occasion, appropriately enough, was the Princeton Bicentenary celebrated in 1947–1948. Albert Tucker, then chairman of the Princeton Mathematics Department, pressed Henkin to take notes of all the talks at the conference on logic, and to help speakers prepare their papers for publication.[101] This put Henkin in direct contact not only with Tarski, but with Kleene, Church, and Gödel as well. This was a golden opportunity, especially working with Gödel, who was so reclusive that Henkin doubts whether he might ever have seen him otherwise. Even more than Church, Henkin recalls that Gödel was exacting about everything, even down to the last dot and period in his papers.

---

theorem, including Henken, Mal'cev, Robinson and Tarski, it was referred to in rather awkward terms by Henkin as "our basic result from logic," for example. Reference to the "compactness theorem" became standard only in the late 1950s; the earliest reference however that I have yet been able to find to the "compactness theorem" as such is an article by Helena Rasiowa, "A Proof of the Compactness Theorem for Arithmetical Classes," Rasiowa 1952.

In her paper, Rasiowa refers to "A simple mathematical proof of the 'compactness theorem for arithmetical classes' stated by Tarski," Rasiowa 1952, p. 8. (There is no reference, however, to where this proof of Tarski's is to be found, or whether Tarski himself at the time called it the "compactness theorem"). Rasiowa adds, however, that the proof given by Tarski at the Cambridge Congress in 1950 is "involved," so presumably her simplified proof is a response to Tarski's Congress paper. In any case, there is no reference to the "compactness theorem" in the published version of Tarski's 1950 Congress paper. Rasiowa also notes that a modification of the algebraic method of proving the Löwenheim-Skolem theorem would also give the compactness theorem (but as Feferman and Tarski point out, what Rasiowa really meant was the Skolem-Gödel theorem; for discussion of the nature of the confusion here, see Feferman and Tarski 1953, pp. 339–340).

This story of the "compactness theorem" is also complicated by the fact that, despite Rasiowa's use of the term in 1952, R. L. Vaught in giving a general overview of results on this subject in the early 1950s, never uses the term "compactness theorem" in Vaught 1954. For a general overview of the history of the compactness theorem, see Dawson 1993.

[100] See Henkin 1950, p. 81.

[101] Although the papers were not printed at the time, individual articles did appear, including Gödel's contribution, which Martin Davis published in his anthology, *The Undecidable.* See Gödel 1965.

The International Congress of Mathematicians: Cambridge 1950

Robinson's promise as a mathematician first came to global attention in 1950. In anticipation of the International Congress of Mathematicians scheduled to take place at Harvard University late in the summer of 1950—the first Congress after the war—Robinson submitted an abstract to the Program Committee based on his dissertation, "On the Metamathematics of Algebra," which he had submitted to Birkbeck College for his Ph.D. in May of 1949.[102] The committee on logic and foundations was so impressed by Robinson's abstract that he was asked to present an invited lecture. Consequently, only a month from celebrating his thirty-second birthday in 1950, Robinson was on his way to the United States, where he would present a half-hour lecture "On the Applications of Symbolic Logic to Algebra." The title of the paper he planned to present summarized not only the subject of his recently completed thesis, but signaled the direction in which Robinson's major mathematical contributions were to develop in the years immediately following the Congress. The paper itself marked the beginning of Robinson's career as a scholar of international reputation.

The 1950 Congress was the first to be held in the United States since the Chicago World's Fair in 1893. In the interim, the status of mathematics in the United States had changed dramatically. No longer was research in American universities done in the shadow of European institutions. On the contrary, many German and other Eastern European mathematicians had left Europe and were now established in the United States. Their presence in American colleges and universities served to invigorate mathematics across the country. As Tarski had said, specifically with logic in mind, the most eminent representatives of contemporary logic were now concentrated in the United States, and consequently it presented "especially favorable conditions for the development of logical thought."[103]

Robinson was about to see for himself. In August he sailed for New York City, and along with several thousand other mathematicians from around the world, arrived in Cambridge, Massachusetts, in time for the official opening of the Congress on Wednesday afternoon, August 30.

No one in Cambridge that summer was blind to the momentous world

[102] The committee appointed on June 3, 1949, to examine Robinson's thesis consisted of Paul Dienes, Max Newman, and George A. Barnard. They accepted the thesis after a *viva voce,* an oral examination, and the degree was officially awarded on October 19, 1949. I am grateful to Ivor Grattan-Guinness for these details from the Senate's unpublished minutes of the Board of Studies for Mathematics, and also from the Senate's printed *Minutes* for 1949, University of London (items 369 and 370).

[103] See Tarski 1965, p. xv.

events of the past year that gave a noticeable political edge to the Congress. On September 1, just as the Congress was getting underway, North Korea opened an all-out offensive against United Nations' forces under General MacArthur along a fifty-mile front just south of Chang-Nyong. Earlier that year, Chinese Communists had seized the U.S. consulate in Peking, although fighting south of Shanghai had not as yet resulted in Chang Kai-Shek's retreat to Formosa. Meanwhile, the Cold War in Europe was escalating to alarming proportions. Churchill's early warnings about the "Iron Curtain" were brought dramatically home in the airlift that managed to keep Berlin alive and independent as all of its supply routes were cut off by the Russians and their East German allies in 1948–1949.[104]

Against such escalating tensions, especially the outbreak of war in Korea, it is little wonder that the Cambridge Congress was not without its share of political problems. As Oswald Veblen, president of the Congress, remarked in his opening address:

> We are holding the Congress in the shadow of another crisis, perhaps even more menacing than that of 1940, but one which at least does allow the attendance of representatives from a large part of the mathematical world. It is true that many of our most valued colleagues have been kept away by political obstacles and that it has taken valiant efforts by the Organizing Committee to make it possible for others to come.[105] Nevertheless, we who are gathered here do represent a very large part of the mathematical world. . . .
>
> I have referred to the political difficulties which have harassed this Congress, but think that if there are to be future international congresses, an even more serious difficulty will be the vast number of people who have a formal, and even an actual, reason for attending. This makes all meetings, even for very specialized purposes, altogether too large and unwieldy to accomplish their purposes.
>
> Mathematics is terribly individual. Any mathematical act, whether of creation or apprehension, takes place in the deepest recesses of the individual mind. Mathematical thoughts must nevertheless be communicated to other individuals and assimilated into the body of knowledge. Other-

[104] Among the many histories of the Cold War and related events that have been written, works consulted here include Fontaine 1968 and Hughes 1976.

[105] Garrett Birkhoff was chairman of the Committee on Transportation Grants, which had the ticklish job of raising money and making sure it was distributed so as to maximize the number of participants. In fact, Birkhoff remembers that the greatest difficulty in organizing the 1950 Congress was to see that people got there. This was at a time when recovery from the Second World War had left most Europeans and other third-world mathematicians without the economic means to cover such expenses themselves, or from other local resources. Birkhoff I-1989.

> wise they can hardly be said to exist. But the ideal communication is to a very few other individuals. By the time it becomes necessary to raise one's voice in a large hall some of the best mathematicians I know are simply horrified and remain silent.[106]

Despite the growing political paranoia, reflected in the dramatic wave of McCarthyism in the United States, only one mathematician from the West was denied a visa for political reasons to attend the Congress, and this was largely because he had failed to notify Congress officials about his difficulties. Only two mathematicians from then-occupied countries failed to get visas. But mathematicians from behind the Iron Curtain "were uniformly prevented from attending the Congress by their own governments, which generally refused to issue passports to them for the trip to the Congress."[107] Instead, the president of the Soviet Academy of Sciences, the physicist Sergey Vavilov, simply cabled the Congress with a terse message: "Soviet Mathematicians being very much occupied with their regular work unable attend Congress. Hope that impending Congress will be significant event in mathematical science. Wish success in Congress activities."[108]

Garrett Birkhoff, who officially opened the first plenary session held in Sanders Theatre on August 30, struck an optimistic note in his capacity as chairman of the Organizing Committee:

> The organizing of a successful International Congress at such a time of political tensions, and after a gap of fourteen years, has had its anxious moments, as many of you know. Your presence here promises that our efforts will be crowned with success."[109]

Particularly successful was the section for logic and philosophy (6), which alone of all the various special divisions of the Congress was permitted four invited lecturers (instead of three). The section was chaired by Alfred Tarski, with L.E.J. Brouwer elected as the section's honorary chairman. The invited lectures were given by Skolem, Tarski, Kleene, and Robinson, along with sixteen other contributed papers. Of the four invited speakers, Robinson was the unknown foreigner making his international debut.[110]

[106] Veblen 1950.

[107] Congress 1950, 1, p. 122.

[108] J. R. Kline, "Secretary's Report," in Congress 1950, pp. 121–123, esp. 122. Also quoted partially in Albers 1987, p. 27.

[109] Congress 1950, 1, p. 123.

[110] The Chairman of each section was allowed to invite not more than three persons to deliver thirty-minute papers in his section. An exception was later made for section 6 on logic and philosophy, in which four invited speakers were permitted. See Congress 1950, 1, p. 138.

Delegates to the Congress were housed in Harvard's dormitories, and most meals were served at the Harvard Union. An opening night reception was held at the Fogg Art Museum, and the following evening (August 31), the Busch String Quartet presented a concert in Sanders Theatre. Teas were given at Wellesley College and at the Harvard Observatory, and everyone was invited to a beer party in Memorial Hall. But the social highlight of the Congress came on Monday evening, September 4, when the famous Wagnerian soprano, Helen Traubel–the American Brünhilde–gave a performance in Boston's Symphony Hall "for which she received a tremendous ovation."[111] On Tuesday evening (September 5), a Congress banquet was given under tents in Sever quadrangle. For the closing evening of the Congress (September 6), a farewell party was sponsored by the director and board of trustees of the Gardner Museum of Art on the Fenway in Boston. As J. R. Kline, the Congress Secretary, summarized the week, the Congress was undoubtedly "the largest gathering of persons ever assembled in the history of the world for the discussion of mathematical research." He was right: the Congress counted a total of 2,316 delegates.[112]

For Robinson, the Congress at Harvard was an opportunity to see old friends and meet new colleagues. Fekete, Motzkin, and Levitzki had all come from Israel, although Fraenkel had not. Fekete and Levitzki were among the official delegates of the Hebrew University, and they also represented the Cercle Mathématique de Tel-Aviv (along with Benjamin Amirà).

Among outstanding logicians at the Congress, Thoralf Skolem was on hand from Norway, as was Andrzej Mostowski from Poland. Of logicians in America, Robinson met Church, Rosser, and Tarski, but Quine was abroad and Henkin was in Cambridge only briefly (and does not recall having met Robinson).[113] Henkin had just been engaged to be married a few weeks before, in Montreal, and had set the date for his wedding only a few weeks after the Congress. Consequently, he was very much preoccupied with things other than logic. As a result, Henkin only spent a few days in Cambridge, principally for a meeting of the Executive Committee of the Association for Symbolic Logic, which he was also serving by then as an editor.[114]

[111] Traubel 1959.

[112] This according to the "Secretary's Report," Congress 1950, 1, pp. 135–145, esp. p. 135.

[113] Nor does Henkin remember exactly when he first met Robinson, although he does have a clear recollection of seeing him at the 1957 International Conference on Axiomatic Method. "Tarski dreamed that up," and commissioned Henkin to be the secretary, doing all of the legwork. "It was the first and only time I ever got a taste of what it would be like to get an ulcer," Henkin I-1989.

[114] An informal meeting of the council was held at 8:00 P.M., August 31, in Cam-

The logic and philosophy sessions began on Friday in Emerson Hall, room 211. Thoralf Skolem delivered the first of the invited thirty-minute lectures, giving some "Remarks on the Foundations of Set Theory." This was followed by papers from Mario Dolcher, Barkley Rosser, Raphael Robinson, Wanda Szmielew, and Alfred Tarski (Szmielew and Tarski presented a joint paper on "Mutual Interpretability of Some Essentially Undecidable Theories").

Robinson's paper was not scheduled until the following Monday. On September 4, Tarski opened the logic and philosophy session at 9:00 A.M. with his invited lecture on "Some Notions and Methods on the Borderline of Algebra and Meta-Mathematics." Much of Tarski's thirty-minute paper covered ground that Robinson was also about to present in his paper. As soon as Tarski was finished, it was Robinson's turn. His thirty-minute invited lecture was devoted to "Applied Symbolic Logic" (this was the title listed in the Congress program; when published, Robinson made it more specific: "On the Application of Symbolic Logic to Algebra").

Robinson began his lecture with a nod to Gödel, Henkin, and Tarski—all of whom had indicated (directly or indirectly) some of the promising connections between logic and mathematics that he also intended to pursue.[115] This was typical of Robinson's style, for he was inevitably generous in acknowledging the research of others. He was careful to make clear the independent discovery by Tarski of the same theorem on algebraically closed fields that he was also about to present to the logic session at the Congress in 1950 (even though Robinson had al-

---

bridge, Massachusetts. Members of the council present included Church, Henkin, Kleene, Meder, Nelson (David), Rosser, and Tarski, with Curry and Skolem as guests. At this meeting it was agreed that Henkin's term as editor be extended for one more year, until January 1, 1952 (Henkin joined Max Black and Alonzo Church as an editor of the *Journal* beginning with vol. 14, 1949. Kleene was added as a fourth in 1950). See Barkley Rosser's report of this meeting, in a letter to members of the Executive Council of the Association for Symbolic Logic, September 5, 1950, p. 1, Tarski Papers, 84/69c, used here courtesy of The Bancroft Library, University of California, Berkeley.

[115] Henkin, for example, used the "compactness theorem" in his dissertation in 1947. In his paper for the *Transactions of the American Mathematical Society,* in 1953, he referred to it obliquely as "our fundamental result from logic." But Henkin did not pick up or use Robinson's metamathematics, and he didn't go much further with model theory, either. Instead, he turned more to algebraic logic and Tarski's theory of cylindric algebras.

When Henkin finally got around to studying Robinson's work in detail, shortly after publication of Robinson's thesis of 1949 as *On the Metamathematics of Algebra* in 1951, he was amazed at how closely Robinson's ideas paralleled his own. But it was clear to Henkin that Robinson was much more concerned with mathematics, and that Robinson was interested in applying the "compactness theorem" to other aspects of mathematical logic than was Henkin. Henkin I-1989.

ready published the theorem himself in 1948 and proven it with full details in his dissertation).[116]

Diagrams and Transfer Principles

Both Robinson's dissertation and his Congress paper considered models and algebras of axioms, but he was especially innovative in his introduction of diagrams and use of transfer principles.[117] All of these techniques were directed towards demonstration of a theorem about algebraically closed fields whose proof by more conventional methods was *not*—as he stressed—at all apparent. Many of the ideas Robinson applied with such success in his later work—especially in his most dramatic creation, that of nonstandard analysis in the early 1960s—were already present in one form or another in his earliest work on the metamathematics of algebra summarized and presented at the Cambridge Congress.

Of all the new ideas in Robinson's Congress paper, however, what he called the "second transfer principle" is noteworthy for its close integration of logic and algebra. For example, in the context of fields of characteristic 0:

> Any statement formulated in terms of the relations of equality, addition, and multiplication, as above, that holds in the field of all complex numbers, holds in any other algebraically closed commutative field of characteristic 0.[118]

In his proof of this theorem, Robinson relied directly upon results of Steinitz's field theory (Steinitz had shown that any two algebraically closed fields of equal characteristic and of equal degree of transcendence are isomorphic).[119]

Robinson concluded his remarks at Harvard, typically even then, with a sense of the historical import of his work. Always interested in the

[116] *Journal of Symbolic Logic*, 14 (1949), p. 74.

[117] Although the word *diagram* may suggest a picture, there was nothing pictorial about Robinson's use of this term in model theory. He introduced the term "diagram" of a mathematical structure to represent the set of sentences that express all possible basic relations between constant symbols representing elements of the structure. Thus, if the structure is the field of rational numbers, the diagram would include such sentences as $2 + 2 = 4$ and $(1/2)(2/3) = 1/3$. Because the diagram is often infinite, and sometimes even uncountably infinite, this represented a radical departure from the way systems of logic had been regarded previously. Leon Henkin, in his dissertation, had introduced the same idea, but had not used the term "diagram." I am grateful to Martin Davis for suggesting this illustration of Robinson's concept of "diagram" to me in an E-mail note of September 7, 1994.

[118] Robinson 1950, p. 689; in Robinson 1979, 1, p. 6.

[119] Steinitz 1910.

history of mathematics, he doubtless derived a certain pleasure from mentioning Leibniz, and the fact that he had now achieved what Leibniz indeed had also anticipated:

> The concrete examples produced in the present paper will have shown that contemporary symbolic logic can produce useful tools—though by no means omnipotent ones—for the development of actual mathematics, more particularly for the development of algebra and, it would appear, of algebraic geometry. This is the realisation of an ambition which was expressed by Leibnitz (*sic*) in a letter to Huygens as long ago as 1679.[120]

This reference to Leibniz is a sign that Robinson was well read, historically, and that he was already familiar with the classics of the history of mathematics (he had, after all, done a reading course in philosophy at Hebrew University on the subject of Leibniz). Leibniz would feature again in Robinson's thinking, in an even more dramatic way, as a catalyst in his best-known work, his creation of nonstandard analysis in the 1960s.

Throughout his Congress lecture, Robinson had struck a number of themes that would later become increasingly prominent in his work. In addition to stressing the novelty of his approach to algebra through logic, he added that while symbolic logic was familiar as a tool in philosophy and epistemology, his own departure—indeed one of his great insights made at the very beginning of his career—was the applicability of symbolic logic to mathematics proper, specifically to abstract algebra. The Congress lecture was only the first of many reflecting Robinson's lifelong dedication to showing that logic was not in fact a dead scaffolding to be used primarily for foundations to shore up the rest of mathematics. On the contrary, Robinson delighted in showing that symbolic logic could be used *within* mathematics to positive, creative effect.

His Congress paper sought to find the solution to "a genuine mathematical problem," as he said, "by applying a decision procedure to a certain formalized statement." Although Robinson was well aware of the practical possibilities of the methods he was developing, he was concerned at the time not with any particular deductive procedure, but with the general relations between a system of formal statements and the mathematical structures described by such a system. In essence, along with Tarski's pioneering work, Robinson's Congress paper was one of the earliest publications of research devoted to model theory.

Robinson and Mathematical Realism

In 1950 the philosophical position Robinson adopted was "a fairly ro-

[120] Robinson 1950, p. 694; in Robinson 1979, 1, p. 11.

bust philosophical realism."[121] By this he meant the acceptance of the full "reality" of any given mathematical structure. Although he was working with formal languages, these were used he insisted merely to *describe* structures, not to justify their "reality" or "existence," which were taken for granted.

This was particularly true in the case of results Robinson presented, formulated from axioms in a formal language, and then related back to particular structures to establish algebraic theorems. By combining the concept of a diagram with the completeness theorem, it was possible to establish theorems, as he liked to stress, "whose proof by conventional means is not apparent." One such proof involved the joint zeros of polynomials over skew fields; others (only mentioned in the Cambridge Congress paper, but developed in his thesis) concerned convex models for which it could be shown, for example, that the union of an increasing chain of models of a convex algebra of axioms K is again a model of K.[122] Here Robinson was especially concerned about the connection between the structures he was working with and the languages used to describe them:

> It was stated at the beginning that we attribute an equal degree of reality to a mathematical structure and to the language within which it is described. Accordingly we may introduce notions which are defined partly with reference to a given algebra of axioms, and partly with reference to its models.[123]

In explaining his "fundamental logical outlook," Robinson realized that although the point of his thesis was mathematical rather than philosophical, he could not avoid taking a stand on foundational issues, especially concerning questions of reality and truth. Above all, Robinson was interested in adopting "as liberal and unfettered a point of view as possible," without having to commit himself to any one school or another. As opposed to the intuitionists, he made it clear that he accepted the axiom of choice, the theory of transfinite numbers, and in formal logic, infinite conjunctions and disjunctions, languages with infinite numbers of propositions, et cetera. Robinson was quick to add that this was not out of any "unbridled desire for generalisation," but that without such concepts it was impossible to formulate many standard, orthodox parts of mathematics, for example, the Archimedean axiom formulated in terms of an infinite disjunction.

121 Robinson 1950, p. 686; Robinson 1979, 1, p. 3.

122 Robinson devoted an entire section of his thesis to "convex systems," chapter 9, Robinson 1951, pp. 116–138; the above theorem is also proved in his *Introduction to Model Theory and to the Metamathematics of Algebra,* Robinson 1963, pp. 80–81.

123 Robinson 1950, p. 693; Robinson 1979, 1, p. 10.

In his Congress paper, Robinson was explicit about the "fairly robust philosophical realism" he adopted, assuming the full "reality" of a given mathematical structure, and using a formal language to describe it without justifying its existence, which Robinson took for granted. Perhaps this was the certainty of youth making itself felt. It may also have reflected the influence of his thesis advisor and Paul Dienes' own allegiance to mathematical realism. But later, Robinson would abandon this variant on a Platonist philosophy of mathematics in favor of a stronger formalism where no reality was attributed to the mathematical concepts in question. Perhaps this change in his philosophical position was simply a reflection of age, experience, and mathematical maturity (for details, see the discussion of Robinson's formalism in chapter 7). In 1950, however, the immediacy and "reality" of his mathematics was uppermost in Robinson's mind.

Model Theory and Algebra

Following his return to England after the Congress, Robinson put the final touches on a revision of his thesis for publication in the prestigious North-Holland series edited by Brouwer, Beth, and Heyting, Studies in Logic and the Foundations of Mathematics. He also published some of the more important results drawn from parts of his thesis in a paper "On Axiomatic Systems which Possess Finite Models," which he submitted to *Methodos* (an Italian quarterly for methodology and symbolic logic).[124]

Here Robinson considered axiomatic systems as sets of statements formulated within the lower predicate calculus, taking algebraic structures as models of systems of sentences. As examples, he first described abelian groups axiomatically, then went on to show how fields, commutative fields, and fields of characteristic p might also be axiomatized. In order to axiomatize algebraically closed commutative fields, ensuring that every polynomial of order n has a root, an infinite sequence of statements was needed. Similarly, defining an axiomatic system for a commutative field of characteristic 0 also required an infinite sequence of statements. At the end of the paper, Robinson again stressed the power he attributed to the axiomatic method he had exploited, which "lends itself readily to the proof of mathematical theorems whose demonstration by 'conventional' mathematical methods would appear to be difficult."[125] This was the same point he had

124 Robinson 1951b. The paper contains a number of typographical and other errors, including the misspelling of Leon Henkin's name. See Henkin's review of the paper in the *Journal of Symbolic Logic*, 200 (1955), p. 186.

125 Robinson 1951b, p. 149; Robinson 1979, 1, p. 331.

made in his Congress paper the year before, and in even greater detail in his thesis. As a practical example he gave the following theorem on the zeros of a set of polynomials at the close of his *Methodos* paper:

> *If a set of polynomials of n variables with integral coefficients* $q_i(x_1, x_2, \ldots, x_n)$ *possesses not less than m different zeros in some commutative field of characteristic 0, then there exists a finite field in which the polynomials* q_i *(whose coefficients are now taken modulo the characteristic p of the field) possess not less than m different zeros.*[126]

The British Interplanetary Society

At the end of 1950 Robinson became a member of the British Interplanetary Society (on December 2). Many of its members were drawn from the faculty and staff of institutions with which Robinson had long been associated, including the Royal Aircraft Establishment at Farnborough and the College of Aeronautics in Cranfield.[127]

The Interplanetary Society, organized in Liverpool at an informal gathering in October 1933, represented "the nucleus of the British movement in rocketry."[128] Its primary interest was "the conquest of space and thence interplanetary travel."[129] Beginning small, with a grand total of 15 members in 1934, its membership in less than three years increased dramatically (to more than 105) by the time the society moved to London in 1937.

Although the Interplanetary Society was dormant throughout the war, a branch in Manchester did its best to remain active, and was subsequently joined by the Astronautical Development Society, a kindred organization founded at Surbiton, Surrey. In 1944, both groups merged to form the Combined British Astronautical Society, with branches in the Midlands, at Farnborough, and at Eccles.[130] After the war, the Combined Society joined with the British Interplanetary Society in 1945 to

126 Robinson 1951b, p. 149; Robinson 1979, 1, p. 331.

127 Fellows paid annual dues of £2.2, members 10s. 2d. Meetings were held fortnightly, and all members received free copies of the society's journal.

128 Cleator 1934, pp. 2–5. Fellows were expected to have a university degree in one of the natural sciences, engineering or medicine; they might also have graduated from a "recognized" technical institution, or they could also be admitted to membership by examination. There were seven sections: astronomy, biology, design, dynamics (headed by M. Woodger), instruments, models, and propulsion. The dynamics section included mathematics, aerodynamics, and orbital mechanics.

129 Cleator 1934, p. 3.

130 "Historical Note" in a pamphlet, *The British Interplanetary Society* (London: n.d., but probably 1946–1947), p. 3.

form a national organization, now called the British Interplanetary Society, Limited.[131]

Although the British Interplanetary Society's purpose, above all, was promotion of interplanetary space flight (at least according to the aims set out in its charter), its vision was actually much broader. It was basically a serious organization of enthusiasts whose interests in aeronautics and rocketry were of the highest order. Often the society jointly sponsored symposia with the Royal Aeronautical Society and the College of Aeronautics.[132] Science fiction writers like Arthur C. Clarke and Olaf Stapledon were also members and contributed regularly to the society's *Bulletin*. At the time Robinson joined towards the end of 1950, the society was busy on various projects, among which was construction of a rocket car, development of new fuels for space travel, and study of ultrashort wireless waves, "which pierce the Heaviside layer and thus enable communication between vessels in space and on earth."

Christmas at Cranfield: 1950

Although the Robinsons did not know it at the time, the Christmas of 1950 would be their last in Great Britain. By then Robinson had not only been promoted to deputy head of the Department of Aerodynamics at Cranfield, but had also been invited to serve as "an independent member of the Fluid Motion Sub-Committee of the Aeronautical Research Council."[133] This appointment was a clear sign of the prominence that Robinson's work was beginning to achieve, especially in the aerodynamics of wing theory and supersonic air flow.[134] At Cranfield,

[131] This brought the total membership by the end of 1945 to slightly over five thousand. In addition to scientists and engineers, the society also attracted writers with an interest in space travel—and science fiction. Among the most prominent of such members were Arthur C. Clarke and Olaf Stapledon. On October 9, 1948, Olaf Stapledon lectured on "Interplanetary Man," published as Stapledon 1948, with lively discussion reported between Stapledon and Clarke, pp. 232–233. The following year Clarke delivered a lecture on "Interplanetary Flight" to the Royal Aircraft Establishment Technical Society, Farnborough (February 9, 1949), subsequently published as Clarke 1949.

[132] Among the more important and better-known of these was the Cranfield Symposium on High Altitude and Satellite Rockets held in July of 1957. Other meetings were convened on space medicine, rocket and satellite instrumentation, space navigation, and even a symposium on liquid hydrogen.

[133] Beginning in June of 1949, his tenure on the Research Council's Subcommittee was to run through the end of May 1952. For his services he was to be paid ten guineas per meeting and provided with first class train fare plus any subsistence allowance as necessary.

[134] In addition to his teaching at the College of Aeronautics, and numerous articles he was publishing on subjects related to aerodynamics, Robinson had also begun to write a book on wing theory. Olga Taussky Todd said she was "amazed to hear that he

Student caricature of Robinson as an angel, with infinite halo, Cranfield, 1950.

Robinson was a popular and highly regarded member of the staff. The students reflected their sentiments that Christmas in the College of Aeronautics' student publication, *Potential*, wherein Robinson was affectionately caricatured as an angel—with an infinity sign for a halo—bearing the legend "location of a point in space!!"[135]

Shortly after the new year, in early February 1951, Robinson received an aerogram from Professor Samuel Beatty, chairman of the Department of Mathematics at the University of Toronto. Beatty wrote to ask Robinson about accepting a position at Toronto as an associate professor:

> Dear Dr. Robinson,
> The President has authorized me to invite you to join our staff in mathematics, starting with the academic year 1951–52, which begins on 1 July, 1951. The title we have in mind is Associate Professor, and the salary we are proposing is $5,600.00. The range of salary for the Associate Professorship is $5300-$6200, which means that the salary proposed is already one-third of the way up the scale.[136]

---

had picked up sufficient knowledge of [aerodynamics] to write a book about it. I myself only worked [during the war] on mathematical problems suggested by aerodynamics," Taussky R–1976. See also Seligman 1979, p. xxi.

[135] *College of Aeronautics Student Bulletin*, winter 1950, p. 27. This was a drawing Robinson liked to keep on his office wall, even at Yale. Renée Robinson I-1980.

[136] Samuel Beatty to Abraham Robinson, February 2, 1951, RP/YUL.

Robinson was needed to replace Leopold Infeld, who had just left Toronto (largely for political reasons due to his close ties to Poland and presumed Communist sympathies). Beatty also explained that it was the president's intention to "re-establish, in due course, Applied Mathematics as a separate Department under its own Head." Robinson would naturally prove instrumental in this development—an opportunity that doubtless appealed to him. He decided to accept the offer, and soon thereafter he received an official letter of appointment from the university's president, Sidney Smith, on March 30, 1951. Having decided to make the move to the University of Toronto, Robinson now found himself in his last semester at the Royal College of Aeronautics.

Robinson's Last Course: Axiomatic Stochastic Processes

In 1951 Robinson's final course at the College of Aeronautics was stochastic processes. As a simple example involving continuous time parameters, he asked his students to consider the mathematical problem of people crossing Waterloo Bridge. Let $f(t)$ be the number of people crossing the bridge from north to south from a given time t onwards. Denote corresponding states by E_n, that is, "n people have crossed." Stochastic processes concerned the full mathematical development of such situations.

In his introduction to the course, Robinson explained that an important part of aerodynamics was related to turbulence, theoretical treatment of which involved statistical arguments. These in turn could be reduced to cases involving time-dependent variables. The theory of such "stochastic processes," he explained, also had applications in areas other than aerodynamics, including (as Robinson emphasized) quality control. Much of this work, he suggested, lay in the future, for as he told his students, it was a subject in its infancy.

The first step was to learn its language. Robinson began with the basics of set theory and Boolean algebra. Then he reviewed probability theory, presenting it axiomatically. The standard reference he recommended (for the axiomatic approach) was A. N. Kolmogorov's *Foundations of the Theory of Probability*, although he realized this was suitable primarily for students who already knew set theory.[137] For the rest of his students he recommended William Feller's *An Introduction to Probability Theory and its Applications*.[138]

Unfortunately, as the semester came to an end, Robinson ran out of time to do more than cursory justice to diverse topics, including appli-

[137] Kolmogorov 1933. See also Batchelor 1990 and Kendall 1990.

[138] Feller 1950.

cations to atomic physics. He did manage a brief discussion of chain reactions as examples of stochastic process involving generations of particles considered over successive time intervals, but was apologetic for not having presented more aerodynamical applications. Even so, as he told his students:

> In conclusion, I will say only that this theory is closely connected with general diffusion problems. Finally although in the time available we have not been able to consider problems of turbulence proper, it is hoped that the course has at least provided a good background for the application of modern statistical methods.[139]

What is remarkable about Robinson's course on stochastic processes is the fact that he again viewed the axiomatic approach in virtually the same way as his earlier applications of symbolic logic to algebra. The value of the axiomatic method in both cases was the same, namely the way in which working axiomatically served to bring clarity, precision, and order to the material in question—all guaranteed to bring new insights. As he promised his students, this would help to "clear up and solve a number of problems whose solution has eluded the grasp of a more naive approach."[140]

Holidays in Europe

The Robinsons always enjoyed traveling, and while there were limited opportunities for doing so immediately after the war, they nevertheless found time to take several summer vacations. Their first such trip was to Ireland, where they went to Dublin and Kilarney.[141] Another summer they visited Scandinavia, where they rented a car for the month. In Sweden they went to Uppsala to meet Alec Young and his wife, and then made a special point of going to Norway to see Thoralf Skolem (who came back to Oslo from his own holiday just to see Robinson). Skolem, who was an elderly man by then, was someone whom Robinson greatly admired and especially wanted to see.

[139] Robinson 1951c, course lecture notes, "Stochastic Processes," p. 58, RP/YUL.

[140] Robinson 1951c, p. 1. In this same spirit, Robinson would surely have been pleased by the considerable power his best-known contribution to mathematics, nonstandard analysis, would later give to the subject of stochastic processes. After Peter Loeb's discovery of how to construct a rich standard measure space from a nonstandard one, numerous important applications were made in various parts of stochastic analysis and mathematical physics. For details, see the survey books by Albeverio et al. 1986; Stroyan and Bayod 1986.

[141] All of the details presented here are based on the extensive collection of scrapbooks, photographs, and slides (amounting to several thousand) in the possession of Mrs. Abraham Robinson, Hamden, Connecticut.

The Robinsons on the island of Ischia and among the ruins of Pompeii.

From July 22 to August 12, 1949, Abby and Renée were in Italy. In Florence they wandered through the Bargello Palace and Museum, admiring the sculpture, saw the galleries of the Uffizi, and toured the Duomo of Santa Maria dei Fiori. They went to Venice, where *La Serenís-sima* was not so serene–it was unbearably hot thanks to a miserable sirocco, and the Robinsons were dismayed to find that they were staying at a vegetarian hotel! Understandably, one snapshot shows Renée and Abby on the Lido in Venice, the two of them in swimwear, enjoying the water and cooling breezes from off the Adriatic. They visited Murano and watched the glassmakers, and explored the Byzantine beauty of San Marco.

Making their way back, they stopped in Verona where Renée took a picture of Abby in the Piazza dei Signori. Further north, they enjoyed a few days hiking in the Italian Alps. From Italy they continued on to Switzerland. In Geneva they visited the Palais des Nations, formerly headquarters of the League of Nations and subsequently home of the European United Nations. There is also a picture of Abby wearing a suit, Renée with hat and gloves, in Lausanne where a friend of Abby's mother had a residence she ran for young girls.

The Robinsons' "Good Bye"

In July of 1951 the Robinsons set off on a "farewell to Europe," during which they spent nearly a month in Italy. This had been carefully planned, for as Renée knew, Abby loved Italy. They stopped in Naples and visited Pompeii, where Renée took a picture of Abby in the Forum. They were greatly impressed by the dramatic sense of history they felt at both Herculaneum and Pompeii, where the colorful frescoes of the Villa of Mysteries made a strong impression. They also took a day boat from Naples to Capri, where they saw the famous Blue Grotto and the imperial Villa Jovis. But on their way to Ischia, unfortunately, Renée was seasick.

In Rome Abby took Renée's picture in the Forum Romanum, while she snapped his in front of the Castel San Angelo (and beside Bernini's famous *Fountain of the Four Rivers* in the Piazza Navona). They went to Tivoli, and saw the magnificent fountains and gardens of the Villa d'Este. They travelled north to Pisa with its famous leaning campanile, to San Gimignano with its many fortressed towers, and to Siena.

Finally, their trip almost over, they again made their way back to England via Switzerland. In the scrapbook Renée made from their photographs remembering this trip, one photograph in particular stands out. It is a picture of Abby, a grand panorama of the Alps unfolding dramatically in the background, the mountains towering majestically above. Under the photograph Renée simply wrote—"Good Bye!"

Passport photographs for Canada, May 1951.

CHAPTER SIX

The University of Toronto: 1951–1957

> Zest for both system and objectivity is the formal logician's original sin. He pays for it by constant frustrations and by living ofttimes the life of an intellectual outcaste. . . . The formal logician gets little sympathy for his frustrations. He is regarded as too rigid by his philosophical colleagues and too speculative by his mathematical friends.
>
> –*Hao Wang*[1]

IN MOVING from London to Toronto, Robinson did not miss a beat, mathematically. Sailing first class from Liverpool on a Cunard liner, the *Franconia,* the Robinsons headed for the New World in August of 1951. Although Renée was seasick most of the time, Abby worked at his usual steady pace, writing three pages a day.[2] By the time they reached North America in early September, he had finished a twenty-five-page manuscript.

AN AXIOMATIC APPROACH TO DIMENSIONAL ANALYSIS

Robinson's latest effort was designed to cast new light upon dimensional analysis:

> It is shown that the main proposition of dimensional analysis, the so-called " -t heorem," can be deduced in strictly mathematical fashion from a set of simple axioms. This provides a firm foundation for the discussion of the physical assumptions involved.[3]

Dimensional analysis, as Robinson explained, was a very useful tool both for physicists and engineers. In his own work on fluid dynamics, the subject arose naturally because real fluids behave in complex ways. In developing satisfactory models for the behavior of fluids, deciding which variables are fundamental is naturally a basic one. Dimensional

[1] Wang 1954, p. 241.

[2] According to George Duff, Robinson's colleague at the University of Toronto, "whenever Robinson was writing a manuscript, he would set aside time at the end of the day. He would keep at it until he had written at least three good pages. This meant foolscap paper, and Robinson wrote in a rather small hand, so three pages was quite a lot." George F.D. Duff, interviewed in his office at the University of Toronto, November 13, 1990; hereafter referred to as Duff I-1990.

[3] Robinson 1951a, a twenty-five-page manuscript "On the Foundations of Dimensional Analysis," dated "*RMS Franconia,* August/September 1951," RP/YUL.

analysis serves to determine the essential relationships between variables, especially for equations that do not depend on any particular system of units of measurement, which are said to be "dimensionally homogeneous."[4]

One of the most important applications of dimensional analysis concerns the relation between a model and its full-scale counterpart. Wind tunnel experiments, for example, would have impressed upon Robinson the importance of predicting forces on a full-scale prototype from the results of force measurements on a particular experimental model. Dimensional analysis was controversial, however, in complex cases for which the dimensional situation was either unclear, misunderstood, or subject to different interpretations.[5]

The axiomatic approach, Robinson insisted, had the advantage of avoiding such debates as whether or not the concept of dimension was "fundamental." Just as in projective geometry, where an axiomatic approach allowed one to take the concept of a straight line as either fundamental, or derivable from other concepts one might prefer to take as fundamental, he found that it was "easy to formulate alternative sets of axioms in which the concept of a dimension is either fundamental or derived."[6] From a purely mathematical point of view, however, he insisted that "there is nothing to choose between them," since the sets of axioms in question were logically equivalent.[7]

Although the axiomatic approach to dimensional analysis was obviously of interest in itself to Robinson, he took its fundamental power to lie in applications. The main practical advantage of dimensional analy-

[4] The early works establishing the importance of dimensional analysis include Buckingham 1914 and Bridgman 1931. See also Esnault-Pelterie 1948.

[5] See, for example, the numerous paradoxes discussed by Garrett Birkhoff in chapter 1 of his *Hydrodynamics: A Study in Logic, Fact, and Similitude*. The second edition contains two chapters devoted to "paradoxes" divided between examples of nonviscous and viscous flow. Birkhoff 1960.

[6] Robinson 1951a, p. 1.

[7] Robinson was not always so indifferent! Several years later, his colleague at Toronto, P. G. (Tim) Rooney, remembers one of the few times he ever saw Robinson in heated disagreement, and it was over the axiom of choice. Rooney and Robinson were talking mathematics. The subject of axiomatic set theory came up and what was the best set of axioms to teach in a course on set theory. Rooney argued for the maximal principle in preference to the axiom of choice (leaving more difficult questions related to the axiom of choice for later in the course). Robinson, however, argued vigorously "that the axiom of choice was simpler and much more natural." Rooney was also surprised that Robinson didn't seem to know the famous shoes-and-socks example in connection with the axiom of choice. "You don't need the axiom of choice," Rooney explained, "to pick one shoe from each pair of an infinite set of shoes—just take the left one. But you can't do this with infinitely many pairs of socks" (unless each sock is already on a foot). Rooney I-1990, with additional details in a letter to JWD of March 28, 1991.

sis was that it enabled the transfer of results from an experiment under one set of conditions to other conditions under which the experiment had not been performed. A good example (according to P. W. Bridgman) was transferring results for flow of a viscous fluid in a tube of fixed diameter to results of flow in tubes of the same form but with different dimensions, or with fluids of differing viscosity. For such transfers of results from one set of conditions to others, it was only necessary to check the values of the dimensionless parameters to make sure they were the same.[8]

Robinson also noted the value of dimensional analysis in demonstrating the equivalence of dimensions that might have been supposed to be different—temperature and energy, for example. The fact that dimensional analysis showed them to be equivalent was not so surprising. It was certainly not, he stressed, the paradox that Lord Rayleigh had taken it to be.[9]

Although Robinson never threw away his paper on the "Foundations of Dimensional Analysis," it remained unpublished at the time of his death in 1974. His decision not to print it may have been due to the fact (evidently unknown to Robinson as he was writing the paper) that Garrett Birkhoff had just written an entire book on the subject.[10]

Hydrodynamics here again provided some especially pertinent examples. Examining in detail the special case of flows with "free boundaries," Birkhoff raised the question of "modeling" and its rational justification. Robinson, too, had long been interested in such questions, beginning with his work at Farnborough which entailed the ever present difficulty of knowing whether the idealized conditions of mathematical theory could provide adequate models for actual, physical experience.[11]

Above all, the most interesting part of Birkhoff's analysis had also occurred to Robinson, namely, the extent to which group-theoretic ideas

[8] Bridgman 1931. [9] Robinson 1951a, p. 24.

[10] Drawing upon the Taft lectures he had given at the University of Cincinnati in January of 1947 (and his experience in many European research laboratories made possible by a Guggenheim Fellowship), Birkhoff analyzed (among other concerns) the numerous "paradoxes" in mathematical physics where "plausible" reasoning had led to incorrect results. See Birkhoff 1960.

[11] One very obvious concern is the difference between fluids of small but finite viscosity and "ideal" fluids having zero viscosity. But as Birkhoff emphasized, this was only part of the problem. Errors of logic and mathematical rigor also came into play, and here dimensional analysis could help to identify those cases in which it was permissible to move from one system to another without encountering difficulties. Among "plausible arguments" that might be useful in physical reasoning (but were mathematically unacceptable), Birkhoff stressed the old adage that "small causes produce small effects." As he was careful to point out, the presence of arbitrarily small terms of higher order could change entirely the behavior of a given solution. See Birkhoff 1960, p. 4.

were applicable to fluid mechanics and questions of dimensional analysis. Birkhoff's book on these subjects was more systematic and thorough than Robinson's comparatively short paper written aboard the *Franconia*, and this no doubt persuaded him to leave it in manuscript. Robinson would immediately have appreciated the comprehensiveness of detail and obvious strength of Birkhoff's analysis, once it had come to his attention.

It is noteworthy that as Robinson was making his way to the New World, where his position at the University of Toronto would call upon his talents almost exclusively as an applied mathematician, his thoughts on the *Franconia* were inspired by questions related to mathematical physics, but treated axiomatically. The axiomatic slant gave his applied interests a characteristically "logical" edge—just as he was beginning to move decisively away from applied mathematics to his surpassing interest in model theory and mathematical logic.

Mathematics at Toronto

When Samuel Beatty (1881–1970) retired in 1952, a year after Robinson's arrival, he had been a member of the Department of Mathematics at Toronto for forty-five years and its Chairman for the last eighteen of these.[12] By then he was part of Canadian academic history itself—the first Canadian Ph.D. in Mathematics (in 1915), having done his graduate work at Toronto with J. C. Fields of Fields Medal fame (Beatty was his only doctoral student). Although Beatty's research was devoted primarily to the theory of algebraic functions, largely to aspects of the Lagrange interpolation formula and the Riemann-Roch theorem, he also taught courses on calculus and complex variables. As soon as Beatty assumed the chairmanship in 1934, his first priority was to strengthen theoretical mathematics.

Seeking to provide a "broader and deeper approach," Beatty hired Richard Brauer in 1935 to cover algebra, and a year later H. S. MacDonald Coxeter to teach geometry. These two appointments, Gilbert Robinson later said, were "Beatty's greatest contribution to mathematics in Canada."[13]

Not to be outdone, John L. Synge, chairman of the Department of Applied Mathematics—was equally interested in expanding what his pro-

[12] Samuel Beatty (1881–1970) also served as dean of the Faculty of Arts from 1936 until his retirement in 1952, when he was elected chancellor of the university (1953–1957). As Gilbert Robinson has said, "in a very real sense he guided Canadian mathematics from the isolation of the 19th century to a significant role in the 20th century," G. Robinson 1971, p. 489.

[13] G. Robinson 1971, p. 489.

gram could offer. He achieved a coup of sorts by hiring the refugee Polish mathematical physicist and friend of Albert Einstein, Leopold Infeld, in 1938.[14] As one of his biographers later wrote, Infeld's life was a "microcosm" of the century.[15] It was the gap caused by Infeld's resignation in 1950 that led to Robinson's appointment. Toronto actually lost two of its applied mathematicians at virtually the same time–A.F.C. Stevenson along with Leopold Infeld.[16] Beatty had already appointed John Coleman in 1949 (a former student of Synge's), and soon added J. A. Steketee to the department as well. The staff in the Department of Applied Mathematics, however, was not at full strength until Robinson was appointed in 1951, along with J. A. Jacobs the following year.[17]

Toronto: September 1951

J. A. Steketee remembers meeting the Robinsons at the train station and driving them across town in his 1929 De Soto. For anyone arriving from London, metropolitan Toronto struck most visitors soon after the war as "an urban disaster":

> congeries of dull, small towns, dangling from a dark entanglement of overhead wires. And in those days, the official social life in the university was stiff with protocol and inhibitions, a trial to [anyone] who had known the gaiety that was always erupting amid the grimness of the wartime British capital. By the fifties, the Toronto environment had begun to change: a genuine city was beginning to emerge with some sense of style and a growing pride and confidence in itself.[18]

It was this "new" Toronto that the Robinsons found most congenial.[19]

[14] In 1943 Synge left Toronto for Ohio State University, where he was to be chairman of the Mathematics Department. In 1948 he returned to Ireland and took a position as senior professor at the Dublin Institute for Advanced Studies. Synge 1972, p. 255. From then on, Beatty headed both departments, the Department of Applied Mathematics becoming once again part of the Department of Mathematics. See G. Robinson 1979, p. 52.

[15] G. Robinson 1968, pp. 123–125. Infeld's political problems affected his entire family, and his two children's Canadian citizenships were revoked. Infeld died in Warsaw on January 15, 1968. In addition to his own autobiography, L. Infeld 1978, see the obituary in the *New York Times*, January 17, 1968, p. 51, col. 2; E. Infeld 1970; and Trautman 1973. Trautman makes no mention of the tremendous political difficulties Infeld faced in Canada due to anti-Communism there.

[16] For biographical details about Stevenson, see Duff 1969.

[17] G. Robinson 1979, p. 57. Robinson is incorrect about A. Robinson having been appointed in 1952; A. Robinson arrived in Toronto the year before, in September of 1951.

[18] Bissell 1974, p. 36.

[19] In fact, it was an exciting time to be in Canada. Toronto had grown dramatically

Relaxing at home, Winnett Avenue, Toronto.

They did not encounter the political problems that had driven Infeld away, but instead were welcomed by their colleagues. They soon discovered the physical beauty of North America as well in the course of many vacations they took throughout Canada, including trips they made to the United States and Mexico in the early 1950s.

The Robinsons soon found a new home on Winnett Avenue, which they were able to buy outright thanks to the success of Renée's work as a fashion designer in London.[20] Although the house was small, it had its own garden, "and when you opened the door it was like entering a new world: cozy, small, attractive. Sometimes, theater groups would meet there, actors and producers together."[21]

Renée, in fact, found that the theater was of increasing interest to her, more appealing and satisfying than the world of fashion. Although she still liked to design her own clothes, by the early 1950s she felt that the industry in general was becoming too commercial, less artistic, and by the time she left England she had grown tired of it all.[22] Instead, her theatrical experience brought her to the attention of the Canadian Broadcasting Corporation, where she was able to make good use of her dramatic talents:

> Soon after having established ourselves [in Canada], as I had an actors' equity card from Vienna and then one from London, I tried to get an audition at the Toronto Radio Station. I was lucky to have one of the most established Canadian actors present, an elderly man who became, more or less, my patron. Soon they asked me to take parts where my accent was either required or did not matter, and I became one of the actors' circle.[23]

After the Austrian actor, Josef Fürst, emigrated to Canada, he and Renée soon became good friends, and they were frequently cast together. In one film they played an Estonian couple, and for two years they were

---

after the war, and by the early 1950s, it was the fastest growing city in North America, with a population well in excess of one million. For general introductions to Canadian history, see Bothwell 1989 and West 1967.

[20] Robinson I-1990.
[21] Robinson I-1983.
[22] Robinson I-1980.
[23] Robinson I-1990.

Renée Robinson, publicity photograph, Toronto.

The Robinsons at home with the actor Josef Fürst.

the von Hohenfelds in a daily radio series, *The Craigs*, in which she played the part of an Austrian noblewoman who had supposedly moved to rural Canada.[24] When the Robinsons left Canada in 1957, Renée was dramatically written out of the script (although she cannot recall whether she met an untimely soap opera death or not).

In addition to radio, Renée made several television appearances, as well as a motion picture in which she played a psychiatrist for a production by the Film Board of Canada, capitalizing again on her Viennese heritage and accent (as she did in the part of a Viennese lawyer in a film made for television).[25] She recorded several of her readings, including

[24] Ross I-1990. Abby, however, would often make tape recordings of her programs. Robinson I-1990.

[25] Robinson I-1980, and Renée Robinson to JWD, March 5, 1994.

one role she had always dreamed of playing—Medea. Another was a comedy, for which she had a special flair. She also made a 78 rpm record with Fürst consisting of excerpts from Shakespeare's *Othello*, in which Renée played Emilia, another role she had always wanted. She was disappointed not to have been given the part opposite her friend Josef Fürst when he was cast as Iago for a CBC television production of the play directed by David Green. (Instead, Green cast his wife as Emilia. This time, Renée knew it was not her accent that was the problem—after all, "If Iago has an accent [Fürst], then why not Emilia?").[26]

At home on Winnett Avenue, Renée would often combine her friends from show business with the mathematicians from the university, "and they enjoyed each other very much." It was just this sort of openness and genial social grace that impressed Robinson's colleague and former graduate student R. A. Ross, who was a guest on several occasions at their home. "Pleasant but small," is how he described it, adding that the Robinsons' generosity in entertaining faculty and graduate students was unusual at the time.[27]

It was a place where Robinson liked to work even after he had come home from the university. And he worked hard, systematically. Often when Renée's sister Sully (by then also a resident of Canada) would stop by to see them in the evenings, Abby would be sitting on the sofa in the living room with his writing pad, working on his mathematics.[28]

Colleagues in Applied Mathematics

The Department of Applied Mathematics was housed in a former private residence at 47 St. George Street. Robinson was given one of the better rooms on the first floor, but it was nevertheless rather sparse:

> The furniture was restricted to some shelves, a filing cabinet, some chairs and a desk in the middle of the room. The desk was usually piled with a disorderly variety of papers leaving a few square feet for his actual writing. He was not the type who carried big businessmen's briefcases around, full with clobber. The work he took home was usually contained in a thin briefcase, a writing pad and some papers, and occasionally a book.[29]

Robinson was a pipe smoker, and there was usually a tin of Players Navy Cut tobacco on hand, and occasionally cigars. His graduate stu-

[26] Robinson I-1990.

[27] Ross I-1990. Tim Rooney also recalls pleasant evenings at the Robinson's home, "the funniest little house, it couldn't have been more than twelve feet wide. Downstairs there was a combined living and dining room, with a little kitchen; upstairs were two bedrooms and a bathroom," Rooney I-1990.

[28] Steketee R-1976.

[29] Steketee R-1976.

dent R. A. Ross remembers that Robinson was always available, but also very disciplined, allocating his time very carefully.

The only occasions on which Ross recalls ever seeing Robinson annoyed was when the phone would ring. There was only a single telephone for the entire building, and it had been installed in a clothes closet just outside Robinson's office. "He seemed to think he was the official telephone answerer—and would have to shout out names from one floor to another whenever the phone rang."[30]

Just downstairs from Robinson, George Duff was assigned an office in the back room on the bottom floor. Duff remembers first meeting Robinson in the summer of 1952. In those days he saw a good deal of Robinson because Duff "sat on his doorstep a lot." Duff liked to discuss his work on partial differential equations, and Robinson was always willing to listen and be helpful.

Duff was a specialist on the subject of differential equations, about which Robinson—to Duff's surprise—also knew a great deal. Duff took it as a reflection, again, of Robinson's wide-ranging abilities that when he was working on a book on partial differential equations, Robinson read over the manuscript and offered several suggestions about how to improve the book, making it easier to follow in various ways for prospective readers.[31] H.S.M. Coxeter was likewise impressed by Robinson's diverse interests, especially his wide-ranging knowledge of geometry, Coxeter's own field.[32] In fact, it was due to his versatility that Robinson soon knew everyone well.

Teaching and Lecturing

In 1952 Robinson was asked to offer a short course of lectures to the National Research Council on "Supersonic Wing Theory."[33] Clearly the quality of his work and his importance in Canada for aerodynamic research were already being recognized. The following year, in 1953, he was invited to lecture at New York University on "Wave Propagation near the Surface of an Elastic Medium," and to present a paper on "Flow around Compound Lifting Units" at a symposium on high-speed aerodynamics at the National Aeronautical Establishment in Ottawa.[34]

[30] Ross I-1990. Similarly, A.J. Coleman recalls the building at 47 St. George Street: "It was a comfortable old building and there were many loud arguments up and down the broken-down stair well." Coleman to JWD, letter of December 3, 1990, RP/YUL.

[31] "Robinson was well appreciated, especially for his versatility, his profundity, his style," Duff I-1990.

[32] Coxeter, too, remembers that as he was working on his *Introduction to Geometry* (New York: Wiley and Sons, 1969), Robinson was helpful in suggesting some problems to be included; Coxeter I-1990.

[33] Steketee R-1976.

[34] Robinson 1953, pp. 26–29.

Meanwhile, his teaching at Toronto was limited primarily to traditional subjects of the basic sort he had been covering at Cranfield: differential equations, fluid mechanics, and aerodynamics. He usually taught a course on partial differential equations in mathematical physics, as well as fluid dynamics in the fourth year applied mathematics program. And he often taught a graduate course on supersonic wing theory, to which he had already contributed a great deal of research on his own while at Farnborough and Cranfield. Robinson also taught some of the "service courses," including calculus, analytic geometry, and differential equations for engineering students.

Administratively, when Irving Pounder became Chairman in 1952, Robinson was asked to serve on a special departmental advisory committee. This was the idea of the University's President Sidney Smith, who had "proposed that there be associated with me, for considerations of matters of policy, a committee consisting of yourself [Robinson] and Professors Coxeter, Webber, Sheppard, and Griffith."[35]

Robinson's First Ph.D. in Applied Mathematics: 1953

Robinson's first graduate student at Toronto was J. F. Hart, who had begun his doctoral work under Arthur Stevenson.[36] His chosen subject was the "Theory of Spin Effects in the Quantum Mechanics of a Many-Electron Atom with Special Reference to Pb^{III}."[37] Mathematically, Hart's thesis was of interest because it calculated for the first time the matrix elements for interconfiguration perturbations within the atom involving mutual spin interactions. This in turn showed that the orbital and spin coefficients were large enough to suggest that spin effects were not negligible, but comparable to electrostatic effects.[38]

Hart's conclusions must have been appealing to Robinson, for again they demonstrated how presumably negligible factors in a physical problem could actually take on considerable importance in reality. This was just the sort of conclusion that Robinson's work in fluid dynamics had impressed upon him time and again.

[35] Pounder to Robinson, July 23, 1952, Robinson's administrative file, Department of Mathematics, University of Toronto.

[36] Ross I-1990.

[37] Hart 1953. Note that Hart is not listed as one of Robinson's graduate students at Toronto in the list of his former students given in Robinson 1979, 1, p. 694.

[38] Hart 1953, pp. 1–2. The relatively large values for the spin effects were due to "large coefficients of kinematic origin and appreciable radial parameters." Thus Hart concluded that the spin effects were comparable in magnitude to the electrostatic ones and could not be neglected. This was in turn helpful in explaining deviations from the normal electron level distributions in the atom as predicted by individual electron-nucleus spin interactions. Hart ascribed such anomalies to interconfiguration perturbations.

Robinson and Aerodynamics

The Cold War insured that Western governments, including Canada, would maintain substantial interest in aeronautics, and Robinson and his graduate students were among those to benefit from increased support and facilities. By the early 1950s there was near universal agreement that gas turbine improvements were essential, and thus both gas dynamics and high-speed aerodynamics were given particular emphasis. The Korean War also served in a direct way to increase military funding, even in Canada, for both applied aeronautical development and theoretical research. Continuing economic expansion after the war also did its share, and the explosive growth of commercial aviation all over the world was an additional factor in maintaining interest in aerodynamics.

Consequently, the University of Toronto was anxious to expand its courses for both undergraduate and graduate students in aeronautics. The undergraduate courses were given as the "aeronautics option" in engineering physics, while graduate studies and research were collectively overseen by a new Institute of Aerophysics (later to become the Institute for Aerospace Studies) with G. N. Patterson as its director. The institute was also supported by the Canadian government with grants from the Defence Research Board (and U.S. agencies as well). Students were enrolled for both the M.S. and Ph.D. degrees. Initially, emphasis was given to the physics of gases, applied aerodynamics, and ballistics, with a special interest in supersonic flight, a field in which Robinson had developed a strong reputation of his own.[39]

The first of Robinson's aerodynamic papers to appear after his move to Toronto was devoted to an analysis of nonuniform supersonic flow.[40] This paper calculated the perturbations produced by a two-dimensional thin airfoil in a nonuniform (i.e., accelerating or decelerating) supersonic flow for the two-dimensional case based on linearized theory. In order to consider the problem of acceleration in isolation, Robinson assumed that the airfoil was placed in a plane of symmetry to the main flow so that the velocity component of the main flow (in a direction normal to the plane) was uniformly zero. Using linearized theory, Robinson then derived formulae for the perturbation potential which involved elliptic integrals (to which approximations could be made for nonuniform streams). These in turn gave correction terms for the case of

[39] Later, rarefied gas flow, blast phenomena, spacecraft orbital mechanics, acoustics, wind loads on buildings, etc., were also added to the areas taught at the Institute. For details, see Green 1970.

[40] Robinson 1953a. The paper had been sent to the *Quarterly of Applied Mathematics* in November of 1951, so the research was actually completed in Cranfield.

a uniform stream, and also provided corrections for both lift and drag.

Robinson's interest in such problems went back to a number of earlier papers where he also dealt with airfoils in nonuniform incompressible flow. The physical significance of this class of problems, Robinson noted, warranted its investigation. He added that one referee had even called his attention to the case of flow around jet vanes, where nonuniform supersonic flow was clearly of practical interest.[41]

On the subject of supersonic airfoil design in general, Robinson once told Ross that he had been one of the pioneers of the subject at Farnborough. Although he had been the first there to work out the details of supersonic flow for delta wings during the war, all of the research was classified. Consequently, he was never appropriately credited for any of these efforts because none of his results were published at the time.[42]

Three Ph.D.'s in 1955: A Bumper Crop

J. A. Steketee had come to Toronto in 1950 from Delft, his interest in aerodynamics having already been stimulated there by J. M. Burgers. Robinson suggested the topic of his thesis on boundary layer transitions, and provided "lively interest and encouragement." Steketee found that due to the usual separate treatment of laminar and turbulent flows, he was unable to formulate the problem as clearly, mathematically, as he would have liked. Instead, he examined several models reflecting various aspects of concrete situations. In particular, he managed to account for "the importance of the velocity gradient," and showed that "some solutions of the perturbation equations can be made to agree with measurements of turbulent fluctuations in a shear flow."[43]

Once again, Robinson had set one of his students upon a problem in which usually neglected features of aerodynamic modeling could not be ignored. As Steketee wrote in the introduction to his thesis, although in many studies of fluid dynamics viscosity might be neglected, "in the neighborhood of walls and of obstacles placed in the flow, where the velocity gradients are generally large, one has to take viscosity into account." It was the problem of transitions at these boundary layers that provided the substance of Steketee's dissertation.

L. R. Fowell's dissertation was devoted to "An Exact Theory of Supersonic Flow around a Delta Wing,"[44] and was concerned primarily with finding a solution for the equations of inviscid supersonic flow in the

[41] Robinson 1953a, p. 183. [42] Ross I-1990. [43] Steketee 1955.

[44] Fowell is also not mentioned in the list of Robinson's graduate students given in Robinson 1979, 1, p. 694, nor is he included in Gilbert Robinson's list of Ph.D. students in the Department of Mathematics at Toronto in G. Robinson 1979, pp. 96–99. But the

particular case of a delta wing. In analyzing flow over the expansion surface he considered two cases for delta wings with supersonic leading edges, and found that below a critical angle of attack a continuous solution might exist, while above this angle the solution was discontinuous. Fowell provided an exact solution for the case of flow over the expansion surface, but in the discontinuous case he only proposed an approximate method.

L. L. Campbell recalls that originally he was "assigned" to Robinson for supervision.[45] Robinson had worked on certain aspects of hyperbolic differential equations during the war, and his report on "Shock Transmission in Beams" dealt with some particular equations governing shock transmission in wings of aircraft.[46] He was especially interested in the initial value problem for hyperbolic partial differential equations in two variables and of general order, and proposed that Campbell try to extend his methods to initial boundary value problems for the same equations. Campbell succeeded in doing so (with appropriate guidance, he adds, from Robinson), and thus produced his thesis.[47]

As Campbell worked on his dissertation, he met regularly with Robinson, who provided "much helpful discussion during the course of the investigation."[48] The principal objective of his thesis was to develop the mixed initial and boundary value problem for a linear hyperbolic partial differential equation of arbitrary order in two independent variables. Campbell was also able to give some results on the mixed problem for a semilinear hyperbolic system of first order equations.[49]

Campbell succeeded in solving the linear, hyperbolic equation of order n (assuming that the unknown function and its first $n-1$ normal derivatives took specified values on a segment of the positive y-axis) by extending a method Robinson had already used to solve the initial value problem for an equation of general order.[50] The method basically re-

---

dissertation on file in the university archives lists Abraham Robinson along with G. N. Patterson as major advisors, and thanks them both in the "Acknowledgement" at the beginning of the thesis "for their guidance and counsel during the course of this project." See Fowell 1955.

[45] L.L. Campbell to JWD, letter of October 30, 1990, RP/YUL.

[46] Robinson 1945.

[47] L. L. Campbell to JWD, letter of October 30, 1990. See as well Campbell 1955.

[48] Campbell 1955, "Acknowledgments."

[49] Campbell's dissertation dealt only with real functions and variables. See Campbell 1955.

[50] Robinson 1950a. One widely used method for solving the initial value problem for an equation of arbitrary order was to solve an equivalent problem for a system of first order equations. This method was also used for solving ordinary differential equations, but introducing the various derivatives of the unknown function z as new variables does

duced to solving mixed problems for a succession of semilinear systems of first order equations. Campbell also solved a mixed problem for the linear hyperbolic equation of order n by using an extension of the Riemann method.

Wing Theory

Robinson's major effort at Toronto in applied mathematics (and the most comprehensive work he ever did in aerodynamics) was his book devoted to *Wing Theory*. This drew upon courses he had taught at Cranfield—and new material he was teaching at Toronto. What Robinson intended to provide was a comprehensive survey of the subject of airfoils in general, designed to serve as a textbook for advanced courses on the subject.[51] As one of Robinson's colleagues at Cranfield put it, "the book took some three years to prepare and it developed into an impressive work of comprehensive scholarship and authority."[52]

Cambridge University Press had imposed a deadline, and Robinson expected to meet it using his famous method:

> by completing three folio-sheets a day, in his own handwriting, during a period of several months. Since he was working so hard he did not allow himself much time for lunch and usually ate his sandwiches at his desk. Only seldom would he appear in the Faculty Union.[53]

The book opens with a two-line aphorism in Greek:

ἔργμασιν ἐν μεγάλοις
πᾶσιν ἁδεῖν χαλεπόν

not take advantage of any of the properties of the characteristic. Instead, the reduction in this case to a first-order system usually introduces characteristic curves which are *not* characteristic curves of the original equation. Robinson's method of reducing the problem to one of integrating a succession of first-order systems overcame both of these difficulties. The resulting first-order systems were in characteristic form and introduced no new characteristic curves. For further details, see Campbell 1955, pp. 93–94.

[51] Actually, it was also intended to reach those "in industry, at the universities, or at research establishments, who are interested in airfoil theory for either practical or theoretical reasons." The major mathematical prerequisites were a "sound knowledge" of calculus and the theory of functions of a complex variable. Those parts of general hydrodynamics needed for the development of airfoil theory were included in chapter 1. Robinson and Laurmann 1956, "Introduction," p. v.

[52] Young 1976, p. 311. According to royalty statements from Cambridge University Press, the book continued to sell throughout the 1960's. Robinson would periodically receive a check every few years for amounts between £10–20 (averaging $30–50). By 1970, sales had dwindled considerably; only one copy was sold—for a payment to Robinson of 4s. Figures based on royalty statements in Robinson's files, RP/YUL.

[53] Steketee R-1976.

Although no translation or source was provided, roughly this may be taken to mean, "in great adversity it is difficult to gain pleasure." Robinson may have intended to allude, in fact, to his experiences during World War II, and the fact that he had come to much of the material contained in the book through the trying years of his war research.

It was E. F. Relf, first principal of the College of Aeronautics, who invited Robinson to write the volume for the Cambridge Aeronautics series.[54] Robinson agreed, and he began to work on *Wing Theory* during his last few years at Cranfield. Only later did he decide to ask his former student from Cranfield, John Laurmann, to join him in the venture as coauthor:

> It so happened [Robinson] and I moved to Canada at about the same time, I to the National Research Council in Ottawa, he to the University of Toronto. It was there that he suggested that I help him with its authorship, and I was of course delighted to accept. Actually, the book was finished a little later when I was at the University of California in Berkeley, finishing my graduate work, and we had some rather complicated mailings back and forth in finalizing the book and in proof reading. I was the one doing the routine stuff in getting properly typed manuscript to the University Press.[55]

Wing Theory covers both subsonic and supersonic airfoil design, considered under conditions of both steady and unsteady flow, and all presented with "that heightened sense of structure and unity that was a characteristic of Abby's work."[56] As the authors explained in their introduction, the basic object of wing theory was:

> the investigation and calculation of the aerodynamic forces which act on a wing, or on a system of wings, following a prescribed motion in a fluid medium, usually air. The theory is based on the assumption that the medium is continuous, and it can be shown that this assumption does not lead to any appreciable errors except at very high speeds or at very low pressures.[57]

[54] The introduction, oddly enough, is dated Toronto, 1953 (December). See Robinson and Laurmann 1956, p. vi.

[55] John A. Laurmann to JWD, December 28, 1990, RP/YUL. Laurmann received his M.S. degree from Cranfield in 1951. He went on to earn a Ph.D. in engineering at the University of California, Berkeley, in 1958. Since then, he has worked as a research scientist or consultant to NASA, Lockheed (Missile and Space Division), the Institute for Defense Analysis, and the National Academy of Sciences, among other institutions. From 1976–1988 he was also senior research associate in the Department of Mechanical Engineering, Stanford University.

[56] Young 1976, p. 311. The book also permitted Robinson to correct some errors in previous publications, especially one formula containing an arithmetical mistake in Robinson 1948, noted in Ward 1952, p. 446.

[57] Robinson and Laurmann 1956, p. 1.

In writing *Wing Theory* Robinson was often torn between conflicting demands: mathematical rigor (involving at times the introduction of subtle "techniques which might be a heavy burden for those who are interested in the subject chiefly for practical reasons") versus presentation of the most efficient methods (as he admitted, for the solution of a particular problem, early and less efficient methods might provide better insight into the topic).

M. J. Lighthill, one of the world's authorities on the subject, regarded *Wing Theory* as "an admirable compendium of the mathematical theories of the aerodynamics of aerofoils and wings. Almost all the important results are referred to, even though there can be only a brief reference to literature in connection with the more difficult topics."[58] But he found the chapter on compressible flow (chapter 4, also the longest) troublesome due to "long stretches of the difficult Hadamard theory for three-dimensional flows being put before the simple two-dimensional theory." Even so, he was impressed by the "overall well-balanced nature" of the account.

Lighthill was more critical of the extent to which the book depended on linearized theory, which relied upon certain restrictive assumptions. Although the assumptions serve to simplify the mathematical analysis, they are not very satisfactory for actual physical applications (and this was especially true for the case of supersonic aerodynamics in compressible flow theory).[59] Although Robinson and Laurmann did try to cover parts of nonlinearized theory, Lighthill was also disappointed that the subject of two-dimensional supersonic airfoil theory was not given more attention. But after reading the book, his judgment remained high, and he regarded *Wing Theory* as "an invaluable introduction to wing aerodynamics for mathematically-minded students, as well as a solid stand-by for purposes of reference for all workers in this and allied fields."[60]

Thus Lighthill neatly sketched the work's major strengths and weaknesses: heavy on mathematics, short on data. This may come as something of a surprise, since Robinson's aerodynamic experience during the war was of a very concrete nature, and at Cranfield he was always interested in wind tunnel studies and the role models played in providing data against which to check his own theoretical work.

Indeed, as Robinson was well aware, models and wind tunnel experi-

[58] Lighthill 1957, p. 529.

[59] For transsonic speeds, the results proved by linearized theory are definitely not reliable. In fact, analysis of shock waves and the particular problem of shock wave boundary layer interaction are well beyond the scope of linearized theory, although these phenomena are very important in determining lift and drag in the transsonic range.

[60] Lighthill 1957, p. 529.

ments only approximated physical conditions in hopes of explaining actual free-flight conditions. There were enough complications in reconciling the two, but at high speeds various nondimensional parameters came into the picture. Two of the fundamentals of aerodynamics—the Navier-Stokes equations and the equation of continuity—provided the fundamental system of differential equations for the motion of an incompressible viscous fluid. But many complications arose in the exact theoretical description of this flow around a given particular body.

As Robinson's own research had shown, for fluids of relatively small viscosity, such as air, approximate theoretical methods were generally sufficient. When checked against experimental data, they usually permitted (with any necessary adjustments) a quantitative explanation and prediction of the behavior of phenomena occurring in a viscous fluid, which was essential for calculating the important aerodynamical problem of the drag of an airfoil.

The Robinson-Laurmann book on *Wing Theory*, however, may have reflected Robinson's deeper and surpassing interest in theory itself, especially the power he believed mathematics had to elucidate the fundamental features of physical nature. In fact, the aerodynamics of wing theory was unusual in the extent to which purely theoretical considerations led to applicable results:

> This branch of the physics of fluids had a remarkable short-term history as one of the few in which theory alone gave immediately engineering applicable results without the need for much empirical amendment. The book on wing theory that I wrote as junior author with Abby (plus I suppose books by others) in a sense represents the close of a chapter in the fulfillment of the theoreticians contribution in this domain—a very neat package. In fact it was this completion of a field of theoretical fluid dynamics that later led me to new fluid dynamic challenges in geophysical application, and a new world of fundamental difficulties in trying to understand atmospheric and oceanic flows.[61]

Thus it was perhaps inevitable that mathematics should dominate the development of *Wing Theory*. Robinson always looked for solutions that were mathematically pleasing, simple where possible yet suggestive or profound. This was evident as early as the research he did at Farnborough, where mathematical sophistication was a notable characteristic of his approach to even the most practical problems he was given. As his former colleague during the war, Alec Young, put it, "[Robinson] had an outstanding ability for understanding in physics

[61] J. A. Laurmann to JWD, December 28, 1990.

and the problems that he dealt with, so that he could apply mathematics to it in a way that was extraordinarily stimulating as well as elegant."[62]

This was all clearly evident in his work with Laurmann on *Wing Theory*. As one of Robinson's major contributions to aerodynamics, and certainly his most comprehensive account of the elements of wing theory, it remained characteristically "Robinsonian" in its deft handling of the mathematics above all else.

Robinson's Last Graduate Student in Applied Mathematics

Robinson's last graduate student at Toronto, R. A. Ross, did not actually finish his dissertation until after Robinson had left Toronto. Nevertheless, the subject of his thesis, "The Waves Produced by a Submarine Earthquake," was inspired by Robinson's own theoretical interests (as well as those of J. A. Jacobs in geophysics). Although Duff took over responsibility as thesis supervisor, Ross duly thanked Robinson at the beginning of his dissertation for having worked with him on the thesis and "suggesting improvements."[63]

Robinson had long been interested in the subject of elastic waves and later suggested the topic to Ross for his dissertation, thanks in part to the proximity of the geophysics group on St. George Street. Since the geophysics section of the physics department was directly next door (where the facilities were a bit more spacious and comfortable than they were at number 47), he sometimes went over for tea in the afternoon, where there was always the chance to talk with Jacobs.[64]

One afternoon, Duff found Robinson at work in his office, reading Lamb's paper of 1904 on the earthquake problem. "You know," said Robinson, "there is nothing in this paper so very deep or outstanding in itself. All the methods and ideas were well known. It is only when it has all been put together, from the beginning to the end, that you have something really fine."[65] This might just as well have been said about Robinson's own work, for he often took familiar subjects or results, but by approaching them in a novel way, with fresh combinations in mind, he would obtain far-reaching and sometimes fundamental conclusions.

As Ross began working on his dissertation, Robinson insisted on seeing him regularly. This was important not only to keep Robinson informed, but to make sure that Ross kept working at a regular pace. This meant that they saw each other at least once a week.

In the classroom, Ross had not found Robinson especially impressive as a lecturer:

[62] Young R-1976.
[63] Ross 1955.
[64] Ross I-1990.
[65] George F. D. Duff to JWD, June 8, 1994.

Alec Young, visiting the Robinsons in Toronto.

> He was well-organized, but even so he was hard to follow. Perhaps he assumed too much. There were also gaps in what he was doing, and you had to fill them in later yourself. For advanced students, or ones willing to do the extra work, this wasn't probably so bad, but it had obvious drawbacks for others.[66]

Usually, Robinson lectured from notes, "but sometimes it seemed as if he was trying to recall something he had done before, but then forgotten." Ross conjectures that Robinson, doing so many things at once, did not always recall exactly what he needed at a given point in the middle of a lecture.[67] When it came to his dissertation, however, Robinson proved an inspiring and encouraging advisor.

What Ross set out to consider mathematically was a challenging physical problem, namely, an underwater earthquake and its resulting effect on the ocean's surface. Representing the earthquake as a radial line source, the earth as a semi-infinite elastic solid, and the sea as a layer of incompressible fluid, he obtained an exact expression for the velocity potential on the surface of the incompressible fluid (under gravity). From this exact solution he went on to give an asymptotic solution valid for large values of time, and for the case in which the line impulse was located on the liquid-solid interface. Ross found that his analysis revealed the wave-like effects of the impulse consisting of disturbances on the fluid surface which were of two types: the one caused by pressure, shear, and Rayleigh waves; the other being (roughly) a gravity-type wave.

[66] Ross I-1990.

[67] This was a tradition in England, Ross points out, where one didn't pay too much attention to lectures. "Instead, you made it up as you went along," Ross I-1990.

Robinson's Last Papers at Toronto on Applied Mathematics

In 1956 Robinson published a paper on the motion of small particles in a potential field of flow, relevant to both the problem of aircraft icing as well as to silting up processes in rivers and estuaries. The approach taken most often to such problems was to compute trajectories of individual particles. Instead, Robinson considered a continuous distribution of small particles and the overall field of flow they would produce.[68]

Later that year he published another paper in which he considered a wave possessing a velocity discontinuity, a problem he approached for wave propagation in an elastic medium with variable properties.[69] In this paper he showed that longitudinal waves do not transform into transverse waves and vice versa. A year later he went on to examine the transient stresses in a beam of variable characteristics subjected to an implosive or concentrated load and investigated the associated variation in the discontinuity in bending moment that can occur across the front of a shear wave.[70]

By the mid-1950s, however, Robinson's interests in applied mathematics had clearly begun to wane. With the completion of his book with Laurmann on *Wing Theory*, he turned his creative energies almost exclusively to mathematical logic, where he was just beginning to develop exciting new ideas, especially in model theory. As his colleague at Toronto in applied mathematics, A. J. Coleman, describes one conversation he had with Robinson at about this time:

> It was in his office, we were seated on either side of his old desk (all desks in the Department of Mathematics were at least 20 and mostly 50 years old!). It was then that I glimpsed for the first time the extent and depth of

[68] Robinson used Stokes' law to determine the motion of an individual particle, assuming that the fluid exerted a drag force on the particle proportional to the vector difference between the velocity of the particle and the velocity of the fluid in question. See Robinson 1956. The paper also calculated the total mass of the particles striking a segment of the contour of an obstacle placed in the field of flow of an incompressible fluid.

[69] Robinson 1957. He noted that H. Jeffreys had already studied problems of reflections and refractions of waves between interfaces of homogeneous layers of an elastic medium, and the loss of energy (negligible) from transverse to longitudinal waves, or vice versa. Robinson approached the problem by a more direct method. Assuming the wave possessed a discontinuity across the wave front, he analyzed the system of ordinary differential equations satisfied by such a discontinuity along the rays. Noting that his conclusions were more precise than Jeffreys', Robinson concluded that it was impossible to create sharp longitudinal waves by a gradual transformation from a sharp transverse wave, and vice versa. Robinson suggested his conclusions might well apply to seismic shocks.

[70] Robinson 1957a.

> his knowledge of fluid mechanics, but I left the conversation with the clear impression that this topic was now of only peripheral interest for him. He was deep into logic . . . and I thought that I had perceived that he had become the purest of pure mathematicians who regarded his mastery of fluid dynamics as a dull necessity forced upon him by the exigency of war work. [But] his heart was now totally caught up with logic. . . .[71]

Colloque de Logique Mathématique: Paris 1952

In August of 1952 Robinson was back in Paris for the second Colloque de Logique Mathématique on "Scientific Applications of Mathematical Logic."[72] All of the sessions took place at the Institut Poincaré, where an opening *réception d'accueil* was held the night before the first sessions actually began.

Professor Albert Châtelet, dean of the Faculty of Sciences of the University of Paris, opened the meeting officially with a brief address at 9:30 on the morning of August 25. A short message of greeting was then read by E. W. Beth from J. Barkley Rosser, president of the Association for Symbolic Logic:

> During the past war it was impossible to maintain the international scientific collaboration which is so necessary for a vigorous development. Since the war the dislocations and suspicions resulting from the war have greatly hampered the resumption of widespread collaboration. We feel that a Colloquium such as the present one will provide a long awaited opportunity for scholars from different countries to associate together again and will provide a stimulus to cooperation and interchange of ideas. The University of Paris is thoroughly to be commended for its action in organizing this Colloquium, and we are happy to hear that it has had a widespread response from scholars of all nations.[73]

Robinson chaired an afternoon session featuring Evert Beth, who spoke on "La logique et le fondement des mathématiques." The following afternoon, Robinson delivered the second plenary lecture (in French) on "L'Application de la logique formelle aux mathématiques."[74] This

[71] A.J. Coleman to JWD, December 3, 1990, RP/YUL.

[72] The first colloquium held in 1950 considered the more conventional topic, "Les méthodes formelles en axiomatique." Robinson had also attended this first meeting, and presented a paper published in the proceedings, "Les rapports entre le calcul déductif et l'interprétation sémantique d'un système axiomatique." See Robinson 1950c, pp. 35–52. This paper is not reprinted in Robinson 1979.

[73] Rosser 1952, p. 14.

[74] Colloquium 1952, pp. 51–63. Robinson's colleague at Toronto, R. A. Ross, remembers overhearing Robinson from time to time speaking French on the phone in his office at 47 St. George Street, "always fluently," Ross I-1990. A brief report of the meeting may

basically continued themes already familiar to those who had heard (or read) his Cambridge Congress paper, or seen his book on the metamathematics of algebra. Once again he offered applications of the generalized completeness theorem to various branches of mathematics, including algebraically closed fields of characteristic 0, the metamathematics of ideals of polynomials (including a metamathematical proof about prime ideals and generic points), and a section on new developments concerning the "automorphism" of a metamathematical ideal and certain groups of such automorphisms.

At the end of his paper, Robinson made the interesting observation that a version of the generalized completeness theorem (his paper's theorem 2.2) implied the axiom of choice. This, however, was based on an error that Henkin spotted and to which Robinson himself referred several years later in his *Théorie métamathématique des idéaux*.[75] In 1956 Robert L. Vaught managed to prove that Robinson's assertion was correct if his theorem 2.2 were amended for a logic with equality (as Vaught noted in the case of Robinson's theorem 2.2, "the logic is that without equality.")[76]

In the discussion following Robinson's paper, George (Djuro) Kurepa responded to *la belle généralisation* of König's infinity lemma that Robinson presented in his paper (§5, on "Application to the Theory of Graphs"). Robinson had extended König's lemma by showing that it held for nondenumerable collections of disjoint (finite) sets. Kurepa noted that he himself had found a particular case of König's proposition which he gave in his thesis on "Ensembles ordonnés et ramifiés," written in Paris in 1935 and published by the University of Belgrade that same year.[77]

A number of the papers delivered at the meeting provoked lively discussion. In response to Robert Feys' paper, "Le probléme des applications de la logique formalisée," Robinson added an historical remark to Feys' quotation of Poincaré to the effect that "Peano has done some nice things." As Robinson pointed out, Poincaré was actually reacting to a statement by Couturat that symbolic logic had given "wings" to

---

be found in the *Journal of Symbolic Logic*, 18 (1953), pp. 95–96. A more detailed summary of the meeting, along with publication of the papers presented, is given in Colloquium 1952.

[75] Robinson 1955, p. 82, n. 28. Robinson had actually finished working on this monograph in 1952, but added the note in January of 1954, explaining the mistake Henkin had found and the changes he had been led to make as a result.

[76] Vaught 1956, pp. 262–263.

[77] Kurepa, in Colloquium 1952, p. 64. For discussion of Dénes König's infinity lemma, a proposition in graph theory that he later applied in topology (to the map-coloring problem) and game theory, see Moore 1980, p. 233. Its role in Gödel's completeness theorem is mentioned on p. 257.

mathematics. Poincaré objected, noting that it had been ten years since Peano had published the first edition of his *formulaire*, and yet, after all that time, symbolic logic still had not flown! Poincaré concluded that instead of giving it wings, logic had only put mathematics in chains!

However great a mathematician Poincaré may have been, Robinson emphasized that he was wrong about logic. What was important, Robinson said, was the way in which logic could be used to characterize certain classes of propositions in a formal manner, after which the value of logic then became clear through its applications. For example, although it was clearly false to say that every proposition that is true of the algebraic numbers is also true of the complex numbers, on the other hand one could say that all true propositions formulated in the first-order functional calculus about the field of algebraic numbers also held for the field of complex numbers. It was the ability of logic to distinguish certain classes of propositions that made it possible to establish such metamathematical theorems.

By specializing results of this kind with the help of model theory, Robinson explained that it was possible in turn to establish other mathematical theorems that were sometimes difficult to demonstrate by "classical" mathematical methods. About this approach in general, Robinson as always was optimistic, and he stressed one of his favorite themes: "it seems to me that one can affirm today that logic has already made positive contributions to mathematics, and that it will also lead to important applications in the future."[78]

Toronto: 1953

Robinson returned to Toronto from his summer in Europe full of ideas. Among the papers he worked on that fall was one coauthored with his colleague, the analyst G. G. Lorentz. Born in Leningrad in 1910, Lorentz had gone to Tübingen in 1944, and emigrated five years later to Canada where he joined the University of Toronto as a lecturer in 1949.[79] In 1953 he went on to Wayne State University, but before leaving Canada, he and Robinson worked together on a paper devoted to summability, a subject for which Lorentz was especially noted. They sent the first version of their article to the *Canadian Journal of Mathematics* in December of 1952.

The Robinson-Lorentz result dealt with regular matrices A and B such that each complex sequence has the core of its A-transform contained in the core of its B-transform. Although in general one could not hope

[78] Robinson 1952, pp. 18–19.

[79] Petersen 1957, pp. 4–5, and Lorentz 1957, pp. 12–16. See also G. Robinson 1979, p. 57 (although Robinson is off by a year in saying Lorentz came to Toronto in 1950).

for a regular matrix C such that $A = CB$ ($CB - A = 0$), b_{nk} 0, R obinson and Lorentz did find that there exists a regular matrix C, c_{nk} 0, such that $CB - A$ is a matrix D for which $\lim_{n\to\infty}\Sigma \mid d_{nk} \mid = 0$. The proof, involving convex sets in Banach spaces, was followed by applications to totally equivalent and core equivalent methods.[80]

Robinson was also invited to visit Cornell University in 1953, where he read a paper on "Applications of the Predicate Calculus to Algebraic Geometry."[81] At the end of the year he attended his first annual meeting of the Association for Symbolic Logic, held at the University of Rochester. He had just been elected to the association's Executive Committee for a three-year term, so the Rochester meeting was also the first in which he participated in an official capacity as well.[82] During the meeting he presented a paper "On Predicates and Algebraically Closed Fields," which subsequently was published in the *Journal for Symbolic Logic.*

In this paper, Robinson noted that many properties of curves, surfaces, and other varieties in algebraic geometry could be formulated in the lower functional calculus as predicates of coefficients of the polynomials defining the varieties in question. Since it was usual to study the properties of varieties in algebraically closed fields, Robinson pointed out how it was of interest to study the general structure of such predicates in terms of algebraically closed fields. The main result established in this paper was the following:

> Let F be a commutative algebraic field of arbitrary characteristic, and let $F' = F[x_1, \ldots, x_n]$ be the ring of polynomials of n variables with coefficients in F. With every predicate $Q(x_1, \ldots, x_n)$ which is formulated in the lower functional calculus in terms of the relations of equality, addition, and multiplication and (possibly) in terms of some of the elements of F, there can be associated an ascending chain of ideals in F',
>
> $$J_0 \subseteq J_1 \subseteq \cdots \subseteq J_{2k+1} \quad k \quad 0,$$
>
> such that
>
> $$V_Q = (V_0 - V_1) \cup (V_2 - V_3) \cup \ldots \cup (V_{2k} - V_{2k+1})$$
>
> for every extension F^* of F which is algebraically closed.[83]

[80] Lorentz and Robinson 1954. [81] Steketee R-1976.

[82] Robinson's election to the Executive Committee for a three-year term beginning January 1, 1953, was announced in the June issue of the *Journal of Symbolic Logic*, 18 (1953), p. 192.

[83] Robinson 1954a, p. 103; Robinson 1979, 1, p. 332. Note that the $V_0, \ldots, V_{2k+1}$ were the varieties of the ideals $J_1, \ldots, J_{2k+1}$ in the coordinate space S_n over F^*, and V_Q was the set of points of S_n satisfying Q. Robinson later improved upon this result in *Complete Theories*, Robinson 1956e. See below, pp. 223–227.

In simple cases, a purely algebraic proof could be given, but for the general theorem, Robinson found once again that "the use of Symbolic Logic is clearly indispensable." Even so, his proof was rather complicated and drew upon a series of preliminary lemmas. Using commutative field theory and the theory of polynomial ideals, he was able to provide a metamathematical proof of the theorem (without having to appeal to any elimination methods).[84]

International Congress of Mathematicians: Amsterdam 1954

The summer of 1954 was another busy one for Robinson, especially since much of it was spent in Europe and Israel where he lectured widely in connection with several international meetings, including the International Congress of Mathematicians held in Amsterdam during the first week of September. Prior to the Congress, Robinson stopped in London where he talked about "Core Consistency" at his alma mater, Birkbeck College.[85] Doubtless he knew that Cooke would be interested in the results he and Lorentz had just recently achieved, but not as yet published.

The Amsterdam Congress was another blockbuster, attracting more than 2,100 participants. Both opening and closing sessions were held in the famed Concertgebouw of Amsterdam, and among the many official receptions, concerts, and parties, one was held in the rooms of the Rijksmuseum. Another featured an evening at the Bellevue, where the Dutch calculating prodigy Wim Klein performed, along with the illusionist Driebeek. An added attraction was the exhibition of works by the local artist M. C. Escher.[86] The official Congress banquet included 1,500 guests, and was elegantly served in the Wintergardens at the Grand Hotel Krasnapolsky.

The University of Toronto sent two official delegates to the Amsterdam Congress, Coxeter and Robinson. Old friends like Fraenkel and Fekete were on hand from Israel, and logic was prominently represented during the Congress by Tarski, who gave an invited hour lecture on "Math-

[84] Robinson did acknowledge, thanks to a reviewer, that Tarski had provided an earlier, more elementary proof of Robinson's theorem. But this relied upon a metamathematical version of the classical theory of elimination in polynomial algebra. Robinson preferred the alternative means, drawing upon the theory of ideals. See Robinson 1954a, p. 104; Robinson 1979, 1, p. 333.

[85] Steketee R-1976.

[86] This was attended by many of the participants, including Donald Coxeter and his Dutch wife, Rien, who remained Escher's friends for the rest of his life. One result of this encounter was the fruitful collaboration of Coxeter and Escher, *M. C. Escher: Art and Science*. See Coxeter and Escher 1986.

ematics and Metamathematics."[87] Doubtless the Congress's most unusual touch was the "traffic system" designed to keep individual sessions running in parallel and on time: "In most lecture rooms there were traffic-lights operated by the chairman in order to keep the speakers to the time allotted. Yellow light meant: you can speak for another two minutes, red light meant: stop."[88]

The Program Committee had decided to sponsor three independent symposia, to run concurrently with the Congress, one of which was to be headed by Arend Heyting on "Mathematical Interpretation of Formal Systems." Robinson was among those specially invited to present a paper, to which was added the small inducement of "a modest compensation in your traveling expenses."[89]

Completeness without Tarski's Elimination Method

The Amsterdam Logic Symposium began on the last day of the Congress, and ran for two days. Robinson decided to speak on "Ordered Structures and Related Concepts," a topic inspired by certain problems raised by Tarski in 1948. Tarski's decision method was well known, whereby it was possible to decide whether or not a statement formulated in the restricted predicate calculus could be deduced from a set of axioms for a real-closed ordered field. As Tarski had shown, this set of axioms was complete.

Robinson had also established similar results on his own by 1950 in the course of writing his dissertation, which he described in his Cambridge Congress paper. Five years later, he was now able to offer a number of refinements. Instead of individual results, he introduced a general metamathematical principle on the basis of which the completeness of certain concepts, including that of a real-closed ordered field, could be deduced without Tarski's elimination procedure.

Tarski had used a generalization of Sturm's theorem and the method of quantifier elimination to show the completeness of a real-closed field, but Robinson established the same result with an entirely different approach based on Steinitz's field theory. Robinson again took the model-theoretic high road, and showed how a general metamathematical principle–model-completeness–could be used to establish the completeness of a general concept (like that of a real-closed ordered field), and

[87] See Congress 1954, vol. 1, p. 392. Tarski's lecture was not published in the Congress *Proceedings,* according to which "no manuscript of this lecture was available [for publication]."

[88] Congress 1954, 1, p. 134.

[89] J. F. Koksma to A. Robinson, February 16, 1954, Robinson's administration file, Department of Mathematics, University of Toronto.

how this could be done without recourse to a cumbersome elimination procedure:

> DEFINITION (MODEL-COMPLETENESS): *If K is a consistent set of statements of the first-order predicate calculus, K is model complete if for every model M of K and for every statement X (containing only relations and constants of M), either X or ~X is deducible from K ∪ N (N the complete diagram of M as given in Robinson's book,* Metamathematics of Algebra*).*

> THEOREM (INCOMPLETENESS): *K is model incomplete if and only if there exist models M and M' of K and a statement X (as above) such that M' contains M, X is satisfied by M' and not by M, and X has the form:*
>
> $$x = (\exists y) \ldots (\exists y_k)\ V(y_1, \ldots, y_k),$$
>
> *where V contains no quantifiers and is a conjunction of atomic predicates and negations of such predicates.*[90]

Among the applications Robinson was able to give, he showed that a set of axioms for the concept of an algebraically closed field is model-complete by proving that if a set of polynomial equations and inequalities is solvable in an algebraically closed extension of an algebraically closed field *M*, then it is already solvable in *M*. These results, and the metamathematical methods Robinson employed to establish them, were typical of the approach he often took in his research throughout the fifties. In fact, one of the main threads running through the purely mathematical research he did at Toronto was to show how *meta*mathematical methods could be used to replace methods Tarski had pioneered so effectively, especially elimination of quantifiers, to establish completeness.

ZÜRICH LOGIC SYMPOSIUM: 1954

After the Congress in Amsterdam, Robinson went to a second meeting, in Zürich, where the International Congress for Philosophy of Science included a special "Colloquium on the Foundations of Mathematics." Although no published record of these proceedings was produced, a draft manuscript among Robinson's papers provides a summary of the remarks he made. Asked to consider the future of symbolic logic (and whether it had reached a dead end where the philosophy of mathemat-

90 Robinson 1955a. The definition and statement of Robinson's "Main Theorem" as given above are slightly modified versions of those to be found in Robinson 1955a, pp. 51–52; Robinson 1979, 1, pp. 99–100.

ics was concerned), Robinson was, as ever, optimistic. But as he cautioned, in trying to make predictions "we are forced to rush in where even the great Hilbert failed."

Basically, Robinson adopted a formalist but pragmatic approach: "We require from our formal framework that it should enable us to decide whether any given mathematical argument is or is not acceptable, but we drop the quest for completeness and consistency." Logic would take a further step forward, he suggested, if it could only resolve the following question: "Is there a specific statement about positive integers, and formulated, preferably, in the first order predicate calculus, such that the truth of the former can be neither confirmed nor denied?"

While in Zürich, Robinson also gave a talk on "Hilbert's Irreducibility Problem" at the Eidgenossische Technische Hochschule, with another on "Completeness Problems in Algebra" for the Mathematical Society in Zürich. From Switzerland he and Renée went on to Jerusalem where he delivered four lectures on "The Metamathematics of Algebra" at the Hebrew University and another on "Mixed Problems for Partial Differential Equations" at the Haifa Institute of Technology.[91]

Embeddings and Model-Complete Systems: 1954

As soon as he was back in Canada, Robinson quickly worked out a brief paper in response to a conversation he had had with B. H. Neumann during the Amsterdam Congress. The result was a "Note on an Embedding Theorem for Algebraic Systems."[92] In 1954 Neumann had published "An Embedding Theorem for Algebraic Systems," and Robinson saw immediately how the basic principle Neumann had used was a direct consequence of the compactness theorem.[93] As Robinson showed, Neumann's results followed easily and directly using the predicate calculus—just one more example of the benefits to be gained in mathematics proper from an appreciation of mathematical logic.

At the end of the year the annual meeting of the Association for Symbolic Logic was held jointly with the American Mathematical Society and the Mathematical Association of America at the University of Pittsburgh. Robinson presented a paper, "On Predicates in Model-Complete Systems," in which he showed (after defining completeness and model-completeness in terms of diagrams) that the concepts of algebraically closed fields and real-closed ordered fields were both model-

[91] Steketee R-1976. [92] Robinson 1955b.

[93] Neumann 1954. Robinson in his article still referred to the compactness theorem as the "extended completeness theorem of the lower predicate calculus." See Robinson 1955b, p. 250; Robinson 1979, 1, p. 345.

complete.[94] He then proceeded to prove that for any model-complete set H, there was a predicate deducible from H that contained no quantifiers, and finished with various illustrations of the theorem.

Théorie Metamathématique Des Idéaux: 1955

It was Paulette Destouches-Février in Paris who persuaded Robinson to write a model-theoretic book on ideals in French for the Collection de logique mathématique.[95] In doing so, he joined such well-known mathematicians as Beth, Curry, Fréchet, Rosser, and Wang, among others, who had already contributed impressive volumes of their own to the series. Destouches-Février had heard Robinson's paper, "Les rapports entre le calcul déductif et l'interprétation sémantique d'un système axiomatique," during the first logic colloquium in Paris in 1950. As Beth said of Robinson's conclusions, "I have to underline the exquisite beauty and the extreme importance of Robinson's results which show in a most successful way the new tendencies in research on foundations reported by van Gramsdonck. I should observe that the results presented by Robinson in a metamathematical form could be transformed into an appropriately mathematical (form) by virtue of the logico-mathematical parallelism discovered by Tarski in 1930."[96]

To French readers, perhaps unfamiliar with the ideas in Robinson's *On the Metamathematics of Algebra*, his new book on ideals must have inspired considerable interest. But actually, for Robinson, it was a transitional work between his thesis and his next significant contribution to model theory, *Complete Theories*, which he had almost finished by the time *Théorie métamathématique des idéaux* appeared in 1955. In fact, the manuscript on the metamathematics of ideals was basically finished in 1952, and Robinson only made minor revisions thereafter.[97]

Although he borrowed considerable material from his thesis, Robinson characterized the new work as broader, deeper, more thorough, simplifying as well some of his earlier algebraic results.[98] Moreover, the book on ideals seems to have been written more with mathematicians in mind, for the austerity (and to mathematicians the unappealing use) of formal logic was much less in evidence. But Paul Halmos, in reviewing the

94 Summary given in the *Journal of Symbolic Logic*, 20 (1955), pp. 200–201. It was the ASL's eighteenth annual meeting, and was held on December 29, 1954.

95 Robinson 1955. According to Robinson's "Avant-Propos," the manuscript was actually finished in 1952.

96 Beth, quoted from remarks made during Colloquium 1950. See Colloquium 1950, p. 51.

97 Robinson 1955. "Avant-propos" was dated Toronto, July 1952, but see p. 82, where n. 28 was added in January of 1954.

98 Robinson 1955, p. 5.

book, complained that "In many outward respects the present work suffers in comparison with the thesis. The typography is bad, often cramped and hard to read, there is a misprint on almost every other page, and the exposition is inadequately motivated."[99] He also warned readers that Robinson's terminology was sometimes unusual (his use of "semi-compact" for "countably compact," for example). Halmos also found Robinson's use of "ideal" itself confusing, and thought the book would have benefited from a systematic use of an interpretation of ideals in terms of Boolean algebra. This, "together with a consistent adherence to the Boolean point of view, would have simplified much of the author's work." Stone's theory of Boolean algebras and their duals (i.e., totally disconnected compact Hausdorff spaces), for example, could then have been applied, and Halmos was surprised that neither in the book on ideals nor in Robinson's *On the Metamathematics of Algebra* was any use or mention made of Stone's work.

It was only at the end of the book on ideals that their role became apparent, in chapter 7 on "Applications" and in chapter 8 on "Transfer Theorems." A familiar application that Robinson again provided here was a model-theoretic version of Hilbert's *Nullstellensatz*: if $p, p_1, \ldots, p_k$ are polynomials in $x_1, \ldots, x_n$ over a commutative field , and if ev ery common zero of $p_1, \ldots, p_k$ in every extension of is a lso a zero of p, then some power of p belongs to the ideal generated by $p_1, \ldots, p_k$. Among the "transfer theorems" discussed in the last chapter, many were familiar from Robinson's earlier work (for example, if a proposition of the first-order functional calculus formulated in terms of addition, multiplication, and equality is true for every commutative field of characteristic 0, then it is true for every field of sufficiently high characteristic).

Ordered Fields and Definite Functions: 1955

Although Robinson's first results using elementary model theory applied to algebra can be traced back to his dissertation and the Congress paper of 1950, five years intervened before he began to publish results that were purely devoted to model theory. But from the late 1940s onward he had been developing a number of main ideas along these lines, especially the technique of diagrams. This method served to exploit the effectiveness of the compactness theorem as used by Henken and Robinson at about the same time in the late 1940s (and implicitly by Mal'cev somewhat earlier).

Robinson had been using both the compactness theorem and the related method of diagrams to produce a variety of results in field theory

[99] Halmos 1955, p. 279.

during the early 1950s, but his major success in applying model theory to algebra came in 1955 with an article "On Ordered Fields and Definite Functions." This was perhaps the most important paper Robinson wrote while he was at Toronto.[100] Published in the prestigious German journal, *Mathematische Annalen*, it was remarkable in part because:

> . . . the results forced one to look at the nascent model theory with a new seriousness. This paper gave a model theoretic proof of Hilbert's 17th problem: that a positive definite real rational function is a sum of squares of rational functions. This had been proved by Artin in 1927, but [Robinson's] new proof had the advantage of giving uniform effective bounds on the number of squares and their degrees (thus answering a question which Artin had raised). However, the main interest of the model theoretic proof lay in its extreme elegance and simplicity.[101]

The problem itself had classic roots. It went back to Emil Artin who had solved Hilbert's problem in terms of formally real fields in 1927.[102] Using model-completeness, Robinson was able to obtain several improvements and generalizations, including bounds on the number of squares and their degrees. At the end of the paper, he asked whether it might not be possible to prove his new results by more traditional methods, without recourse to model theory. His answer was negative. Although he had tried to do so, Robinson insisted that he could find no way to establish even minimal conditions without using the predicate calculus. This was indeed impressive, and a signal that model theory had significant advantages to offer, even to pure mathematics. Robinson's paper of 1955 was clearly another integral part in his overall strategy to show the power and generality of model theory in applications to algebra.

One reader who responded immediately to Robinson's paper on definite functions was Georg Kreisel, who was particularly impressed with what Robinson had done and took the trouble to tell him so. At the time Kreisel was at the Institute for Advanced Study, and wrote a detailed letter to Robinson from Princeton in January 1956, saying that he had read Robinson's paper on definite rational functions "with great interest."[103]

Kreisel explained that he, too, had succeeded in applying an exten-

[100] Robinson 1955c.

[101] Kochen 1976, p. 314. This was all reminiscent of what Robinson had said of Lamb's 1904 paper: "All the methods and ideas were well known. It is only when it has all been put together, from the beginning to the end, that you have something really fine." George F. D. Duff to JWD, June 8, 1994.

[102] For a survey of work on Hilbert's seventeenth problem, see Pfister 1976, pp. 483–489.

[103] Georg Kreisel to A. Robinson, January 25, 1956, RP/YUL.

sion of the ϵ-substitution method to Artin's original proof, but that his bounds were "awful":

$$\lambda(2, \alpha) \leq 2^{2^{2^{2^{2^{2^{\alpha}}}}}}.$$

Kreisel saw no way, however, of using Robinson's proof to get better bounds, and asked Robinson if he had thought about this? Kreisel seems to have answered this question himself, for several weeks later he wrote to Robinson again, noting that "On further study of your paper I find that the explicit determination of the bounds l and m (not only for coefficients in real-closed fields, but also, e.g., for the rationals), is very much simplified by your approach." Even so, trying to find an explicit determination of the bounds for arbitrary f seemed to Kreisel "very much like hard work" (but not so bad as when applied to Artin's original proof where the successive elimination of square roots was involved).[104]

Mathematical Logic at Toronto: A. H. Lightstone's Thesis

Although Robinson's teaching at Toronto had been exclusively devoted to introductory courses and applied mathematics (with the single exception of a course on foundations of mathematics in his final year), he nevertheless managed to attract the attention of a small group interested in logic.[105] Among his first students was A. H. Lightstone, who completed his dissertation with Robinson in 1955. Meanwhile, Paul Gilmore had come to Toronto as a postdoctoral student to work with Robinson, and Elias Zakon arrived from the Technion in Haifa to spend a year's sabbatical with Robinson during his last year in Canada.

Lightstone was born in Ottawa in 1926, received his B.S. from Carleton University in Ottawa and his M.A. at the University of New Brunswick, Canada. It was Robinson, of course, who suggested the subject of Lightstone's dissertation: "Contributions to the Theory of Quantification." In his preface, Lightstone was pleased to thank Robinson "for many stimulating hours spent together during the course of the investigation."[106]

The "investigation" was prompted by a noticeable asymmetry in the restricted predicate calculus, which as Lightstone observed, permitted only one kind of variable to be quantified. In the Hilbert-Ackermann

[104] Georg Kreisel to A. Robinson, February 7, 1956, RP/YUL. It should be noted that in a follow-up paper less than a year later, Robinson made a significant advance over Artin's paper as well as his own (both depended upon field extensions involving square roots). By introducing the idea of semimodule, Robinson found that he could avoid square roots entirely. See Robinson 1956a.

[105] Ross I-1990. Because he was already finished with his requirements, Ross did not take the course.

[106] Lightstone 1955.

version, there were three basic variables: sentential, individual, and predicate. Lightstone, in his thesis, introduced a variant calculus designed to include all of the various classes of variables on which the calculus was based. Consequently, the axioms and rules of formation (of well-formed formulae, WFF) did not have to distinguish between relations and individuals. However, a distinction was drawn between types of variables in determining WFFs obtained from a finite sequence of integers.

With slight variations, Lightstone's "Symmetric Predicate Calculus" (SPC) was applicable to the restricted predicate calculus, the extended predicate calculus, and even to the many-sorted calculus. Lightstone also showed that SPC was consistent and complete "by the usual methods." As an application of SPC, Lightstone considered the axioms of incidence in projective geometry, "with a view to proving the principle of duality in a formal fashion." In fact, Lightstone may well have been echoing Robinson in declaring that the "usual" proof of duality "takes no account of the underlying logical calculus."

The second part of Lightstone's thesis was devoted to "Syntactical Transforms." Using his SPC notation, he gave various sets of rules—"syntactical transforms" (mappings of WFFs into themselves; Skolem's normal form is an example of a syntactical transform). Lightstone's aim was the development of transforms under which provability was invariant. The thesis was especially interested in various kinds of continuity—continuity at zero, uniform continuity, and a weakened form of topological continuity at a point. Lightstone observed that the notation of SPC suggested yet another kind of transform which he called "duplication" (i.e., each variable or functor appearing in a formula was replaced by two symbols of the same type). Not only was provability invariant under this transform, but duplication was also useful in developing the transform for uniform continuity.

Lightstone brought his thesis to a close with a number of applications, all of which were very Robinsonian in character. Among these, he showed that if a WFF holds in every abelian group, then a transform of the formula (expressing the uniform continuity of any functor appearing in the formula) holds in every ordered, completely divisible, abelian group. For a system of equations with integral coefficients and with parameters from an algebraically closed integral domain of characteristic 0 with a solution in every such integral domain (including those without a unit element), the solution is continuous in the parameters at zero (i.e., the solution can be made arbitrarily small by choosing suitably small parameters).[107]

[107] See Lightstone and Robinson 1957. In a note at the beginning of the paper, the authors indicated that the results reported in the article constituted a part of Lightstone's thesis.

Hilbert's Irreducibility Theorem

In addition to his work with Lightstone, Robinson was also pleased to be working with Paul Gilmore, who came to Toronto in 1953 as a postdoctoral student to work with Robinson on mathematical logic. Gilmore was a Canadian who had graduated in 1949 from the University of British Columbia with honors in mathematics and physics after having served a three-year term in the Royal Canadian Air Force. As a graduate student at Cambridge University, on a two-year scholarship from St. John's College, he became interested in logic. After completing the work for his M.A. at St. John's in 1951, he went to Holland to continue his studies in logic with E. W. Beth and A. Heyting. Two years later he was awarded a doctorate in mathematics from the University of Amsterdam. It was in his second year, supported by a scholarship from the National Research Council of Canada, that he heard about the availability of NRC postdoctoral fellowships:

> After four years away I wanted to return to Canada. It was probably Professor Beth who suggested Abby's name to me. When I applied for the NRC fellowship, I was aware of his work in logic and algebra. My application for the fellowship stated that I wanted to use my knowledge of intuitionistic mathematics to further Professor Robinson's work. One of my first memories of Abby is his response to my application when we met in Toronto to discuss what I would do during my two year fellowship. He spoke with some amusement about how my proposed research was more suitable for the purposes of the application than for a serious program of research. Such honesty and openness made his compliments all the more valuable.[108]

From 1953 to 1955 Gilmore was a university research associate in mathematics at the University of Toronto. He later taught briefly at Pennsylvania State University before going on to the T. J. Watson Research Center in Yorktown Heights, where he was employed by IBM. After supervising the group working on combinatorial mathematics, he returned to Canada in 1977 to head the Department of Computer Science at the University of British Columbia.

Working together at Toronto, Gilmore and Robinson succeeded in providing an elegant model-theoretic proof of Hilbert's irreducibility theorem, namely, that if a polynomial $F(X_1, \ldots, X_n)$ in n variables over an algebraic number field K is irreducible, then there is an irreducible polynomial in $X_1, \ldots, X_m$ $(0 < m < n)$ obtained from $F(X_1, \ldots, X_n)$ by specializing $X_{m+1}, \ldots, X_n$ to a suitable set of values in K.[109]

[108] Paul C. Gilmore to JWD, letter of February 8, 1991, RP/YUL.
[109] Hilbert 1892, Gilmore and Robinson 1955.

Their joint result arose, not from Gilmore having read Robinson's earlier result presented at the CNRS Colloquium in Paris in 1950, but from another paper Robinson had given to Gilmore for his comments:

> Abby gave me a draft of a paper he had written and asked me to read it and suggest any corrections or improvements. I could not find any errors in the paper, but I was able to generalize his main theorem. When I described my generalization to Abby, he became very excited and enthusiastic. He rushed us to the library to consult *Math Reviews* and other references which would suggest applications of the extended theorem. Abby listed the papers he wanted me to read and suggested the results we could obtain, and the collaboration was started. He left the writing of the paper pretty much to me, although he suggested changes and improvements to my drafts.[110]

Drawing on earlier papers dealing with Hilbert's irreducibility theorem by K. Dörge and W. Franz, they extended a metamathematical result Robinson had already published (in the paper he presented at the first CNRS Colloquium in Paris in 1950), whereby not only Hilbert's irreducibility theorem followed, but also two other theorems: one related to results established by Dörge, the other a new result on irreducibility due to Gilmore and Robinson for fields K with a valuation j in an ordered field W.[111]

Again, the main significance of the Gilmore-Robinson paper was the way it showed how algebraic results of substantial and recognized content could be obtained from symbolic logic thanks to Robinson's new methods. The success of an "elegant model-theoretic proof of Hilbert's irreducibility theorem" was itself an important achievement. In Simon Kochen's words, "The possibility now seriously arose that model theory could be a new method of attack on purely algebraic questions. This hope has been amply borne out in a number of cases where the only proof of algebraic results to date is the model-theoretic one."[112]

Beth's Theorem: 1955

At the end of the year Robinson attended the annual meeting of the Association for Symbolic Logic at Boston University. Following a joint session with the American Philosophical Association in the morning, the ASL held a separate meeting of its own in the afternoon. Robinson presented a paper "On Beth's Test in the Theory of Definitions," the

[110] Gilmore to JWD, February 8, 1991, RP/YUL.
[111] For details, see Gilmore and Robinson 1955, p. 487; Robinson 1979, 1, p. 352.
[112] Kochen 1976, p. 314.

purpose of which was to demonstrate once again the ease and directness of model-theoretic arguments. Given $X = Y(F, G_1, \ldots, G_m)$ in the lower predicate calculus with atomic predicates $F = F(x_1, \ldots, x_n)$ and $G_1, \ldots, G_m$, he took X to define F *implicitly* (semantically) in terms of $G_1, \ldots, G_m$, if the equivalence

$$(x_1) \ldots (x_n)[F(x_1, \ldots, x_n) \equiv F'(x_1, \ldots, x_n)]$$

was deducible from the conjunction

$$Y(F, G_1, \ldots, G_m) \wedge Y(F', G_1, \ldots, G_m).$$

In this case, according to Beth's test, X also defines F *explicitly* (syntactically), that is, there is a formula $Q(x_1, \ldots, x_n)$ containing only $G_1, \ldots, G_m$ such that

$$(x_1) \ldots (x_n)[F(x_1, \ldots, x_n) \equiv Q(x_1, \ldots, x_n)]$$

is deducible from X.[113]

In order to obtain this result, Beth had originally used a complicated analysis of derivations in the style of Gentzen. Robinson bypassed this analysis with a model-theoretic lemma:

> *Let K be a complete set of sentences (of the lower predicate calculus) and let* X_1 *and* X_2 *be two sentences such that all individual constants and atomic predicates which are common to* X_1 *and* X_2 *occur also in K. Then if* X_1 / X_2 *is deducible from K, either* X_1 *or* X_2 *must be deducible from K.*[114]

Robinson was candid in admitting, however, that the proof of this lemma was not as straightforward as one might suppose. Nevertheless, Beth's theorem followed as a nearly direct consequence in only a few steps.

Robinson published a more complete version of his model-theoretic proof of Beth's theorem in a paper Arend Heyting communicated to the Dutch Academy of Sciences giving "A Result on Consistency and its Application to the Theory of Definition." This paper contains what has come to be well known as Robinson's "Consistency Theorem," which he used to establish the general consistency result from which Beth's theorem followed.

At about the same time, William Craig was also interested in proving Beth's theorem on definability, although his work on the subject was virtually independent of what Robinson had done:

[113] Beth 1953. He also discusses his theorem, as well as proofs by Robinson, Craig, and Lyndon in Beth 1959, p. 293.

[114] See the report of Robinson's paper in the *Journal of Symbolic Logic,* 21 (1956), pp. 220–221.

> Our exchanges about his joint-consistency theorem and my interpolation theorem were not very extended. Apparently each of us originally was mainly interested in providing a proof of Beth's theorem on definability. It took both of us some time, I believe, to realize that the respective theorems that we used in the proof of Beth's results were of intrinsic interest. (Lyndon's strengthening of the interpolation theorem to obtain results on closure under homomorphisms may have been the main factor). The equivalence of Robinson's theorem and mine, for first-order logic, which only then became an issue, also did not come out until some time after our papers had been published.[115]

The relation between Beth's theorem, Robinson's consistency theorem, and Craig's interpolation theorem are now standard topics in any introduction to metamathematics or model theory. In fact, the search for extensions satisfying one or another of these has been a driving force behind a good deal of work in abstract model theory.[116]

In bringing his paper on definability to a close, Robinson also presented an analogous (and simpler) argument to show that the two notions of semantic and syntactic definability coincide for the sentential calculus, along with a generalization of Beth's theorem for n predicates $F_1, \ldots, F_n$.[117]

Persistence and Completeness: 1956

Considering Robinson's accomplishments in 1955, it is hard to believe that he could have done even more in the following year. Yet he began the new year in January with a visit to the University of Montreal, where he presented two lectures meant to provide an "Aperçu metamathématique sur les nombres réels."[118] Later, he also published two papers that echoed ideas he had pioneered in 1951, inspired this time by a paper of Leon Henkin's.

In "Two Concepts from the Theory of Models," Henkin posed the problem of characterizing syntactically the statements X which are persistent with respect to a given set K (of sentences of the predicate calculus). The idea of persistence was due to Robinson, who introduced it in his first book, *On the Metamathematics of Algebra*. Henkin noted that "Robinson's theorem [on the finite persistence property] can be ex-

[115] William Craig to JWD, December 18, 1990, RP/YUL. See also Craig 1957 and Lyndon 1959.

[116] See, for example, Keisler 1971 and the articles by Makowski and Shelah 1979. See also 4.6, 18.4, and 19.1 in Barwise and Feferman 1985. I am indebted to William Craig for this bibliographic detail. Craig to JWD, June 14, 1994.

[117] Robinson 1956b.

[118] See item [41] in the "Bibliography" of Robinson's works in Robinson 1979, 1, p. 577.

pressed by saying that the class of abelian groups and the class of commutative fields both have the finite persistence property."[119] Henkin went on to determine what was required for a statement X to be absolutely persistent. Robinson, in response to Henkin's paper, was able to prove a more general theorem:

> *In order that the statement X be persistent with respect to the set K it is necessary and sufficient that $X \equiv Y$ be deducible from K for some existential statement Y (i.e., it is in prenex normal form and contains no universal quantifiers).*[120]

Robinson had submitted his note on Henkin's paper to the *Journal of Symbolic Logic* in April of 1955, but continued to think about the problem. In less than a year he had come up with a longer paper, "Completeness and Persistence in the Theory of Models," which was published in 1956. Here the concept of persistence was extended from its earlier meaning in terms of statements in the lower predicate calculus to include predicates as well. Robinson also added two new concepts of completeness: precompleteness and *n*-completeness. The paper itself took up various themes, all of them related to completeness, including 1-completeness and restricted completeness, as well as persistence. An entire section of the paper was devoted to tests for completeness, including various necessary and sufficient criteria. For example:

> THEOREM: *In order that a set of sentences K be model-complete, it is necessary and sufficient that for every model M of K, the set $K \cup N$ is 1-complete, where N is the diagram of M.*[121]

This theorem led directly, Robinson noted, to the model-completeness test he had presented at the Amsterdam Colloquium in 1954, but which he was now able to prove using very different methods.[122] All of this, in fact, looked ahead to the many examples he had systematically worked out as applications of the test for his next major publication, *Complete Theories*.

[119] Henkin 1956, p. 28.

[120] Robinson 1956c.

[121] Robinson 1956d, p. 24. Robinson gave several other necessary and sufficient conditions for K to be model-complete: namely, that for every model M of K, $K \cup N$ is precomplete (theorem 4.7), and that K be precomplete (theorem 4.8). Consequently, the notions of model-completeness and precompleteness are equivalent. But, alternatively, Robinson also gave examples to show that 1-completeness and precompleteness did not imply each other, and that completeness did not imply precompleteness. Finally, Robinson brought his paper to an end with a necessary and sufficient condition for model-completeness stated in terms of Henkin's related concept of G-completeness.

[122] Robinson 1955a.

Complete Theories: 1956

Robinson spent much of 1956 overseeing *Complete Theories* through press. This volume gathered together many of his recent results, most of which were less than twelve months old.[123] The new book was built around the idea of model-completeness which provided an ingenious approach for determining the conditions under which various theories were complete and decidable. He also gave a new proof of Tarski's classic result that the field of real numbers is decidable (a problem to which he returned again in 1971, after he had developed the concept of Robinson forcing).

Robinson sent *Complete Theories* to North-Holland in 1955, again for the Dutch series Studies in Logic and the Foundations of Mathematics.[124] Having heard some of his new ideas during the logic symposium held in conjunction with the International Congress the year before, the editors of the Studies series invited Robinson to elaborate, and he did.[125]

Upon publication, *Complete Theories* was hailed as "a milestone in the development of model theoretic algebra."[126] In addition to model-completeness, Robinson developed the related concept of model-completion (a model-completion of K is a theory $K' \subset K$ such that every model of K can be embedded in a model of K' and the union of K' with the diagram of any model of K is complete). The theory of algebraically closed fields, for example, is a model-completion of the theory of integral domains. Not all theories, however, are model-complete–neither the theory of differential fields nor that of formally real fields has a model-completion. But if a model-completion does exist for a given theory, then it is unique.[127]

One of the leitmotifs of Robinson's career was his continuing examination of the abstract properties of algebraically closed fields. Although various approaches may be taken to this subject (including the question

[123] Robinson 1956e, "Preface."

[124] Robinson signed the contract for *Complete Theories* to appear in North-Holland's series, Studies in Logic, on May 2, 1955, RP/YUL.

[125] The paper, "Ordered Structures and Related Concepts," was presented during the Symposium on the Mathematical Interpretation of Formal Systems, held in conjunction with the International Congress of Mathematicians meeting in Amsterdam in 1954, RP/YUL.

[126] J. Keisler, "Preface to the Second Edition," in Robinson 1977, p. vi.

[127] In 1969 Eli Bers (who Keisler has called Robinson's "fourpartite colleague") improved upon the concept of model-completion with the idea of model-companion. This is a more general concept in that it concerns a theory $K' \subset K$ such that every model of K can be embedded in a model of K' where K' is model-complete. This is an important generalization of Robinson's idea because (for example) the theories of formally real fields and differential fields do have model-companions (but not model-completions).

of categoricity in power of algebraically closed fields of fixed characteristic), Robinson's approach improved upon Tarski's method of quantifier elimination, which thanks to Robinson's generalization of model-completion could now be avoided in favor of subtler and more elegant tools.

The basic idea behind *Complete Theories* is comparable to an abstract version of Hilbert's *Nullstellensatz*, one of the most fundamental results in the theory of algebraically closed fields.[128] Once the nature of Robinson's insight is understood, the idea can be widely applied to a number of other important examples. What Robinson succeeded in doing was to show how a general theory could be developed that applied to all model-complete theories. In the course of doing so, he was able to give a short, elegant proof of the completeness of real-closed fields, as well as other completeness results that had been obtained previously using various other methods.

The book begins, after expected preliminaries, with a discussion of model-completeness (chapter 2).[129] After explaining the idea of persistence, Robinson introduced his well-known test for model-completion, namely the important theorem 2.3.1:

> *In order that a non-empty consistent set of statements K be model-complete, it is necessary and sufficient that for every pair of models of K, M and M', such that M' is an extension of M, any primitive statement Y which is defined in M can hold in M' only if it holds in M.*[130]

Robinson then considered various examples (in chapter 3) of model-complete groups and fields. He began with simple examples of certain types of abelian groups (e.g., completely divisible torsion-free abelian groups with at least two elements), and then went on to other examples, including algebraically closed fields, real-closed fields, fields with valuation, integral domains with valuation, and modules (groups with operators).

In the next chapter he took up the problem of finding conditions under which model-completeness entailed completeness. Here he introduced another new concept, that of a prime model, which led naturally to the "Prime Model Test":

> PRIME MODEL: *A structure M_0 is said to be a prime model of a set of statements K if M_0 is a model of K and if every model M of K con-*

[128] For further evaluation of this approach, see Keisler 1979, p. xxxiii.

[129] This chapter also introduces related notions of "partial completeness" and "partial model-completeness," along with corresponding versions of the model-completeness and prime model tests. R. L. Vaught has pointed out, however, that the only example Robinson gives here for partial completeness is wrong. See Vaught 1960, p. 176.

[130] Robinson 1956e, p. 16. A WFF Y is *primitive* if it is of the form $Y = [(\exists y_1) \ldots (\exists y_n) Z(y_1, \ldots, y_n)]$, where Z is a conjunction of atomic formulae or negations of same.

tains a partial structure M' (i.e. M is an extension of M'), such that M' is isomorphic to M_0. If K contains any constants then these shall correspond to themselves in the isomorphism.[131]

THE PRIME MODEL TEST: *Let K be a model-complete set of statements which possess a prime model M_0. Then K is complete.*[132]

The prime model test, combined with his results on model-completeness, provided, as Robinson said, "a very effective tool for the investigation of concrete cases."[133]

Robinson used the prime model test to establish the completeness of a wide variety of different groups and fields (in section 4.3 of *Complete Theories*). Although it may be fairly simple to show that K has a prime model, it is usually not so simple to show that it is model-complete. Here Robinson's success reflected his considerable facility, not only in his handling of metamathematical arguments, but also in his thorough understanding of the basic algebraic concepts involved.[134]

In chapter 5, a number of direct applications were given, beginning with an example Robinson had discussed at length with the topologist Beno Eckmann. Using topological methods, Eckmann had proved a variation of this theorem for polynomials with coefficients in the field of complex numbers:

5.1.1. THEOREM: *Let*

$$5.1.2.\quad \begin{cases} p_j(x_1,\ldots, x_n) = 0 \quad j = 1,\ldots, n,\ n \text{ odd}, \\ x_1 p_1(x_1,\ldots, x_n) + x_2 p_2(x_1,\ldots, x_n) + \ldots + x_n p_n(x_1,\ldots, x_n) = 0 \end{cases}$$

be a system of polynomial equations with coefficients in a field of characteristic zero. Then the system possesses at least one solution other than $x_1 = x_2 = \ldots = x_n = 0$ (within some extension of the field).[135]

[131] Robinson 1956e, p. 72. For example, the field of rational numbers is a prime model of the set of axioms K for the concept of a commutative field of characteristic zero. Note that no prime model exists if the characteristic of the field is not specified.

[132] Robinson 1956e, p. 74.

[133] Robinson 1956e, p. 74. Robinson also gave a corresponding test for partially complete sets.

[134] At the Cornell Summer Institute in 1957, Tarski and Vaught presented a paper later published in *Compositio Mathematica*, in which they also considered model-completeness. Their paper showed that the idea could be formulated conveniently in terms of arithmetical extensionality, and that the model-completeness of the theories of algebraically closed and of real closed fields could both be established, of course, by elimination of quantifiers, a persistent theme in Tarski's work.

[135] Eckmann 1942. See also Robinson 1956e, p. 90.

Basically, what Robinson did was to translate the general idea of this theorem into a statement X of the lower predicate calculus. Since the statement held for the field of complex numbers (as Eckmann had shown), Robinson could apply a result he had established earlier in *Complete Theories* (namely theorem 4.3.8, that the elementary theory of algebraically closed fields of specified characteristic is complete) to conclude that it must also hold for all other algebraically closed fields of characteristic 0. This proves Eckmann's theorem directly, since every field of characteristic 0 can be embedded in an algebraically closed field of characteristic 0.

This very elegant, simple example demonstrates nicely the power and great appeal of the new kind of results Robinson was able to achieve in *Complete Theories.* It also led him to the subject of transfer principles, which were to assume an increasingly important role in virtually all of his future work.

Robinson explained transfer principles as follows:

> By a *transfer principle* we mean a metamathematical theorem which asserts that any statement of a specified type which is true for one particular structure or class, is true also for some other structure or class of structures. Thus the proposition that a particular set of axioms K is complete may be expressed in the form of a transfer theorem since it amounts to the assertion that any statement which is defined in K and which holds in one particular model of K, holds also in all other models of K.[136]

This notion of transfer principles also harkened back to his thesis and other ideas put forward in *On the Metamathematics of Algebra*.[137] A good example was Eckmann's theorem. Robinson was able to prove this based upon the metamathematics of transferability: a result that holds in one case for an algebraically closed field of characteristic 0 must then hold in all such cases, including that for which Eckmann's theorem was formulated.

Robinson went on in this chapter to provide a stronger description for algebraic varieties than he had given on this subject in his earlier paper, "On Predicates in Algebraically Closed Fields."[138] But now he added the condition that dimension $V_i >$ dimension V_{i+1}. This was a significant improvement, and showed that the length of the chain of varieties in question cannot exceed $n + 2$ (i.e., $2k < n$).[139]

The final chapter of *Complete Theories* (chapter 6) was devoted to "syntactical transforms," a subject familiar from the paper Robinson had written with his graduate student Harold Lightstone on this same sub-

[136] Robinson 1956e, p. 92.

[137] See, for example, Robinson 1950 and Robinson 1951.

[138] See above, Robinson 1954a, pp. 208–209.

[139] Robinson 1956e, p. 106.

ject.[140] After defining syntactical transforms, Robinson explained how they permitted passage from a phrase like "for all *x* . . ." to "for all *x* which belong to *R* . . . ," or from "there exists an *x* . . . " to "there exists an *x* which belongs to *R* . . . " In mathematics, Robinson observed that "assertions on the *continuity* or *boundedness* of the solution of certain problems can be obtained from the assertion of the mere *existence* of a solution by this kind of modification."[141]

Details for the continuity transform were given in the paper with Lightstone (and indeed constituted a significant part of Lightstone's thesis). In *Complete Theories,* Robinson instead took up the syntactical transform concerned with *boundedness.* Among other examples, he then applied this to results of the following sort: If a statement *X* which asserts the existence of elements with a certain property holds in all completely divisible torsion-free abelian groups (with at least two elements), then a correlated statement asserting the existence of such elements as bounded functions of the parameters involved holds in all ordered groups of the same type.[142]

Complete Theories was an important book for Robinson—and for model theory. It was also a thorough embodiment of his own characteristic approach to the metamathematics of algebra. The book ably demonstrated how adept Robinson had become at taking concepts from algebra, interpreting them from a model-theoretic point of view, and then finding the significant generalizations—the important metamathematical properties of theories themselves that could be used, in turn, to shed new light on various parts of algebra. This was a method that Robinson had honed to perfection.

The Erdös-Gillman-Henriksen Problem: 1956

In addition to the publication of *Complete Theories,* Robinson also wrote a provocative two-page paper offering a solution to a question considered earlier by Erdös, Gillman, and Henriksen: "Is a nondenumerable real-closed field—in particular if it is nonarchimedean—characterized by its type of order as an ordered set?" Robinson answered this question negatively by specifying two such fields with no order-preserving isomorphism between them.[143] This paper is particularly noteworthy in light of nonstandard analysis, which Robinson would develop with such brilliance and success four years later during a year he spent at Princeton.

Robinson took up several interrelated problems in this paper, begin-

[140] Lightstone and Robinson 1957.
[141] Robinson 1956e, p. 108.
[142] Robinson 1956e, p. 125.
[143] Robinson 1956f.

ning with the fact that while any isomorphism of a real-closed field is order-preserving, the converse is false. Considering two independent real transcendentals t_1, t_2 over the rational field R, their real-closures $R(t_1)$ and $R(t_2)$ are similarly ordered, but not isomorphic. Erdös, Gillman, and Henriksen had established a partial converse, but left open the question for a nondenumerable real-closed field, especially in the nonarchimedean case.[144] Robinson was interested in this question, not only because, as he said, "of the distinguished parentage of the problem," but because it also underscored the importance of the Erdös-Gillman-Henriksen paper in which the existence of order-preserving isomorphisms was ensured under stronger conditions.

Robinson was able to show that the answer to this question about order type was indeed negative, and did so with a simple example.[145] Considering the field of fractional power series R_k^*, then R_k^* is real-closed, nonarchimedean, and nondenumerable. R_1^* and R_2^* are similarly ordered but not isomorphic. As Robinson showed, there can be no order-preserving isomorphism between R_1^* and R_2^*.

The Canadian Summer Research Institute: Kingston 1956

In the summer of 1956 Robinson was among a group of mathematicians taking part in the annual Canadian Summer Research Institute at Queens University, Kingston, Ontario.[146] At the end of June he and Tim Rooney, who were staying at the same rooming house, went out one evening together, to a nearby coffee shop. To Rooney's surprise, "he

[144] Erdös 1955.

[145] Robinson 1956f. Robinson, however, misstated what he was about to establish at the beginning of his paper, saying that he would "specify two algebraically isomorphic real-closed non-archimedean and nondenumerable ordered fields, such that no order-preserving isomorphism exists between the two fields," Robinson 1956f, p. 908; Robinson 1979, 1, p. 559. However, if any two real-closed fields are algebraically isomorphic, then the isomorphism necessarily preserves order (squares will map to squares, exactly the positive elements in a real-closed field). Salma Kuhlmann has taken up this question in her dissertation, producing a counterexample of her own to the Erdös-Gillman-Henriksen question, namely, by producing for every cardinal k, 2^k real-closed fields isomorphic as ordered sets but not as ordered fields, proposition 9.6 in Kuhlmann 1991. See also Kuhlmann and Alling 1994. I am indebted to Salma Kuhlmann for bringing Robinson's error in this matter to my attention during the recent meeting on "Nichtstandard Analysis und Anwendungen" held at the Mathematisches Forschungsinstitut, Oberwolfach/Walke, Germany, January 30–February 5 1994; Kuhlmann I-1994.

[146] The inspiration for such meetings had come from Ralph Jeffrey, head of mathematics at Queens University. Summer Institutes proved to be one of the Canadian Mathematical Society's "most successful projects," according to G. Robinson 1971, p. 490. The idea was to provide a stipend for everyone who participated, supporting three months of summer research at Queens University. The institutes, in fact, were so successful that they continued for some twenty-five years; Duff I-1990.

paid for my coffee, explaining that as of midnight he would be promoted to Professor—which means that it was June 30—and this was his way of celebrating."[147]

Robinson spent most of the summer working on several papers, one of which was later entitled, "Some Problems of Definability in the Lower Predicate Calculus."[148] Actually, this reflected more work on Hilbert's seventeenth problem, applying the idea of model-completeness to field extensions. Robinson was again interested in whether or not there was a uniform bound to the number of squares required to express a totally positive element of a finite algebraic extension of an ordered field as a sum of squares of elements of the extension. The answer Robinson gave was "yes," and in working out the details of his proof, he was led, as he later said, to a relativization of model-completeness.[149]

Another product of Robinson's concentration that summer was a joint paper written with his just-graduated student, Harold Lightstone. Together they worked on the metamathematics of algebraically closed fields, specifically, "On the Representation of Herbrand Functions in Algebraically Closed Fields."[150]

Elias Zakon

Back in Toronto, Robinson began a new collaboration with a colleague who had come to Canada specifically to study with him on mathematical logic. This proved to be the beginning of a long and productive mathematical friendship. Some months earlier, Elias Zakon had taken (as he admitted) the bold step of writing to Robinson:

> I have to apologize in advance for the liberty I am taking in writing this letter. I hope, however, that you will find it possible to consider my request favorably. I am a senior lecturer of mathematics at the "Technion," Israel Institute of Technology. My scientific interests are in the field of set theory, logic and foundations of mathematics. So far I have published several papers, mainly on transfinite numbers. Two of my results, modest as they are, are cited in H. Bachmann's book *Transfinite Zahlen,* Zürich, 1955,

[147] P. G. Rooney to JWD, March 28, 1991.
[148] Robinson 1957b.
[149] This idea had already been presented in his paper on ordered structures at the Amsterdam Symposium in 1954, and was included in his book on *Complete Theories*. See Robinson 1957b. Robinson acknowledged suggestions that Harold Lightstone had made to help improve the presentation. Moreover, just as the paper was about to appear, Robinson learned that Kreisel had published a notice of work he was doing on bounds of degrees of polynomials. Kreisel 1957. Robinson therefore added a reference to Kreisel's notes in his own paper. See Robinson 1957b, p. 395.
[150] Lightstone and Robinson 1957a.

> pp. 52 and 98, footnote (see also *Mathematical Reviews,* September 1953 and May 1954).
>
> In the next academic year I shall have my "sabbatical leave," and I intend to use it for advanced study and research in the field that interests me. The Institute for which I work ("The Technion") is willing to help me by allocating an appropriate stipend, so as to enable me to study abroad, at a university or institute where I could obtain proper facilities and scientific guidance. On my part, I should like very much to spend the period of my leave at Toronto, so as to have the opportunity and the privilege of your most competent guidance, and I should be very grateful to you for agreeing to supervise my work. I wish to add that it is not my intention to become a candidate for a degree, and I should prefer to study as a guest or a visiting research fellow. Of course, I understand that my intended research work will not place any financial obligations on the University. . . .[151]

At the time, Zakon was forty-eight, having studied at the Friedrich Wilhelm University in Berlin (1926–1930) and at the Stefan Batory University in Vilna, Poland (1930–1934). In 1950 he obtained a position at the Haifa Institute of Technology and by 1954 had been advanced to the position of senior lecturer of mathematics. Among his "recent" publications, Zakon mentioned two that had appeared in the *Mathematical Quarterly* (*Riveon Lematematika*) published in Jerusalem, namely, "Left Side Distributive Law of the Multiplication of Transfinite Numbers," and "On the Relation of 'Similarity' between Ordinal Numbers."[152] He had also published a monograph, "On Fractions of Ordinal Numbers,"[153] and a three-volume *Collection of Exercises in Higher Mathematics* that was used as a basic textbook in the engineering departments at the Technion.[154] Meanwhile, he was preparing a new textbook on calculus, also intended for use at various institutes of technology in Israel.[155]

Robinson was suitably impressed, and agreed to invite Zakon to Toronto. He even sent a memorandum to Pounder, urging that Zakon be offered "something better than the formal standing of 'occasional student' (e.g., the title of 'research fellow'). It is understood that this would not involve the University of Toronto in any financial obligation."[156]

Indeed, over the next decade and more, Robinson and Zakon worked

[151] E. Zakon to A. Robinson, May 14, 1956, in Robinson's administrative file, Department of Mathematics, University of Toronto.

[152] Zakon 1953 and 1954. [153] Zakon 1955. [154] Zakon 1954a.

[155] E. Zakon to A. Robinson, June 19, 1956, in Robinson's administrative file, Department of Mathematics, University of Toronto.

[156] A. Robinson to I. R. Pounder, July 1, 1956, in Robinson's administrative file, Department of Mathematics, University of Toronto.

together on several papers which they jointly published, the first on "Elementary Properties of Ordered Abelian Groups," and later, "A Set-theoretical Characterization of Enlargements."[157] Zakon, in fact, remained in Canada and soon found a position at Essex College, Assumption University of Windsor, Ontario.

An Offer from Israel

Late in 1956 Robinson received an enquiry from the Hebrew University, a tempting invitation to accept the chair in mathematics held by his former teacher, Abraham Fraenkel, at the Einstein Institute. The opportunity to return to Jerusalem, and above all to devote his energies full time to pure mathematics and logic, proved irresistible. On January 9, 1957, Robinson wrote to the chairman of mathematics at Toronto, I. R. Pounder, about his decision:

> Dear Professor Pounder,
>
> Several months ago I was asked by a senior member of the Faculty of Science at the Hebrew University, Jerusalem, whether I was interested in an appointment at that University. I replied that I intended to give serious consideration to any offer and, at the same time and subsequently, [have] kept you informed of the gradual development of this matter.
>
> I have now received an official letter from the President of the Hebrew University and have decided to accept his offer of a professorship in Mathematics. Accordingly, I herewith give notice that I wish to resign my position at the University of Toronto at the end of the present academic year.
>
> I may add that this has not been an easy decision for me. I regret the fact that I shall have to sever my connections with the University of Toronto, and I am also sorry that I have to impose on you the task of finding somebody to take my place. However, I have no doubt that in view of its great academic reputation the University of Toronto will have no difficulty in choosing a man who will carry out my duties as well as (or better than) I.[158]

Robinson's last semester at Toronto was a busy one, during the course of which he accepted invitations to participate in two major conferences, a summer institute in logic at Cornell University and the Sixth Mathematics Seminar organized by the Canadian Mathematical Congress (scheduled for August in Edmonton, Alberta). Robinson was asked to give a series of lectures on mathematical logic meant to introduce

[157] Zakon and Robinson 1960 and 1969.

[158] A. Robinson to I. R. Pounder, January 9, 1957, in Robinson's administrative file, Department of Mathematics, University of Toronto.

the subject in a serious way throughout the country. There could have been no more fitting end to the five years he had spent in Canada.

Before setting off for the summer, however, Robinson must have been surprised to receive a letter from Sidney Smith. Still president of the University of Toronto but about to depart for an official position in the Canadian government as secretary of state for External Affairs,[159] Smith wrote in June with the happy news that as of July 1, 1957, Robinson's salary would increase to eight thousand dollars annually.[160] Apparently, he knew nothing of the letter Robinson had sent to Pounder in January, resigning his position. Ironically, as Robinson later told R. A. Ross, had the salary increase come just a little sooner, he could never have *afforded* to leave Toronto.[161]

Summer Institute in Logic: Cornell 1957

The idea for an institute was first raised by Paul Halmos. In September of 1955 he took it upon himself to write to Edwin Hewitt, who was then chairman of the Summer Institutes Committee of the AMS. Halmos suggested that the time had come for an institute on mathematical logic, possibly as early as 1957:

> You may recall that in our conversation in Ann Arbor you mentioned two very reasonable necessary conditions that a subject must satisfy in order to be eligible for consideration by the committee. The first was that the subject be a live one, with something happening in it that would make an extended conference worth while, and the second was that it be not an obvious recipient of support from the many industrial and governmental sources that other subjects (such as statistics and partial differential equations) can tap nowadays. The various disciplines usually grouped together under the name of symbolic logic (foundation problems, set theory, recursive functions, applications to algebra, algebraic methodology in logic, etc.) certainly satisfy both these conditions.
>
> In regard to the non-availability of other support, I think little need be said. Although logic is one of the oldest subjects of mathematical interest and although I am convinced that its continued study is of tremendous mathematical value, the subject is not such as to capture the imagination

[159] When Smith left the university for this new position in the fall of 1957, Robinson sent a warm note of congratulations from Jerusalem. Smith replied, thanking Robinson for his congratulatory letter "on my assuming the portfolio for External Affairs," and hoping their paths would cross again soon. Sidney Smith to A. Robinson, Ottawa, October 15, 1957, RP/YUL.

[160] C. E. Higginbottom, bursar and secretary to the board of governors, University of Toronto, to A. Robinson letter of June 27, 1957, RP/YUL.

[161] Ross I-1990.

of an admiral of the navy or a tycoon of industry. In regard to liveliness, much can be said. Several books on logic have appeared in the last few years—e.g., Kleene, Rosenbloom, Rosser, and the entire Dutch series. Other books (not mere expositions) are in progress—e.g., Church and Myhill. The sudden blossoming of these books is just one visible sign of energetic activity; I am sure that an opportunity for an exchange of ideas over an extended period would bring to light many others.[162]

A few weeks later, Henkin and Tarski wrote jointly to Hewitt, agreeing that "we are very far from being in the position of those branches of mathematics which can hold forth hope for quick application to military and industrial problems." More to the point, they addressed the accelerating growth of the subject itself:

In addition to the many books which are appearing in this field there may be cited the recent appearance of two new journals devoted exclusively to mathematical logic—*Zeitschrift für Mathematische Logik und Grundlagen der Mathematik* and *Studia Logica*—as well as the increasing number of general mathematical journals which are publishing articles in this domain. If further evidence be needed we may mention the fact that during the past two years the Association for Symbolic Logic has held meetings in September to supplement its regular annual meetings (in December), while in Europe conferences in logic have been held last year (in Marburg) and this year (in Paris), drawing their personnel from several countries.

There is one further point which perhaps deserves particular mention. In part because much of the work in logic is published in special journals, there are some mathematicians who are not familiar with the many directions in which this field has recently developed. These mathematicians have the feeling that logic is concerned exclusively with those foundation problems which originally gave impetus to the subject; they feel that logic is isolated from the main body of mathematics, perhaps even classify it as principally philosophical in character.

Actually such judgments are quite mistaken. Mathematical logic has evolved quite far, and in many ways, from its original form. There is an increasing tendency for the subject to make contact with the other branches of mathematics, both as to subject and method. In fact we would go so far as to venture a prediction that through logical research there may emerge important unifying principles which will help to give coherence to a mathematics which sometimes seems in danger of becoming infinitely divisible.[163]

[162] Halmos to Edwin Hewitt, September 13, 1955, Tarski papers, 84/69c, Bancroft Library, University of California, Berkeley.

[163] Leon Henkin and Alfred Tarski to Edwin Hewitt, September 26, 1955, Tarski papers, 84/69c, Bancroft Library, University of California, Berkeley.

By the spring of 1956, Hewitt's committee had approved the idea for an institute, and serious planning began. It was decided that the meeting would be held the following summer at Cornell University, where the institute would be jointly sponsored by the American Mathematical Society and the National Science Foundation. A committee was formed to decide on speakers, and discussion began on the list of participants.[164] As letters started to circulate among Halmos, Kleene, Quine, Rosser, and Tarski in the spring of 1956, Robinson's name, surprisingly, did not come up.

It was again Paul Halmos who first mentioned this in a circular letter of May 28, 1956. "By the way," he wrote, "what about G. Birkhoff and (second afterthought) A. Robinson?"[165] Several weeks later, Rosser also mentioned Robinson in a letter he sent to everyone on June 18, in which he transmitted the following information based on a letter he had received from Maurice L'Abbé at the University of Montreal:

> L'Abbé said that it had turned out to be premature to do much in logic in Canada for the summer of 1957. He hopes that logic will be a major focus of the Seminar planned for 1959; to prepare for this it has been arranged for Abraham Robinson to give a series of lectures entitled "Introduction to Mathematical Logic" at the Seminar for 1957. This Seminar will be held at Edmonton, at the University of Alberta, from August 12 to the end of August.[166]

Twice within the space of a month, not only had Robinson's name been mentioned, but prominently as the figure chosen to introduce mathematical logic on a wider scale to Canada. Only a few days later, Tarski again wrote to everyone on the committee (June 22):

> I am somewhat surprised by the fact that the name of A. Robinson has been mentioned so far in but one letter, and then only as a "second afterthought." Personally, I appreciate his results and ideas very highly, and each of his publications which have appeared so far has further enhanced my appreciation. . . . Except for Henkin and Robinson, none of the logicians in [Julia Robinson's] age group who have [*sic*] been mentioned in our correspondence has made a more (or even equally) substantial contribution.[167]

164 The schedule, when finalized, ran for the entire month of July, from July 1 through August 2, 1957.

165 All of the following material regarding the Cornell Summer Institute for 1957 is drawn from memorandums and correspondence among the Tarski papers, 84/69c, folder 5.13, Bankroft Library, University of California, Berkeley.

166 Rosser to Halmos, Kleene, Quine and Tarski, June 18, 1956.

167 Alfred Tarski, June 22, 1956, letter in the Tarski papers, 84/69c, Bancroft Library, University of California, Berkeley.

Although none of this explains why Tarski failed to mention Robinson's name previously, he now gave Robinson full credit (and even admitted that he admired Robinson's work), saying that he "appreciated his results." This was all it took to get Robinson's name onto everyone's list.

The Cornell Meeting

As Yiannis Moschovakis has so aptly described the first AMS Summer Research Institute in Symbolic Logic, the "list of participants is an incredible who-will-be-who of young logicians and its proceedings are full of exciting new beginnings."[168]

All of the talks presented at the Cornell meeting were divided among four sections. The first was devoted to logic and computing machinery; the second to intuitionism, recursive functions, and related topics; the third to logic and algebra, model theory, and related topics; and the fourth to many-valued logics, combinatory logics, and set theory. Robinson contributed papers in three of these four divisions; he was scheduled to give his first lecture on July 10 (in section 3)—a paper on "Relative Model-Completeness and the Elimination of Quantifiers."[169]

Model-Completeness and Elimination of Quantifiers

Most early proofs of the decidability or completeness of mathematical theories were based on methods of eliminating quantifiers, as Tarski had done in his well-known results. The level of difficulty of this approach was directly related to the mathematical properties of the particular system in question. Under the best of circumstances, when there was an effective method for carrying out the elimination of quantifiers, one obtained a useful (if laborious) decision procedure, and in some cases a completeness proof as well. But as Robinson observed, "its realization may still require considerable ingenuity [as in Tarski's paper of 1948 which considered the case of real-closed ordered fields], so that in general it represents a suggestion, or piece of good advice, rather than a systematic method."[170]

In hopes of better understanding how the method of eliminating quantifiers worked, Robinson probed the conditions under which a given predicate could be reduced to a predicate free of quantifiers. Drawing

[168] Moschovakis 1988, p. 346.

[169] Robinson 1957c. Published by the same title in a more extensive version as Robinson 1958.

[170] Robinson 1957c and 1958. For the relevant work by Tarski to which Robinson referred, see Tarski 1948.

on the first Hilbert-Bernays *e*-theorem, Robinson sought to "link up," as he put it, the elimination method with model-completeness. First he explained relative model-completeness drawing upon "Some Problems of Definability in the Lower Predicate Calculus."[171] When applied to the theory of real-closed fields (for which an elimination method had already been worked out by Tarski),[172] this method could be used to avoid complications like square roots and equations of odd (and possibly high) degree which otherwise arose in trying to eliminate existential quantifiers (at least in the approach Tarski took). Relative model-completeness made it possible to avoid such complications, and enabled Robinson to show the existence of elimination procedures for a wide class of theories, including 1) integral domains and algebraically closed fields, and 2) commutative ordered rings and real-closed ordered fields.

Unfortunately, Robinson was only able to establish the *existence* of elimination methods, but this did not in turn provide "an effective elimination procedure for the general case." Nevertheless, he had been able to establish that the completeness of a mathematical theory entailed the existence of an elimination method.

Metamathematics of Field Theory

Two weeks later, on July 29, Robinson spoke again, this time on "Applications to Field Theory."[173] Explaining a metamathematical result, that the theory of algebraically closed fields of specified characteristic is complete, he indicated how this might be taken as a precise formulation of the "Lefschetz principle"—that any theorem (of a certain type) which holds in the field of complex numbers holds in any other algebraically closed field of characteristic 0.

As an example of how one might reduce a given nonelementary concept to an equivalent elementary concept, Robinson considered the problem of expressing certain properties of an ideal J in a polynomial ring as a property of the coefficients of a basis for J. He had already considered a related problem in his *Théorie métamathématique des idéaux*, namely, the question of providing bounds for the index r and the degrees of the polynomials $q_i(x_1, \ldots, x_n)$ whose existence was asserted by Hilbert's *Nullstellensatz*:

> *If $p(x_1, \ldots, x_n)$ vanish for all joint zeros of the polynomials $p_i(x_1, \ldots, x_n)$, then for some r and $q_i(x_1, \ldots, x_n)$, $p^\rho = \quad q_i p_i$.*

[171] Robinson 1957b.
[172] Tarski 1948.
[173] Robinson 1957d; not published in Robinson 1979.

In his book on ideals, Robinson proved this theorem first for algebraically closed K and then for general K. Again, he mentioned Kreisel's success in showing that the bounds in question could be determined as primitive recursive functions of n and the degrees of the p_i [see above, pp. 213–214 and p. 216].

Robinson also wanted to show in his Cornell paper how familiar results from the lower predicate calculus might be extended to languages with various nonelementary features. Suppose, for example, one wants to make a statement about polynomials with coefficients in a given field K. Phrases like "for all polynomials" or "there exist polynomials" involve quantifiers and, in turn, infinite conjunctions and disjunctions expressing quantified statements. A number of interesting problems thus arose concerning rings of polynomials and formal power series which Robinson left as open questions.

For example, given two elementarily equivalent fields K_1 and K_2, were the corresponding rings of polynomials $K_1[t]$ and $K_2[t]$ or the rings of formal power series $K_1\{t\}$ and $K_2\{t\}$ also elementarily equivalent? Robinson noted that if the polynomial rings were elementarily equivalent, then the same would follow for the fields of rational functions $K_1(t)$ and $K_2(t)$.

Theorem Proving: Man Versus Machine

Two days later, Robinson changed gears again, and gave his third and last lecture at the Cornell Symposium, this time in section one on "Proving a Theorem (As Done by Man, Logician, or Machine)."[174] This paper seems to have been inspired by one given earlier in the institute by Herbert Gelernter (then at IBM).[175] Robinson had been talking about machine proofs with Beth, Gilmore, and Kreisel during the conference, and had some constructive suggestions to make of his own. Basically, Robinson drew a distinction between a theorem-proving machine and how a logician or mathematician might approach the same problem—differently.

First of all, Robinson was critical of books on the strategy of mathematical proofs because he found that they usually overlooked mathematical logic entirely. Convinced that logic had much to contribute to heuristics, and in response to those who saw logic as simply a formal tool, he posed a straightforward question: "Can Mathematical Logic do

[174] Robinson 1957e; not published in Robinson 1979.

[175] Gelernter had spoken on July 19 about "Theorem Proving by Machine," Gelernter 1957. In his paper he described a program designed to enable a machine to prove theorems of elementary geometry with the help of some "built-in 'artificial intelligence,'" rather than by an automatic decision procedure.

more than provide a notation for the detailed formulation of a proof on a computer?"[176]

Confident that the answer to this question was "yes," Robinson used an approach inspired by Herbrand to indicate why.[177] Assuming that a proof devised by a logician for some proposition *A* might well be unnecessarily long and quite beyond the capacity of a then-contemporary computer, "the question therefore arises how to replace the systematic method . . . by a less systematic method which, however, may give us a reasonable chance of finding a relatively short sub-conjunction of *A* which is contradictory":

> Now some reflection shows that this is precisely what Man (i.e. the working mathematician) has been doing intuitively for a long time when trying to prove a mathematical theorem. For example, to prove that the sum of two sides of a triangle is greater than the third side, $AB + BC \geq AC$, we construct a single auxiliary point C' (on AB produced, so that $BC' = BC$) and we then make use of a previously established theorem. Now this point C' may be regarded as a Herbrand functor of the initial A, B, C and hence . . . we are in a position to write down a suitable sub-conjunction A′.[178]

Without wishing to exclude other possibilities, Robinson suggested that an efficient theorem-proving program might be based on Herbrand's procedure rather than on the standard predicate calculus.[179] The computer would have to be instructed to select small numbers of elements to produce sub-conjunctions. These in turn would suggest new patterns for the construction of sub-conjunctions until the desired contradictory sub-conjunction was found.[180]

Although the Cornell Symposium was serious business and a great suc-

[176] Robinson 1957e, p. 350.

[177] Herbrand 1930. This was Herbrand's Ph.D. Thesis (Univesity of Paris), chapter 5 of which was translated as "Investigations in Proof Theory: The Properties of True Propositions," in van Heijenoort 1967, pp. 525–581.

[178] Robinson 1957e, p. 351.

[179] P. C. Gilmore did exactly this a few years later, in his article, "A Proof Method for Quantification Theory: Its Justification and Realization," Gilmore 1960. The basic idea involves examining the negation of a valid formula, which, if inconsistent, will be detected. But during execution of the program, the multiplication method used by Gilmore produces very large numbers of conjunctions. Even so, a program written for the IBM 7094 "data procesing machine" was able to produce proofs for "moderately complicated sentences" in less than two minutes. Martin Davis and Hilary Putnam succeeded in cutting down the number of ground clauses that Gilmore's method had employed. Even so, most valid formulas in first-order logic resisted proofs by computer in a reasonably brief period of time. This was eventually solved by a major breakthrough made in 1965 by J. A. Robinson, who introduced a "resolution principle." For further refinements since Robinson's work, see the discussion in Chang and Lee 1973, esp. pp. 45–66.

[180] Robinson 1957e, pp. 351–352.

cess, it was not all work and no play. There were informal parties and outdoor picnics. For the Robinsons, it was doubtless a nostalgic time as well, knowing they were about to leave North America. The Cornell meeting made it possible for them to see a number of colleagues and old friends, many of whom they might not see again for some time. A good example of the congenial mood of the meeting was the closing "Recessional" composed for the end of the symposium by "P.R.H. Anonymous":

> If you think that your paper is vacuous,
> Use the first-order functional calculus.
> It then becomes logic,
> And, as if by magic,
> The obvious is hailed as miraculous.[181]

This certainly justified the claim Halmos made to his colleagues as they first began to work on the logic colloquium that he was a logician *humoris causa*. He also noted that "Every subject has its lunatic fringe; logic has, perhaps, more of it than many other subjects."[182]

The Edmonton Summer School: 1957

Abby and Renée arrived in Edmonton in time for the fifteen lectures Robinson had promised to give as an introductory course on mathematical logic for the Sixth Mathematics Seminar sponsored by the Canadian Mathematical Congress (August 12–31, 1957). The success of the meeting, which was held at the University of Alberta in conjunction with the first seminar of the Canadian Association of Physicists (the Theoretical Physics Division), was reflected in the number of participants—approximately 190 (including wives and children). Almost everyone was housed in student dormitories.

The seminar opened officially on Monday afternoon, August 12, with welcoming remarks from the president of the university, the mayor of Edmonton and the Alberta minister of education. In addition to conferring honorary degrees upon Professors H.S.M. Coxeter and Eugene Wigner, a garden party "in glorious weather" was held on the lawn of the president's home on Saturday afternoon, where "the colorful saris of the Indian ladies present provided a picturesque touch."[183] In addi-

[181] Paul Halmos, in Cornell 1957, 3, p. 432. In "Errata," published in vol. 3 of the Cornell *Proceedings*. It was noted that in line 1, "vacuous" should have been replaced by "ridiculous."

[182] Halmos, September 18 and December 4, 1956, Tarski papers, 84/69c, Bancroft Library, University of California, Berkeley.

[183] "The Canadian Mathematical Congress Seminar in Edmonton," *Canadian Mathematical Bulletin*, 1 (1958), p. 62.

Participants at the Canadian Mathematical Congress, August 1957.

tion to other social gatherings, receptions, and dinners, the staff of the Rutherford Library also prepared "an exhibit of old books, pictures and other items of historical and mathematical interest."

Due to the relatively short amount of time at his disposal for an introductory course on mathematical logic, Robinson limited his lectures primarily to the propositional and predicate calculi (first order). The text he used followed basically this same plan, namely the Hilbert-Ackermann *Principles of Mathematical Logic* (1950). Theorems were proven selectively, but with enough detail to give everyone a sense of what the methods and approach were like. When Robinson reached the end of his course, with the Löwenheim-Skolem theorem, he admitted that he had only reached "the end of the beginning."[184]

Of the many advances mathematical logic had made in the previous decade alone, Robinson mentioned higher order predicate calculi (in which relations can be quantified and may appear as arguments of other relations), questions of constructivity, algorithms and decision methods, recursive functions and predicates (closely related, he stressed, to the theory of computers), and many-valued logics. Finally, "in addition to the pursuit of various ramifications at the top, we may wish to investigate more closely the logical and philosophical foundations of the entire subject."[185]

Preparing to Leave Toronto

After their summer of institutes, seminars, summer schools, and travels, the Robinsons returned briefly to Toronto, to pack and say their

[184] Robinson 1958a, p. 41.

[185] Robinson 1958a, p. 42.

Farewell party with students and friends from the faculty, at the Robinson's home in Toronto, Summer 1957.

Good-byes to close friends, including the Coxeters and Luxemburgs.

good-byes. Renée recalls that just before they left, Dean Beatty said to her, "This isn't one man, but two." This was doubtless more than a reference to the long hours Robinson spent in his office and the steady, regular, tenacious manner in which he worked. Indeed, while he had been at Toronto, Robinson had excelled in two important areas: both applied and pure mathematics. The department gave a big going-away party for them at which Beatty, genuinely sorry to lose Robinson, made a "wonderful speech."[186]

Of the many expressions of gratitude Robinson received as he was about to leave Toronto, one he always kept was a warm letter from Mary Cooper, corresponding secretary of the university's organization for Friendly Relations with Overseas Students (FROS). This group's mission was to welcome foreign students to Canada and (according to its letterhead) "provide opportunities for mutual understanding and appreciation."[187] Robinson was an active member of the Toronto Committee, and under its chairman Professor James Ham (later, the president of the University of Toronto), served several terms as vice chairman of the Executive Committee. As Cooper wrote:

> At our Annual Meeting it was the unanimous wish of the Committee that I should express to you our warm appreciation of your interest and sup-

[186] Renée Robinson I-1980.

[187] Cook 1961. In 1965 FROS changed its name to International Student Centre of the University of Toronto. See the *Varsity* (student newspaper, Univerity of Toronto), July 1, 1965.

> port while you have served on the Toronto Community Committee of F.R.O.S.
>
> We would like to send warmest good wishes to you and Mrs. Robinson as you take up your new post in Jerusalem. We shall miss you in Toronto, but shall always count you among the good friends of F.R.O.S. wherever you are.[188]

Among the photographs the Robinsons took after five happy and productive years in Toronto were several prompted by a few last parties or gatherings of friends. One was of a small group of students, including Vichien Verapanish and Nizam Ahmed Qazi, who gave Robinson a beautiful wooden box, dated August 6, 1957, with greetings carved on the lid in both Hindi and Arabic.

Two of Robinson's colleagues, Wim Luxemburg and Donald Coxeter, were also invited, along with their wives, for a final farewell before Abby and Renée sailed for Europe. The picture taken on that occasion provides a fitting group portrait, a reminder that as Robinson prepared to leave Toronto, the decision to do so had not been easy. In giving up his position at the university, he was giving up more than a job, indeed congenial colleagues and good friends as well. What awaited them in Israel, a country in a state of war with most of its neighbors following the Suez Crisis of 1956, was anybody's guess.

[188] Mary Cooper to Abraham Robinson, May 27, 1957, RP/YUL.

CHAPTER SEVEN

The Hebrew University: Jerusalem 1957–1962

> The Hebrew University may fill the part in the restored Jewish nation, which the Temple filled in the Second Commonwealth of the Maccabees. . . . There is something in the atmosphere of Jerusalem which makes man dream, and see visions of a better and peaceful humanity.
>
> —*Norman Bentwich*[1]

Israel: The Political Background

The Israel of 1957 was not the Palestine that Robinson had known as a boy in the 1930s. The crucial moment for Eretz-Israel had come in November of 1947, when the United Nations adopted its historic resolution in favor of establishing a Jewish State in Palestine. Only a few months later, on May 14, 1948, the British gave up control and withdrew from the country.[2]

Meanwhile, without the British, full-scale conflict erupted between Jews and Arabs. The War of Independence that began on May 15 found the new country in a seemingly defenseless position, threatened by five hostile armies converging on Israel—the Lebanese from the north, the Syrians from the northeast, the Iraqis and Transjordanians from the east, and the Egyptians from the south. The army of Transjordan attacked Jerusalem directly on May 15, and the key sector of Sheikh Jarrah was captured shortly thereafter. This in turn cut off Mount Scopus, including the Hebrew University and the Hadassah Hospital, from the rest of the modern Jewish city.

With a combination of "tenacity, boldness, improvisation, and luck," the Israelis managed to withstand the initial phase of this assault, and were successful in blocking the main thrust of the Arab invasion.[3] In less than a month, the United Nations Security Council called for a four-week truce that began on June 11. Although the Egyptians in the south had cut off the Negev from Israel, their advance on Tel Aviv had

[1] Bentwich 1961, p. 164.

[2] The Hebrew University 1969, p. 7. The United Nations resolution actually established two states, one Jewish and one Palestinian Arab, as Moshé Machover stressed in a letter to me of October 22, 1992.

[3] Safran 1981.

failed. Despite a second outbreak of hostilities on July 9, another cease-fire ten days later led to several months of negotiations. By the time fighting began again in mid-October, only the Egyptians and Israelis were involved.

The cease-fires, in fact, were not observed. The Israelis feared that the United Nations was prepared to surrender both Jerusalem and the Negev to Transjordan. By concentrating their forces largely in the south, the Israelis were eventually able to drive the Egyptians from the Negev and into the Sinai. When the British demanded that the Israelis withdraw, the Egyptians agreed to an armistice. Israel, too, decided that the moment had come to end the war—largely because signing the documents for peace would ratify Israel's sovereignty, acknowledging that it was indeed an independent state. Over the next few months Israel agreed to separate treaties with each of its Arab neighbors, thereby securing the legitimacy it had long anticipated.[4] But for this it also paid a heavy price, and not only in the lives lost during the conflict itself.

While the War of Independence assured the survival of Israel, it in turn created Jordan out of Transjordan, and left Egypt in control of the Gaza Strip. Jerusalem, instead of being internationalized as the UN had hoped, was instead divided between Jordan and Israel, with the Hebrew University and all of Mount Scopus cut off from the Israeli part of the city. But there was also a very strong, positive effect of the War of Independence that was largely psychological: "The experience of the war indelibly stamped a sense of unity and common destiny on the psychic fiber of all those who partook of it; and by the time it had reached its last stages, virtually every corner of the country had experienced it."[5]

Jerusalem, however, was another matter. The Jewish quarter of the Old City had been destroyed and its residents driven out.[6] With the Arabs in control of Sheikh Jarrah, there was only one road connecting the Jewish part of the city with Mount Scopus, but this was perilous territory:

> The attacks on the road by Arab guerrilla fighters and Arab Legionnaires made communicating between the Hebrew University and Jewish Jerusalem impossible. On the other hand, the stout defenses maintained by the University's students and the Haganah, foiled every attempt by the Arabs to capture Mount Scopus. Arab bombings, however, threatened to reduce the magnificent University buildings to rubble, destroying priceless equipment and the library. To avert destruction and loss of life, both Arabs and Jews agreed to demilitarize the whole area under the United Nations.[7]

[4] Safran 1981, p. 60.

[5] Safran 1981, p. 62.

[6] Bentwich 1961, p. 124.

[7] The Hebrew University 1955.

As the *Bulletin of the American Friends of the Hebrew University* put it in 1955, one of the great sacrifices Israel had to make for its independence was the loss of the buildings of its only university. More than ever, Israel needed trained professionals in every field, and the university was called upon to provide them under the most difficult of circumstances.

At first, despite its isolation, some faculty and students nevertheless tenaciously made their way once every two weeks up to Mount Scopus. But vehicles on the single road through Sheikh Jarrah were easy targets for snipers and bomb throwers. Even before the War of Independence, in fact, on April 13, 1948, an unusually large convoy was on its way to the university and the Hadassah Hospital when an Arab attack killed seventy-seven in the group, including some of the university's most outstanding scholars and scientists.[8]

That summer Mount Scopus was officially placed under United Nations' supervision. A year later, among the terms to which Israel and Jordan agreed in signing their armistice of April 3, 1949, was a provision that called for "the resumption of the normal functioning of the cultural and humanitarian institutions on Mount Scopus and free access thereto."[9] Jordan, however, permitted neither free access nor the normal functioning of the university or other facilities. Instead, until the Six-Day War which would reunite all of Jerusalem in June of 1967, Israeli access to Mount Scopus was limited to only a small number of police and civilian caretakers. Every two weeks the armoured convoy was permitted to take provisions and a change of personnel. Otherwise, since the founding of Israel, for nineteen years (including the period when the Robinsons were in Jerusalem), the university on Mount Scopus was a ghost town, empty, unused, fighting a "steady battle against mold and decay."[10]

The Suez-Sinai War: 1956

The year before the Robinsons arrived in Jerusalem, Israel was again at war. The trigger this time was the United States and a smoldering crisis over the Suez Canal. The first overt signs of trouble came soon after the

[8] The Hebrew University 1969, pp. 7–8.

[9] The Hebrew University 1969, p. 9.

[10] Lachman 1956, p. 2. Upon prior application, a handful of Jewish visitors were allowed once every two weeks to make the trip from Mandelbaum Gate, the one frontier station, to Mount Scopus where they could spend two hours before the convoy returned. The trip was made in an old truck, stripped down and completely covered with armor plate. Inside it was dark, ventilation poor, with only a small slit across the window in front of the driver's seat to see what was directly in front. Robinson went several times, but Renée never did. Renée Robinson to JWD, letter of March 25, 1991.

American government reversed its decision to help finance construction of the Aswan Dam, which prompted Egypt's president, Gamal Abdel Nasser, to retaliate by nationalizing the canal. Built by a French consortium and owned predominantly by French and British shareholders, its operation was overseen by the British until they relinquished control a few years earlier.[11]

While the French and British mobilized to counter Nasser's move to seize the canal, Israel was on the alert as well. Concerned not only by escalating Egyptian-sponsored Fedayeen attacks, the Israelis feared that Jordan was at last on the brink of falling completely under Nasser's control. And Nasser, confident in his popularity among Arabs, seemed ready to declare war against Israel.

In a calculated measure, the Israelis launched a strike against Egypt on October 29, followed by a joint Franco-British assault the next day. Within the week Israeli forces had occupied the entire Sinai Peninsula, although French and British troops had only managed to occupy Port Said and a small area to the south. In a surprising show of solidarity, the United States and Soviet Union together led the United Nations to negotiate an agreement whereby the French and British would withdraw as soon as a UN Emergency Force was in place. Simultaneously, the Israelis agreed to withdraw from Egyptian territory. With assurances that they would be guaranteed shipping rights through the Gulf of Aqaba, and that the Gaza Strip would not be used as a base for future Fedayeen attacks, Israeli forces relinquished the areas they held in the south.[12]

Although Israel gained nothing tangible from the Sinai-Suez War, the psychological and political advantages it realized were substantial. Above all, the war resulted in nearly a decade that proved free of overt conflict, although tensions remained. The war did serve both to preempt Nasser's hopes of unifying Arab forces against Israel and cement relations with France, which proved crucial to Israel's procurement of military equipment. Above all, UN forces on duty in the Sinai significantly enhanced Israeli security in the south.

The Return to Israel: Summer 1957

After the flurry of summer meetings and going-away parties in Toronto,

[11] On July 19, John Foster Dulles, U.S. secretary of state, announced the withdrawal of U.S. guarantees for a substantial loan from the World Bank to build the Aswan Dam. A week later, on July 26, Nasser nationalized the canal. Two years earlier, England ended its military presence, and in June of 1956 the last British soldiers left Egypt. But the British, who also renounced their part of the loan for the Aswan Dam, were additionally threatened by the fact that the canal was vital to their economy. On these events, see Creighton 1976, pp. 273–274.

[12] Safran 1981, pp. 356–357.

Renée and Abby finally departed for Israel in September. Abby was looking forward to their trip by sea to England, and Renée to being back in Europe. The crossing was uneventful, but being back in England gave both Abby and Renée a chance to see old friends—they were met in Southampton by Hermann Jahn and his wife, then proceeded to London to see the Abramskys and of course Abby's former colleagues at Birkbeck as well as the College of Aeronautics in Cranfield. Only months before, Robinson had been awarded a D.Sc. degree from the University of London based upon his many publications in aerodynamics and mathematical logic. The degree was officially approved on July 17, in recognition of his growing reputation.[13]

The Robinsons decided to make their way leisurely to Israel from England, motoring across Europe in order to revisit some of their favorite places. They bought a new car, a yellow Vauxhall, which took them back through the countryside of France where they enjoyed the champagne country of Reims, the pâtés and white wines of Alsace, continuing on to Einstein's birthplace, Ulm, then to Austria, Salzburg, the great monastery at Melk, and finally a beautiful drive through the Wachau with its dramatic vineyards to Vienna. They went to the Prater, visited the Burgtheater, the Opera, the Kunsthistorisches Museum, and Schönbrunn—it all brought back memories from both of their childhoods. Abby took Renée to Weidling, where his aunt, the widowed wife of Isak Robinsohn, still lived in the home where Abby and Saul had spent so many happy vacations as children.

From Vienna they drove south, to Yugoslavia, where Robinson had been invited to give some lectures arranged by his old friend Djuro Kurepa. This also gave them a chance to enjoy the dramatic scenery along the Dalmatian coast—the medieval austerity of the island of Trogir, the magnificent remains of Diocletian's palace at Split, and a dramatic view from the balcony of their hotel, the Excelsior, over the walled city of Dubrovnik. Unfortunately, on their way to Belgrade, the drive took considerably longer than anticipated:

> Finally they arrived at the university, and Abby left the car to learn where and when his lecture was to be held. As it turned out, he was already late and his listeners were waiting. He sampled opinion as to the most appro-

[13] In support of his application for the higher doctorate in science, the D.Sc., Robinson submitted copies of most of his publications. Max Newman, W. R. Dean, and George Temple were appointed examiners, and they recommended the award on February 20, 1947. The official approval was made by the University Senate the following July. I am grateful to Ivor Grattan-Guinness for checking the University of London's unpublished minutes of the Board of Studies for Mathematics and the printed *Minutes* of the University Senate, 1957 (items 2453 and 4138), where details concerning Robinson's D.Sc. degree may be found.

The Robinsons and their yellow Vauxhall, Israel, Summer 1957.

priate language and began at once in French. After a while, someone from the institute was finally sent out to the car to tell Renée what had happened and to bring her in.[14]

Renée sat in the back of the auditorium, another of the few times (except for special occasions) when she was actually present for one of Abby's lectures. In Belgrade, Kurepa gave the Robinsons his apartment for a few days while he took his family to visit relatives, and this gave Renée and Abby a chance to see the city at their leisure.[15]

From Yugoslavia the Robinsons drove through Macedonia to Greece where they made a small tour of Peloponnesus before stopping in Athens. They climbed up to the Acropolis and had their photograph taken together in front of the Erechtheum, just as Abby had done with his mother and Saul on his way to Palestine in 1933.

Upon reaching Israel, Renée and Abby spent a few nights in Haifa where they stayed with his brother Saul and his wife Hilde. Although Abby was delighted to see his brother, a few days later he and Renée left for Jerusalem where they hoped to be settled before the beginning of the fall term. To their dismay, they arrived to find that none of their belongings, shipped from Canada, had been delivered. Worse yet, there were problems with their accommodations. As Robinson explained in his first letter back to Canada from Israel:

> There has unfortunately occurred a hitch in our housing arrangements and we still live in a boarding house while our things are in storage. Otherwise everything is all right, the weather is fine (like late October in Toronto) and the last thing that people worry about is the fact that there is a border a few hundred yards away from where I am writing this letter.[16]

[14] Seligman 1979, p. xxiii.

[15] Renée Robinson, I-1982.

[16] Robinson to Pounder, January 1, 1958, Robinson's administrative file, Department of Mathematics, University of Toronto.

The boarding house was run by an old friend of Robinson's mother, from Breslau:

> It had a fine view of the city and counted among its residents a famous archaeologist with whom Abby could enjoy discussing ancient Israel and the places he was visiting. After three months they found a pleasant apartment with a good view. Abby's real preference would have been for something closer to the soil and the people, a house of the type the Arabs lived in.[17]

The Hebrew University at Terra Sancta

Like the unsettled state of the Robinsons' temporary quarters in a small pension, the Department of Mathematics at the Hebrew University was doing its best to survive in a number of buildings in town. Actually, Robinson's office and the department's classrooms were in a building at the northern edge of the King David Hotel.[18]

Cut off from its old campus, the Hebrew University was desperately short of space and operated from a patchwork of make-do facilities scattered throughout west Jerusalem. The former college attached to the Franciscan Monastery of Terra Sancta was rented for many years for administrative offices and to house what there was of the university's library—most of which was trapped on Mount Scopus. Only essential books, trucked out sporadically with the bi-monthly convoys, were on hand for student and faculty use. Although the beatific figure of a Madonna is incongruously perched high atop the building, there was an inscription in Hebrew over the front door indicating that part of this unlikely looking structure was indeed a section of the Hebrew University.[19]

Conditions were "makeshift" at best:

> Students stand in crowded classrooms to study the Humanities, Science, Medicine, Law, Agriculture and Education. From the third floor of Terra Sancta, one can see the Judean Hills, rich in Biblical history . . . Through

[17] Seligman 1979, p. xxiii. The apartment was in the Talbieh quarter. Renée remembers the archaeologist as L. A. Mayer (1895–1959), who was librarian and keeper of the records of the Department of Antiquities under the British mandate and later (from 1932 onwards), professor of Near Eastern Art and Archaeology at the Hebrew University's Institute of Oriental Studies. He was also well known for his excavations. After his death, the Museum of Islamic Art and Culture was built in Jerusalem in his honor, as was the building next to it, which was designed as a residence for elderly academics. See "Professor L. A. Mayer—1895–1959. *In Memoriam,*" *Israel Exploration Journal,* 9 (1959), pp. 147–148.

[18] Machover I-1991.

[19] The Hebrew University 1956, p. 2.

> another window one can see a transit work camp which houses thousands of Israel's new citizens . . . The past, the present, and the future is involved in the scope of the Hebrew University and of its students there.[20]

The university—fully realizing the need for centralized facilities—finally made the decision to build a new campus. At first the government had suggested a site in the Judean hills close to the ancient Roman fortress of Castellum, but this was regarded by university authorities as too remote. Instead, they preferred an empty site in the Jewish area close to the heart of Jerusalem on Givat Ram, the "Lofty Hill." This was on the western outskirts of Jerusalem, only a mile and a half from the residential and business center of the city, and was considered ideal.[21]

On June 2, 1954, the site was officially dedicated, and building was immediately begun. In the words of Norman Bentwich:

> The site does not command the magic view of Scopus, but has its own austere majesty. It is in the folds of the hills of Judea, hidden away both from the Holy City with its medieval walls and from the sprawling modern suburbs of our day.[22]

Teaching at the Hebrew University

Robinson's teaching that first year was divided between undergraduate courses (one on linear algebra, the other on hydrodynamics), and one advanced course on logic. As he wrote to Irving Pounder at Toronto:

> I hope you received the postcard which we sent you just before crossing the Austrian/Yugoslav border. We completed the remaining stretch of our trip without incident and arrived here about two months ago. My duties for this year include (i) a course in linear algebra, with an introduction to groups, rings, fields (a 1st year course to about 160 students), (ii) Hydrodynamics (about 60 students), (iii) a Seminar in Mathematical Logic (about 30 students). The high attendance figures under (ii) and (iii) are, to some extent, explained by the fact that many courses are attended by students of different years, beginning with the second year.[23]

The logic course was actually a seminar that Robinson gave along with Abraham Fraenkel. Moshé Machover, an advanced student who attended the seminar, found that it opened up a whole new world.

[20] Kent 1951, p. 7.

[21] Bentwich 1961, p. 52.

[22] Bentwich 1961, p. 53.

[23] Robinson to Pounder, January 1, 1958, Robinson's administrative file, Department of Mathematics, University of Toronto.

Machover had already finished all of his course work, and in 1956 was inducted into the army. In considering a topic for his M.Sc. thesis, Fraenkel had suggested that he do something in combinatorial set theory. While standing guard duty at night, Machover had plenty of time to think about the subject, but despite concentrated effort, eventually realized he was "getting nowhere."[24] Fortunately, Robinson's arrival in Israel proved to be a turning point in Machover's career. Before Robinson joined the staff, no one had taught mathematical logic at the Hebrew University. Consequently, the seminar Robinson offered in his first semester drew many interested students. Machover (having been transferred by the army to a clerical position in Jerusalem) was among them.

The course was devoted primarily to the *Eutscheidungsproblem*. Church's revised *Introduction to Mathematical Logic* had just appeared, and some of the students used it for the prerequisites in logic needed for the seminar.[25] The book caught Machover's interest, and as soon as he had finished reading it, he went to Fraenkel to ask if he might change the subject of his thesis from set theory to logic–and write it with Robinson. Fraenkel agreed, giving Robinson his first graduate student in Israel.

The seminar also examined Tarski's classic paper on decision procedures for elementary algebra and geometry. This included Tarski's method, using Sturm's theorem, for the elimination of quantifiers (see chapter 6). At the end of the paper, a number of open problems were raised, and Robinson suggested Machover try to obtain a decision procedure for algebra with the addition of exponentiation.[26]

As Machover began working more closely with Robinson, Robinson would often suggest that they go to a nearby neighborhood café where they could enjoy some pastry and talk:

> He didn't get me interested in the Tarski open problems, perhaps they were too difficult for a M.Sc. thesis. Then Robinson gave me a paper by Mostowski on model theory, on properties preserved under various kinds of products. Ultraproducts were being invented in Poland at the time, but I missed this, although I did get results on direct products.[27]

Machover recalls that Robinson was a rather shy person, always extremely kind, always ready to make helpful suggestions. He even gave Machover a copy of his book on the metamathematics of algebra–but was so unclear about it–both in what he said and how he said it–that Machover was never sure if Robinson meant it as a gift, for Machover to

[24] Machover I-1991.

[25] Church 1956, "a revised and much enlarged edition of *Introduction to Mathematical Logic* . . . which was published in 1944 as one of the *Annals of Mathematics Studies.*"

[26] Tarski 1948, pp. 43–45, 57. E-mail from Moshé Machover to JWD, March 3, 1994.

[27] Machover I-1991.

keep, or not. He kept the book, and Robinson never asked about it again.

Evaluating Arab Schools: Robinson's Interests in Pedagogy

One of the first assignments Robinson accepted upon his arrival in Israel, apart from his teaching, was to visit local Arab schools and evaluate the effectiveness of classes offered in mathematics. After his first such visit, he recognized immediately the severity of the problems facing Arab schools, and his reports to the Ministry of Education strongly emphasized the need for improvement in basic mathematics education. In this he became a strong advocate for educational reform:

> He was particularly interested in getting Arab students to the university, and he visited Arab schools all over the country. . . . He had a special interest to get the same treatment for Arabs and to get more of them to the university.[28]

After visiting the government high school in Nazareth, for example, in mid-February of 1958, the picture he painted was grim:

> On February 13th I visited the (governmental) high school in Nazareth and attended three lessons [12th grade calculus and integration, 10th grade algebra and quadratic equations, 9th grade geometry, congruence theorems, etc.]. . . . All three [teachers] impressed me that they are serious and devoted teachers. The discipline and the activity were very good and it seems that the majority of the class knew the study material.
>
> My main negative criticism concerns the school apparatus. For instance: the teaching is done in buildings which are far from each other and the laboratories need equipment. But in particular one must point out the sorrowful situation about the books they have for their study. It is a strange mixture of English books, old books from Egypt and notes prepared by the teachers. Besides the fact that preparing the notes (in a hurry?) is a heavy burden for the teachers, such a mixture would possibly have a bad influence on the study and on the general atmosphere and would put the pupils in a more difficult situation than their friends in the Hebrew schools, whose books are more compatible with the study plan and matriculation examinations.
>
> It is clear that this problem needs handling as soon as possible and that one must devote special effort to its solution.[29]

[28] Renée Robinson interviewed by Lynn Steen; transcription of tape recording, RP/YUL.

[29] Robinson to Birenbaum (in Hebrew), the Hebrew University, dated February 28, 1958, RP/YUL.

A month later, Robinson visited another high school in Taibé. He observed two classes (eleventh and twelfth grades) and found "a large percentage of indifferent pupils." This may have been due to the unfortunate timing of Robinson's visit—as he noted, his impression that the level of the school was lower than Nazareth "was explained to me by the fact that my visit took place in the Ramadan month." Again, Robinson stressed the need for suitable books in the report he filed.[30]

Renée has especially vivid memories of one trip they made to Nazareth, where two students were delegated to show her the local sights, including the Church of the Nativity, while Abby undertook his classroom observations. Afterwards, on their way back to Jerusalem, they stopped at a small Arab village where one of Abby's university students lived. "He notified his family that we would be coming and the whole village was already expecting us. It was a rather primitive village. It was getting quite dark before we could leave."[31] In fact, they were invited to the student's home—very simple, shared with livestock, and without electricity. But the hospitality was gracious; they were served rich Arab coffee and shown off with pride to the other villagers.[32]

Back in Jerusalem, Robinson began to think about what might be done to improve the quality of teaching in Arab schools where mathematics was concerned. As a gesture towards helping, he offered to give a special course designed specifically for teachers at such schools. Lack of interest or official indifference apparently thwarted this effort, as a memorandum from Schmuel Shalmon, in the Ministry of Education, suggests:

> Dear Sir [Robinson],
>
> I enquired about the mathematics course and we have reached the conclusion that in the best case six people would participate and practically no more than four are expected. Obviously it is impossible to have this course. . . . I am using this opportunity to thank you for your deep concern about what is done in Arabic education and for your goodwill to help it advance and we hope that you will continue these efforts in the coming years.[33]

Robinson's interest in Arab students was by no means limited to those in secondary schools. Although there were few Arab students at the

30 Robinson to Birenbaum, April 16, 1958, RP/YUL. During Ramadan Muslims are expected to fast during the day, which might well account for the "indifferent pupils" Robinson observed.

31 Renée Robinson to JWD, letter of August 4, 1983, RP/YUL

32 Seligman 1979, p. xxiii.

33 Schmuel Shalmon, Ministry of Education, to Abraham Robinson, June 1, 1958 (in Hebrew), RP/YUL.

Hebrew University, he had a reputation for being a sympathetic listener and adviser.[34] This was due in part, no doubt, to the fact that unlike many professors who appeared at their offices for only a very limited number of hours each week, Robinson was in his office nearly every day, his door always open to students and colleagues alike:

> Professor Robinson devoted great attention to every single student. He devoted himself entirely to this work. He was ready to give many hours to the solution of problems that a student might raise. Every student who sought him out found that he would listen to their problems as if these were the most important of all. . . . and every student also had the feeling that [Robinson] was paying absolute attention to them. He sought every possible way to help each student, but without giving up fundamental requirements. Even when a student received a negative answer from him, one could tell that the student was grateful for the attention he was given.
>
> In particular, Professor Robinson worked devotedly to support the minority students—Arabs and Russians and the other immigrants, and he always took into account the special situation in which these students found themselves.[35]

As for the Arab students, Robinson not only offered as much help and counsel as he could, but he regularly invited them to his home, offering his own brand of warm hospitality that was typical of the relations he sought to establish with many of his students. As Renée explains, "Sometimes, Arab students came to our home. They looked no different from other students, and so I seldom knew who was Jewish and who was not. They all wanted advice from Abby." Perhaps Michael Rabin summarized it best when he told Renée that "Abby was always interested in closing gaps in society, whether socio-economic or racial. . . . We consider Abby's stay in Jerusalem as a highlight in the history of our Mathematics Institute."[36]

This did not mean, however, that Robinson was equally sympathetic to pan-Arab nationalism and the threat this posed to Israel as a nation. As he confided in a letter to Irving Pounder:

> As for the new Arab State which you mention in your letter (now new Arab States in the plural), I was at first inclined to reply that any federation of kindred nations is to the good, provided it lives on good terms with its

[34] In Robinson's last year at the Hebrew University, for example (1961–1962), out of 7,442 students enrolled, only 95 were Arab (or Druze). The Hebrew University 1963, p. 18.

[35] Mrs. Lipschitz, secretary to the Teaching Commission, read through all of the faculty protocols in order to verify the statements and information given here. Quoted from a transcript of her account, hereafter cited as Lipschitz R-1975, RP/YUL.

[36] Renée Robinson to JWD, letter of August 4, 1983, RP/YUL

> neighbours. This reply still stands, but at the same time it appears that the two federations regard each other only as mutual instruments of aggression. It is a pity that the only thing that all Arab states can agree on is their dislike of Israel. There is nothing personal in these remarks, I have met a number of Arabs and, as expected, they are quite as nice as anybody else.[37]

The Robinsons' First Year in Israel

By January of 1958, the Robinsons had found an apartment not far from the university in Talbieh, and soon were settled in their own home. Every room had a balcony, and a direct view as well to the university campus taking shape on Givat Ram. As Robinson described it to Pounder:

> I have hung the crest of the U of T in my study. As appears from this, we have recently settled down in a nice four-room apartment from which parts of the campus can be seen, about two miles away. The new building of the Department of Mathematics will be ready in June.[38]

Abby was not altogether happy with their location, however, despite the convenience of being near the university. He would have preferred to live in the Old City, in the historic center, had it not been cut off from the Jewish part of Jerusalem. He was happiest just walking the streets of the ancient quarters. As Renée puts it, "he loved every stone."[39]

Nevertheless, their apartment was within a reasonable walking distance of the university, and was sufficiently large to accommodate guests comfortably. More often than not, Robinson usually drove to his office rather than walk, even after his office was relocated to the new campus on Givat Ram. That first winter, when temperatures plunged, there was even snow–just enough for Renée and Abby to build a snowman, although it melted within hours.

Now that he was back in Israel, Robinson could look forward to seeing his older brother on a more or less regular basis. Saul was in Haifa, and from time to time Renée and Abby would drive down to the coast to pay short visits to Saul and Hilde, who lived on the slopes of Mount Carmel. Often Saul and Abby would have their breakfast on the balcony of the Robinsohn's apartment, looking out to sea. The two brothers loved to talk, not only politics but philosophy, and sometimes they avidly discussed aspects of education in which they were both interested. Occasionally they would take the time to visit local archaeological sites,

[37] Robinson to Pounder, March 13, 1958, Robinson's administrative file, Department of Mathematics, University of Toronto.

[38] Robinson to Pounder, March 13, 1958.

[39] Renée Robinson to JWD, letter of August 4, 1983, RP/YUL.

as they did on one occasion to Roman Caesarea, not far from Haifa, where Abby took Saul's picture on the waterfront against the ruins.

Renée and Abby in Acre, Spring 1958.

Given the Robinsons' penchant for travel, it is not surprising that their yellow Vauxhall took them everywhere. The car, however, was not without its problems, and once a wheel came loose and went rolling on ahead of them! This happened late on a Friday, so they had to abandon the car, take the last bus to Jerusalem, and worry about retrieving the Vauxhall the following day.

On Friday evenings, among the more pleasant aspects of life in Jerusalem was an invitation on occasion to the Fraenkels' home, where Renée and Abby were often included among the guests:

> There was a big table, and Fraenkel would sing, hitting all the wrong notes for he was not musical, but he would sing anyway. Sometimes Abby would sing too.[40]

Renée paints an especially lively picture of Fraenkel:

> He was the strangest man—with a little pointed head. He went swimming every day. He also stood on his head every day, doing exercises. He was a mountaineer. When he was nearly 70 he went to Switzerland for his "last hike." He died of a heart attack after coming home from swimming, but lived long enough to see Abby installed in his place.[41]

Fraenkel was especially helpful to Renée in making a prerecorded program for Toronto Radio. She wanted to do a report on the Yemenites, and Fraenkel took her "all around to help put the show together." He was especially helpful in accompanying Renée to a Yemenite synagogue for one of their Friday services, which she would not have been able to do on her own.[42] Renée had already done a documentary piece on the Druze, the religious sect concentrated mostly in neighboring Lebanon and Syria, but with small numbers in Israel. The Druze, who she explained permitted no intermarriage or conversion to their faith, were consequently isolated, cut off from the rest of the world, and therefore a subject of considerable interest for her program.[43]

[40] Renée Robinson I-1980.
[41] Renée Robinson I-1980.
[42] Renée Robinson to JWD, March 5, 1994.
[43] Hitti 1928 and Tarcici 1971.

Givat Ram: The "Vision Splendid"

As the new Hebrew University campus was coming to life on Givat Ram, Norman Bentwich referred to it as the "vision splendid."[44] The Robinsons were in Jerusalem just in time to see that vision become reality.

The new campus was officially opened with a ceremony coinciding with Israel's tenth anniversary. By then, twenty-five major buildings had been completed. Next to the seven-story administrative center, the Sherman Building, was the half-domed university *aula* or assembly hall. Just outside was Weizmann Square, a stone court planted with beds of flowers and boasting a pool with reeds from Lake Huleh and a mosaic floor from a Byzantine synagogue near Beth-Shaan.

This higher level gave way to a lower tier of buildings which led down past the Institutes for Jewish Studies, Humanities, and Oriental Studies to Physics, Organic and Inorganic Chemistry, and finally, to the site of Manchester House for Mathematics and Theoretical Physics. Further along was an open-air theater, the applied sciences building, workshops, and various laboratories. Below these were physics and botany, prominent thanks to a planetarium dome giving "architectural interest to the landscape."[45] Actually, the only "daring architectural monument" was the synagogue, which dominated the students' quarter at the end of a tree-lined avenue named after Selig Brodetsky.

Students and the Rewards of Teaching

Student life at the Hebrew University was by no means easy. As a staff correspondent from the *Baltimore Sun*, Louis R. Rukeyser quoted one faculty member on the subject of students:

> They know how desperately this country needs trained professional leaders. And they're very conscious of the important role students have played throughout the struggle for national independence.[46]

As a result, teaching was especially rewarding, not only because of the service it represented to the new country, but because of the determination of the students. Most of them worked very hard, "in fact, too hard in many cases." As one faculty member observed, "To the Israeli, there is no limit to human endurance; anyone who admits to fatigue is a traitor."[47]

[44] Bentwich 1961, p. 57.

[45] Bentwich 1961, p. 54.

[46] The Hebrew University 1961, p. 11.

[47] Samuel 1958, p. 2. Samuel taught in the Public Administration program in the Eliezer Kaplan School of Economics and Social Sciences. The article he wrote for the *Bulletin* was based on material he also prepared for the *Jewish Vanguard* of London.

Although this was perhaps an exaggeration, it was only part of the story. Another reason why teaching at the Hebrew University was taken so seriously was that most students were older and more mature than the average college student elsewhere, having completed their required two-and-a-half-years (two for women) of military service before entering the university. It comes as no surprise that students in Israel were also extremely well disciplined. "They, too, feel impelled by the prevailing Israeli patriotism to do their very best."[48]

The Department of Mathematics

As Norman Bentwich explained in writing a history of the Hebrew University, mathematics held a special place in Jewish intellectual tradition, and hence it was an especially important part of the university:

> Mathematics has always been a favoured field of the Jewish genius, perhaps because of its abstract character. Almost from the foundation of the Hebrew University, the Institute of Mathematics has counted in the academic world.[49]

The interesting link that Bentwich pursued between the Jewish spirit and its seeming penchant for mathematics was attributed to its abstractness:

> One branch of the sciences pursued [at the Hebrew University] promises to be particularly fruitful, if we may judge from the Jewish experience of the last century and of this century. In that period the Jewish genius has shown a remarkable power of contributing to the mathematical and physical interpretation of the universe. Something in those studies seems to be connected with their spiritual heritage and stimulates their creative thought. Planted in their own environment, it is to be expected that they will develop to the sum of knowledge in these architectonic sciences. For every historic people can develop its quality best in its own environment. From the age of the Bible the genius of the Jews lay in their profound sense of abstract unity. . . .[50]

This suggestion of the ethnic character of Jewish mathematics may now seem rather startling, especially in light of the distinction between Jewish and so-called Aryan mathematics drawn in Germany only a few decades earlier, but with a distinctly anti-Semitic objective. Such attempts to characterize a "special" Jewish temperament in mathematics was by no means unique, however, to the Nazi period, and had been given

[48] Samuel 1958, p. 3.
[49] Bentwich 1961, p. 78.
[50] Bentwich 1961, p. 161.

prominent expression by no less a mathematician than Felix Klein in 1893:

> It would seem as if a strong naive space-intuition were an attribute preeminently of the Teutonic race, while the critical, purely logical sense is more fully developed in the Latin and Hebrew races. A full investigation of this subject, somewhat on the lines suggested by Francis Galton in his researches on heredity, might be interesting.[51]

Subsequently, what may have seemed a harmless characterization in the nineteenth century became a blunt instrument in the twentieth, when Nazi mathematicians like Ludwig Bieberbach and Theodor Vahlen proselytized on behalf of a racially pure *Deutsche Mathematik*.[52] That Bentwich, writing in 1960, should also have singled out a special Jewish proclivity for abstract mathematics is thus all the more remarkable. And yet the truth of the matter perhaps lay more with Jewish tradition and custom than with any particular genetic disposition, ethnically, for mathematics.

Robinson's official faculty photograph, the Hebrew University.

Whatever the reasons, the Department of Mathematics at the Hebrew University, from its inception, always enjoyed a strong reputation internationally. The faculty, officially founded during E. Landau's brief tenure in Jerusalem, had grown into a formidable enterprise by the time Robinson returned in 1957. In addition to Fraenkel, who continued teaching until 1959, the faculty included Shmuel Agmon, Shimshon Amitsur, Robert Aumann, Aryeh Dvoretzky, Shaul Foguel, Yitzhak Katznelson, and Michael Rabin, with slightly more than a dozen lower-ranking teachers.[53] Half of the senior members were originally from Germany or Eastern Europe, while the others were Sabras, born in Israel. Virtually all had visited or taught in the United States, and some had also spent varying amounts of time at universities in Western Europe.

[51] Klein was writing specifically about intuitions of space. See Klein 1922, vol. 2, pp. 225–231; esp. p. 228. This same idea was also developed in Klein 1926, part 1, pp. 114–115.

[52] Lindner 1980. See also Mehrtens 1980.

[53] The Hebrew University 1969.

The International Congress of Mathematicians: Edinburgh 1958

That summer, Robinson was back in Europe for the International Congress of Mathematicians in Edinburgh. The Robinsons took their time getting to Scotland, driving from Israel in the Vauxhall, spending several days in Belgium, visiting Louvain, Bruges, Ghent—and especially "Expo '58" in Brussels. In Edinburgh, they went to the usual tourists sites, enjoyed the sounds of pipers in the park, visited Edinburgh castle, saw the John Knox House, and made a trip through the Firth of Clyde.

The Edinburgh Congress was jointly sponsored by the city and University of Edinburgh, along with the royal societies of both Edinburgh and London. It was officially opened at a morning plenary session in the university's McEwan Hall. After several brief messages of welcome, Professor W.V.D. Hodge was elected president of the Congress, by acclamation, whereupon he read a message of greeting from H.R.H. The Prince Philip, Duke of Edinburgh, Patron of the Congress, who wrote from Buckingham Palace.

In his "Presidential Address," Professor Hodge echoed concerns many Congress participants had heard before at both of the previous postwar congresses:

> In recent years there has been a steady growth in the number of symposia held, many with the support of the International Mathematical Union. These symposia have done excellent work in advancing research in special fields. But this is not enough.[54] It is essential for the well-being of mathematics that there should be periodic gatherings attended by representatives of all branches of the subject, and this for several reasons: in my personal opinion, the most important reason is that gatherings such as this serve as an invaluable safeguard against the dangers of excessive specialization.[55]

It was in hopes of being comprehensive that the Program Committee "planned our meetings so as to pass under review all the main developments in mathematics and to try to get things into perspective." The one-hour speakers were picked "as a team so that a continuous spectrum will be presented and they have been asked to make their lectures broad surveys of recent developments. In this way, it is our hope that their contributions will present a general survey of all that is important

[54] Hodge also noted that symposia were closed meetings, by invitation only, whereas Congresses were "open to all mathematicians," and were especially important therefore to younger mathematicians, where they could "meet and listen to the leaders in their subject," Hodge, in Congress 1958, p. lii.

[55] W.V.D. Hodge, "Presidential Address," in Congress 1958, pp. l–li.

in modern mathematics, and that when our *Proceedings* are published, they will form a focus from which many of the developments of mathematics in the next few years may begin."[56]

The Edinburgh Congress was notable, in part, for the emphasis given to the "youngest child of mathematics, the science of computing." Hodge noted that it had presented mathematicians with "many fascinating new problems, and we have considerably enlarged the amount of space given to this subject."[57]

As soon as the Fields Medals were awarded—to Klaus Roth and René Thom—the lord provost declared the inaugural session closed, whereupon everyone stood for the national anthem of the United Kingdom. Congress sessions began immediately, including invited lectures (19 one-hour lectures, 37 half-hour lectures) and 604 fifteen-minute communications. Robinson had elected to give one of the fifteen-minute talks in the first of eight sections devoted to logic and foundations.

On Thursday afternoon, a garden party was held for all members of the Congress on the grounds of Lauriston Castle. Saturday evening, August 16, an informal dance took place in McEwan Hall, with a team from the Scottish Country Dance Society on hand to demonstrate traditional dances. Instead of the usual banquet, however, a Congress reception was held at the Royal Scottish Museum, on the last evening of the Congress.

There were also chamber music recitals, an evening of Scottish song and dance, and a program of Scottish films at the Gateway Theatre. An excursion on Sunday offered a steamer cruise down the Clyde from Glasgow to the island of Bute, ending at Gourock, which Renée and Abby chose to join. Others went to Loch Lubnaig and Loch Earn. And on two evenings, sight-seeing by bus of the city of Edinburgh was provided as well. Book displays were organized, as was an exhibition of mathematical typography by the Monotype Corporation. And both the Scottish National Library and the Edinburgh University Library mounted special exhibitions of mathematical books.

Decision Methods

Robinson's short communication was presented in section 1, "On a Problem of Tarski's." The problem to which he alluded was mentioned in a footnote in Tarski's classic monograph, "A Decision Method for Elemen-

[56] Hodge, in Congress 1958, p. li.

[57] Hodge, in Congress 1958, p. lii. Hodge admitted, however, that due to limitations, only the mathematical side of computing, and not its engineering aspects, were included at the Edinburgh Congress.

tary Algebra and Geometry."[58] There Tarski raised the question of providing a decision procedure for elementary sentences (expressed in terms of equality, order, addition, multiplication, and a unary predicate $A(x)$), which are true in the field of real numbers when $A(x)$ is satisfied only by the *algebraic* real numbers.

Robinson solved this problem using model-theoretic methods by specifying a complete set of axioms K^{**} for all of the above-mentioned relations, including $A(x)$, for which the real numbers were a model. K^{**} was then shown to be complete (in the sense that for any sentence X of K^{**} either X or $\sim X$ was a theorem of K^{**}). Proving the completeness of K^{**} was by no means easy or direct, and depended upon an argument establishing the model-completeness of a closely related set K^*, which required nearly half the paper and was, in itself, as Paul Gilmore later described it, "ingenious."[59]

Models and Sigma-Persistence: Spring 1959

The move to Jerusalem and the demands made upon Robinson's time in the first year must have been substantial. Considering the additional, welcome distraction of the celebrations commemorating the tenth anniversary of the State of Israel and the inauguration of the new Hebrew University campus in Givat Ram in the spring of 1958, it is little wonder that there was a temporary hiatus in Robinson's publications (although material completed for publication before Robinson left Canada continued to appear).

It was not until September that Robinson completed his first article since moving to Israel, namely, the paper he had given at the Edinburgh Congress on Tarski's problem. The following spring, before leaving Jerusalem for the summer, he finished another paper. This one he sent to Arend Heyting, who communicated "Obstructions to Arithmetical Extension and the Theorem of Łoś and Suszko" for publication at a meeting of the Dutch Royal Academy of Sciences on June 27, 1959.[60]

Robinson's argument depended upon an idea he had already used successfully (see chapter 6), namely persistence, and σ-persistent sentences—those for which the union of every increasing sequence of models of K is again a model of K. The Łoś-Suszko Theorem established that if a set of sentences K is σ-persistent then K is equivalent to a set K'

[58] See Robinson 1959 and n. 21, p. 57, of Tarski 1948 for the problem in question.

[59] Although K^{**} was not model-complete, K^* was constructed in such a way that it would be. For details, see Robinson 1959, pp. 184–201. For Gilmore's evaluation of the proof, see *Mathematical Reviews,* 22 (1961), no. 3690.

[60] Robinson 1959a.

of class *UE*.[61] Robinson was able to generalize this result by introducing a relativization to a set of sentences K_0: If a set of sentences H is σ-persistent with respect to a set of sentences K_0, then there exists a set H′ of class *UE* such that H is equivalent to H′ with respect to K_0.

The proof Robinson developed depended upon the concept of an "obstruction" (an extension *M′* of *M* is said to obstruct *M* if there does not exist an extension of *M′* which is an arithmetical extension of *M*). Whereas Łoś and Suszko had established their result for the pure first-order predicate calculus, Robinson's generalization held for arbitrary axiomatic theories formulated in the predicate calculus.

Azriel Levy

Among Robinson's graduate students at the Hebrew University was Azriel Levy, who had just finished writing his master's thesis with Abraham Fraenkel on "The Independence of Various Definitions of Finiteness," when Robinson arrived in Jerusalem.[62] After completing the Hebrew version of his doctoral dissertation, Levy went to the United States as a Sloan Fellow of the School for Advanced Study at the Massachusetts Institute of Technology, where he prepared an English version of part of his thesis, along with several other papers, for publication.[63]

Levy's dissertation, "Contributions to the Metamathematics of Set Theory," was completed in 1958 under the joint supervision of both Fraenkel and Robinson. Gödel's concept of constructability, introduced in lectures in 1938–1939, had already been generalized by András Hajnal a few years later.[64] Joseph Shoenfield, independently, also established the conditional independence of the axiom of constructability and related axioms, all of which he explained at the 1957 Cornell Symposium where Levy (and presumably Robinson as well) first heard of them and began to think about extending Shoenfield's results even further. This eventually became the subject of Levy's thesis.[65]

The axiom of constructability (the axiom that all sets are constructable, $V = L$ in Gödel's notation), implies both the axiom of choice and the generalized continuum hypothesis. Shoenfield dealt with the problem

[61] A set of axioms is said to be of class *UE* if it is comprised of sentences in prenex normal form in which no existential quantifier is followed by a universal quantifier. See Robinson 1959a, pp. 489–490; in Robinson 1979, 1, pp. 160–161.

[62] Results from the thesis were published in a paper Levy finished in June of 1957. See Levy 1957.

[63] Levy 1960a.

[64] Hajnal 1956. Constructability had been introduced by Gödel in his lectures on the continuum hypothesis, published as Gödel 1940.

[65] Shoenfield 1957 and 1959. Levy published some of his own results, even prior to finishing his thesis, as Levy 1957.

of the independence of $V = L$, for which Levy obtained somewhat stronger results of his own.[66] The method used by both Hajnal and Shoenfield, as well as by Levy, started with a model of set theory in which $V = L$ is not true, obtaining from it another model of set theory in which $V = L$ is not true, while the axiom of choice and one or another form of the generalized continuum hypothesis are true.

Applied Mathematics

In the spring of 1959 Robinson was invited to teach a course the following semester on fluid dynamics at the prestigious Weizmann Institute at Rehovot, which supported primarily research along with some graduate teaching. Although Chaim Weizmann had been a founding member of the Hebrew University in 1918, he was instrumental in establishing the Daniel Sieff Research Institute in 1934, from which the Weizmann Institute eventually evolved. In fact, for reasons that were both personal and professional, Weizmann chose to pursue his own research (in chemistry) at the institute, which was created expressly for him in Rehovot next to his own home.[67]

While the Hebrew University was his first love, Weizmann was disappointed by certain features of its development, particularly in the sciences. Friction between Weizmann and the university's first president, Judah Magnes, produced as Norman Bentwich put it, "a soreness which could not readily be healed."[68] Consequently, a good deal of fundamental research in the sciences was deliberately pursued separately, elsewhere.

The course Robinson agreed to offer on fluid dynamics at the Weizmann Institute reflected his reputation as an expert and his experience in teaching similar courses at Cranfield, Toronto, and the Hebrew University. It was doubtless due to similar respect for his book on *Wing Theory* that Robinson was asked to contribute the article on "Airfoil Theory" for Wilhelm Flügge's *Handbook of Engineering Mechanics*. Flügge was professor of engineering mechanics at Stanford University, and asked Robinson to write about wing theory in the *Handbook's* part 7 on fluid mechanics.

Such projects are never as easy as they seem. They require the synop-

[66] Shoenfield used weakened forms of the Generalized Continuum Hypothesis (there is an ordinal α_0 such that $2^{\aleph_\alpha} = \aleph_{\alpha+1}$ for $\alpha \geqq \alpha_0$) and of the Axiom of Constructibility (every set of integers is constructible). See Shoenfield 1959, p 537.

[67] For details of Weizmann's life not only as a scientist, but as a Zionist leader, president of the World Zionist Organization and first president of Israel (1949–1952), see Weizmann 1949 and Carmichael 1963.

[68] Bentwich 1961, p. 73.

tic overview of an expert who knows how to pare away the nonessentials, giving the skeleton of the subject in a way that is both coherent and meaningful. Robinson somehow managed to reduce the fundamentals of airfoil theory to twenty-four pages, and as A. D. Young has said of the piece, it represents "a remarkable demonstration of concentration without loss of clarity; in the twenty-four pages allotted to him the essentials are all to be found."[69]

Due to space limitations, Robinson had to narrow his focus to steady conditions in a fluid medium assumed to be inviscid and incompressible. This meant no supersonic cases were considered, nor did he try to include the theory of wind tunnel interference or how measurements derived from models required correction to adequately approximate free-flight conditions. Robinson's article gave greatest emphasis instead to lifting line theory, in which he considered various types of wings and explained how to calculate pressure distributions and other aerodynamic forces.[70]

After Robinson left Toronto for Jerusalem, he only published two more items on aerodynamics. One of these was the brief article for Flügge's *Handbook*, while the other appeared as a joint work (written with A. E. Hurd) in 1968. This was a study on flexural wave propagation in nonhomogeneous elastic plates, an evident sequel to his earlier work on beams.[71]

But as A. D. Young has said of Robinson's apparent declining interest in applied mathematics:

> Those of us who had worked with him in those fields know how closely he continued to follow developments and read all important publications and he took a keen interest in the careers and work of his former students and colleagues. One felt that the applied mathematician in Abby was never far away, and that some exciting development offering the challenge of a structural unity to be uncovered could yet again bring the applied mathematician into action.[72]

Indeed, this was to prove the case during Robinson's last years at Yale, when nonstandard analysis found applications in such diverse areas as mathematical economics and theoretical physics. As Simon Kochen has suggested, the connection between Robinson's abilities in pure and applied mathematics was doubtless of deep and consequential significance:

> Almost one-half of his papers and one book are devoted to aerodynamics and structures and his last paper on the latter subject was written as re-

[69] Young 1976, esp. p. 312.
[70] Robinson 1962.
[71] Hurd and Robinson 1968.
[72] Young 1976, p. 312.

cently as 1968. I think, in assessing Robinson's mathematical outlook, one would ignore such a large and integral part of his work at one's peril.

I believe that the thread that runs through all his work lies precisely, in fact, in this aspect: that also as a mathematical logician, his viewpoint was that of an applied mathematician in the original and best sense of that phrase; that is, in the sense of the 18th and 19th century mathematicians, who used the problems and insights of the real world (that is, physics) to develop mathematical ideas. To logicians, it is the world of mathematics which is the real world.[73]

The Italian Mathematical Union: Naples 1959

Among the results Robinson published during his first years at the Hebrew University, local differential algebra occupied an important place. Here he continued to be interested in Joseph Ritt's work, one aspect of which was of special concern.[74] Robinson focused his attention on the problem of initial and boundary values, something Ritt's work had not included. Robinson liked the results he was able to achieve, again thanks to his model-theoretic approach, and decided to present some of his new ideas at a meeting of the Union of Italian Mathematicians in Naples, to which he had been invited in the summer of 1959.[75] Basically, this paper gave an algebraic theory for the solution of initial value problems for systems of algebraic differential equations.

In the detailed version of Robinson's argument published the following year, he considered a localized differential ring R, together with a ring homomorphism $H{:}R \rightarrow R_0$. Beginning with elements whose values "at a given point" were represented by a homomorphic mapping into an ordinary ring, he then considered ideals, polynomial ideals, and corresponding varieties, all of which led up to a consistency condition for a system of algebraic differential equations with given initial values.

On the subject of localized differential fields and rings, Robinson went on to consider S-perfect ideals, a theory of ideals in localized differential rings, extension and polynomial adjunction, regular localized differential rings, and reflexive localized differential rings. At the end of the paper he was pleased to acknowledge help one of his students, Shlomo Halfin, had given him in developing some of his ideas on the subject. Later they wrote two joint papers on the related subject of local differential algebra.[76]

[73] Kochen 1976, p. 313.

[74] Ritt 1950.

[75] Robinson 1959b; a detailed version of what Robinson only sketched in this abstract of his paper was published as Robinson 1960.

[76] Halfin and Robinson 1960 and 1963.

Foundations of Mathematics: Warsaw 1959

Following the meeting in Naples, Renée and Abby drove to Poland, where an international symposium on foundations of mathematics was held in Warsaw during the first week of September. Abby was pleased that this also gave them an opportunity to revisit places he remembered from his childhood. He took Renée to see Breslau (now Wrocław in Polish), and showed her the first school he attended as a boy in Waldenburg (now Walbrzych).[77]

A particularly moving moment was their visit to the local Jewish cemetery in Waldenburg, where Robinson's father was buried. The graves were overgrown with weeds, but with the help of an old caretaker, Robinson managed to find the headstone marking his father's grave. The old porter could still speak German, but he was the only one—everyone else working at the cemetery spoke only Polish.[78]

Although Robinson was the single participant from Israel at the Warsaw meeting, old friends were on hand, including Lorenzen from Hamburg, Février from Paris, Kreisel from England, Beth and Heyting from Holland, Bernays from Switzerland, and Henkin, Kleene, Tarski, and Vaught from the United States. The general theme of the symposium was infinitistic methods in the foundations of mathematics, with an emphasis on nonconstructive methods, including inaccessible cardinals, infinitistic rules of proof, higher classes of number-theoretic, and function-theoretic hierarchies of predicates. Robinson spoke on "Model Theory and Non-Standard Arithmetic," a subject notable for its relation to his unexpected discovery of nonstandard analysis just two years later.[79]

Robinson began by noting that the lower predicate calculus provided a framework both for studying the incompleteness and undecidability of arithmetic theories as pioneered by Gödel, Kleene, and Church, and for what had come to be known as model theory (and earlier the applied predicate calculus or metamathematics of algebra) as developed by Mal'cev, Tarski, Henkin, Łoś, and of course, Robinson himself.

The first to try putting the two together was Skolem in an early paper on nonstandard models of arithmetic. Later, "more subtle constructions" were due to Łoś, Frayne, Scott, Tarski, Kochen, and Rabin. One of the most impressive applications of nonstandard models was advanced in a paper by Ryll-Nardzewski in 1952, who showed that Peano's axioms as represented in the lower predicate calculus could not be replaced by a finite system.[80] An especially important example of applications of non-

[77] Seligman 1979, p. xxiv. [78] Renée Robinson I-1980.

[79] Robinson 1959c.

[80] Ryll-Nardzewski 1952. Mostowski also proved this result, but without the use of nonstandard models, in Mostowski 1952.

standard models of arithmetic was their use in proving Hilbert's classic irreducibility theorem (and certain extensions of the theorem as well).

To make clear what he meant by a model-theoretic approach to arithmetic, Robinson explained:

> Generally speaking, the task of model theory is the investigation of the relation between sets of sentences (or, used here synonymously, *axioms*) on one hand, and sets of mathematical structures on the other. A typical problem of model theory is as follows. Let K be a set of sentences in the LPC [lower predicate calculus] such that the class of models of K is closed under union of monotonic subclasses. A set K which satisfies this condition is called σ-persistent. How can we characterize the sets K which have this property?[81]

Robinson stressed that especially important here were the concepts of completeness and relative model-completeness. Considering the examples of algebraic, formally real, and differential fields, and their relative model-completions, each was model-complete. Thus the theories of commutative fields, of formally real fields, or of differential fields of characteristic 0 were especially amenable to metamathematical treatment.

This was unfortunately not the case with arithmetic, where "the situation is far less satisfactory":

> Indeed, while the axiom systems of standard algebra seem to be particularly suitable for the discussion of the model-theoretic properties of the structures satisfied by them, especially as regarding problems of extension and intersection, the contrary is true for arithmetic.[82]

For example, let K_0 denote a particular formulation of Peano's axioms for arithmetic of natural numbers in the lower predicate calculus. Any model of K_0 was said to be a weak model of arithmetic. A particular model of K_0 is of course given by the natural numbers themselves, the "standard" model, which Robinson denoted M_1. Any weak model contains M_1 as a submodel. The set of all sentences with extralogical constants (including the sentence asserting the existence of the number 0, and those expressing the relations of equality, successor, sum, and product) which hold in M_1 was denoted by K_1. Then $K_1 \supseteq K_0$. As Robinson pointed out, K_0 was recursive but not complete, while K_1 was complete but not recursive (and not even recursively enumerable). Any model of K_1 was said to be a *strong* model of arithmetic. Robinson cautioned that while K_1 was complete, it was not model-complete (due to the existence

[81] Robinson 1959c, p. 266; Robinson 1979, 1, p. 168.
[82] Robinson 1959c, p. 266; Robinson 1979, 1, p. 168.

of predicates of arithmetic definable in the lower predicate calculus but not decidable).

In turning to the question of the relations between different nonstandard models, answers for even the simplest cases were often difficult. Consider a weak (or strong) model M where M' and M'' are two other weak (or strong) models contained in M. Is the intersection of M' and M'' again a weak (or strong) model? The answer (as it turned out) was negative. It was Dana Scott who pointed this out to Robinson (probably at the meeting) with an argument first established by Georg Kreisel and drawing upon a theorem of Łoś and Suszko.[83]

Robinson, not surprisingly, was concerned by the messiness of it all:

> Abnormal results of this kind seem to indicate that it is foolhardy to leave the well-tended fields of algebraic model theory on one hand, and of recursive functions theory in arithmetic on the other to venture into the barren deserts of a model theory of (non-standard) arithmetic. Nevertheless, if one may be permitted to make a suggestion or prophecy, just this seems to be a natural task of mathematical logic for a number of years to come, comparable to the erection of a comprehensive edifice of non-euclidean geometry in the last century and perhaps no less rewarding.[84]

But it was the business of a mathematician, he said, "not to preach but to prove," and what Robinson had in mind was to present two kinds of results bridging model theory and nonstandard arithmetic. The first concerned an extension theory for nonstandard models of arithmetic. Among results of this type, Robinson proved that every Skolem model was a simple extension of its prime model. He also showed that the Skolem models were the only simple extensions of the prime models.[85]

When it came to understanding the structure of extensions in arithmetic, however, much was to be gained by the "remarkable method" used by Mostowski and Ehrenfeucht in their work on the automorphisms of strong models. Robinson also stressed the *algebraic* relations between fields generated by strong models that had recently been investigated by Elliott Mendelson.[86]

[83] Kreisel 1952. The reference to M_1 and M_2 in Robinson 1959c, p. 271 (Robinson 1979, 1, p. 271), are misprints; what Robinson clearly meant were intersections of weak or strong models in M.

[84] Robinson 1959c, p. 281; Robinson 1979, 1, p. 173. Considering all of the recent work on models of arithmetic in the last decade or so, Robinson's remarks were indeed prophetic (Martin Davis to JWD, in an E-mail note of September 7, 1994).

[85] See theorems 2.33 and 2.34, Robinson 1959c, p. 285; Robinson 1979, 1, p. 187.

[86] The procedure of adjunction of elements used by Ehrenfeucht differed from the method Robinson used, since Ehrenfeucht had presupposed an arbitrary choice of Herbrand functions for all the existential quantifiers which occur in sentences of K. See Ehrenfeucht 1956, Mendelson 1957, and Mostowski 1955.

Turning to field theory and a characterization of the natural numbers, Robinson considered the axioms for an algebraically closed field H_0, K_1 denoting the set of sentences which held for the natural numbers. Taking $N(x)$ as a one-place relation, K_2 resulted from relativizing the sentences of K_1 with respect to $N(x)$.

Using Rabin's definition of a computable structure (adapted to notation for relations $S(x, y, z)$ instead of Rabin's use of functors $x + y = s(x, y)$ for the sum), he defined $H_1 = H_0 \cup K_2 \cup \{N(0)\}$. Robinson then focused most of his attention on properties of H_1, and proved in particular that H_1 was complete.[87] This, he noted, meant that every two models of H_1 were elementarily equivalent. Considering models M of H_1 for which the substructure M' coincided with the standard model (of the natural numbers) M_1, Dana Scott had raised the question of whether or not all models M of this kind were elementarily equivalent (M an algebraically closed field and M' a strong model of arithmetic determined by $N(x)$). To answer this question, it was necessary to introduce nonstandard models of arithmetic.[88]

At the end of his paper, Robinson explored the extent to which corresponding results held for rational numbers (instead of natural numbers).[89] Likewise, he considered whether or not there were counterparts in which algebraically closed fields might be replaced by real-closed fields (although, as he showed, in such cases completeness failed).[90]

Robinson as Chairman

When Abraham Fraenkel retired as chairman of the Department of Mathematics in 1959, it was Robinson who assumed his position—and soon found himself in the midst of university policy-making.[91] In fact, he had already served for a year on the faculty's Teaching Commission, to which he was appointed in June of 1958 and over which he presided in 1959.[92] As George Seligman has put it so well, "his keen interest, his conscientiousness, his availability, and his wisdom made him a favorite for committee work and other responsibilities, and had him marked for more administrative duties in the future."[93]

Robinson's capacity for patience and wise counsel had already won him universal respect, especially during several student strikes. Because

[87] Theorem 3.9, Robinson 1959c, p. 291; Robinson 1979, 1, p. 193.

[88] Robinson 1961, p. 293; Robinson 1979, 1, p. 195.

[89] Robinson 1961, p. 294; Robinson 1979, 1, p. 196. According to Levy, Robinson always preferred relations to functors.

[90] Robinson 1961, p. 297; Robinson 1979, 1, p. 199.

[91] For biographical details concerning Fraenkel, see van Rootselaar 1972, pp. 107–109.

[92] Lipschitz R-1975.

[93] Seligman 1979, p. xxiv.

student political and social activism was high, student rallies were frequent, sometimes disruptive, as in the case of one protest held in the cafeteria that Robinson was called in to settle.[94] But the major administrative responsibility Robinson inherited was a thorough overhaul of the university's curriculum, including a complete restructuring of the courses and requirements offered by the Department of Mathematics. This included the introduction of a new B.Sc. degree, which had not previously been offered.

Under the university's old system, an M.Sc. degree was conferred upon submission of a major thesis after successful completion of a battery of written examinations (consisting of ten separate papers in the major subject, in addition to written examinations in two minor subjects), as well as a final oral examination.[95] This was conducted in the old European tradition whereby any faculty member in any part of the university might participate. Students who were able to complete only a part of the requirements, that is, a major with only one minor, or three minors, were entitled to a B.Sc. degree.

As Mrs. Lipschitz, secretary of the natural sciences faculty at the time, described the situation:

> At this time the faculty was expanding to an enormous extent. . . . The number of students who had finished the *Abitur* and who were endeavoring to be admitted to the faculty was steadily increasing. Then there was only one university. On the other hand, steps had to be taken to see that as few students as possible would drop out in the course of their studies. . . .
>
> To fulfill all these goals, the faculty of Natural Sciences introduced the B.Sc. as a requirement, while the students who presented good marks, would still be able to get the M.Sc. This difference had in principle great significance.[96]

Robinson, as chairman of mathematics, made a number of fundamental organizational improvements in this scheme, largely to accommodate a continuously increasing number of students applying for admission. This meant hiring new teachers, increasing budgets, and thoroughly reorganizing the entire educational structure of the department. It was hoped that introducing the three-year B.Sc. as a formal requirement would enhance the curriculum, even if it meant revising the courses faculty would have to offer.

As a result, teaching hours were reduced and students were no longer required to take a final examination at the end of their studies. Instead,

[94] Renée Robinson, cited in Seligman 1979, p. xxiii.
[95] Moshé Machover to JWD, October 22, 1992.
[96] Lipschitz R-1975.

Robinson lecturing, the Hebrew University.

at the end of each course, students were to be examined, and at the end of their third year, with satisfactory completion of their requirements, they would then receive the B.Sc. degree. Implementing these reforms, however, which were undertaken with thorough discussion and reconsideration of all the department's various courses—required two years.

The new program began in the academic year 1959–1960, and at first it was not at all clear whether it would succeed or not. The fact that it did was in no small measure due to Robinson's continuing efforts, for he was the "living spirit" behind the new curriculum:

> He worked tirelessly on this, and developed creative thoughts all handled with great patience, and he did so resolutely and intensively, for he had to convince all of the professors of the desirability of this, to prepare them [for the changes], all of which required hard work and steadfastness, especially in the face of opposition and complaints that arose.
>
> It should be noted that at the time, the Natural Sciences Faculty (like all of the other Faculties of the Hebrew University in Jerusalem), worked according to the old system of the European universities. This meant that a Professor had his own teaching assistants under him, and developed his own specialty wholly apart from what anyone else was doing. There was no cooperative work in the departments, and everyone did in fact whatever he wanted to do. The coordination between specialties, the cooperation of different departments, the introduction of democratic methods—all these were the innovations that Professor Robinson brought about here. Of course this wasn't easy, since there were so many opposed to it. . . .[97]

[97] Lipschitz R-1975.

Differentially Closed Fields

When Robinson returned to Israel after his travels during the summer of 1959, he put the finishing touches on a paper that had taken shape using some of the ideas he had presented at the Warsaw Symposium, this time applied to differentially closed fields. In October he sent a paper off for publication in the *Bulletin of the Research Council of Israel.*[98] The results Robinson presented, inaugurating in effect the subject of differentially closed fields, was easily the most important work he accomplished while at the Hebrew University.

Algebraic field theory and algebraic geometry are closely related to algebraically closed fields as extensions of commutative fields. E. R. Kolchin had already raised the question of whether corresponding extensions exist for differential fields, but he did so in terms of an extension in differential algebra to the concept of a universal algebraic extension in ordinary field theory.[99] Robinson was interested in a slightly different concept, corresponding to that of an algebraically closed field, and it was this that he called a differentially closed field.[100]

Seidenberg showed that to any finite system S of differential equations over a differential field M, it was possible to associate a condition C on the coefficients which is necessary and sufficient for S to have a solution in some extension M' of M. A differential field M was then said to be differentially closed if each S has a solution in M itself whenever the corresponding condition C is satisfied.

Using Seidenberg's elimination techniques, Robinson discussed how it was possible to give a model-completion for the axioms of differential fields. The differentially closed fields could then be taken as models of the "closure" axioms associated with this completion. As George Seligman has said, reflecting the views of Angus Macintyre, "it would be appropriate to say that [Robinson] *invented* differentially closed fields."[101]

Robinson made clear the significance of differentially closed fields in section 3 of his paper, where he gave a number of applications including a theorem on the specialization of parameters and a proof of the existence of a constructive version of J. F. Ritt's *Nullstellensatz.*[102] The

[98] Robinson 1959d.

[99] Kolchin 1953. Kolchin's universal extension was of infinite degree of *differential* transcendence over the ground field, just as in ordinary algebra, a universal extension is of infinite degree of transcendence over the ground field.

[100] It is also worth noting that Kolchin's universal extensions are differentially closed, but the converse is not in general true.

[101] See Seligman 1979, p. xxiv.

[102] Ritt 1950. This result had also been obtained by Seidenberg, using a different method. See Seidenberg 1956.

paper came to a close in section 4, where he provided a test for the existence of such model-complete extensions.

Beginning with several facts about ordinary commutative fields, Robinson explained various properties of differentially closed fields:

2.1 *Every field M can be embedded in an algebraically closed field, M'.*
2.2 *Given M, there exists [an algebraically closed] field M' such that all elements of M' are algebraic with respect to M. Moreover, M' is determined uniquely (up to isomorphism) by M, although this does not carry over to the differential case.*
2.3 *With M and M', every finite system of polynomial equations and inequalities which possesses a solution in some extension of M, possesses a solution also in M'.*

This last property, in fact, was the essential feature of algebraically closed fields.[103]

Algebraically closed fields also possessed a number of significant metamathematical properties, among them: if K' denotes the set of axioms for the concept of an algebraically closed field of characteristic 0, K' is recursively enumerable, complete, model-complete, and model-complete relative to the set of axioms for fields of characteristic 0.

In considering the counterpart of an algebraically closed field in differential algebra, Robinson proceeded axiomatically, beginning with a set of axioms K_D for a differential field of characteristic 0. This was done by adding to the relations of K_0 the relation $D(x, y)$ (where $y = x'$), along with the appropriate axioms for the existence of the derivative and the basic rules for addition and multiplication. The first of these axioms, for example, was given by:

$$(x)(\exists y)(z)[D(x, y) \wedge [D(x, z) \ldots E(y, z)]],$$

that is, the result of differentiation is unique, in the sense of equality within the system.

Robinson focused attention on a related set of axioms K_D', an extension of K_D containing the basic relations for equality, sum, product, and differential, with no individual constants. This was model-complete relative to K_D, and every model of K_D could be embedded in a model of K_D'. Thus K_D' was intended to serve as an axiomatic counterpart of K' for algebraically closed fields. Robinson termed a model of K_D' a "differentially closed differential field." In his construction of K_D', Robinson was closely guided by Seidenberg's elimination theory in differential algebra for differential fields of characteristic 0.[104]

[103] Robinson 1959d, p. 114; Robinson 1979, 1, p. 441.

[104] For details, see Robinson 1959d, p. 117; Robinson 1979, 1, p. 444.

Among applications, Robinson included a theorem (also provable by purely algebraic means) that if M_1 and M_2 are two extensions of a differential field M_0 which have no elements in common (other than the elements of M_0), then M_1 and M_2 can be embedded in a joint extension M_3. Since this theorem is true only for differential fields of characteristic 0, Robinson acknowledged there could be no strictly analogous corresponding theory for differentially closed fields for $p \neq 0$. A second application concerned predicates satisfied by independent transcendentals in a differentially closed field, and a third determined the bounds on the orders and degrees of polynomials in Ritt's *Nullstellensatz*.[105]

Robinson brought his paper on differentially closed fields to a finish with a test for the existence of model-complete extensions using results on σ-persistence due to Łoś and Suszko, as well as the methods he had developed himself for dealing with differentially closed fields.[106] Thus he concluded with a very general formulation of his main idea:

> Let K be a non-empty, consistent, and σ-persistent set of sentences. In order that there exists an extension K' of K which contains no additional relations or individual constants such that every model of K can be embedded in a model of K' and such that K' is model-complete relative to K– it is necessary and sufficient that every primitive predicate $Q(z_1, \ldots, z_k)$ which is defined in K, $k \quad 0$, possess a test $Q'(z_1, \ldots, z_k)$.[107]

New Light on Ordered Abelian Groups: Research with Elias Zakon

Although it had been easier working with Elias Zakon when they were both in Canada, Robinson continued their collaboration after he left Toronto and in 1960 their joint results on "Elementary Properties of Ordered Abelian Groups" were published in the *Transactions of the American Mathematical Society*.[108] The paper Zakon and Robinson worked on together was initially inspired by the complete classification of abelian groups (using elementary properties that could be formalized in the lower predicate calculus) as conceived by Wanda Szmielew in 1955.[109] It

[105] See Ritt 1950, p. 7.

[106] Łoś and Suszko 1957.

[107] Theorem 4.1, Robinson 1959d, p. 124; Robinson 1979, 1, p. 451. Q' was said to be a "test" for Q if Q' was an existential predicate defined in K, and if for every model M of K and elements $a_1, \ldots, a_k$ of M, $Q'(a_1, \ldots, a_k)$ held in M if and only if $Q(a_1, \ldots, a_k)$ held in some extension M' of M which was a model of K.

[108] Zakon and Robinson 1960. After coming to Toronto to study with Robinson during his sabbatical year from the Technion in Haifa, Zakon stayed in Canada, having obtained a permanent position at Essex College, Assumption University of Windsor, Ontario, Canada.

[109] Szmielew 1955.

was apparently Szmielew's paper that originally gave Robinson and Zakon the idea of providing a similar classification for *ordered* groups. Their joint paper provided a classification by elementary properties for all archimedean ordered abelian groups (and for a more general class as well which they called regularly ordered groups).

Although archimedean groups could not be formalized in the lower predicate calculus, Robinson and Zakon were able to show that every regularly ordered group was elementarily equivalent to some archimedean group. As it turned out, this meant that the concept of a regularly ordered group was the elementary counterpart to that of an archimedean group. This, in turn, "throws a new light on the significance of that concept" (i.e., that of a regularly ordered group). Zakon took up the purely algebraic issues related to regularly ordered groups in a separate paper of his own.[110]

Sabbatical at Princeton

The Hebrew University encouraged sabbaticals, faculty exchanges, and international meetings as means of insuring that its faculty was always at the forefront of research in all fields. Indeed, "in no other university is the practice of the sabbatical year more faithfully observed by the staff."[111]

In 1960 Robinson took advantage of this policy to spend a year away as a visiting member of the faculty at Princeton University. In anticipation of Alonzo Church's sabbatical leave from Princeton in 1960–1961, Robinson had been invited to spend the year as a visiting professor in the Department of Mathematics. This meant more packing and planning than usual as Renée and Abby set off that August for the United States, beginning with a short voyage from Haifa to Genoa via Rhodes and Cyprus, which they made with Ernst and Luise Straus.

Straus and Robinson had known each other since their school days together in Rehavia. Straus went on to study mathematics and physics, obtaining his Ph.D. from Columbia University for a thesis on unified field theory in 1948. From 1944–1948 he was Einstein's assistant at the Institute for Advanced Study, and together they wrote several papers on relativity and gravitational fields. Subsequently, Straus joined the faculty at UCLA, but in 1955–1956 he returned to the institute in Princeton as a member for a year.[112] In 1960, he was again on leave from UCLA to spend the year in Jerusalem. This gave Straus and Robinson a chance to become reacquainted, and Straus even attended

[110] Zakon 1961.

[111] Bentwich 1961, p. 151.

[112] Institute for Advanced Study 1980, p. 360.

one of Robinson's courses: "I remember the rather complicated-to-understand seminar in Logic that he gave there to which I went with not too much success and the very pleasant way the department was run and how everybody enjoyed the lack of excessive administration that marked his chairmanship."[113]

From Cyprus Renée and Abby continued on to Athens and then to Genoa. From there, they drove to Monte Carlo, where they had arranged to meet his colleague from the Hebrew University, Yehoshua Bar-Hillel and his family, for a trip together through southern Europe, including a tour of Spain and the south of France. The Robinsons finally returned to Genoa where they boarded the *Leonardo da Vinci* for their passage to New York.[114]

The ship called at a number of ports along the way, including Gibraltar and the Azores before crossing the Atlantic. Robinson usually worked in the mornings, but in the afternoons they enjoyed sunning on the pool deck. After dinner, Renée and Abby would often spend the evenings dancing. Renée always appreciated the fact that on such occasions Abby liked to dress up. From time to time he enjoyed "being elegant," and the ship provided the first opportunity to wear his tuxedo since leaving Canada.

Stanford University: Summer 1960

The Robinsons had arranged to rent a faculty home for the year they were in Princeton. They were scarcely settled in New Jersey, however, before they flew to California, where the 1960 International Congress for Logic, Methodology, and Philosophy of Science was to be held at Stanford University from August 24 through September 2. Sixty-three papers were eventually published as a record of the invited addresses arranged by the Program Committee.

Opening ceremonies were held at Stanford on Wednesday morning, August 24, with Alfred Tarski presiding. J. E. Wallace Sterling, Stanford's president, offered a warm welcome. S. C. Kleene greeted everyone in his capacity as president of the Division of Logic, Methodology, and Philosophy of Science (of the International Union of History and Philosophy of Science), as did Alan T. Waterman as Director of the U.S. National Science Foundation.

That afternoon, in a session chaired by Abraham Fraenkel, Robinson delivered the first invited plenary address on "Recent Developments in Model Theory."[115] The rest of the week was devoted to papers, invited

[113] Straus R-1976.

[114] Renée Robinson to JWD, letter of May 29, 1991.

[115] Robinson 1960a.

The Robinsons with Harry Wolf and Robert Cohen among other colleagues, at the Studio Inn, Palo Alto, August 1960.

lectures, and a number of special symposia. The weekend was given over to local sight-seeing, with a tour of San Francisco on the following Wednesday. Thursday evening the Congress banquet was held at a local restaurant in Palo Alto, Rickey's Studio Inn.

Robinson's hour address was a fitting opener for the Congress, and provided a broad survey of model theory as it had developed in the decade since he, Tarski, and Henkin had first begun to exploit its potential in the late 1940s. Much of the paper was devoted to Mal'cev's pioneering work in the theory of models, which only recently had come to the attention of logicians working in the West, although the fundamental ideas had already been published (in Russian) in the early 1940s.

Rather than try to give a comprehensive introduction to the subject, Robinson took a different approach. What he wanted to do was to give a feeling for how some of the pioneering results in model theory "take on a new complexion when viewed in the light of more recent developments, and I shall emphasize some trends and methods which appear to be destined to play an important part in the near future."[116]

Always interested in historical questions, Robinson noted that the fundamental ideas of model theory and the methods to prove such classics as the extended completeness theorem of the lower predicate calculus, the compactness theorem for arithmetical classes, or the principle of localization—originally formulated within different frameworks but later seen to be basically equivalent as far as applications were concerned—appeared in the work of two of the pioneers of modern logic: Löwenheim and Gödel (e.g., Gödel's completeness theorem of the lower predicate calculus). Choice functions, as used by Skolem, Hilbert, and Herbrand, went back to the twenties. Of special interest to Robinson was Skolem's construction of a nonstandard model for arithmetic, the

[116] Robinson 1960a, p. 60; Robinson 1979, 1, p. 12.

"legitimate forerunner" of ultraproducts which Robinson also discussed in the Stanford lecture.[117]

Following a broad analysis of Mal'cev's work, Robinson returned to a topic basic to model theory–the construction of models, more precisely structures or relational systems subject to given conditions, either in terms of axioms in the lower predicate calculus, or in connection with other structures introduced previously. The mere existence of a model for any consistent set of axioms proved sufficient to yield subsequently interesting results. Of all such constructions, Robinson emphasized the best known–that of the reduced direct product, or ultraproduct.

Skolem gave the first example of what Robinson termed a reduced subdirect product in order to establish the existence of nonstandard models of arithmetic. A general definition was given by Jerzy Łoś in 1954.[118] Subsequent interest in ultraproducts was stimulated by a brief abstract published by Frayne, Scott, and Tarski a few years later,[119] and by abstracts or papers by Chang, Keisler, Kochen, and Rabin.[120] Among important advances, it turned out that model-theoretic results based upon the extended completeness theorem (or one of its equivalents) could also be achieved using ultraproducts.

In emphasizing the role of valuations, Robinson noted that the ultraproduct construction was really a special case of a limit type of construction which led to a proof of the principle of localization at the same time. Although he mentioned the results of Frayne, Keisler, and Kochen relating elementary equivalence to ultrapowers, he did not go into details. Instead, he suggested that in the future, further study of different constructions yielding models of a given consistent set of sentences would prove to be of increasing importance.

In closing, Robinson returned to some basic notions and results of meta-algebra.[121] Mal'cev's work served nicely as a typical example of how direct applications of model theory led to concrete algebraic results. Other questions, Robinson confessed, were not so direct and could only be considered from the standpoint of metatheory. As an example, he mentioned the problem of bringing together different notions leading to quotient structures (normal subgroups and ideals), which he had already treated successfully using a model-theoretic approach in his *Théorie métamathématique des idéaux.*[122] Likewise, bringing the "intuitively

[117] Robinson 1960a, p. 61; Robinson 1979, 1, p. 13.

[118] Łoś 1955, pp. 98–113.

[119] Frayne, Scott, and Tarski 1958, pp. 673–674. See as well the comprehensive paper by Frayne, Morel, and Scott 1956.

[120] Keisler 1960, Kochen 1961, and Rabin 1959.

[121] Section 7, Robinson 1960a, pp. 76–78; Robinson 1979, 1, pp. 28-30.

[122] Robinson 1955.

analogous" concepts of algebraically closed fields and real-closed fields together provided another good example of the power of model theory. It was this last example that brought his Stanford lecture to a conclusion with a flourish.

Robinson's most recent research, in fact, touched on exactly these problems. Taking algebraically closed fields as a special part of commutative field theory, to solve an algebraic problem whose parameters belonged to a field M, there was no need to go beyond the algebraic closure of M itself. Similarly, given an ordered field M, in many cases it was sufficient to consider nothing more than the real closure of the field itself. Moreover, instead of considering the algebraic, or real, closure of M, any algebraically or real-closed extension would do.

Consider, for example, a set of axioms K for a commutative field of specified characteristic, and another set of axioms K' for an algebraically closed field (or possibly a real-closed field). Using Seidenberg's elimination theory, Robinson had earlier shown how it followed that if K were a set of axioms for a differential field of characteristic 0, then it possessed an extension to a suitable theory of differentially closed fields K'.[123] Unfortunately, it was not always possible to assign to every model M of K a model M' of K' which was an extension of M comparable to that of the algebraic closure (or of the smallest real-closed extension) in the sort of examples Robinson had given.

Somewhat ironically in retrospect, since Robinson was only months away from one of the greatest discoveries of his career as a logician, namely nonstandard analysis, he apologetically excused himself for not having included the subject of nonstandard models. At the time he had in fact been considering nonstandard models of arithmetic (rather than analysis) but added, in a remarkably understated way that could only be appreciated a decade later: "there is no doubt that [the subject of nonstandard models] will continue to grow vigorously." He was less apologetic, however, about having overlooked infinite languages, another important branch of model theory developed by Tarski and his school. Acknowledging that here a "spectacular success was scored not long ago," he could add that "Fortunately, Tarski is here to tell you about it himself."[124] And Tarski did.[125]

Embedding Problems

Among the problems Robinson was working on during his first semester in Princeton, discussions with two of his colleagues, Dov Tamari

[123] Robinson 1958b.

[124] Robinson 1960a p. 78; Robinson 1979, 1, p. 30.

[125] See Tarski 1960.

and Michael Rabin, had set Robinson's mind to thinking about several embedding problems which resulted in a paper he sent to *Fundamenta Mathematica* in December of 1960.[126] Łoś had shown that not every semigroup with cancellation could be embedded in a group, and similarly, not every ring without zero divisors could be embedded in a (skew) field. Although tests to determine whether or not a semigroup could be embedded in a group had been worked out by Mal'cev, Lambek, and Tamari, the corresponding problem for rings and skew fields had only been dealt with in particular cases. In considering these cases, Robinson found that a model-theoretic approach could be used to prove the existence of a test to solve this problem.

Dov Tamari had already shown that a Birkhoff-Witt ring with a finite number of indeterminates over a commutative field could be embedded in a skew field.[127] He was also able to show that the same still holds if the ring is locally finite. Robinson was able to use his methods to bridge the gap between Tamari's two results.[128]

Michael Rabin suggested the last of the embedding theorems which Robinson proved for any extension K of the theory of rings without zero divisors:

> *Let* $\{M_n\}$ *be a set of models of* K *where* n *ranges over the non-empty index set* $I = \{n\}$ *and let* M *be the strong direct sum of* $\{M_n\}$. *Let* M' *be a subring of* M *which does not contain any zero divisors. Then* M' *can be embedded in a model of* K.[129]

First News of Nonstandard Analysis: Washington, D.C. 1961

During his first semester at Princeton, Robinson had been thinking about nonstandard models of arithmetic, particularly Skolem's work, and one day, as he walked into Fine Hall, the idea of nonstandard analysis suddenly flashed into his mind! This was to prove the beginning of something entirely new, even revolutionary, in mathematics.

It was one of those fortuitous coincidences—for Robinson had also

[126] Robinson 1961a. Previously, Robinson had already found model theory useful in deriving various embedding theorems (which B. H. Neumann had done conventionally in 1954). Robinson found these theorems also followed as a direct result from the extended completeness theorem, and in fact could be produced even "more directly by means of the predicate calculus." See Robinson 1955b and Neumann 1954.

[127] Tamari 1953.

[128] See Robinson 1961a, pp. 460–461, in Robinson 1979, 1, pp. 476–477. Work on conditions under which a ring (without zero divisors) could be embedded in a skew field had been considered as early as 1933 by Øystein Ore. For discussion of later results, including applications related to Ore's work by A. W. Goldie on the universal enveloping algebra of a finite-dimensional Lie algebra, see Jacobson 1962, chapter 5, "Universal Enveloping Algebras," pp. 151–197, esp. pp. 165–166.

[129] Theorem 3.4, Robinson 1961a, p. 461, in Robinson 1979, 1, p. 477.

just been invited to give the plenary address for the silver anniversary meeting of the Association for Symbolic Logic. The lecture he gave was indeed special, entirely appropriate for such an important occasion, for he used it to make the first public announcement of his latest discovery, included as part of his presentation on "Non-Standard Arithmetics and Non-Standard Analysis."

That year the ASL meeting was held at the Willard Hotel in Washington, D.C. (on Tuesday, January 24, 1961), where the American Mathematical Society and the Mathematical Association of America were also holding their annual joint meeting.[130] Robinson's paper heralded his new idea: how to give rigorous foundations to the calculus with infinitesimals. The paper was also communicated almost immediately by Arend Heyting to the Dutch Royal Academy of Sciences in April, and was subsequently published in the *Proceedings* of the Academy.[131]

Beginning with a reference to Skolem's well-known work on nonstandard models of arithmetic, Robinson explained how Skolem had shown the existence of proper extensions of the natural numbers N which possessed all of the properties of N formulated in the lower predicate calculus. Such extensions were known as (strong) nonstandard models of arithmetic.[132] The question Robinson now posed: What happened if the same approach were taken to the real numbers R_0?

Taking K_0 to be the set of sentences formulated in the lower predicate calculus in terms of the elements of R_0 and all relations definable in R_0, all elementary statements about functions could be expressed within K_0. Robinson then introduced his nonstandard models of analysis—models of K_0 which were proper extensions of R_0 with respect to all the relations and individual constants in R_0. Taking R^* to be any nonstandard model of analysis, elements of R_0 were termed "standard" elements. The infinitesimals comprised a maximal ideal M_1 in the ring M_0 of finite elements of R*. The existence of such models followed, as he had shown as early as 1951, from a familiar application of the extended completeness theorem of the lower predicate calculus.[133] But nonstandard models of analysis could also be constructed in the form of ultrapowers, as Simon Kochen had done.[134]

130 That afternoon, S. C. Kleene also gave an invited address on "Foundations of Intuitionistic Mathematics." For details on both, see David Nelson's summary of the meeting in the *Journal of Symbolic Logic*, 25 (1960), p. 93.

131 Robinson 1961. For a helpful, very readable introduction to nonstandard analysis, see the article in *Scientific American*, Davis and Hersh 1972. Upon reading this article, Robinson sent a postcard—showing a statue of d'Alembert—to Martin Davis, "complimenting me on the article." Martin Davis to JWD, E-mail note of September 7, 1994.

132 Skolem 1934.

133 Robinson 1951.

134 Kochen 1961.

Kochen's method was especially helpful because of the insights it suggested into the structure of nonstandard models and how (or to what extent) it was possible to single out certain distinguished models in particular. As Robinson explained:

> It is our main purpose to show that these models provide a natural approach to the age old problem of producing a calculus involving infinitesimal (infinitely small) and infinitely large quantities. As is well known, the use of infinitesimals, strongly advocated by Leibnitz (*sic*) and unhesitatingly accepted by Euler, fell into disrepute after the advent of Cauchy's methods which put Mathematical Analysis on a firm foundation. Accepting Cauchy's standards of rigor, later workers in the domain of nonarchimedean quantities concerned themselves only with fragments of the edifice of Mathematical Analysis. We mention only du Bois-Reymond's Calculus of infinities and Hahn's work on nonarchimedean fields which in turn were followed by the theories of Artin-Schreier and, returning to analysis, of Hewitt and Erdös, Gillman, and Henriksen. Finally, a recent and rather successful effort of developing a calculus of infinitesimals is due to Schmieden and Laugwitz whose number system consists of infinite sequences of rational numbers.[135]

Robinson's approach was conceived in such a way as to yield a proper extension of classical analysis. Thanks to the precision of the framework in which he was working, standard properties of specific functions including the trigonometric functions, Bessel functions, and relations in general, simply carried over to nonstandard counterparts in R^*. Because his system also contained infinitely small and large quantities, it was possible to "reformulate the classical definitions of the infinitesimal calculus within a calculus of infinitesimals and at the same time add certain new notions and results."[136]

Although not generally appreciated at the time, Robinson indicated that the ultrapower construction coincided (in certain special cases, at least) with the construction of residue fields in the theory of rings of continuous functions. Additionally, there were also connections between these theories and Robinson's version of nonstandard analysis. Without giving details, he promised to have more to say about these connections in his forthcoming *Introduction to Model Theory and to the Metamathematics of Algebra*, which was soon to appear in the Studies in Logic and the Foundations of Mathematics series from North-Holland.

[135] Robinson 1961, p. 433; Robinson 1979, 2, p. 4. The major drawback to the Schmieden-Laugwitz approach was that it included zero divisors and was only partially ordered. Thus many basic results of the calculus, for example, required modification if they were to be treated in the Schmieden-Laugwitz fashion.

[136] Robinson 1961, p. 433; Robinson 1979, 2, p. 4.

The first task of Robinson's paper was to introduce nonstandard analysis and the general subject of nonarchimedean fields. As Robinson explained the crux of his method:

> Syntactically, or linguistically, our method depends on the fact that we may enrich our vocabulary by the introduction of new relations, such as $R_0'(x)$, $M_0'(x)$, $M_1'(x)$ which define R_0, M_0, M_1, in R^*. (Note that the singulary relations just mentioned are, provably, not definable in terms of the vocabulary of K_0). We are therefore in a position to reformulate the notions and procedures of classical analysis in nonarchimedean language. Since all the "standard" results of analysis still hold we may make use of them as much as we please and we may therefore carry out our reformulation either at the level of the fundamental definitions (of a limit, of an integral, . . .), or at the level of the proof or, finally, by introducing nonstandard notions into a result obtained by classical methods.[137]

The last section of Robinson's paper briefly sketched some examples. Taking s_n to be a standard sequence, that is, a function defined on the natural numbers N and taking values in R_0, the definition could be extended automatically to the nonstandard positive integers. Taking s to be a standard number, Robinson offered the following straightforward definition:

> 3.1 *s is called the limit of s_n if and only if s_ω is infinitely close to s for all infinitely large positive integers ω. In symbols:*
>
> $$(\omega)[N'(\omega) \wedge \sim R_0(\omega) \supset |s - s_\omega| \in M_1].$$ [138]

Similar definitions were given for limit points and Robinson provided an alternative nonstandard proof for the Bolzano-Weierstrass theorem. Cauchy's necessary and sufficient conditions for convergence were introduced in a nonstandard version, and for functions of a real variable, nonstandard conditions were formulated for such concepts as limit, continuity, uniform continuity, and derivatives. Using the notation $a =_1 b$ to mean that a was infinitesimally close to b, that is, that $b - a$ was infinitesimal, Robinson expressed the nonstandard version of the derivative as follows:

> 3.5 *$f(x)$ has the derivative f_0 at the standard point x_0 (f_0 a standard number) if for all infinitesimal η:*
>
> $$\frac{f(x_0 + \eta) - f(x_0)}{\eta} =_1 f_0 .$$

[137] Robinson 1961, p. 434; Robinson 1979, 2, p. 5.
[138] Robinson 1961, p. 435; Robinson 1979, 2, p. 6.

As Robinson noted, various "standard" results of the differential calculus, including Rolle's theorem, could be established with remarkable ease using nonstandard analysis. Due to the brevity of the paper, Robinson was not able to go into detail, although he did offer some preliminary observations:

> We touch only briefly upon integration and remark that, up to infinitesimal quantities, Cauchy's integral and Riemann's integral can be defined by means of a partition of a given standard interval into an infinite number of subintervals of infinitesimal length combined with the formation of the usual sums such as $(x_n - x_n - 1)y_n$. The nonstandard definition of the Lebesgue integral appears to be more intricate and has not been carried out in detail so far.[139]

Robinson went on to speculate about the important applications he anticipated from nonstandard analysis, especially in cases where infinitesimals had a natural role to play. As an obvious example, Robinson pointed out that much of the classic work in differential geometry and analytical mechanics had been done using a vague notion of infinitesimals. It had always been "a matter of general belief that all this work could, if necessary, be rewritten to conform to the rigor of contemporary mathematics but nobody would think of carrying out this task. It is therefore not without interest that we may now justify the use of infinitesimals in all these problems directly."[140]

To give an idea of the possibilities he had in mind, Robinson offered some details for the osculating plane of a skew curve. He also considered the n-parametric Lie group G, where passage to R^* naturally extended it to a wider nonstandard group G^*. Consequently, the "infinitely small" transformations in G^* could now be interpreted in Robinson's new, nonstandard way. He noted that the nonstandard transformations constituted a genuine subgroup G' of G^*, but did not go into any more detail (except to say that this could all be taken much further).

Robinson also introduced nonstandard integrals. For example, taking the Riemann integral, the nonstandard definition "preserves the properties of an integral to the extent which they can be expressed in the lower predicate calculus, e.g., approximation by sums of the form $(x_n - x_{n-1})y_n$."[141]

Nonstandard approaches to the Dirac delta function, Green's formula, and Green's identity were all suggested, along with interesting results to

[139] Robinson 1961, p. 436; Robinson 1979, 2, p. 7.

[140] Robinson 1961, p. 437; Robinson 1979, 2, p. 8. For the problems arising with Lebesgue integration, see Robinson's letter to W.A.J. Luxemburg, May 13, 1962, RP/YUL, quoted below, pp. 341–342.

[141] Robinson 1961, p. 438, Robinson 1979, 2, p. 9.

be expected from examining Volterra's formula for the solution of the two-dimensional wave equation. As for classical applied mathematics, Robinson noted that "it would of course be natural to reword the usual statements about particles of fluids and about infinitesimal surfaces and volumes in terms of the present (i.e., nonstandard) theory." But instead, he only hinted at this, preferring to consider a specific example from fluid mechanics with which he was especially familiar, one that gave rise to some interesting conceptual difficulties:

> It is the assumption of boundary layer theory, e.g. for flow round a body or through a pipe, that the equations of inviscid flow are valid everywhere except in a narrow layer along the wall. It is found that the thickness of the layer, d, is proportional to $R^{-1/2}$ where R is the Reynolds number, supposed large. Within this boundary layer, the flow is determined by means of the boundary layer equations which are obtained by simplifying the Navier-Stokes equations of viscous flow. However, when solving these equations, it proves natural to suppose that the boundary layer is *infinitely thick* [*sic*].[142]

But consider the case of a straight wall along the x-axis. Solving the boundary layer equations for boundary conditions at $y = 0$ and $y \rightarrow \infty$ raised a "conceptual difficulty" that interested Robinson, since this was incompatible with the assumption of the smallness of d.

This difficulty could be resolved quite nicely by supposing that the inviscid fluid equations held for all positive standard values $y > 0$ while confining the influence of viscosity to values of y belonging to $O(\delta)$, δ infinitesimal (in which case the Reynolds number R was infinite). Introducing $y' = y/\delta$, it was then possible to derive and solve the boundary layer equations for $0 < y' < \infty$, a region in which the flow had not been previously defined. As Robinson suggested, "there are other problems in continuous media mechanics that should be amenable to a similar analysis."[143] But he went on to add:

> In reality it is of course not true that the region in which viscosity is effective may be regarded as infinitely thin. It can in fact be seen with the naked eye both in certain laboratory experiments and in every day life. Thus, the above model is intended only as a conceptually clear picture within which it should be easier to discuss some of the more intricate theoretical questions of the subject such as the conditions near the edge of the layer.
>
> For phenomena on a different scale, such as are considered in modern

[142] Clearly Robinson meant "infinitely thin." The emphasis in the original is his. See Robinson 1961, p. 439; Robinson 1979, 2, p. 10.

[143] Robinson 1961, p. 439; Robinson 1979, 2, p. 10.

physics, the dimensions of a particular body or process may not be observable directly. Accordingly, the question whether or not a scale of nonstandard analysis is appropriate to the physical world really amounts to asking whether or not such a system provides a better explanation of certain observable phenomena than the standard system of real numbers. The possibility that this is the case should be borne in mind.[144]

Clearly Robinson's mind was racing to think of all the myriad possibilities nonstandard analysis held out–in theory and in practice–in a wide variety of applications. But all of that was still in the future, especially as Robinson was about to take up a new position at UCLA.

First there were other developments from a mathematical point of view that demanded Robinson's attention. Perhaps he felt it was necessary to establish nonstandard analysis with full legitimacy on purely mathematical terms–not only for logicians but especially for mathematicians–before going on to explore its usefulness in applied mathematics. In any case, after completing the year at Princeton, Robinson spent several months supported by a grant from the National Science Foundation at the University of California, Berkeley, where he worked on nonstandard arithmetic and an accompanying nonstandard language.

The main purpose of Robinson's research at this point was to discuss the syntactical properties of his nonstandard language, and to consider two types of truth definition–internal and external–for the nonstandard language itself. The value of these results became clearer over the next few years as the elements of nonstandard analysis began to unfold. Robinson's approach, however, marked an important departure from previous efforts to develop nonstandard arithmetics by using the standard lower predicate calculus to discuss a nonstandard model of arithmetic (as had Henkin, Scott, and Tarski in the late 1950s).

Fraenkel's Seventieth Birthday: Robinson on Constructing Models

On February 12, 1961, Abraham Fraenkel celebrated his seventieth birthday. His colleagues in Jerusalem had organized a collection of essays to be presented to him in honor of the occasion, and Robinson had been one of the volume's editors. Due to his commitments in Princeton that year, however, he had to commemorate the event *in absentia,* although he contributed a paper of his own, "On the Construction of Models."[145]

Fraenkel was fond of saying that Robinson, his best-known student, was also proof of the exception to the old adage that mathematicians

[144] Robinson 1961, pp. 439–440; Robinson 1979, 2, pp. 10–11.
[145] Robinson 1961b.

did their best work when they were very young.[146] At the time of the Fraenkel *Festschrift*, Fraenkel was seventy and Robinson just over forty years old.

For the *Festschrift*, Robinson chose to write a broad, expository paper on the subject with which he was most closely identified: model theory. In doing so, he intended to demonstrate in a quiet yet unmistakable fashion—one of his trademarks—the power of several new results he had only recently achieved.

The paper began by explaining the basic aims of model theory: to determine the relations between systems of sentences (often of the lower predicate calculus) and mathematical structures satisfying these systems. Thus given a set of sentences (which might be finite or infinite in number), a major problem of model theory was to find a structure (or structures) M satisfying the sentences in question. It was well known that according to the extended completeness theorem of the lower predicate calculus, if K is a syntactically consistent set of sentences of the lower predicate calculus, then K possesses a model M. In pioneering applications of this theorem, Mal'cev showed that it could be replaced by a purely semantic, "general principle of localization," namely, that "K possesses a model if and only if every finite subset of K possesses a model." Indeed, Mal'cev's localization principle is what is now simply known as the compactness theorem. Another type of model construction was the method used for obtaining a reduced direct product of a given set of models, a construction related to Skolem's method for obtaining nonstandard models of arithmetic.

What Robinson aimed to do in the Fraenkel anniversary paper was to "formulate and prove a fundamental lemma which can be made to play a central role in independent proofs of the several theorems and constructions mentioned above."[147] The fundamental theorem in question was a total valuation lemma (lemma 2.1). Through a corollary in which the valuations were indexed over the set of integers $N = \{n\}$, it was possible to prove the extended completeness theorem of the lower predicate calculus: If K is syntactically consistent then it possesses a model.[148] Thanks to the valuation lemma, Robinson could establish the extended completeness theorem without recourse to either transfinite induction or to the maximal ideal theorem for Boolean algebras, both of which had been used in earlier derivations of the theorem.[149] Noting that the

[146] Fraenkel 1967, p. xxx.

[147] Robinson referred to Skolem 1955 and Łoś 1955. He also mentioned in particular results published by Frayne, Scott, and Tarski on reduced products, as well as recent papers by Kochen on filtration systems, and by Rabin on arithmetical extensions.

[148] Theorem 3.2, Robinson 1961b, p. 210; Robinson 1979, 1, p. 35.

[149] In fact, the compactness theorem is equivalent to the assertion that every Boolean

general principle of localization and the compactness theorem were "very nearly two versions of one and the same result," Robinson gave another application of the valuation lemma, namely, the compactness theorem (although without, as yet, calling it such): *If every finite subset of K possesses a model then K also possesses a model.*

Robinson's final applications concerned model constructions involving ultrapowers, requiring first a special case of the valuation lemma. As Robinson concluded, this was the fundamental property of ultrapowers:

> *In particular, if all M_n are copies of the same M_0, then M_0 is called an ultrapower of M. In that case, all sentences which hold in M_0 (and which may include individual constants of M_0) hold also in M, i.e. M is an arithmetical extension of M_0.*[150]

THE UNIVERSITY OF CALIFORNIA: LOS ANGELES BECKONS

In the spring of 1961 the Robinsons drove across the United States, picnicking as they went. They retraced parts of earlier trips across South Dakota to visit the Badlands and Dinosaur National Park, and returned to the Rockies where Abby's camera caught Renée against the snow in a summer dress, freezing she remembers in the unexpected chill of the cold mountain air. They drove through Nevada and stopped at Lake Tahoe (more touristic than was to their liking) before eventually reaching their initial destination: Berkeley.

While in California they made a point of visiting the missions, including San Miguel, Santa Barbara, and San Juan Capistrano. It was on one of his trips south to Los Angeles that Robinson was invited to join the faculty at UCLA.

Even before leaving Israel for the United States in 1960, Robinson seems to have been thinking seriously about the possibility that the time had come to look for another position. As early as 1959, the University of Toronto made it clear that it wanted Robinson back:

> Dear Professor Robinson:
>
> The purpose of this letter is to ask you if you would think at all kindly of the notion of returning to the University of Toronto. I am not thinking particularly of an appointment directed to Applied Mathematics, but sim-

---

algebra contains an ultrafilter, which is in turn a version of the maximal ideal theorem, since filters and ideals are duals. I am grateful to Mel Fitting for calling my attention to this fact, established in Theorem 5.2 of Bell and Slomson 1969, p. 104. See also Robinson 1963, p. 50, problem 2.6.2.

[150] Robinson 1961b, p. 216; Robinson 1979, 1, p. 41.

ply a Professorship in Mathematics, with a free hand to offer what you like. Indeed, the distinction between Pure and Applied seems to be diminishing here. Perhaps the fact that we are all housed together in Baldwin House contributes to this. This is only a temporary arrangement, but when we move into the new Arts building a year or so from now, we will stay together. . . .

Yours sincerely, D. B. DeLury, Chairman[151]

Robinson replied as soon as he received DeLury's letter, expressing his interest in the offer but explaining that he could not accept for at least a year:

Dear Professor DeLury,

Thank you very much for your letter of November 12. I certainly have not forgotten Toronto and I am glad to see that, in a very definite sense, Toronto has not forgotten me. I particularly appreciate the terms of appointment contained in the first paragraph of your letter, which are suited ideally to my present outlook.

It so happens that a short while ago I was approached by Princeton University with the offer of a visiting professorship for 1960–61, to take the place of one of their senior people, who will have his sabbatical leave next year. My reply was positive and this means that I shall not be available before the beginning of the academic year 1961. For that date I wish to give your offer my very serious consideration. However, since this refers to a time which is still more than a year and a half away, it might be best to defer a final decision until the end of 1960 when I shall presumably be in the States and may even be able to visit Toronto. I appreciate of course that in the meantime you do not remain committed to the offer contained in your letter. However, if for some reason I come to a definite decision at an earlier date I shall not hesitate to write to you.[152]

DeLury responded on December 1, congratulating Robinson on his appointment at Princeton, adding that he hoped this would indeed provide an occasion for Robinson to visit Toronto sometime during the year. He also added a word of explanation about the position he had in mind: "When I wrote to you, I was not thinking necessarily of the year 1960. If, any time during the next year, you decide to have a go at Toronto, I shall be delighted to hear it and I have no doubt that I can attend to the arrangements here."[153]

[151] DeLury to Robinson, November 12, 1959, Robinson's administrative file, Department of Mathematics, University of Toronto.

[152] Robinson to DeLury, November 24, 1959, Robinson's administrative file, Department of Mathematics, University of Toronto.

[153] DeLury to Robinson, December 1, 1959, Robinson's administrative file, Department of Mathematics, University of Toronto.

More than a year passed before Robinson again wrote to DeLury, but the following letter makes clear that by February 1961, Robinson was definitely interested in the possibility of returning to Toronto:

> This is just to let you know that I am giving very serious consideration to your offer. However, I felt that before saying anything more definite I ought to discuss the matter and timing of my release with the authorities of the Hebrew University, and I have written to them accordingly. As soon as I have clarified the situation in this respect, I shall write to you again.[154]

Toronto was equally serious about hiring Robinson, so much so that Claude Bissell, then president of the university, also wrote to him (in a letter marked "Personal") to say that he hoped their negotiations would result in Robinson's return to Canada: "I was very happy to hear that you have been having talks with Professor DeLury, and I earnestly hope that they result in your decision to return. Christine and I send you and Renée our warmest regards. . . ."[155]

The Decision to Go to UCLA

While Robinson was mulling over the prospects of academic life back in Toronto, he was invited to lecture in the Philosophy Department at UCLA. According to Angus Taylor, who had met Robinson only the year before, in Israel:

> The next time I saw Robinson was in Los Angeles in the Spring of 1961, when he was visiting the UCLA Philosophy Department. The idea of a joint appointment for Abby in Mathematics and Philosophy was beginning to take shape. This idea was brought to a successful conclusion later that Spring when Abby agreed to come to UCLA at the beginning of the 1962–63 academic year. I was chairman of the Mathematics Department from 1953 to 1964, with the year 1961–62 out for a sabbatical in Europe, so my return from sabbatical coincided with Abby's coming to UCLA.[156]

Not only was Robinson favorably impressed by southern California, but the faculty there was equally impressed with him. Soon an official offer was made of a joint professorship in philosophy and mathematics. Robinson promptly accepted the offer, and phoned his old friend Claude Bissell to say that he was sorry, but had decided not to return to Toronto after all. He tried to break the news to DeLury by letter in as straightforward a way as possible:

[154] Robinson to DeLury, writing from Princeton, February 21, 1961, Robinson's administrative file, Department of Mathematics, University of Toronto.

[155] Bissell to Robinson, April 10, 1961, RP/YUL.

[156] Angus Taylor in a letter to George Seligman, November 29, 1976, RP/YUL.

> I can assure you that this has not been an easy decision for me since both my wife and myself have only fond memories of our years in Toronto. Indeed, my decision is not due to any general preference but rather to the concentration of good people working in my field at UCLA and elsewhere in California. While I appreciate that you will now wish to make alternative arrangements for filling the available vacancy in Toronto, I hope that my personal relations with all our friends at the University of Toronto will remain unaffected.[157]

DeLury wrote back on April 21, saying that he understood the grounds for Robinson's decision. He also did his best to emphasize that the position at Toronto was one that would be created especially for Robinson, whenever he might wish to accept it. There were no time limits or other constraints. "Please do not think of the post I suggested to you as a vacancy that must be filled. Rather, it is a new post, to be created for you if you should choose to come back to Toronto. Therefore, I ask you to take the view that it will still be open if your decision should turn you in our direction."[158]

By then, however, Robinson had made up his mind in favor of UCLA. Doubtless the appeal of inheriting Rudolf Carnap's chair was a powerful factor, as was Renée's unhappiness with their situation in Israel, the lack of stimulating opportunities for her, and the sense of cultural isolation she increasingly felt after the excitement and glamour of living in London, even during the war, and the pioneering spirit of Toronto in the postwar years.

Although Israel was alive with pioneering energy of its own, it was a nation at war with virtually all of its neighbors, and this was something she could never forget, especially in a divided Jerusalem. Above all, Renée could find little to engage her. Considering the situation in Israel in the 1950s, there was really no place for her skills either as an actress or fashion designer. Theater was concentrated in Tel Aviv, and in Jerusalem, even the opportunities for her interests in broadcasting were minimal:

> In Jerusalem, only programs for children were produced. The head of the children's department at the radio station was very interested to try me and we made a record telling a story for children. But the radio people who listened to it said that my accent would irritate the children, that they would concentrate on that rather than the story. I spoke Hebrew well

157 Robinson to DeLury, April 18, 1961; Robinson's administrative file, Department of Mathematics, University of Toronto.

158 DeLury to Robinson, April 21, 1961, Robinson's administrative file, Department of Mathematics, University of Toronto.

enough, but the Sabras talk very fast and with a certain list. So that was the end of that.[159]

Consequently, while in Israel Renée turned her artistic talents to painting, sculpture, and crafts.

Renée Robinson, Jerusalem, 1960.

California, however, offered not only the allure of Hollywood but the sophistication of Beverly Hills where fashion and the arts were prized. As a world capital for both television and film, Los Angeles would offer new opportunities that Renée might pursue if she wished.[160] For Abby, UCLA offered a tremendous opportunity to advance the fortunes of mathematical logic. Working in both Departments of Philosophy and Mathematics was ideally suited to his interests at the time. Moreover, the proximity of a good number of the world's leading logicians throughout the University of California system must have been equally enticing.

Before Robinson left Berkeley that summer, Clark Kerr, president of the University of California system, wrote to make his appointment official. It was agreed that he would be given a nine-month contract at UCLA, beginning July 1, 1962, with a salary (step 3) of $14,556.[161]

Around the World

Rather than return to Israel via the familiar transatlantic route, Renée and Abby decided to make an around-the-world tour, which included as much lecturing as it did sight-seeing in Hawaii, Japan, Hong Kong, Thailand, Cambodia, Nepal, and India. From San Francisco, after saying their farewells at Berkeley, they flew to Honolulu to enjoy several days on the beaches of Waikiki.

From Hawaii they continued on to Japan where Robinson spent several weeks lecturing in Tokyo, Kyoto, and Nagasaki. In addition to visiting colleagues like Sigekatu Kuroda, Akira Kobori, and Gaisi Takeuti, the Robinsons saw many of the historic and natural sights of the coun-

159 Renée Robinson to JWD, letters of May 29, 1991, and March 5, 1994.

160 Renée Robinson I-1980.

161 Kerr to Robinson, June 26, 1961, RP/YUL.

Abby and Renée in Kyoto.

try, with special trips to Nagoya and Nikko. Especially impressive was the Tosho-gu Mausoleum of Tokugawa Ieyasu, intended as a monument to the deified founder of the Tokugawa shogunate and ornately decorated with rich reliefs and brilliantly colored with lavish use of gold and lacquer.

The *shinkansen* "bullet train" sped them from Tokyo past Mount Fuji to Kyoto, where they visited magnificent temples and went to a special tea ceremony–Renée never forgot the bitter taste of green tea. The Robinsons were invited to the Kuroda's home, where they were served western sandwiches. And one morning, sitting on the floor of their room at the Miyako Hotel, they had a fish soup breakfast! In their Japanese robes, it all seemed quite fitting, if exotic.

In Kyoto the shimmering vision of the Golden Pavilion was unforgettable, and provided a dramatic contrast to the austerity of the Zen Buddhist temples with their formal rock gardens and intricate geometric patterns. Always drawn to the countryside, Renée and Abby especially enjoyed visits to Yomoto Spa and the falls at Lake Chuzenji. One morning they set out on a hiking trip with Professor Tarasaka to Hakone, which ended abruptly when they heard a storm warning for an approaching typhoon. Back in Kyoto, everything that could be nailed down was, and the Robinsons sat in their hotel room with the wind howling outside. Fortunately, as they followed the progress of the storm on television, the typhoon eventually moved in another direction.

From Osaka they flew to Hong Kong, made a tour of the New Territories, and visited the harbor at Aberdeen where thousands of sampans and floating restaurants constituted a veritable city on water. Although

Robinson did not give any lectures, Li Ping and his wife did show them the sights of street life in Kowloon, which Renée thought was very busy but excessively noisy. She did find some Chinese silk that she liked, and bought it to use later for a dress that she designed when they were back in Israel.

In Cambodia (Kampuchea) they stayed in Phnom Penh, where their hotel overlooked the river and they could watch people-powered trishaws plying the boulevard from the window of their room. The Buddhist monks on the streets with their shaved heads and orange robes all seemed to Renée like the Hare Krishna groups that they had seen in San Francisco, but as Abby reminded her, "these are *real* ones!"[162]

The Robinsons' primary goal in Cambodia was to visit the famous ruins at Angkor Wat and Angkor Thom. Although in disrepair, the temples were still very much artistic masterpieces. Among the most memorable sights were the many sculptures, including a long row of two-faced Buddhas–one side the benevolent face, the other side the evil one–and an extraordinary "elephant gallery."

In Thailand Robinson was invited to lecture through the auspices of a high-ranking naval captain who had spent a year in Israel studying at the Hebrew University, and was in fact one of Robinson's former students. What impressed Renée at the time was the fact that he had two

[162] Renée Robinson I-1980.

Abby at one of the entrances to the Royal Palace, Bangkok, Thailand.

Renée and Abby, visiting Angkor Thom by trishaw in Cambodia.

wives, both sisters. He sent a chauffeured car to drive Abby to the lecture, and later took both Renée and Abby sight-seeing. They visited the ornate Grand Palace complex, constructed in the eighteenth century, where in addition to the many temples and shrines, the famous image of an Emerald Buddha is revered.

From Bangkok the Robinsons flew to Calcutta, then Muzaffarpur (in Bihar, not far from Patra) where Renée and Abby were greeted with garlands of marigolds and jasmine. Robinson was scheduled to deliver a lecture at the university, where they had tea. Hundreds of students were present for Abby's talk, after which he and Renée were invited to the home of one of the professors to meet his family.

Proceeding to Katmandu, they found enticing environs in Nepal for long walks against the majesty of the Himalayas. As guests of the president of Tribhuvan University, the Robinsons stayed in a special guest house where the mosquito nets were not especially reassuring—"there must have been ten years of blood on them," exclaimed Renée! While in Nepal they visited many of the local Hindu sites, and at one stop Abby had his picture taken on the steps of a beautiful temple, surrounded by Nepalese children. Renée was impressed by the local crafts and had two hand-embroidered hats made for her in Katmandu.

They did a lot of hiking, observed the daily routine of the local people, and especially enjoyed the carved wooden façades of the shops and houses. Renée barely avoided disaster, however, when she slipped on cow dung and landed with a terrible thud on her right arm. To their surprise, the only available doctor had studied in Los Angeles. He took Renée to a pharmacy, upstairs in one of the old buildings, and gave her some tablets to relieve the pain (although even a year later her hand continued to bother her).

Abby in front of a Stupa, with local children, Katmandu, Nepal.

Back in India, they went to New Delhi, and made a point of visiting Agra and the Taj Mahal at night, its ethereal beauty unforgettable in the moonlight. A former student from the University of Toronto served as their guide to

Renée and Abby at the Taj Mahal, Agra, India.

the site, which offered a serene combination of gardens, fountains, and reflecting pools dominated by the sleek white marble of the mausoleum.

From Agra they went on to Bombay where Robinson had arranged to spend a few days at the Tata Institute. There too he gave a lecture, and was invited to return whenever his travels might next bring him to India. All in all, the trip proved to be a remarkably rich combination of university visits and Asian cultures. As the Robinsons made their way back to Israel, Abby had even begun to think about writing a novel, and actually began to draft the first chapter of a manuscript that might well have turned into an international thriller had not mathematics reclaimed his attention.

Back in Israel: New Students and Old Friends

Having agreed to accept Carnap's chair at UCLA, Robinson found that his last year in Israel was a difficult and nostalgic one. He made the best of his time, seeing as much of the country as he could, as if taking it all in for one last time. When Calvin Elgot visited from IBM, Robinson took him to Wadi Ara. Another visit to see a Druze village produced a memorable photograph, in which Abby appears wearing dark glasses against a background of blooming fruit trees.

Among the dissertations Robinson supervised that year, the one he directed as sole advisor was Shlomo Halfin's "Contributions to Differential Algebra."[163] Halfin's thesis was concerned primarily with local differential algebra and ideals, subjects in which Robinson himself was also keenly interested. Amram Meier's thesis, also completed in 1961–1962, was devoted to "Analytic Continuation by Summability and Relations between Summability Methods." Robinson served as senior supervisor of his thesis, which was also read by Dr. Amnon Jakimovski.[164]

Although Shmuel Schreiber apparently worked for a time with Robinson as well, he did not finish his dissertation on "Loop Rings" until well after Robinson had left the Hebrew University for UCLA. Nevertheless, when Schreiber finished his dissertation, Robinson was listed along with A. S. Amitsur and R. Artzy as one of his thesis advisors.[165]

Nonstandard Analysis Begins to Take Shape

Robinson spent much of the year working on a revision of his first book, *On the Metamathematics of Algebra*. This time he aimed to provide not only a concise overview of model theory (with all of the latest developments produced in the decade following the book's first publication in 1951), but to present the basics of nonstandard analysis as well. Many mathematicians had been intrigued by Robinson's early reports of nonstandard analysis, and the book would finally make an introductory description widely available. In a lengthy letter to W.A.J. Luxemburg written in May of 1962, Robinson described his own progress, and history, in developing the subject. As he said, he was now "living in a nonstandard world":

> Dear Wim,
>
> I regard it as a great compliment that an analyst of your caliber has been willing to invest so much effort in Nonstandard Analysis and I was very pleased to see the result. As you know from my paper I regard the ultrapowers as special cases of the sort of models to which my theory applies but I quite agree with you that to many analysts the (apparently) more abstract arguments of logic [~~appear~~] seem strange. As you anticipated, your chapter II is new to me. Your remark about the prime ideal

[163] Halfin 1962. [164] Meier 1962. [165] Schreiber 1963.

theorem versus the axiom of choice is relevant in view of Halperin's result that one is indeed weaker than the other. As it happens, I inserted a similar remark in the paper mentioned below. I also found some of your examples very interesting.

My own history in the field of Non-standard Analysis in recent months has been as follows. Having written up the sort of thing described briefly in the paper you know, I rested on my laurels for some time. Having done so I discovered a non-standard way of proving the existence theorem for $dy/dx = f(x, y)$, $f(x, y)$ continuous, by proving in this way Ascoli-Arzelà. This is the best result included in my new book in "Studies in Logic" which is mentioned in my paper. More recently I discovered that I can use my methods in Complex Function Theory, where it serves to replace, and to improve upon, the usual normal family methods. In this way I have obtained several new results in the theory of Picard-Julia on the behavior of a function in the neighborhood of an essential singularity, and on the zeros of complex polynomials. At the same time I found that it is natural and convenient to broaden the basis of the theory by relying on a higher order calculus. The new [~~model~~] theory also can be translated into the language of ultraproducts but as you know I find logic more congenial, and at least in some cases really simpler. My paper on the subject is now being printed in a form similar to that of your notes and I shall send you a copy of it as soon as it is ready. Incidentally, I sent you a reprint of my "Non-s[tandard] Analysis" in the *Ind[aginationes] Math[ematicae]* some time ago. The use of a higher order language also settles the problem of the introduction of the Lebesgue integral by Non-s[tandard] Analysis, a problem which I, and apparently you also, found a little difficult to cope with in an orderly manner. For some time now I have been thinking about problems in Functional Analysis but so far as I can see our activities also may intersect there. Altogether, so far as my standard duties permit, I am now living in a non-standard mathematical world, so you will forgive me if I reply only briefly to your query regarding my method in "On the Construction of Models." First of all, you are quite right in thinking that a special case of my lemma can be identified with Rado's (in the Canad[ian] Journal 1950 ca.).[166] This was pointed out to me by Büchi at Stanford,[167] and I mentioned it in a written version of a lecture which is due to be published.

If I am not mistaken, your lemma is rather like one I proved last year in Berkeley in order to apply my method to a two-valued logic. Dana Scott then showed that similar things can be done by Tychonoff's theorem. I then generalized my method to infinite valued logic to a point which I

[166] Rado 1949.

[167] Büchi 1962; Büchi was then at Michigan, but gave a paper at the Stanford Congress in 1960.

(and I think my audience also) found so confus[~~ing~~]ed that I gave up in disgust.

Wishing you all the best in your noble endeavor. Regards to Erdélyi. Renée and I are sending you and Trudy our kindest regards. Yours ever, Abby.[168]

[P.S.] Congratulations on your promotion.

Although it was not published until 1963, Robinson completed his revised version of *On the Metamathematics of Algebra* in 1961. He explained that the earlier book had been out of print for some time, and that, in the intervening decade, "the subject had developed vigorously."[169] Rather than reissue *Metamathematics*, or attempt to update it, Robinson decided an entirely new work was called for. Although nearly half of the contents was the same, much of the material in the opening chapters was presented in a different way, often with improved or simplified terminology, arguments, and proofs. The second half of the book was almost entirely new.

What Robinson had first written in the 1950s about noncountable languages, properties of classes of structures that were closed under extension or intersection, the completeness of algebraically closed fields, and his method of diagrams—these were new ideas that required more justification and explanation than was needed ten years later, when such material was well established. Robinson regretted, however, that one important aspect of *Metamathematics* had met with less success, namely, "the suggestion that numerous important concepts of Algebra possess natural generalizations within the framework of the Theory of Models." Not only did this approach provide "a certain unity of outlook," but it produced new results—the most important criterion in Robinson's view—as the concept of a differentially closed field more than adequately demonstrated.[170]

Consequently, the revised book was retitled *Introduction to Model Theory*, and its most significant new element—for which it became immediately well known—was its presentation of his new idea of nonstandard analysis. As he put it rather modestly, "This is a new application of Model theory which provides an effective calculus of infinitesimals and which appears to have considerable potentialities."[171]

The promise—and first basic results—of Robinson's model-theoretic approach to defining infinitesimals had already been sketched in his

[168] Robinson to Luxemburg, May 13, 1962, RP/YUL.

[169] Robinson 1963, p. v. The original, *On the Metamathematics of Algebra,* was first published by North-Holland in 1951. Robinson finished writing the new book in December of 1961, in Jerusalem.

[170] Robinson 1963, p. vi.

[171] Robinson 1963, p. vi.

brief article on "Nonstandard Analysis" in 1961.[172] From the beginning he acknowledged other work bearing on the subject, and gave credit to Carl Schmieden and Detlef Laugwitz for their "recent and rather successful effort of developing a calculus of infinitesimals."[173]

The major drawback, he added, to what Schmieden and Laugwitz had accomplished was the fact that their method (using infinite sequences of rational numbers) included zero divisors and was only partially ordered. This meant that many basic results of the calculus required modification before they could be treated using the Schmieden-Laugwitz approach. In general, Robinson characterized these earlier results as "fragments of the edifice of Mathematical Analysis," whereas he had succeeded in providing a "proper extension of classical Analysis":

> That is to say, the standard properties of specific functions (e.g. the trigonometric functions, the Bessel functions) and relations, in a sense made precise within the framework of the Lower predicate calculus, still hold in the wider system. However, the new system contains also infinitely small and infinitely large quantities and so we may reformulate the classical definitions of the Infinitesimal calculus within a Calculus of infinitesimals and at the same time add certain new notions and results.[174]

The book opened with chapters on the lower predicate calculus, algebraic theories (including Mal'cev's theory of normal chains), the basic concepts and methods of model theory, model-consistency, and an entire chapter on completeness, along with tests for model-completeness.[175] Chapter 6 of *Introduction to Model Theory* dealt with generalizations of algebraic concepts, for example, polynomials, bounded predicates, algebraic predicates, convex systems, and separability, all of which were taken with little modification from *Metamathematics*. Likewise, chapter 7 on the metamathematical theory of ideals drew from his book of 1955, *Théorie métamathématique des idéaux*. Chapter 8, devoted to the metamathematical theory of varieties, ended with a discussion of Hilbert's seventeenth problem.[176]

[172] Robinson 1961.

[173] See the passage quoted more extensively above, p. 283, from Robinson 1961, p. 433; Robinson 1979, 2, p. 4. By the time he wrote *Introduction to Model Theory*, the work of Schmieden and Laugwitz had taken on greater significance, and he now emphasized that their work "deserves special mention." See Robinson 1963, p. 271.

[174] Robinson 1961, p. 433; Robinson 1979, **2**, p. 4.

[175] The theory of model-completeness was developed in detail in Robinson 1956e; for his theory of relative model-completeness, see Robinson 1957b. Both works are discussed above, chapter 6, pp. 223–227 and p. 229.

[176] This included Robinson's improvements inspired by Artin-Schreier, incorporating material he had already published in the mid-1950s. See Robinson 1955c, 1956a, and 1957b. For Robinson's earlier work on Hilbert's seventeenth problem, see above, pp. 215 and 229.

The last chapter, oddly, was not trumpeted as "nonstandard analysis." This was simply included as one of several "selected topics." Others included elimination of quantifiers, direct products, and ultraproducts.[177] Only the last three sections of the book's final chapter were devoted to nonstandard analysis itself, and this material went no further than applications to the theory of functions of a real variable. The results were basic, but gave a good sense of how nonstandard methods could be used in place of standard arguments.

Robinson explained his approach in a no-nonsense way. He constructed a strong nonstandard model of arithmetic using ultrapowers to produce a proper elementary extension of the natural numbers. Similarly, a nonstandard model of analysis arose naturally as an ultraproduct of R_0, the set of all real numbers. Applications followed almost immediately. After proving a theorem involving partial derivatives, he added, "we may also justify in this manner the use of infinitesimals as they occur in classical Differential Geometry and in Analytical Mechanics."[178]

Robinson then introduced standard and quasistandard functions, explaining that "quasi-standard" functions can be used to represent generalized functions such as Dirac's delta function, "but we shall not follow up this possibility here."[179] This passing reference to delta functions may have been included because they had served as a major source of inspiration for an independent and possibly rival version of nonstandard analysis, first suggested by Schmieden and Laugwitz, and extended with considerable vigor by Arthur Erdélyi at Caltech in the early 1960s—at just the same time, in fact, that Robinson was doing his own work at the Hebrew University in December of 1961.

Among additional applications Robinson gave, for example, of nonstandard analysis at work was a nonstandard proof of the existence of a solution for a differential equation of the first order with a given initial value. Although to a large extent this followed the standard proof (originally due to Peano), instead of applying the Ascoli-Arzelà theorem, Robinson used nonstandard arguments to achieve the same sort of compactness result in nonstandard terms.

Saying Good-Bye: Leaving Israel Again

Leaving Israel was not so difficult for Renée—as the American Friends of the Hebrew University put it in 1959, "Jerusalem is one of the qui-

[177] Robinson credited Łoś with formulating the idea of ultraproduct, referring to Łoś 1955. The approach Robinson took was that from his own "On the Construction of Models," Robinson 1961b.

[178] Robinson 1963, p. 265.

[179] Robinson 1963, p. 267.

eter world capitals."[180] Actually, "Renée was not very happy in Jerusalem. Living on a border with a hostile neighbor often gave her a feeling of being cooped up in a garrison town."[181]

But leaving Israel was more difficult for Abby, if only because of his deep attachment to the country both intellectually and emotionally. His leaving also brought some repercussions as well—and these were to weigh, sadly, upon Robinson for the rest of his life. Above all, for Robinson who was always of an open and generous spirit, it must have hurt, personally, when some of his colleagues in Israel proved less than sympathetic to his decision to leave. Even though he returned to Israel on numerous occasions, his relationship with some of his former colleagues was never quite the same.

Indeed, there is a special terminology reserved for Jews who emigrate from—or decide to leave—Israel. While Renée and Abby were hailed as new *olim* (a term meaning to go up or ascend) upon their arrival in 1957, those who left Israel were called *yordim*—descenders. In the case of Jews leaving Israel, the term had a sharply pejorative, negative sting. Abby, once he made clear his intentions to leave Israel, was subsequently something of an outcast in the minds of more than one of his former colleagues.

It may be difficult to understand this fully unless one is an Israeli, rooted in the history and the land that Robinson himself loved more than any other place on earth. Nathan Rotenstreich, rector of the Hebrew University, conveyed something of the commitment Israelis expected, especially of those in the university: "By its very nature, an institution of higher learning is intimately concerned with the great movement of national liberation."[182]

This visceral identification of the Hebrew University with Israel itself explains how some might have felt that leaving the university was tantamount to abandoning the nation at precisely that moment when it was struggling in its early years to survive. This feeling was present from the very beginning, in fact, when the thirteenth Zionist Congress passed its resolution establishing the Hebrew University, several decades before land and buildings were to make the vision a reality. The university, in the words of the Congress resolution, "is destined to constitute a central force in the national and political life of the land of Israel."[183]

[180] It went on to add, "There is practically no night life," The Hebrew University 1959, p. 5.

[181] Seligman 1979, p. xxv.

[182] Tsvi Hermann Schapira, *Das Projekt einer Jüdischen Hochschule*, quoted in an address given by the rector of the Hebrew University at the ceremonial opening of the academic year 1965–1966: Rotenstreich 1966, p. 6.

[183] Rotenstreich 1966, p. 7.

Judah Magnes, the founding president, put it as follows in opening the Institute of Jewish Studies in 1924:

> The Jewish people needed a university just because it was not like other peoples. They were dispersed through the world, speaking all the languages; and nevertheless, they were a specific unity, but threatened by assimilation. The University should be the hearth of a spiritual revival in the land of Israel. . . .[184]

The repeated emphasis upon the ties which bound the university and Israel together, their past, present and future, was reflected yet again, decades later, in a poignant statement by Rabbi Bernard M. Casper, dean of students: "Hebrew University is much more than an institution of learning. It has come to symbolize all that we are and everything we hope to be."[185]

Despite those who were displeased with his decision, it was not for any lack of national feeling or identity with Israel on Robinson's part that he had decided to leave. On the contrary, as Renée often said, he "loved . . . every stone." She once told George Seligman "how every stone in the city had its special appeal to him, how he loved to walk the streets of Jerusalem and to breathe in the historical air of its places."[186]

Whatever others might think, Robinson always regarded the Hebrew University as a special part of Israel and of himself–in the words of Norman Bentwich, the "hub of the things of the mind and spirit for Israel and for Jewry." The majesty and beauty of the "vision splendid" was something Robinson always remembered, and carried with him wherever he went, whether he was in Israel or not. Indeed, as Bentwich described the Hebrew University, it was a powerful symbol, a "visible beacon of light, a people's Menorah."[187]

[184] Judah Magnes, quoted in Bentwich 1961, pp. 158–159.

[185] B. M. Casper, quoted in Scofield 1959. See as well "Hebrew University Rises Again," Samuels 1959.

[186] Seligman 1979, p. xxiii.

[187] Bentwich 1961, p. 57.

CHAPTER EIGHT

UCLA and Nonstandard Analysis: 1962–1967

> Mathematics is a jungle, the Jungle of the Infinite.
>
> —*Ian Stewart*[1]

> Mathematics is not a deductive science—that's a cliché. When you try to prove a theorem, you don't just list the hypotheses, and then start to reason. What you do is trial and error, experimentation, guesswork.
>
> —*Paul R. Halmos*[2]

THE POSSIBILITY of bringing Abraham Robinson to the Los Angeles campus of the University of California arose during the visit he made to southern California in the spring of 1961, when he was working briefly at Berkeley while on leave from the Hebrew University. Although he was invited by the Mathematics Department, the Philosophy Department was just then considering ways to replace Rudolf Carnap (who was about to retire), with the hope of luring Alfred Tarski from Berkeley. Robinson, having expressed his interest in UCLA, offered an attractive alternative, one that would also serve to continue the department's strong reputation in mathematical logic. As R. M. Yost (then chairman of philosophy, soon to be replaced by E. A. Moody) wrote to Robinson in early April of 1961:

> Dear Professor Robinson:
>
> Professor Angus Taylor and I were delighted to learn, just before your visit to UCLA, that you were willing to consider the possibility of joining us permanently in 1962–1963. After an informal consultation, we concluded that by pooling our prospective resources in position and money, we could in all probability offer you a joint appointment in our two departments.
>
> My own department, no less than the Department of Mathematics, would be highly pleased and honored if you should decide to come to Los Angeles. Our own understanding is that if you should accept a joint appointment at UCLA, you would offer at least one course per semester in the Philosophy Department. Out of concern for the interests of my department, I should indicate to you the kinds of courses that would be of special value to us. We feel that our graduate students, especially those who

[1] Stewart 1991, p. 75.

[2] Halmos 1985, p. 321.

intend to go on for doctorates, should in some degree become acquainted with the most important developments in logic that have been made during the last generation.

But most of these students will never attain the technical proficiency to master these subjects. We therefore hope that you would be willing to offer an advanced undergraduate course in which such important ideas as deducibility and completeness are presented discursively to those of our serious students who do not intend to specialize in logic. We hope also that you would be willing to offer a graduate course in our department in which topics in logic are presented to students who, though fairly well trained in the techniques of mathematical logic, have not had a complete undergraduate background in mathematics. From my conversations with Professor Taylor, I have the strong impression that the courses you would offer in his department could presuppose as much mathematical background as you wish.

Meanwhile, Professor Taylor and I will press arrangements for your appointment as vigorously as we are able so long as you do not signal us to stop.[3]

Writing in support of Robinson's appointment was Alonzo Church, who endorsed Robinson's reputation as among the very best, internationally, despite the fact that he was still comparatively young (Robinson was forty-two), yet already well known as a major figure among logicians:

There is no question that Robinson is a man of the highest standing in scholarship and research, and I have no hesitation in recommending him to you in the strongest terms in this regard. He has contributed to various branches of mathematics, but his most important work and that with which I am most familiar is in mathematical logic and in the important and rapidly developing intermediate field between logic and abstract algebra. Indeed the credit for originating the application of modern logic to algebra must be shared equally between Robinson and Tarski, who introduced the method independently of each other at about the same time. (It happened in my case that I first heard of this important idea from Robinson, though I afterwards learned that similar work was being done by Tarski.)

Your request for as definite an estimate of Robinson as possible led me to ask myself, just who and how many of those who have done original work in mathematical logic I would rank as superior to him. Naturally it is difficult to make such comparisons final—opinions will inevitably differ from person to person—and it is moreover especially to the philosophical

[3] R. M. Yost, Jr., to Robinson, Princeton University, April 12, 1961, Robinson's administrative file, Department of Philosophy, UCLA.

side of mathematical logic, in view of the fact that his own original work is entirely in mathematics. Nevertheless I go on to give you my estimate for what it is worth. In the United States there are just four men in this field whom I would rank certainly ahead of Robinson, and a fifth as possibly so. The five names are Rudolf Carnap, Kurt Gödel, S. C. Kleene, W. V. Quine, Alfred Tarski. Possibly three or four others have to be put as about equal, or nearly so. And among the half dozen ablest of the younger men in this field whom I know, perhaps one or two will in the end surpass Robinson, but this is presently speculative.

Turning to Europe I know again of about five men who might be ranked ahead of Robinson. These include the younger Markoff in Russia, certainly, and possibly Novikoff (though here I have some doubts), Mostowski in Poland, Skolem (in Norway) and Bernays (in Switzerland) have to be mentioned on the basis of past accomplishments rather than present activity. I would rank Specker (in Switzerland) below Robinson, though I know that some others think differently. The same thing applies to Georg Kreisel.

Briefly, if I rank Robinson among the first fifteen men in this field throughout the world, I am being safely conservative. If I put him among the first ten, the proposition becomes doubtful but can definitely be argued.[4]

The fact that the University of Toronto was also interested in making Robinson a handsome offer did not hurt his case at UCLA (see chapter 7). In fact, this gave additional urgency to the letter Taylor and Yost now wrote together asking the administration to approve Robinson's appointment with all dispatch:

On behalf of the Departments of Mathematics and Philosophy we propose that Dr. Abraham Robinson be appointed jointly as Professor of Mathematics and Professor of Philosophy, effective July 1, 1962. We propose that the position for this appointment be supplied, initially, by the position to be vacated by Professor Carnap on his retirement, June 30, 1962. The necessary salary is estimated to be about $16,000. This would then require a small augmentation of the salary presently paid to Professor Carnap. The Mathematics Department undertakes to contribute the amount of .5 FTE from new positions made available to it at the earliest possible time after 1962, thus releasing .5 FTE to the Philosophy Department for other use.

We shall presently describe the qualifications of Dr. Robinson. But first it is necessary to point out the need for extraordinary dispatch in consid-

[4] Alonzo Church to Angus Taylor and R. M. Yost, April 12, 1961, Robinson's administrative file, Department of Philosophy, UCLA.

ering this recommendation. Professor Robinson, now serving as Visiting Professor at Princeton, will return to the Hebrew University in Jerusalem for the year 1961–1962. But he is strongly interested in obtaining a position in America after that. The University of Toronto is trying to induce him to go there, where he was during 1951–1957. His interest in UCLA and his potential availability for a position of the type we propose, was not discovered until a few months ago. Now we know that, if he is to avoid making a commitment to Toronto in the very near future, we must be able to give him definite and strong assurances that we shall make an effort to bring him to UCLA, and that our efforts have a good chance of success. We must be able to do this at the beginning of the week of April 17. Accordingly, a truly exceptional effort must be made to act swiftly in this case. We believe the case justifies such efforts, for we have in Dr. Robinson a man of unusual ability and distinction.

Professor Robinson is a versatile and powerful scholar. At the age of 42 he has reached a position of world eminence as a mathematical logician. We think he may safely be ranked among the best ten men in this field. We have excellent verbal estimates of his standing from our own colleagues, several of whom (in both departments) are thoroughly acquainted with the names and works of leading logicians. With this letter we are sending a letter from Professor Alonzo Church about Professor Robinson. We are confident that ample supporting appraisals can be obtained from other outstanding scholars. . . .

Professor Robinson was at UCLA as a guest of the Mathematics Department just before the Spring Vacation, and at that time he gave two lectures. He also held extended discussions with members of the two departments on matters of professional interest. He showed himself to be a fine lecturer; both brilliant and gracious, and a man of broad intellectual interests. If we can add him to our faculty, we shall in so doing perform a notable strengthening of our two departments, and we shall also greatly please certain of our colleagues who are especially admiring of and interested in Robinson's work. This is an important matter in the case of Professors Montague and Chang, who have felt just a bit isolated for lack of more intense activity in mathematical logic at UCLA.

It may also be appropriate to point out that Professor Robinson's appointment would be only the third tenure appointment in Philosophy at UCLA in twenty-three years. The Department of Philosophy had at one time hoped to replace Professor Carnap with a distinguished scholar in the philosophy of science. After a vigorous canvass, extending over a period of several years, the Department has now reached the conclusion that there is no possibility of finding a suitable candidate of this sort. Hence, the Department now feels that this proposed action of appointing Professor Robinson jointly in Mathematics and Philosophy is the best opportunity which can be expected of retaining a distinguished scholar in a field that is at least closely allied to Professor Carnap's.

We respectfully urge your prompt action toward carrying out our recommendation.[5]

Within a week the necessary paper work had been processed, and Robinson promptly accepted the offer subsequently extended. Yost, upon hearing this news, immediately sent off a telegram to Princeton:

I have just heard the good news from Vice-Chancellor Sherwood. My Colleagues and I are delighted that you have decided to come to UCLA. With Cordial Regards, R.M. Yost.[6]

Following an exchange of telephone conversations, it was agreed that Robinson would accept a step 3 professorship.[7] In support of his appointment, letters were solicited from Tarski, Heyting, Rosser, and Bernays. By the end of June, the appointment papers had finally passed all the way up the administrative chain of command, and on June 26 the final, official letter confirming Robinson's appointment was sent. UCLA Chancellor Franklin D. Murphy wrote personally to Robinson:

to express our great satisfaction at the approval by the Board of Regents of your appointment as Professor. I hope that you will see fit to accept this appointment and join the distinguished and growing UCLA family.[8]

The Robinsons *en Route*

When Abby and Renée left Israel late in the summer of 1962, they made a brief stop in Europe, in part to spend a few days in Munich with his brother Saul and wife Hilde. Saul was now based in Hamburg, where the Robinsohns had moved in 1959 (when he was named director of the UNESCO Institute for Education).

Initially, the UNESCO project was intended to serve as an educational resource center "to help Germany overcome her intellectual isolation which

[5] Angus E. Taylor and R. M. Yost, Jr., writing jointly to Chancellor Franklin D. Murphy, April 13, 1961, Robinson's administrative file, Department of Philosophy, UCLA.

[6] Copy of telegram sent April 18, 1961, to Abraham Robinson from R. M. Yost, Robinson's administrative file, Department of Philosophy, UCLA.

[7] See the letter of May 11, 1961, written jointly by Taylor and Yost to Chancellor Murphy. This represented a compromise of sorts—it was not the highest salary that might have been negotiated, but definitely the most expedient. For further details about Robinson's later advancement to step 5 (after having been denied a step 4 promotion!), see below, pp. 324–326.

[8] Franklin D. Murphy to Abraham Robinson, June 26, 1961, Robinson's administrative file, Department of Philosophy, UCLA.

Saul and Hilde Robinsohn.

The Robinsons in formal attire, aboard the *S.S. France* in 1964.

resulted from the war period and National-Socialism."[9] Founded in 1952, the institute aimed to promote "informal international exchange of ideas." Over the years, however, its goals were sharpened as it sought to stimulate studies on comparative education on a broader scale.[10]

Robinsohn's appointment as director of the institute provided a "dynamique impulsion."[11] Indeed, in 1961, as a widely recognized authority on comparative education, Robinsohn helped to organize a special international meeting at the University of London's Institute of Education. He was also instrumental in founding the European Com-

[9] Gal 1962, pp. 11–12.

[10] The institute was also responsible for publishing the *International Review of Education*, founded in 1955 within three years of the Institute itself. See Gal 1962, p. 11.

[11] Gal 1962, p. 4.

parative Education Society, in which he assumed a leading role from the start.

When Renée and Abby saw Saul in the summer of 1962, he had just returned from an important meeting in France, where a group of the society's officers had met at the Centre International d'Études Pédagogiques, in Sévres, for three days in May. Because the Society hoped to work in close cooperation with the newly formed Comparative Education Division of UNESCO and the UNESCO Institute for Education in Hamburg:

> Members of the Committee were therefore extremely pleased when Dr. Robinsohn of the Hamburg Institute agreed to serve not only (as at present) on the Constitution Committee but as a member of the Executive Committee.[12]

Saul's best news for Abby and Renée in Munich, however, was the fact that his position with UNESCO in Hamburg, initially only a temporary appointment, had just been renewed.[13]

From Munich, Renée and Abby continued on to Italy where they embarked on the *Leonardo da Vinci,* first class, bound for New York. Abby reveled in the crossing, the daily routine of deck-chair reading and before-dinner cocktails, the pleasure of Italian cuisine, fine wines, and midnight buffets. While Abby sometimes dressed for dinner, black tie, Renée took the opportunity to wear her evening dresses and, occasionally, a mink stole that she had brought with her for cool evenings on deck. But in less than a week the *Leonardo da Vinci* docked in New York, and the real world was back in focus. They off-loaded the Vauxhall that had served them so well in Israel and across so much of Europe, and set off for the West Coast.

Adjusting to Southern California

The City of Angels—sprawling, smoggy, crisscrossed with freeways—stood in sharp contrast to the history and tradition of Jerusalem. UCLA, however, was set on rolling hills once part of a Spanish land grant, the Rancho San José de Buenos Ayres. Ironically, the campus itself had a distinctly Italian feeling. This was due entirely to the setting, which reminded UCLA's first architects of northern Italy, and thus "a Romanesque or Italian Renaissance style of architecture was adopted, featuring red brick, cast stone trim, and tile roofs."[14] In fact, many of the early buildings at UCLA were based on designs patterned after churches and universities in Bologna, Milan, and Verona.

[12] Holmes 1963, p. 419.
[13] Seligman 1979, p. xxiv.
[14] Stadtman 1967, p. 330.

Consequently, the campus offered a pleasant combination of Italianate brick and stone buildings, grassy lawns, and landscaped sidewalks. But this was no hilltop town. When Robinson arrived in late summer of 1962, the registration issue of the UCLA student newspaper, the *Daily Bruin*, ran a banner headline: "Record Number to Enroll"—nearly twenty thousand.[15]

> UCLA has 19,000 students, taught by 3000 professors, housed in 42 major buildings. It has 62 departments, 12 schools and colleges and several research categories enclosed within its 400 acres. . . . The University is BIG. . . . Many feel she is too big and is growing too fast. Some students say that they feel like a number with the word "taught" stamped and processed on their perforated card. . . .[16]

Indeed, finding a parking place was a major effort. As the *Bruin* declared, to no one's surprise, "First Day Auto Jam Clogs Parking Lots."[17] At first, Robinson did not have to worry about driving, for he and Renée lived not far from campus. By the end of the semester, however, they had found a serene home high in the Santa Monica mountains, only twenty minutes from campus:

> At first we rented a house, but within three months we bought this beautiful place in Mandeville Canyon. I heard about it and when I saw the place that was it. It also had a beautiful pool. Many film stars lived in this canyon—Robert Taylor was our next-door neighbor. We had lots of friends coming to swim, and Tarsky, when he visited us, said "your house is nearly more famous than you are."[18]

It was not long before Renée had converted one of the rooms into a working studio where in addition to the printmaking she had done in Israel, she took up a new interest:

> I had an enormous large window and, quite often, wild animals would come down from the hills and look in; if I were inside, we would both stand and look at each other. I became a ceramicist soon, and enjoyed the work very much. It was not long before I began to exhibit, and won prizes.[19]

[15] 19,922 students were expected in the day sessions alone. See the issue for Tuesday, September 11, 1962, p. 1.

[16] *UCLA Daily Bruin*, Tuesday, September 18, 1962, p. 2.

[17] According to *Time Magazine*, "In a rueful moment President [of the University of California] Clark Kerr once defined his three main headaches as "sex for the students, athletics for the alumni and parking for the faculty." Quoted from Hamilton and Jackson 1969, p. 146.

[18] Renée Robinson to JWD, letter of March 3, 1992.

[19] Renée Robinson to JWD, letter of March 5, 1994.

The Robinson's home in Mandeville Canyon, in the Santa Monica mountains, Los Angeles, California.

Robinson at UCLA: Philosophy and Mathematics

The interaction of mathematics, logic, and philosophy had a long history at UCLA that Robinson doubtless found attractive:

> First [there was] the brief encounter with Bertrand Russell. Though he stayed for one year only (1939–40), the impact of his presence has lingered over the years. In 1938, Professor Hans Reichenbach came to the LA campus from the University of Istanbul. His arrival accelerated the rate of good studies in the [Philosophy] department and marked the beginning of a curricular tradition, with special emphasis on studies in logic and the philosophy of science, that has continued to the present. After Reichenbach's untimely death in 1953, Professor Rudolf Carnap joined the department, lending his distinction to the advancement of this program.[20]

Robinson, as Carnap's successor, was expected to further enrich and

[20] Stadtman 1967, p. 359.

extend this tradition. Ernest A. Moody, having recently assumed the duties of department chairman, made a special point of welcoming newcomers at the first department meeting in September, Robinson included.[21] The department also lost no time in putting him to work. In preparing questions for the written qualifying examinations that year, Robinson was naturally asked to serve as a consultant for questions on logic.

It was the habit of the Philosophy Department to give a cocktail party for its new members (along with visiting, junior, and emeriti colleagues), and this was held at Moody's home on September 29. Another party was scheduled separately for graduate students, assorted deans and guests at the home of Donald Kalish. Over the years Kalish and Robinson became good friends, and for a time they worked together on the UCLA Vietnam Day Committee.

In philosophy, however, Robinson invested a considerable amount of time in committee work. Mathematics demanded less, due primarily to its larger staff, but even there he was a member of the Committee on Graduate Studies and several subcommittees.[22] His joint appointment, however, meant administrative double-duty.

Philosophy, because it was smaller, worked as a committee of the whole, and met weekly. Department records show that Robinson was faithful in attending almost every meeting. In his first year alone, he accumulated a variety of responsibilities–including service on a committee reconsidering requirements for dissertation plans and oral qualifying examinations.[23] He also served on the Carnap Prize Committee, and was responsible for a committee report on advising graduate students, written jointly by Yost and Robinson, which concerned not only qualifications for the M.A. degree, but outlined necessary revisions for the UCLA *Graduate Manual* as well.[24] In the years to come, both the importance and amount of Robinson's committee work increased sub-

[21] Minutes of the Philosophy Department Meeting for September 13, 1962, archives of the University of California, Los Angeles, record series no. 411, "Philosophy Department, Chairs' Correspondence, 1930–1987." Cited hereafter as Philosophy Department Minutes.

[22] When Robinson joined the faculty in 1962, the number of part- and full-time students in mathematics had risen to almost three hundred; nearly twenty doctoral degrees were awarded annually. The number of tenured faculty members was about twenty-eight, with approximately twenty assistant professors, lecturers (and several distinguished visitors thrown into this number each year). I am grateful to Angus Taylor for correcting the figures here from the less-accurate account to be found in Stadtman 1967, p. 355. Angus Taylor to JWD, November 22, 1992.

[23] Philosophy Department Minutes, meeting of January 18, 1963.

[24] Staff contributions paid for the Carnap prize which was given for the best student essay. See Philosophy Department Minutes for March 7 and 28, and April 23, 1963. The report on graduate student advisement and degree requirements was submitted on May 2.

stantially. What impressed his colleague David Kaplan at the time was the fact that Robinson not only seemed to have an unlimited capacity for taking on administrative responsibilities without apparent detriment to his scientific work, but that he always seemed available to "play." He remained, however, "very conscious of time."

Teaching at UCLA

In the autumn of 1962 Robinson offered a seminar in logic, Math 260, and an introduction to philosophy of mathematics, followed that spring with another logic seminar, Phil 271.[25] During 1963–1964, he again offered no lower division courses, but in the first semester taught introduction to modern logic in the Philosophy Department, and in the second, philosophy of science. This included not only a general survey, but "philosophical analysis of the concepts and laws of natural sciences."[26]

In the Mathematics Department, however, he offered a year course on set theory, which according to the UCLA catalogue covered axiomatic set theory, including relations, functions, cardinal and ordinal numbers, finiteness and identity, infinite arithmetic, partial-, simple- and well-orderings, the axiom of choice, the continuum hypothesis and its consequences, inaccessible numbers, and results on independence and consistency.[27]

The next year, just prior to his taking a leave of absence, Robinson offered a graduate seminar on applications of mathematical logic to analysis (in the fall), followed that spring by a lower division course on logic (the first half of which David Kaplan had taught), as well as a graduate seminar, foundations of mathematics.[28]

Kaplan, also a newly appointed member of the UCLA Philosophy Department, regarded Robinson as a cultured, European "gentleman" who wanted to participate from the beginning in everything at UCLA related to logic.[29] This included the logic colloquium, founded by Chen-Chung Chang and Richard Montague. Sessions alternated between departments—even numbered sessions were held in the Mathematics Department, odd numbers in Philosophy—and the emphasis was usually very mathematical.

In philosophical discussions, Kaplan found Robinson "a very fair

[25] UCLA 1962, pp. 451, 453.

[26] UCLA 1963, pp. 470, 471.

[27] UCLA 1963, p. 409.

[28] UCLA 1964, pp. 408, 466, 471.

[29] Kaplan was appointed to the UCLA faculty in 1962, but didn't receive his doctoral degree until 1964. He remembers how unnerving it was when Robinson decided to sit in on his logic seminar, although he recalls that Robinson "was very courteous . . . a warm, charming person with very deep philosophical interests," Kaplan I-1991.

player," and was impressed that in discussing philosophy of logic, Robinson never tried to use technicalities to win a point. "He talked philosophy the way philosophers did. He was a brilliant man who thought it through, philosophy being an *a priori* subject."[30] Robinson and Kaplan often discussed philosophical issues, in fact, of mutual interest:

> One issue we discussed a lot was the objectivity of truth conditions. I argued that we could move into a meta-language in which we could say something about the interpretation of the meta-language, but Robinson felt the meta-language itself was open to interpretation. Consequently, how could we ever get truth conditions for the object language metamathematically? According to Robinson, everything was open to interpretation.[31]

Robinson's openness in discussions with his colleagues extended to a generally informal style that he brought to his teaching. Donald Kalish remembers a lively example of his adroitness in the classroom during a course Robinson was teaching using Kleene's book, *An Introduction to Metamathematics*. Kalish, along with his colleague Richard Montague, were attending Robinson's lectures regularly that semester (a not uncommon practice among faculty in the Philosophy Department).

Montague liked to insist upon drawing every subtle distinction, making sure that every assumption was made explicit. Other logicians were not as obsessed with such detail—as Quine once said of Reichenbach, he played "fast and loose" with notation in particular.[32] Robinson, however, was interested primarily in establishing results—as many as he could, and one incident in particular reveals the very different philosophy that distinguished Robinson's approach from that of his colleague Montague:

> During Robinson's course, Montague kept asking questions about Kleene's *Introduction*—he wanted to "polish the book." Robinson was more interested in getting results out, stimulating students. Robinson, as politely as he could, but also as firmly as he could, turned to Montague at one point and said: "There is no question that you know your Stephen Kleene, but you don't know your Emily Post."
>
> Montague—confused as if he hadn't heard correctly—said "Did I miss some theorem?" Robinson explained that he was not interested in every quotation mark *à la* Tarski.[33]

[30] Kaplan I-1991. [31] Kaplan I-1991.

[32] W. V. Quine, review of Hans Reichenbach, "Elements of Symbolic Logic," in the *Journal of Philosophy*, 45 (1948), pp. 161–166, esp. 164. Specifically, Quine complained that "the new notations . . . play fast and loose with variables with a perversity which must be seen to be believed." Reichenbach replied to Quine's review in the same volume of the journal, pp. 464–467.

[33] Kalish I-1991.

This was simply a matter of conflicting styles, but on at least one occasion Robinson found himself playing the role of mediator. William Lambert was in the middle of a presentation, which Montague again interrupted:

> I remember getting caught in the crossfire once when I was presenting a class report; I referred casually, as we will, to "THE algebraic closure of a field." The late Richard Montague was in the room and proceeded to pick nits as only he could. Robinson, of course, was not at all into that particular sport. Things got tense for a while (I retired behind the nearest large piece of furniture!) but all worked out—with Abraham, despite his obvious annoyance, the peacemaker.[34]

As his colleagues realized, "Montague was a man of powerful will as well as intellect, and when his views on educational or philosophical questions were in conflict with those of his colleagues, personal clashes could sometimes ensue."[35] His life came to a tragic end in 1971, when he was murdered "at the hands of persons still unknown." But in the words of his colleagues Furth, Chang, and Church (in a brief *éloge*), he was recognized as a philosopher "of uncommon brilliance." Those who knew him best also valued "his qualities of humor, of sympathy, and of unshakable personal loyalty; the many friendships he formed with his colleagues were strong, uninterrupted, and deeply valued."[36]

Generalized Limits and Linear Functionals: Spring 1963

The twenty-eighth annual meeting of the Association of Symbolic Logic was held on Saturday, January 26, during the annual AMS and MAA joint meeting at the University of California, Berkeley, where Robinson presented a paper "On Some Topics in Nonstandard Analysis." This provided yet another informal introduction to the subject, but with a particular emphasis, for Robinson went on to explain how one could "give proofs of old and new results in the theory of functions of a complex variable. One can also define a notion of summability which links the Banach-Mazur limits with the theory of Toeplitz matrices."[37]

This was all spelled out at much greater length in a paper Robinson wrote over the next few months. The article he sent to the *Pacific Jour-*

[34] W. M. Lambert, Jr., to JWD, January 8, 1991.

[35] Furth, Chang, and Church 1974, pp. 68–70. Montague was forty years old when he was murdered in 1971, "in his home in Los Angeles, at the hands of persons still unknown at this writing. A man of uncommon brilliance and versatility, he packed into his tragically brief existence greater achievement than most can expect in a lifetime."

[36] Furth, Chang, and Church 1974, p. 70.

[37] *Journal of Symbolic Logic*, 27 (1962), pp. 475, 477.

nal of Mathematics in March of 1963 was devoted to generalized limits and linear functionals, and offered full details for many of the results he had covered only briefly during the ASL meeting at Berkeley.[38]

Robinson was concerned with continuous linear functionals F on the space m of all bounded sequences $\sigma = \{s_n\}$ of real numbers, where $F(\sigma) =$ standard part of $\sum_{n=1}^{\infty} a_n s_n$, where the a_n were elements of a nonstandard model R of the field of real numbers, and $\sum_{n=1}^{\infty}$ was taken over R. Not only did he establish necessary and sufficient conditions under which a linear functional $F(\sigma) = \lim_{n\to\infty} s_n$ for every convergent sequence s, but by using ultraproducts, he was also able to show that there was a nonstandard model R such that every continuous linear functional on the space of all bounded sequences s had the form $F(\sigma) \geq 0$ (for all nonnegative σ), and that $F(\{s_n\}) = F(\{s_{n+1}\})$.

Traditionally, using infinite matrices in summations had proven very useful in both real and complex function theory. In considering how Toeplitz matrices in particular figured in such summations, Robinson distinguished Toeplitz-limits (termed the very concrete matrix methods of summation) from other types of generalized limits such as Hahn-Banach and Mazur-Banach limits. Although Hahn-Banach limits might be more efficient, in general it was not always possible to use them in place of Toeplitz-limits. Nonstandard analysis, however, made it possible to "bridge the gap between the Hahn-Banach limits on one hand and Toeplitz-limits on the other hand."[39] In doing so, Robinson had to derive generalized limits related to the matrix methods but efficient for all bounded sequences. These limits, defined in terms of linear forms with nonstandard coefficients, satisfied the conditions of Hahn-Banach and Mazur-Banach limits.

Model Theory and the Metamathematics of Algebra

Robinson's *Introduction to Model Theory and to the Metamathematics of Algebra* appeared shortly after he arrived in Los Angeles, in 1963.[40] For anyone following Robinson's work in model theory over the decade since the Cambridge Congress in the early 1950s, there was little in the *Model Theory* book that was new, although one reviewer did recognize that it was "the first attempt to write a connected exposition of the new subject of model theory." As Erwin Engeler explained, "The main body of the work consists of rewritten versions of the author's main contributions to the subject, which are brought into a smooth and eminently readable sequence. Since these contributions are almost dense in the

[38] Robinson 1964.

[39] Robinson 1964, p. 270; in Robinson 1979, 2, p. 48.

[40] Robinson 1963.

domain, there results as complete a survey as can be expected at this time from any single author."[41]

But not everything in *Model Theory* was old news, rehashed or reworked for the sake of the book. Above all, there was a new proof of the completeness theorem of first-order logic (based on a simple combinatorial lemma), and three sections that were entirely new on what was fast becoming Robinson's best-known creation, nonstandard analysis. This too was the first systematic treatment that Robinson had given of the subject, and it was the avenue by which many mathematicians first learned of the strength and importance of the new subject—which were not at all apparent from the all-too-brief paper that had first introduced nonstandard analysis in the *Proceedings* of the Dutch Academy of Sciences in 1961. (For a discussion of the contents of the book, written while Robinson was still at the Hebrew University, see above, chapter 7, pp. 300–302.)

In addition to the wealth of technical material presented in *Model Theory*, Robinson found it "a matter of historical interest" that in the standard existence proof given by Peano of a solution for a first-order differential equation with given initial value:

> Peano employed an early version of symbolic logic. He justified the use of his notation by the remark that the complete development of the argument in terms of an ordinary language would be excessively complicated.[42]

This again reiterated, with historical emphasis, one of the most valuable reasons Robinson could advance for applying logic to mathematics, namely, the significant economies it frequently offered. Robinson was now doing his part to continue and strengthen this tradition. Indeed, he could not conceal his delight at the prospects nonstandard analysis seemed to offer. As his old friend and colleague at UCLA, Ernst Straus put it, nonstandard analysis "reached its next flowering in those years at UCLA":

> And you could see how much he enjoyed and how excited he was about this possibility. He kept quoting Fraenkel, who said "Nobody achieves a good mathematical result after the age of 30," and he was very pleased that his best ideas came to him, at least by Fraenkel's standard, in an advanced age.[43]

[41] Engeler 1964. In addition to finding the book weak on ultraproducts and ultralimits, he also thought more might have been done on connections between model theory, topology, and cylindric and polyadic algebras.

[42] Robinson 1963, p. 270.

[43] Straus R-1976.

Berkeley International Symposium on Model Theory: 1963

Robinson spent part of his first Californian summer at Berkeley, participating in the International Symposium on the Theory of Models which ran from June 25 through July 11. The meeting was designed to "bring together from all countries, for review and exchange of ideas, scholars who have done significant work in the area of research in the foundations of mathematics generally known as 'the theory of models.'"[44] Robinson had served as a member of the Organizing Committee, co-chaired by Henkin and Tarski, although most of the administrative work was shouldered by Dana Scott, secretary for the Organizing Committee. This proved to be no small task, considering the size of the meeting: one hundred sixty participants from seventeen different countries.

Although the symposium was dedicated to Thoralf Skolem, who had died in March, the published *Proceedings* included a brief opening paragraph about Evert Beth who had died in April of the following year. Appropriately, the editors voiced their appreciation for the many contributions Beth had also made over the years to the theory of models. The symposium, it was hoped, would help to give the new and somewhat disorganized field some focus. As a preliminary notice explained:

> To date no comprehensive treatise on [model theory] is available or (as far as we know) planned. The many contributions to this domain are widely scattered and some are difficult of general access. Hence, a meeting among the workers in this field, resulting in a published volume, would seem to be particularly valuable at this time.[45]

Indeed, the published *Proceedings* not only provided a comprehensive overview of model theory at the time, but offered an extensive bibliography on the subject and an opening "Foreword on Terminology." This, it was hoped, might help to establish some sort of uniformity throughout the field. The need for this was due not only to the newness of model theory as a discipline, but to discrepancies in terminology used by different authors: "In an effort . . . to give this volume greater unity the editors decided to *suggest* to authors, on an experimental basis, some uniform terminology and notation."[46] There followed a list of recommendations, not only of basic terminology (for models, rational structures, universe, language, sentence, grammar, formula, term, etc.), but also for symbols and even the proper typefaces and fonts that should be used for names, variables, et cetera.

[44] Berkeley 1963, p. vii.

[45] See the announcement describing the meeting in the *Journal of Symbolic Logic,* 27 (1962), pp. 128–129.

[46] Berkeley 1963, p. xiii.

The Berkeley Symposium itself was an ambitious and arduous one, spanning a seventeen-day period and two weekends, when excursions were planned in addition to various informal gatherings, including lunches, cocktail parties, and dinners. There were thirty-five one-hour invited addresses, among which was a lecture by Robinson on "Topics in non-Archimedean Mathematics."[47] Appropriately (for a meeting sponsored by the International Union for the History and Philosophy of Science), he began with some brief historical remarks about Leibniz and the history of the calculus:

> Leibniz wished to base the differential and integral calculus on a number system that included infinitely small and infinitely large quantities. More precisely, he regarded the new numbers as ideal elements which were supposed to have the same properties as the familiar real numbers and stated that their introduction was useful for the *art of invention*.[48]

Leibniz and his followers were never able to justify this approach, and so the attempt was largely "abandoned" by later generations that preferred the surer foundations for the calculus laid down by Cauchy and Weierstrass. Robinson was pleased to explain how he had succeeded where Leibniz had not, thanks to the methods of model theory. Inspired by the work of Thoralf Skolem, he was able to produce a model of the real numbers that legitimately contained both infinitesimals and infinitely large, nonstandard numbers, and he was anxious to show mathematicians as well as logicians just how useful nonstandard analysis could be.

With exactly this in mind, Robinson's paper offered a dazzling array of applications whereby model-theoretic methods were applied to analysis. For example, he explored the theory of complex functions and the analytic theory of polynomials, proving Rouché's theorem and a "far-reaching" generalization of Kakeya's theorem on the location of zeros. Although the tools were nonstandard, the results, including Kakeya's theorem as a special case, were standard: "Note that this is a classical theorem," Robinson insisted, "without reference to any nonstandard extension of the real or complex numbers."[49]

Equally of interest was a brief section on applications to topological spaces, with a very simple nonstandard version of the Heine-Borel theorem that a closed, bounded set in n-dimensional Euclidean space is compact (based on the nonstandard result that an arbitrary topological space $^{\circ}M$ is compact if and only if all points of its nonstandard extension $^{\circ}M^{*}$ are near-standard).[50] Another section devoted to normed linear spaces

[47] Robinson 1963b.

[48] Robinson 1963b, p. 285; Robinson 1979, 2, p. 99.

[49] Robinson 1963b, p. 291; Robinson 1979, 2, p. 105.

[50] Robinson 1963b, pp. 294–295; Robinson 1979, 2, pp. 108–109.

applied nonstandard methods to infinite-dimensional separable Hilbert space, and sketched how nonstandard methods could be used to develop spectral theory in particular.

Robinson closed his Berkeley paper with a few methodological remarks. While acknowledging that W.A.J. Luxemburg and Gaisi Takeuti had succeeded in producing nonstandard models of analysis using ultraproducts ("this has the advantage of making the theory more concrete in appearance"), he preferred to rely upon the formal properties of the models $^{\circ}M^*$ "rather than on their mode of construction." In fact, what he chose to emphasize were not ideal elements or structures but *methods*:

> It may be said that in spite of the model-theoretic garb of our method, it really represents an extension of the *deductive procedures* used in various branches of mathematics. And, if we wish to pass to a more philosophical level of discussion, we may well argue that the operation of the existence of standard models as opposed to non-standard models, e.g., of arithmetic, has a meaning only inasmuch as it implies the adoption of certain rules of deduction which distinguish the standard model from any non-standard model that is under consideration at the same time.[51]

Robinson as Consultant to IBM

Within months of his arrival at UCLA, Robinson signed a contract to work part time for IBM. In a letter of November 7, 1962, he was formally appointed an "independent consultant" to help develop a new programming language, the appointment beginning January 1 and running through December 31, 1963.[52] His first lengthy period spent working for IBM came the following summer, when he and Renée drove crosscountry in the Vauxhall. Along the way, they stopped to enjoy the breathtaking beauty of the Grand Canyon.[53]

At IBM Robinson worked mostly with Calvin Elgot at the Thomas Watson Research Center in Yorktown Heights, New York. He and Renée lived in the small town of Katonah, where they enjoyed three pleasant months living on a private estate where they had a cottage to themselves:

> There was a big pool between the main house where the owners lived and us. The next year we stayed there a much shorter time and the third sum-

[51] Robinson 1963b, pp. 297–298; Robinson 1979, 2, pp. 111–112.

[52] Letter from IBM Watson Research Center, Yorktown Heights, New York, dated November 7, 1962, RP/YUL. Under the terms of this agreement, Robinson was to consult at the rate of $250 per day with two summer months at $2000 per month, with travel and living expenses to be covered as well.

[53] Renée Robinson to JWD, letter of March 3, 1992.

Abraham Robinson with Oskar Werner aboard *HMS Queen Elizabeth*.

mer we took a house much closer to IBM in Yorktown Heights, which was fine. Our next-door neighbors were a couple of opera singers. She was very well-known, but had already retired. He still went to the Metropolitan Opera, but mostly to sing in the chorus and play small parts.[54]

While the Robinsons were away in New York, they rented their home in Los Angeles to Oskar Werner:

We rented our home in Mandeville Canyon to Oskar Werner, who was in Hollywood to film "The Ship of Fools." His mother-in-law, the famous French actress Annabella (who was married for nearly ten years to Tyrone Power), took one look and said: "It is very much like the Werner's home in Lichtenstein." So they took it!

The Werners only had one big pool party when the film was finished (Oskar did not like to be around actors very much). Later when the film was about to open, there was a big première, to which we were invited with all of the Hollywood stars. After seeing the film we all went to the Beverly Hills Hotel where a red carpet had been laid out and a gala reception took place well into the night.

We stayed in touch with Annabella and the Werners for quite some time. In Paris we visited Annabella in her penthouse, where the photos of her husband with whom she lived in Hollywood for ten years were all around.[55]

That first summer Robinson produced two articles at IBM, one written with J.H. Griesmer and A.J. Hoffman "On Symmetric Bimatrix Games," and a paper of his own presenting "Some Remarks on Threshold Functions."[56] The latter considered n-variable truth functions whose

[54] Renée Robinson to JWD, March 3, 1992.

[55] Renée Robinson to JWD, March 3, 1992, and March 5, 1994.

[56] Griesmer, Hoffman, and Robinson 1963; Robinson 1963c. According to C. C. Elgot, Robinson's paper on threshold functions was never sent out for publication; Calvin C. Elgot, quoted by H. J. Keisler in his "Introduction" to Robinson 1979, 1, pp. xxxvii.

arguments ranged independently over 0 and 1 with functional values also equal to 0 or 1. Those representable as linear forms with real arguments were "threshold functions." Determining whether a given truth function was a threshold function was relatively straightforward, reducing in fact to the consistency of a finite system of linear inequalities, but determining the overall properties of threshold functions (as opposed to other truth functions) was not so simple, although Robinson showed how this problem might be attacked from a logical point of view.

Promotion: Spring 1964

The initial effort to advance Robinson in rank with a corresponding increase in salary came from the Philosophy Department. On January 7, Moody (now chair) wrote to recommend a "merit increase" along with Robinson's promotion to a level 4 appointment effective retroactively to January 1, 1964. Moody explained that since coming to UCLA, Robinson had published "an important book in the area of the foundations of mathematics and logic (*Introduction to Model Theory and to the Metamathematics of Algebra*), as well as eight articles of high merit." Moody added that:

> Professor Robinson is presently engaged in writing an additional volume to be published in the same (North-Holland) series, concerned with the philosophical and mathematical implications of non-standard models of arithmetic and of analysis. These publications, as well as the papers he has read at scientific meetings, and as guest lecturer at universities here and abroad, testify to his continuous creative work in the field of logic and mathematics, in which he ranks as one of the top five to ten men in the world.
>
> At UCLA he has made valuable contributions to our graduate program in philosophy, as well as in mathematics, and has served conscientiously and very efficiently on departmental committees, and on the [University-wide] Committee on Educational Policy [see below, pp. 333–342]. Despite the fact that his name is primarily associated with mathematics, he has impressed the members of my department with his breadth and penetration as a philosopher—not only as a philosopher of mathematical logic, but as a philosopher of science and as one with great insight and familiarity with the Greek and the modern philosophical tradition. We value him very highly in our department, not only as a scholar, but as a person of exceptionally sound judgment in matters of educational policy.[57]

[57] Ernest A. Moody to Chancellor Franklin D. Murphy, January 7, 1964, Robinson's administrative folder, Department of Philosophy, UCLA.

This letter seems to have fallen on deaf ears, for Robinson was not given the promotion. Consequently, at the end of May, Moody joined forces with Angus Taylor. This time, the Philosophy and Mathematics Departments together urged that Robinson be advanced to level 5, skipping the level 4 promotion altogether:

> With the full support of the tenured staff of my department, and in concurrence with the request being made by Professor Angus Taylor, Chairman of the Department of Mathematics, I am writing to ask that a merit increase to the Step V professorial level be granted to Abraham Robinson, Professor of Philosophy and Mathematics, to be effective July 1, 1964. This adjustment is, in my opinion and that of my colleagues, fully warranted by Professor Robinson's internationally recognized status and creative achievements in mathematics and logic, and by the great distinction which his presence on the UCLA faculty lends to our two departments and to the University and its Los Angeles campus.
>
> As one of the most original and creative minds of the present generation, in his field of study and investigation, Robinson stands at the top level of the profession. There are perhaps eight or ten men in his field, throughout the world, who exceed him in eminence, but most of these are men whose main creative work lies in the past, whereas Robinson has achieved comparable eminence while still in his forties, and is doing new and original work in several important fields. . . .
>
> At UCLA, in both the Philosophy and Mathematics Departments, he is contributing greatly to the reputation of this campus as a foremost center for the study of mathematical logic, and has attracted graduate students of ability from other parts of the country. I might add that Professor Robinson has taken an active and highly responsible part in university service, serving on the Statewide Committee on Educational Policy, of which he is to be chairman for 1964–1965. His excellent judgment, and penetrating grasp of university and departmental problems, are very highly valued by his colleagues in philosophy and mathematics, and by those with whom he has worked on university committees.
>
> At the time when his original appointment was negotiated, in April 1961, it was felt appropriate by the chairmen of both departments to request what was at that time an overscale salary of $16,000 per year; but Professor Robinson, in order to save the extra time that would have been needed for processing an over-scale appointment, accepted appointment at the Step III level, which was then the top regular professorial level.[58] At the time this appointment became effective, in 1962–1963, both Professor Taylor and I felt that the level of the appointment had been unduly low for

[58] According to the application for "Merit Salary Increase," Robinson earned $15,100 at the professor 3 level. At step 4 level, his salary would have increased to $17,400.

> a man of Robinson's distinctions, but we decided to wait, perhaps mistakenly, for the end of his first year of service before requesting an upward adjustment. When, in January of 1964, we were informed that some funds were available for merit increases, for higher level staff, to be made retroactive to January 1 of this year, we put in a request for such a retroactive increase for Professor Robinson to the Step IV level, in the belief that such a request would be more easily granted than one to the fifth step, even though we were in agreement that Robinson's stature, as a man at the top of his profession, should be recognized by early adjustment to our top regular step. When our January request was not approved, the urgency of making the adjustment at the Step V level became very apparent to us, and on that account the present request is being made at a somewhat tardy date.
>
> There is no question in my mind of Robinson's qualification for the top regular step, or of his ability to command a salary of that level, or higher, at other major institutions in the United States. If confirmation of this opinion is desired, from distinguished men of the field outside the university, Professor Alonzo Church, of Princeton, or Professor G. H. von Wright, of the Academy of Finland, would be well qualified to judge of Robinson's work and reputation, not merely from the point of view of mathematics, but from that of philosophy. From the point of view of fair and appropriate relations of salary level, commensurate with relative professional standing and achievements, holding *within* the two departments in which Professor Robinson serves, Professor Taylor and I are in full agreement that the requested adjustment of Professor Robinson's position to the fifth step of the Professorship is both warranted and needed. I earnestly hope that very special consideration will be given to this request, with as little delay as possible, in order that the best interests of our campus, and of the two departments concerned, may be served, and a very valuable and distinguished man given the recognition by UCLA that he could so easily find elsewhere.[59]

This time, the promotion was approved, and by the end of July the university officially advanced Robinson with "Merit increase to Professor V."[60]

That summer Abby and Renée again drove cross-country to Yorktown Heights, where Abby worked with Elgot at the IBM Watson Research Center. At the end of the summer the Robinsons set off for Europe, where Abby had been invited to attend a series of meetings which they

[59] Ernest A. Moody to Chancellor Franklin D. Murphy, May 28, 1964, Robinson's administrative file, Department of Philosophy, UCLA.

[60] Foster H. Sherwood to Donald Kalish, July 27, 1964, Robinson's administrative file, Department of Philosophy, UCLA.

interspersed with visits to some of their favorite cities. This provided yet another opportunity for a sea voyage to Europe, as well as their first chance to visit Israel since leaving two years earlier.

Robinson, Bernstein, and the Invariant Subspace Problem: 1964

When the Robinsons disembarked in Southampton, they went directly to Bristol for a meeting of the Association of Symbolic Logic held at the university from July 13–17. The meeting was a NATO Advanced Study Institute, which included a symposium on axiomatic set theory at which Robinson gave a one-hour invited address on "Non-Standard Analysis."[61]

The title itself gave no hint of the remarkable result Robinson would unfold during his talk. The Bristol meeting was the ideal occasion, in fact, for an announcement that he knew would create a stir. Robinson was pleased to explain that he and Allen Bernstein had together just solved the invariant subspace theorem (in the case of polynomially compact operators) for Hilbert space.[62]

In fact, this was the first truly remarkable mathematical achievement using nonstandard analysis. As C.-C. Chang recalls, it "instantly rocked the mathematical world, so to speak, although many said a standard proof would soon turn up." Indeed, within months Paul Halmos found one, "but once you know something is true, it is easier to find other proofs. Major credit must go to Robinson."[63] As Halmos himself recalls:

> The polynomially compact business was one of the exciting mathematical events of the 1960s. The first published proof that compact operators always have non-trivial invariant subspaces is due to Nachman Aronszajn and Kennan Smith; it appeared in 1954. Smith pointed out, I might also say complained, that the proof was "tight". It left no room for modifications and generalizations; it proved exactly what it was designed to prove, no more. It did not, for instance, help to prove that a square root of a compact operator also had to have non-trivial invariant subspaces.[64]

Aronszajn had explained his proof to Halmos on a restaurant napkin. Halmos "cherished it, and along with many others I kept trying to 'loosen' it so as to be able to apply it more broadly—but all to no avail." Robinson's result, therefore, showing that square roots of compact operators did

[61] See the report of the Bristol meeting in the *Journal of Symbolic Logic*, 29 (1964), pp. 218–228, esp. p. 222.

[62] Although Robinson announced the result at the Bristol meeting in 1964, he waited until after Bernstein had completed his thesis before they jointly published the proof. See Bernstein and Robinson 1966.

[63] Chang I-1991.

[64] Halmos 1985, p. 320.

have invariant subspaces, came as "a surprise, a jolt." Robinson and Bernstein had indeed succeeded in "loosening" the Aronszajn-Smith technique.[65]

When Robinson sent Halmos a preprint of the paper he and Bernstein had written, Halmos wrote back from the University of Michigan to say how impressed he was with their result, which resolved "certainly a very difficult problem."[66] The theorem in question:

> *If p is a non-zero polynomial, T a bounded linear operator on a complex Hilbert space H, and $p(T)$ is compact, then there is a closed subspace L of H, distinct from* $\{0\}$ *and H, which is invariant under T.*[67]

Robinson and Bernstein proved this by establishing a relation between the operator T and an associated operator on a nonstandard Hilbert space H, restricting the operator T_ω to a subspace H_ω of nonstandard dimension w in H^*. A chain of invariant subspaces for T_ω then implied the existence of corresponding invariant subspaces for T. Basically, Robinson and Bernstein then followed Aronszajn and Smith in showing that at least one of these subspaces was nontrivial.

Although Halmos soon found a standard proof avoiding the direct appeal to nonstandard analysis, he admitted that his proof was basically nothing more than a reworking of the insight Robinson and Bernstein had already provided, but recast in standard terms:

> The Bernstein-Robinson result inspired me to re-study that technique, and I thought I saw what made it work. I abstracted the principal property that did the trick and called it quasitriangularity.[68]

But Halmos had reservations as well:

> However admirable some of the accomplishments of the logicians may be, the mathematician doesn't need them, cannot use them, in his daily work. The logic proofs of mathematics theorems, all of them so far as I know, are dispensable: they can be (and they have been) replaced by proofs using the language and techniques of ordinary mathematics. The Bernstein-Robinson proof uses non-standard models of higher order predicate languages, and when Abby sent me his preprint I really had to sweat to pinpoint and translate its mathematical insight. Yes, I sweated, but, yes, it was a mathematical insight and it could be translated.[69]

[65] Halmos 1985, p. 320.

[66] P. R. Halmos to Abraham Robinson, letter of June 19, 1964, RP/YUL.

[67] Bernstein and Robinson 1966, p. 421; Robinson 1979, 2, pp. 88.

[68] Halmos 1985, p. 320.

[69] Halmos 1985, p. 204. His standard proof was published in the same issue of the *Pacific Journal*, immediately following the Bernstein-Robinson paper. Halmos 1966.

Robinson was to return to this dilemma concerning the ultimate value and significance of nonstandard versus standard methods in a most emphatic way several years later.

Following the Bristol meeting, Renée and Abby crossed the English Channel, rented a car, and drove through Holland to Germany, where they saw Saul and Hilde in Berlin. Saul had just accepted a position as director of the newly founded Institute for Educational Research in the Max-Planck-Gesellschaft in Berlin. His experience as an educator uniquely qualified him for this responsibility, since the institute was to be concerned especially with matters of curriculum and comparative education.[70]

After buying a beautiful, wooded piece of property in Grunewald, on the outskirts of Berlin, Saul and Hilde decided to design their own home, from the foundations to the light fixtures. When Abby and Renée arrived in the summer of 1964, the new house was still under construction, so they stayed in a hotel.[71] Renée remembers in particular driving with Saul and Hilde from West Berlin across the border into East Berlin, where the two couples saw the major landmarks. The Berlin Wall had only existed for three years, since 1961, and it was a chilling experience spending even a few hours in the recently isolated communist capital of the Deutsche Demokratische Republik.

After leaving Berlin, Abby and Renée continued on to Vienna, where Renée was pleased to show Abby the inside of the Musikvereinsgebäude, which also housed the conservatory where she had first studied acting. They went to the opera, to Schönbrunn—in fact, they revisited all of the most famous places in Vienna. And Robinson even found time to stop in at the local IBM office.[72]

From Vienna the Robinsons traveled south and undertook a brief tour of the Greek isles, making their way from Mikonos to Delos, Santorin (with its beautifully white city of Thera set atop the volcano), and finally Rhodes. It was then but a short sail to Haifa, and soon thereafter, they were back in Jerusalem.[73]

"Formalism 64"

Considerable effort went into organizing that summer's International Congress for Logic, Methodology, and Philosophy of Science, hosted in Jerusalem by the Israel Academy of Sciences and Humanities. The meeting, initiated by the Israeli Association for Logic and Philosophy of

[70] Noah 1972, p. 405.

[71] Renée Robinson to JWD, letter of March 3, 1992.

[72] See Bristol 1964, p. 222.

[73] Renée Robinson to JWD, letter of March 3, 1992.

Science, was inspired largely by the highly successful example of the Stanford Congress in California two years earlier. Robinson's hour address, "Formalism 64," offered his most recent reflections on the nature and foundations of mathematics.

In the 1950s, working under the influence of his teacher Abraham Fraenkel, Robinson seems to have been satisfied with a fairly straightforward philosophy of Platonic realism. But by 1964, his philosophical views had undergone considerable change, which was reflected directly in the title he gave his Jerusalem paper.[74] Indeed, Robinson chose his title deliberately—to indicate that his views on the foundations of mathematics were in many important respects close to those of Hilbert and the more or less traditional formalists, but that by 1964 significant changes were nevertheless in order.

Hilbert, for example, could not have known in the 1920s that his own particular brand of formalism, which sought to prove the consistency of mathematics within the bounds of an axiomatized formal system, was "doomed to failure."[75] This only became apparent in the early 1930s thanks to the famous series of papers by Kurt Gödel in which he showed, among other things, the existence of formally undecidable propositions in arithmetic. Moreover, foundations of mathematics had also been affected by several other more recent developments, especially by Paul Cohen's proof of the independence of the continuum hypothesis in 1963, and debate over various axioms of infinity that were an increasingly popular topic for discussion.

In characterizing his paper, Robinson described it "as a personal statement of a point of view arrived at over a number of years." He had, in fact, developed his philosophical principles carefully—and empirically—based upon his experience as both an applied and pure mathematician. Consequently, he was aware simultaneously of both practical and theoretical issues affecting any foundations for mathematics, which he reduced to two main principles, one descriptive, the other "denotic" (or prescriptive):

(i) Infinite totalities do not exist in any sense of the word (i.e., either

[74] Georg Kreisel, in reflecting on Robinson's paper a year later, quipped that he could not tell whether Robinson had meant 1864 or 1964: "The number '64' is supposed to refer to 1964, but the paper might just as well have been written in 1864 since it presents only a crude version of formalism. It neither takes into account Frege's classical critique of that kind of formalism, nor, worse still, the massive studies of the one genuine attempt to pursue formalism, namely Hilbert's program, nor even Hilbert's analysis of adequacy conditions," Kreisel 1971, p. 190, n. 2. Although Kreisel was right about Frege, Robinson did discuss Hilbert, although not to the extent that Kreisel thought necessary. For a more positive response to Robinson's "Formalism 64" see Cohen 1971.

[75] Robinson 1964a, p. 229; in Robinson 1979, 2, p. 506.

Robinson at the International Congress for Logic, Methodology, and Philosophy of Science, Jerusalem, 1964.

really or ideally). More precisely, any mention, or purported mention, of infinite totalities is, literally, *meaningless*.

(ii) Nevertheless, we should continue the business of Mathematics "as usual," i.e. we should act *as if* infinite totalities really existed.[76]

Realizing that his first principle was most at odds with Platonism, Robinson described briefly the position of Kurt Gödel, whom he regarded as the "outstanding platonist of our time." But he found it impossible to agree with Gödel, and objected that:

> I feel quite unable to grasp the idea of an actual infinite totality. To me there appears to exist an unbridgeable gulf between sets or structures of one, or two, or five elements, on one hand, and infinite structures on the other hand or, more precisely, between terms denoting sets or structures of one, or two, or five elements, and terms purporting to denote sets or structures the number of whose elements is infinite.[77]

[76] Robinson 1964a, p. 230; Robinson 1979, 2, p. 507. Nearly ten years later, Robinson recalled the major points of "Formalism 64" as follows: "(i) that mathematical theories which, allegedly, deal with infinite totalities do not have any detailed meaning, i.e. reference, and (ii) that this has no bearing on the question whether or not such theories should be developed and that, indeed, there are good reasons why we should continue to do mathematics in the classical fashion nevertheless." Robinson added that nothing since 1964 had prompted him to change these views, and that in fact "well-known recent developments in set theory represent evidence favoring these views." See Robinson 1973a, p. 42; in Robinson 1979, 2, p. 557.

[77] Robinson 1964a, p. 231; in Robinson 1979, 2, p. 508.

However, by saying that any reference to an infinite totality was *meaningless*, Robinson was not advocating any sort of mathematical nihilism—all he meant was that any such reference was meaningless "in the sense that its terms and sentences cannot possess the direct interpretation in an actual structure that we should expect them to have by analogy with concrete (e.g., empirical) situations. This is not to say that such a theory is therefore pointless or devoid of significance."[78]

Due to what Robinson termed the "bifurcation" of set theory—a reference to the complementary Gödel-Cohen results admitting the consistency of both Cantorian and non-Cantorian set theories—formalists were forced, he believed, to conclude that "the entire notion of the universe of sets is meaningless (in the sense indicated by our first principle)."[79]

The second of Robinson's principles, that one should act *as if* infinite totalities existed, was in its turn directly at odds with intuitionism (along with various types of finitist, constructivist, and operationist schools of thought). More generally, despite their ultimate very serious differences, Robinson noted that "the starting point of Intuitionism is, like my own, the rejection of the naive notion of an infinite totality."[80] But there any further similarity ended abruptly. Instead, Robinson found his philosophical sympathies closest to those of the logical positivists, who were also pragmatic in their approach to the development of formal theories. But here Robinson's views as a formalist took a novel turn, on the level of *form* itself, where syntax was the heart of the matter. Unlike logical positivists, Robinson refused to believe that the syntax of a language was always a matter of choice:

> It seems to me that the rules of logic and of certain parts of Arithmetic which are *used* in the analysis of a formal theory are by no means arbitrary. For example, even when considering some non-standard, e.g., three-valued, logic, the logician ultimately assumes that a given concrete situation either obtains or does not obtain, and that these possibilities are mutually exclusive. In other words, at this point the logician actually *uses* two-valued propositional logic in syntactical terms, he assumes the law of contradiction, not-(X and not-X), and the law of the excluded middle, X or not-X.
>
> Thus it appears to me that there are certain basic forms of thought and argument which are prior to the development of formal Mathematics.[81]

[78] Robinson 1964a, p. 231; in Robinson 1979, 2, p. 508.
[79] Robinson 1964a, p. 231; in Robinson 1979, 2, p. 508.
[80] Robinson 1964a, p. 231; in Robinson 1979, 2, p. 508.
[81] Robinson 1964a, p. 237; in Robinson 1979, 2, p. 514.

In 1964 Robinson was still a decade away from his most mature development of a philosophy of mathematics, but in his Jerusalem paper he offered a basic outline of his own brand of formalism. It was a pragmatic philosophy based in part on the knotty problem of the infinite—and of course his own very recent experience with the controversial subject of infinitesimals. Here his foundational views could be traced back to Leibniz, whose characterization of infinite concepts in mathematics as *fictiones bene fundatae* seemed right to Robinson.[82] More intensive consideration of these questions, especially how history and philosophy of mathematics together might support a satisfactory foundations of mathematics, came nearly a decade later in an interdisciplinary seminar on mathematics and philosophy that he offered jointly at Yale with Stephan Körner in the fall of 1973 (see chapter 9).[83]

Enjoying Israel: Summer 1964

Apart from the Congress and his usual mathematical preoccupations, Robinson took advantage of being back in Israel that summer to visit some of his favorite places, all of which he and Renée recorded in dozens of photographs. Among these is a picture of Abby on Mount Zion in Jerusalem, and another taken at Akko, an Arab city on the coast just north of Haifa. In 'En Hod, an old Arab village, Abby and Renée were just in time to witness a colorful Arab wedding, which Abby photographed. One weekend they went to Zefat (Safad), a popular artist colony high in the mountains. Saul and Hilde were also in Israel, and there are pictures not only of Lotus Street, where Saul and Hilde lived, but others taken at Bet Oren (this just south of Haifa), where Abby and Saul were photographed together in front of a colorful bank of flowers.

There is also a photo of Abby on his way to the Dead Sea, with a cave high in the background where as a student he once stood guard with the Haganah, and another taken of Abby with Bar-Hillel, basking in the sun.

Robinson, Educational Policy, and the Berkeley Free-Speech Movement: 1964–1965

In the fall of 1964, as chairman of the University-wide Educational Policy Committee, Robinson also had a seat on the university's important Academic Council, whose primary function was to advise the president of the university, Clark Kerr. The council was peripatetic, meeting on a different campus each month (generally on a Wednesday). Robinson's

[82] Körner 1979, p. xlii.

[83] Seligman 1979, p. xxxi.

first meeting convened on September 23 at 10:00 A.M. in the Small Regents' Conference Room at Berkeley, with Angus Taylor presiding as chairman.[84] Once introductions and preliminary discussion of major issues facing the council that year were over, President Kerr joined the meeting at 11:00 A.M. Kerr briefly reviewed the university's academic plan for the coming year which included new campus sites in both northern and southern California.[85]

A major concern from Berkeley to Los Angeles was the impending change to a quarter system. This would also occupy Robinson's own committee on educational policy later in the year. All of these administrative issues, however, were soon overshadowed by what later, euphemistically, came to be known as the Berkeley Free Speech movement (FSM).[86]

At first this was a conflict limited to the Berkeley campus. The catalyst at work, set against the background of the Civil Rights movement and student opposition to the escalating U.S. prosecution of the war in Vietnam, was a confrontation between the Berkeley administration and student political activists over a sudden reversal of University policy regulating political activity on campus. Previously, a small strip of sidewalk just outside Sproul Hall Plaza—the Student Union was on one side, the central administration building, Sproul Hall, was on the other—had served as a meeting place where "clubs set up folding card tables, duplicated literature, other publications, collected funds, sold bumper stickers, buttons, etc."[87]

Without warning, at the beginning of the fall semester, "it was officially announced by the campus Administration that the customary practice would be discontinued." This, in turn, "triggered the opening moves in what soon became a confrontation between a relatively small group of students and the campus Administration, followed by an expansion of turbulence that threatened to engulf the campus. The eventual effects were felt through the whole University."[88]

[84] The following information about meetings of the Academic Council during the period of Robinson's service is drawn from the minutes of the council's meetings, kept as part of the Academic Senate Records, University of California Archives, used here courtesy of The Bancroft Library, University of California, Berkeley. The "Minutes of the Academic Council, 1964–64" are catalogued as CU-9, vol. 123.

[85] Minutes of the Academic Council Meeting for September 23, 1964, pp. 39–41.

[86] For a general overview of the Free Speech Movement, see the "Chronology of Events: Three Months of Crisis," compiled by the editors of *California Monthly*, the alumni magazine of the University of California, Berkeley, which provided a day by day listing of events from September 10, 1964, through January 4, 1965. Reprinted in Lipset and Wolin 1965, pp. 99–199.

[87] Draper 1965, p. 27.

[88] From the draft of a book by Angus E. Taylor, the first chapter of which details the

Berkeley's change in policy was enforced on October 1. It was nearly noon when two deans, along with campus police, approached the largest of the student tables. This was for the Congress of Racial Equality (CORE), manned by Jack Weinberg, a former graduate student in mathematics who had dropped out the year before. Weinberg was promptly escorted to a waiting police car, and then the trouble began.[89] The police car was immediately surrounded by students:

> On October 1 and 2, 1964, several hundred students at the University of California's Berkeley campus held a police car captive for thirty-two hours, until administration leaders of the university agreed to negotiate a series of grievances. The prolonged conflict that emerged from the encounter of the newly formed "Free Speech Movement" with the university administration convulsed the campus for most of the year, leading to mass arrests of students, a general strike, the involvement of the entire faculty in the dispute, the removal of some administration officers, and a continuing atmosphere of crisis and distrust.[90]

As Calvin Trillin described subsequent events for the *New Yorker*:

> During the fall semester, the free-speech controversy demanded the attention, and often the full-time participation, of a large number of Berkeley students, administrators, and faculty members; it involved an unprecedented use of mass action by students and two potentially disastrous confrontations between hundreds of students and hundreds of policemen; and it eventually produced a situation in which a distinguished university of twenty-seven thousand students nearly came to a halt—a situation that the chairman of the Emergency Executive Committee of the Academic Senate called, with little disagreement from anybody who spent the fall at Berkeley, "one of the critical episodes in American higher education."[91]

Even so, as Angus Taylor recalls, "from all that I know, a great deal of campus activity went on with minimal interruption. The attempted strike had little effect. The situation was worrisome, of course."[92]

As the political situation at Berkeley temporarily stabilized, the Academic Council held its second meeting on the Davis Campus, Wednesday, October 14. Meanwhile, Robinson's Educational Policy Committee had met to consider the university's academic plan for the year:

---

events at Berkeley in 1964–1965; quoted here with his permission as communicated to JWD in a letter of June 13, 1994.

[89] For details, see Draper 1965, chapter 11, "The Police Car Blockade Begins," pp. 39–41.

[90] Heirich 1971.

[91] Trillin 1965, in Miller and Gilmore 1965, pp. 256–257.

[92] Angus Taylor to JWD, November 22, 1992.

Professor Robinson reported that the University-wide Educational Policy Committee had studied the [Academic] Plan and reported to the Administration on it. The reaction of this committee, he stated, was on the whole favorable to the Plan. The council discussed the Plan and raised some questions concerning the items which deal with "Proposed New Programs" on the several campuses. The following motion was passed: That the portions of the Plan relating to the several campuses, particularly those concerning "Proposed New Programs," be referred to the campuses for review and comment, and that with this proviso the Council go on record as approving the Plan.[93]

A month later, the Academic Council returned to Berkeley for its November meeting. While the subject of "Student Political Activity on University Campuses" was the fourth item on the day's agenda, it did not receive protracted discussion and inspired no resolutions for council action of any kind.[94]

Among Berkeley students, Mario Savio, a graduate student in philosophy, emerged as an impressive spokesman for the movement. In part, he blamed the impersonal nature of the university and its attitude towards students for what had happened:

If you are an undergraduate still taking non-major courses, at least one of your subjects will be a "big" lecture in which, with field glasses and some good luck, you should be able, a few times a week, to glimpse that famous profile giving those four- or five-year-old lectures.[95]

Indeed, the FSM took as its "war cry" the phrase printed on every student's computerized registration punch card: "Do Not Fold, Spindle, or Mutilate."[96]

Throughout November renewed debate and the seeming inability of the university's administration to come to terms with the FSM problem prompted the Berkeley faculty to take action of its own. On December 8 it sought to resolve the controversy by setting guidelines for acceptable student political activities on campus. Simultaneously, it was hoped the university would begin to address deeper problems which were thought to underlie the conflict in the first place.[97]

[93] Minutes of the Academic Council meeting for October 14, 1964, pp. 42–44.

[94] Minutes of the Academic Council meeting for November 18, 1964, pp. 45–47.

[95] See the "Introduction by Mario Savio," pp. 1–7 of Draper 1965, esp. p. 4.

[96] Ironically, it was not much later that the FSM found itself using punch cards and borrowing the university's IBM computer to track all of its complicated legal matters and the people involved in resolving the conflicts that had arisen between students, the university, and police. See Trillin 1965, in Miller and Gilmore 1965, p. 258.

[97] As John Searle viewed this outcome, it was the *faculty* "and not the students nor the administration nor the regents [who] solved the crisis and brought the present atmo-

In response to the Berkeley faculty meeting of December 8, Angus Taylor decided that the Academic Council finally had to "get into the action":

> During the several days that followed I did much telephoning, to Kerr, to members of the Council, and to Regents' chairman Carter, whom I called at the request of Kerr. On December 9 I went to Davis to attend a meeting of the Committee on Educational Policy [Robinson's Committee]. Two resolutions in support of President Kerr and his actions were adopted. . . .
>
> My thought at this time was to get the Council to suggest to the Regents that they make a serious study of University policy and find a way to get out of the current predicament without a showdown between the Regents and the Berkeley faculty. By Saturday night of that week I had completed first drafts of the two documents that were later worked over and approved by the Council for transmittal to Kerr and the Regents [at their December meetings at UCLA]. On Sunday I called in Robinson and Holzer, the two other members of the Council from UCLA. They helped me to improve both documents. On Wednesday began the job of getting the Council to work on the drafts.[98]

The stage was now set for the most critical meeting of the Academic Council that year, which convened for three days at UCLA, Wednesday through Friday, December 16–18, 1964. When Angus Taylor called the first meeting to order at 10:10 A.M. in the Regents' Room, all members of the council were present. This time, there was only one item on the agenda. The entire meeting was concerned "exclusively with the problems raised by the recent student unrest on the Berkeley Campus."[99] Within days the council had agreed upon its official "Report to the President by the Academic Council," dated December 17, 1964, which among other things warned:

> The problem will not be solved by proclaiming slogans, whether they be slogans in favor of law and order, or in favor of freedom of speech. A solution will not be reached merely by having some segments of the University present a list of demands. Nor can the proposals espoused by a single Division of the Academic Senate provide a complete solution, when it is not clear that those proposals are acceptable in their entirety by other Divisions.[100]

---

sphere of normal peace and freedom to the Berkeley campus." See Searle 1965, p. 92.

[98] Angus Taylor to JWD, November 22, 1992.

[99] Minutes of the Academic Council meeting for December 16–18, 1964, pp. 48–52, esp. p. 48.

[100] The entire "Report to the President by the Academic Council," dated December 17, 1964, was appended to the minutes of the Council's December 16–18 meeting, pp. 51–51d.

The council was well aware that as debate on the so-called Free Speech movement spilled over from Berkeley to other UC campuses, concern had focused in part on the regents' decision in November that "under certain general conditions, students, faculty members, and University employees could use University facilities for organizing or financing 'lawful off-campus action, not for unlawful off-campus action.'" The FSM (along with some faculty members at Berkeley) had attacked implied limitations they associated with the word "lawful." According to minutes of the Academic Council that December at UCLA:

> Some of those who are interested in promoting the cause of civil rights by means of demonstrations contemplate the possibility that they may engage in actions which may be held to violate the criminal law (perhaps in the South, perhaps elsewhere) in the course of their demonstrations or other activities. They do not wish to be prevented merely on this account from planning their activities on campus. Others worried that the University may wish to impose discipline on students as a result of such illegal actions taken off-campus. . . .
>
> The present issue is whether the Regents' regulations, as revised on November 20, 1964, are permissive enough to give to the students, staff and faculty of the University their full constitutional rights with respect to on-campus political activities.[101]

Out of such concerns the Academic Council managed to forge what amounted to an article, or articles, of faith at its December meeting:

> What is needed, then, is time and good faith and cooperative endeavor. The Academic Council has faith in the President and the Board of Regents—faith that they desire full freedom within the law for students and faculty on the campuses of the University.[102]

Although this all seemed quite positive, there was also a darker side to what was happening in California—alienation, anger and collapsing morale, felt by students and faculty alike:

> The Council also recognizes that some of the unrest among students is traceable to a feeling that the University is a huge corporate enterprise run by remote administrators and geared to the mass production of research and of candidates for degrees. The present situation has produced tremendous soul-searching on the whole issue of impersonality and inaccessibility.[103]

[101] Minutes of the Academic Council meeting for December 16–18, pp. 51–51d.

[102] Minutes of the Academic Council meeting for December 16–18, "Appendix," p. 51b.

[103] Minutes of the Academic Council meeting for December 16–18, "Appendix," p. 51c.

Robinson could not have helped but harbor similar thoughts himself. Meanwhile, the Academic Council eventually reached perhaps overly idealistic conclusions:

> The Council especially urges that faculty members and administrators take a greater personal interest in students. Discussions with individual students should be encouraged concerning both personal and University problems.[104]

Robinson expressed his own views, largely in keeping with those of the Academic Council, in a letter to one of the university's regents, T. R. Meyer, who was chairman of the Regent's Special Commission to review university policies. Meyer had written to Robinson towards the end of January, asking for his opinion concerning resolution of the FSM question for the university at large. Robinson replied with a thoughtful letter of February 3:

> Thank you very much for your letter of January 27, 1965. Before offering my comments on the points raised by you, I wish to recall that two University-wide committees, of which I am a member, have made unanimous statements and recommendations concerning the matter under discussion. While, in part, these gave expression to our desire to reestablish mutual trust between the several sections of our University community, they also dealt with certain questions of principle arising out of recent events at Berkeley. In addressing myself to the more specific problems detailed by you, I speak only for myself. . . .
>
> 1) There should be no restriction upon content of on-campus speech and advocacy. . . .
> 2) The University should have the right to prevent on-campus activities which lead clearly and directly to off-campus unlawful action. Moreover, the University should have the right to discipline students who persist in organizing such activities. . . .
> 3) The University should not impose discipline for off-campus activities except where these activities have been followed by a conviction in court and it has become evident that the presence of the student on campus represents a menace to the members of the University community. . . .
> 4) There should be clear rules of procedure in disciplinary cases, including the right of appeal. . . .
> 5) There should be no wide variations on questions of principle between the campuses. . . .
> 6) As I understand it, any student (or group of students) has the right to lay his suggestions or grievances before members of the administration

[104] Minutes of the Academic Council for December 16–18, "Appendix," p. 51b.

> or the faculty. An alert administration or faculty might want to associate the students with the work of the University in study groups (e.g., on teaching methods) and the like. All this can be done under the present policy; I do not recommend any major changes in this policy.
>
> I am aware of the difficulties involved in the interpretation of some of the terms used here (e.g., the distinction between advocacy and action, the identification of the crimes which imply that the continued presence of their perpetrator on campus represents a menace to the University community). In any case, we can, as an intellectual exercise, imagine situations which would compel the University to tighten its regulations or to relax them further. The suggestions which I have outlined in response to your request are intended to represent a reasonable approach to the real situation with which we are faced at this time.[105]

This letter certainly reflects Robinson's restraint, as well as his reasonableness, but it also conceals his concern for the general lack of respect he perceived on both sides—on the parts of administration and students alike—for traditional academic values. As Ernst Straus recalls:

> I remember also his relatively strong adherence and respect for tradition of every kind. Not only the Jewish tradition, but the academic tradition and the academic courtesies. He was distressed at the onset, what was called "the free speech movement" in Berkeley and the various insults to academic tradition and to academic standards that he noticed in them. I must say, much before the rest of us did, he foresaw in it a general anti-intellectual tendency that most of us, I think, at the time did not perceive.[106]

Indeed, as Angus Taylor knew from his close association with Robinson that year:

> Abby was not a defender of the FSM. He empathized with the interest in the civil rights movement that motivated many of those who joined in FSM demonstrations, but he did not approve of the tactics of the leaders of the FSM and he knew that the controversy was not really about free speech but about the use of the University as a base for action and organizing for action.[107]

After its December three-day marathon, the next meeting of the Academic Council was something of an anticlimax. This time, Robinson drove down to San Diego, where Angus Taylor called the group to or-

105 A. Robinson to Theodore R. Meyer (chairman of the Special Committee to review university policies), February 3, 1965, RP/YUL.

106 Straus R-1976.

107 Angus E. Taylor in a letter to JWD, June 13, 1994.

der at 10:00 A.M. on January 20 in Bonner Hall, UCSD. The first item under discussion was a report on the budget, and the picture was bleak. No money was available for new research projects or to continue expansion programs in the libraries. Major reductions were to be made in capital improvements, while budgetary limitations "affect also the possibilities of faculty salary increases." Kerr pointed out that the university's salary levels had dropped behind those of its peers (Harvard, Yale, Columbia, Princeton, and Michigan), but this only resulted in a motion by the regents to increase out-of-state student fees.[108]

The second item on the day's agenda concerned the advisory role Senate committees should assume in approving proposals for new research institutes. This had been referred to Robinson's committee on educational policy, and he duly reported that the chairman of the University-wide Committee on Research had been invited to attend meetings of his committee on educational policy at which proposals for new institutes were being considered.[109] Before any formal action could be taken on the question of new programs under consideration, however, the question was deferred pending a report of the coordinating Committee on Graduate Affairs.

In February Robinson drove up to Santa Barbara for the council's next meeting. Most of the morning was spent going over routine business, but after lunch the subject of research institutes again came up:

> Professor Raleigh stated that the Coordinating Committee on Graduate Affairs would like to be informed about actions taken regarding new institutes. Professor Robinson said that he would ask the University-wide Committee on Educational Policy to allow him to invite the Chairman of the CCGA to meetings in which the Educational Policy Committee discusses new research institutes.[110]

On the subject of appointments at the newly established campus at Santa Cruz, Robinson moved that the Academic Council advise the president that appointments at the university be reviewed by ad hoc committees. The committees would be nominated at Santa Cruz and would report to the chancellor through a local advisory committee. The motion was passed unanimously.[111]

In March the Academic Council met on the Berkeley campus, but in April it was back at UCLA. Part of this meeting was devoted to discussion of Regent Meyer's report on "University-wide Regulations relating to Student Conduct, Student Organizations, and Use of University Fa-

108 Minutes of the Academic Council meeting for January 20, 1965, pp. 52–55.
109 Minutes of the Academic Council meeting for January 20, 1965, p. 53.
110 Minutes of the Academic Council meeting for February 17, 1965, p. 58.
111 Minutes of the Academic Council meeting for February 17, 1965, p. 58.

Abby and his red sports car, Los Angeles.

cilities," which also took up a good deal of time at the council's subsequent meeting in May, when a revised version of the Meyer Committee report was again on the agenda.[112] This meeting was held at the university's Riverside campus, after which Robinson stayed on for another day to meet with his educational policy committee, about which he reported at the council's last meeting for the year, in June. This farewell session was held in Millberry Union on the San Francisco campus. The subject of Robinson's last report as Chairman of the Educational Policy Committee concerned Senate authority over the curriculum:

> Professor Robinson presented a proposed resolution, and explained the connection between his proposal and the discussion which occurred at the Regent's Educational Policy Committee meeting in Riverside on May 20, 1965.[113]

Robinson was not quite finished with his remarks when President Kerr arrived at 10:45, whereupon discussion shifted to yet another draft of "Regulations Concerning Student Conduct." Kerr also distributed a draft document intended to delegate to each chancellor "the authority to take final action on all appointments including appointments to tenure ranks. . . ." After discussion, the council voted unanimously to endorse Kerr's efforts to encourage administrative decentralization. It was recommended, however, that local, university-wide agencies of the Academic Senate consider the resolution and comment before a final decision was reached.[114]

[112] Eventually, the council voted to express its "favorable view" of the Meyer report in its "present form." See Minutes of the Academic Council meeting for May 19, 1965, p. 66.

[113] Minutes of the Academic Council meeting for June 16, 1965, p. 68.

[114] Minutes of the Academic Council meeting for June 16, 1965, p. 68. Angus Taylor recalls "the proposed delegation to Chancellors of 'authority to take final action, etc.,' was not an unconditional delegation. Further on in the draft it was made clear that it would not include authority to make tenured appointments at salaries over the top of the approved salary scale." Angus Taylor to JWD, November 22, 1992.

Numbers and Ideals

Despite the demands of his teaching, department meetings and service as chairman of the University-wide Educational Policy Committee and member of the Academic Council, Robinson still found time for mathematics. In April of 1965 he finished the preface of a brief, college level textbook on *Numbers and Ideals: An Introduction to Some Basic Concepts of Algebra and Number Theory*, which he duly sent off to Holden-Day, his publisher in San Francisco. In little more than one hundred pages, Robinson hoped to stimulate interest in the simplicity and beauty—as well as the power—of abstract algebra when applied to relatively familiar concepts like the integers. Above all he wanted to show how basic ideas in modern algebra, including groups, rings, primes, and ideals led to very general yet very strong results.

Because he felt mathematics was becoming increasingly difficult to teach at the undergraduate level, Robinson wrote this book as a possible remedy. But he also wanted to stimulate receptive minds to think like mathematicians, and believed it was "highly desirable to introduce the student at the earliest possible moment to the modern methods which will enable him to appreciate, and later to take part in, contemporary research."[115]

As he stressed, the power of the algebraic methods was largely a product of their generality. But he was concerned that there might be a danger in introducing students to overly abstract mathematics without giving them a sense of the concrete cases which gave rise to the abstractions in the first place. As Robinson cautioned, "the beginner who is familiar only with a general theory and not with the concrete cases which are its roots and branches misses much of the beauty of the subject and may well be at a disadvantage when embarking on original research."[116]

Thus Robinson sought to "bridge the gap" (one of his favorite expressions) between abstract methods and concrete applications. He did so by focusing on a central theme—the theory of ideals in algebraic number fields, specifically quadratic number fields. Beginning naturally with the integers, he explained groups and rings, went on to consider primes, the Euclidean algorithm, prime factorization, and integral domains. The

[115] Robinson 1965. Initially, the book was praised as "leisurely and clear" but faulted for offering fewer than four problems per chapter: "this is especially regrettable in view of the author's stated objective of making abstract ideas concrete; surely copious problems would further this aim." W. J. LeVeque, *Mathematical Reviews*, 32 (1966), no. 79. Robinson apparently took this criticism to heart, for almost immediately Holden-Day issued a "new printing with added problems"—noted explicitly on the paperback cover. The new printing was also dated 1965.

[116] Robinson 1965, p. v.

subject then advanced to fields, quadratic fields, congruences, and isomorphisms. Algebraic integers in quadratic number fields were the subject of an entire chapter, which included discussion of units and specifically units in $J(\sqrt{2})$.

The subject of primes, along with factorization and divisibility in integral domains, gave Robinson the chance to discuss a "memorable error." As Robinson good-naturedly noted:

> Mathematicians, even good mathematicians, are not infallible. Every now and then they make mistakes. Some mistakes are trivial and best forgotten. There are, however, also "interesting" mistakes which occur because some theorem which was known to be true in familiar cases was taken for granted also in some new and unfamiliar case. Such a mistake occurred in the work of the mathematician E. E. Kummer (1810–1893). Kummer, who was one of the first to consider algebraic integers tried to solve a famous problem in number theory [Fermat's Last Theorem] by making the *unwarranted* assumption that the *unique* factorization theorem [for ordinary integers] is always true in rings of algebraic integers.[117]

Robinson explained the difficulty Kummer faced by demonstrating the now well-known fact that unique prime factorization fails to hold for algebraic integers. A simple counterexample makes this clear, since for $J(\sqrt{10})$, $6 = 2 \cdot 3 = (4 + \sqrt{10})(4 - \sqrt{10})$. Here, in the ring in question, 2, 3, $4 + \sqrt{10}$ and $4 - \sqrt{10}$ are all primes. Robinson then went on to provide the happy ending to the story, namely, that when confronted with his mistake, Kummer found a way of "saving" the unique factorization theorem, even for algebraic integers. Although Robinson did not provide details for Kummer's subsequent theory of "ideal" numbers, he did add that Dedekind's later concept of an *ideal* achieved similar ends. This, however, proved to be of far-reaching significance not only for number theory, but in many other parts of mathematics, and even in mathematical physics.[118]

When it came to elaborating the subject of ideals, Robinson included the finite ascending chain condition, homomorphisms, and both prime and maximal ideals. The last two chapters of *Numbers and Ideals* concerned rings of polynomials, their roots, and a test for algebraic integers. The last chapter, which Robinson entitled "Paradise Regained," was devoted to the factorization of ideals, which also provided a brief discussion of the problem of factorization of ideals in special cases.

At the very beginning of the book, Robinson explained that like all scientists, mathematicians were "driven by a lifelong desire to discover new worlds":

[117] Robinson 1965, p. 62.

[118] Robinson 1965, pp. 62–63.

> Reasons for doing so vary. [Mathematicians] may have a practical purpose in mind. Or they may require some knowledge in a new field in order to solve an old problem. Or, like mountain climbers and explorers, they may want to tackle the unknown simply "because it is there." For a pure mathematician the world to be explored is in his own mind, although the field of application of his work may be in physics or technology. It follows that it is sometimes hard to say whether the mathematician invents a new subject, or discovers it.[119]

In *Numbers and Ideals*, Robinson wanted to convey a sense of discovery, but also to stimulate interest by closing with several suggestions for further consideration. He urged his readers to investigate what they had just learned, in hopes they might find the joy of invention on their own, by trying their luck with the case of ideals for integral domains, rather than for quadratic fields which they had presumably learned from the book. But ultimately, the most important lesson in *Numbers and Ideals* was an aesthetic one: "Here we shall be content if we have managed to show you some of the beauty of the subject in a concrete case."[120]

Nonstandard Analysis: Spring 1965

It was also in April that Robinson put the finishing touches on his most ambitious book to date, *Nonstandard Analysis*, which he dedicated to Renée. This time, reflecting his optimism, Robinson chose an aphorism for the book from Voltaire's *Micromégas*:

> Je vois plus que jamais qu'il ne faut juger de rien sur sa grandeur apparente. O Dieu! qui avez donné une intelligence à des substances qui paraissent si méprisables, l'infiniment petit vous coûte autant que l'infiniment grand.[121]

The book's frontispiece, when published a year later, was taken from Euler's *Introductio in analysin infinitorum*–a graphic sign that Robinson saw himself working in an important historical tradition.[122] Like Euler, who had utilized infinitesimals in a wide range of important applications to both pure and applied mathematics, Robinson expected no less of nonstandard analysis.

In his preface, Robinson reprised a leitmotif of his career: how mathematical logic could benefit mathematics proper. This time it was model theory providing "a suitable framework for the development of the Dif-

[119] Robinson 1965, pp. 62–63. [120] Robinson 1965, pp. 62–63.

[121] "More than ever I see that we must not judge anything by its apparent size. O Lord, who have given intelligence to beings which seem so contemptible, the infinitely small costs You as little as the infinitely great," Voltaire, *Micromégas,* trans. Fowlie 1960, p. 33. The French is cited in Robinson 1966, p. vii.

[122] Euler 1748.

ferential and Integral Calculus by means of infinitely large numbers." Once again, he had found a way to make metamathematics work for mathematics in a profound and original way. Moreover, this was not vague metamathematics. As W.A.J. Luxemburg expresses it, "Robinson believed that if you have an idea that something exists, provide a model. Don't talk about infinitesimals vaguely in a metamathematical sense—define them explicitly."[123]

The purpose of Robinson's book was to show how infinitesimals "appeal naturally to our intuition":

> It is shown in this book that Leibniz' ideas can be fully vindicated and that they lead to a novel and fruitful approach to classical Analysis and to many other branches of mathematics. The key to our method is provided by the detailed analysis of the relations between mathematical languages and mathematical structures which lies at the bottom of contemporary model theory.[124]

Robinson introduced his readers to what was needed of mathematical logic for the rest of the book—including first and higher order theories, as well as the compactness theorem. Following the basics of nonstandard arithmetic and analysis, Robinson then proved some basic theorems from classical differential geometry, using infinitesimals "in the spirit of l'Hôpital's *Analyse des infiniment petits*" (of which Robinson actually owned a copy). As he was pleased to admit, "after the removal of some glaring and frequently attacked inconsistencies, the method used in these texts can be put on a firm foundation."[125] The last chapter of *Nonstandard Analysis* went even further. In examining history of the calculus from his new perspective, Robinson argued that nonstandard analysis required an entirely new history, and urged a revolution of sorts in the historiography of mathematics as a result (see below, pp. 349–355).

On the subject of a nonstandard approach to nonmetric topological spaces, Robinson highlighted a "striking characterization of compact spaces," along with applications. An entire chapter was devoted to the theory of functions of a real variable, as well as Lebesgue measure and Schwartz distributions (including discussion of the local value of a distribution).

Robinson's new book also devoted an entire chapter to complex nonstandard analysis. This was based almost entirely on research completed in 1962 for the Air Force Office of Scientific Research while Robinson was still in his last year at the Hebrew University.[126] Among applications were the analytic theory of polynomials (dealing with the location of

[123] Luxemburg I-1991.
[124] Robinson 1966, p. 2.
[125] Robinson 1966, p. 2.
[126] Robinson 1962a.

Leonhard Euler, *Introductio in analysin infinitorum*, frontispiece, Lausanne, 1748.

zeros of polynomials in the complex domain), and the theory of exceptional values of entire functions (including Picard's theorems and Julia's directions). "However, it is even more significant that the theory of normal families is replaced by certain generalized functions provided in a natural way by our approach."[127]

Another chapter dealt with the theory of linear spaces, including normed spaces, Hilbert space, and a nonstandard spectral theory of compact operators. Here Robinson included the important result he and Bernstein had only just published in the *Pacific Journal* on the invariant subspace theorem for complex Hilbert space (see above, pp. 327–329). Robinson also praised Bernstein's dissertation, which went beyond their joint result to investigate the spectral theory of self-adjoint bounded linear operators using nonstandard analysis. Bernstein

[127] Robinson 1966, p. 3. Much of the material in this chapter had also just appeared in a paper Robinson contributed to the Nevanlinna *Festschrift*, Robinson 1965a. For discussion, see below, pp. 358–360.

also showed how the invariant subspace theorem could be generalized to arbitrary Banach spaces.[128]

Equally natural, Robinson found, was the application of nonstandard analysis to topological groups, especially Lie groups. As he explained, the infinitesimal neighborhood of the unit element of a group also constituted a group, and was the intuitive counterpart of the infinitesimal group for a given topological or Lie group. This was followed by an investigation of variational principles applied to various problems using nonstandard methods—among them an adaptation of the classical proof of Riemann's mapping theorem, as well as the Dirichlet integral method in potential theory. Several problems from hydrodynamics, including boundary layer theory using infinitesimals, were also treated from a nonstandard point of view (whereby a counterpart of the de Saint-Venant principle in the theory of elasticity was given).

Based upon the impressive array of results Robinson exhibited in *Nonstandard Analysis*, he hoped that soon others might begin to explore the possibilities that nonstandard analysis offered. Although finding the right nonstandard interpretation for a given classical theory might not always be easy, he was sure that once found, "the subsequent reformulation and development of the theory can be a rewarding experience."[129]

Even theoretical physics, Robinson hoped—especially areas "afflicted" with divergence problems—might benefit from the methods of nonstandard analysis. This was as true for modern physics as it was for classical results; and although he had not given any examples of applications beyond classical applied mathematics, he was confident that this was probably due to "limitations of the author and not of the method."[130] It was not long, however, until Robinson did pursue nonstandard applications to quantum field theory in research undertaken with Peter Kelemen (see chapter 9).

In addition to simplifications in the development of certain parts of classical function theory resulting from nonstandard analysis, Robinson offered new results on the zeros of complex polynomials and on the behavior of analytic functions in the neighborhood of an essential singularity as evidence that his new methods were indisputably worthwhile, leading as they did to proofs of new theorems. This was especially true of the proof he and Bernstein had found for the invariant subspace theorem, which was already well known in mathematical circles and helped to create, overnight, understandable interest in Robinson's new book.

[128] Robinson 1966, pp. 200–201.
[129] Robinson 1966, p. 5.
[130] Robinson 1966, p. 5.

Nonstandard History of the Calculus

Robinson's last chapter, with obvious pleasure, was devoted to history of the calculus. "The fact that the more recent writers in this field were convinced that no such theory can be developed effectively, colored their historical judgment. Thus, a revision has now become necessary."[131] Indeed, although infinitesimals were "late in returning to the mathematical scene," as Robinson put it, he was confident that they were destined to reestablish themselves. He was especially sensitive to the injustice done Leibniz, he felt, by later commentators who treated him severely, not because his views were inherently untenable, but "because they were writing in the light of later developments which happened to go in a different direction."[132] As Robinson could now argue, Newton and Leibniz had *not* gone in the wrong direction. If anything, they had blazed a trail on which Robinson now proposed to venture considerably further–but with more precision and finesse.

As for Leibniz, Robinson was basically interested "in getting into Leibniz' mind."[133] Indeed, he liberally peppered his mathematical conversations with names like Newton and Leibniz. He had an exceptionally historical view of the calculus, and often stressed its origins and the relevance of nonstandard analysis for a proper understanding of what Newton and Leibniz had actually accomplished. As C.-C. Chang found especially remarkable, Robinson spoke of Newton and Leibniz "as if he knew them."[134]

In bringing historians of the calculus to task, Robinson was particularly critical of Carl Boyer's *The Concepts of the Calculus*.[135] The history of mathematics was on shaky ground, Robinson felt, if it chose to pass judgment on earlier theories based upon currently fashionable prejudices. Nonstandard analysis cast a new light on the history of the calculus, and Robinson was interested to see how it might appear if reexamined without assuming that infinitesimals were wrongheaded or at all lacking in rigor.

Although Leibniz only settled on a more or less consistent view of infinitesimals during the last twenty years or so of his life, Robinson characterized his belief in infinitesimals as:

> a useful fiction, adopted in order to shorten the argument and to facilitate mathematical invention (or, discovery). And if one so desires, one can always dispense with infinitely small or infinitely large numbers and revert to the style of the mathematicians of antiquity by arguing in terms

[131] Robinson 1966, p. 4.
[132] Robinson 1962a, p. 120.
[133] Kaplan I-1991.
[134] Chang I-1991.
[135] Boyer 1959.

> of quantities which are large enough or small enough to make the error smaller than any given number.[136]

Robinson applauded Leibniz's refusal to assert, however, the reality of infinitesimals. This he based on a letter Leibniz sent to d'Angicourt in 1716, in which "Leibniz considered it impolitic to upset the Marquis de l'Hôpital who was one of his followers but who, unlike himself, believed in the reality of infinitesimals."[137]

Whatever the status of infinitesimals ontologically, Robinson realized from the moment he discovered nonstandard analysis that it raised an important historical question. As he posed it first in 1962:

> One may ask, if Leibniz' ideas are indeed tenable, how is it that they degenerated in the eighteenth century and suffered eclipse in the nineteenth? The reason for this can hardly be found in the absence of a consistency proof for his system. Leibniz was aware of a need for justifying his procedure and he did so by relying on the fact that "*tout se gouverne par raison*"—hardly a consistency proof in our sense. However, even in the twentieth century no mathematician is known to have changed his profession because Gödel showed that no conclusive proof for the consistency of Arithmetic is possible. Nor can it be said that a theory of infinitesimals is less intuitive than the δ, ϵ-procedure and its ramifications and developments.
>
> The reason for the eventual failure of the theory is to be found rather in the fact that neither Leibniz nor his successors were able to state *with sufficient precision* just what rules were supposed to govern their system of infinitely small and infinitely large numbers.[138]

There is a major difference, however, between Gödel's result and Leibniz's imprecision regarding the infinitesimal calculus, for Gödel's work never cast any doubts on the validity of arithmetic per se—although it did dash hopes that the consistency of arithmetic could ever be *proven*. On the other hand, the dual problems of the infinite and infinitesimals had always plagued mathematicians and philosophers alike because they always seemed to be involved with inconsistencies and paradoxes. While there were no good grounds to doubt the consistency of arithmetic, there were historically long-standing reasons to be suspicious of infinitesimals, for reasons as well known to the Greeks as they were to both Berkeley and Nieuwentijt. It was the nagging worry

[136] Robinson 1966, p. 261, and (with only minor differences) Robinson 1962a, p. 121.

[137] Robinson 1962a, p. 122. Robinson quoted Leibniz's letter without reference to Dangicourt (or the fact that Leibniz's letter was written in 1716) in Robinson 1966, p. 263. For Leibniz's letter to Dangicourt, see Leibniz 1789, 3, pp. 499–502.

[138] Robinson 1962a, p. 123. See Robinson 1966, p. 266, for a different approach to this same question.

that contradictions could not be explained away that made their case very different from Gödel's.[139]

In explaining the major shortcoming of Leibniz's attempt to handle infinitesimals, Robinson suggested that "In fact, what was lacking at the time was a formal language which would have made it possible to give a precise expression of, and delimitation to, the laws which were supposed to apply equally to the finite numbers and to the extended system including infinitely small and infinitely large numbers as well."[140]

On this point, however, nonstandard analysis was somewhat tricky, especially when it came to definitions and the distinction to be drawn between internal and external statements, on which the status of nonstandard infinitesimals hinged. As Robinson once noted:

> In our own theory the answer to the question whether Archimedes' axiom is true not only in R but also in R^* is unambiguously, yes–and no![141]

The answer was "yes" if by Archimedes' axiom one meant the sentence in K formalizing the statement, "For every pair of real numbers a and b such that $0<a<b$ there exists a natural number n for which $na > b$." This sentence holds in R and therefore holds for R^*, as well. In R^*, n can simply be taken as an infinite natural number.

But if Archimedes' axiom is taken to mean that for any pair of real numbers a and b, $0<a<b$, there is a natural number n such that

$$\underbrace{a + a + a + \ldots + a}_{(n \text{ times})} > b,$$

then this postulate was *not* fulfilled in R*.[142]

What Robinson's nonstandard analysis neatly avoided was the problem for which Berkeley had condemned Newton–namely that his infinitesimals were the "ghosts of departed quantities," or the equally problematic treatment in the Leibnizian calculus as ratios of infinitesimals. It did so by dealing with equivalence instead of equality. Nonstandard analysis was concerned only with the *equivalence* of two elements $a \cong b$ which did not require that $a = b$, but only that a was *infinitely close* to b, i.e., that $a - b$ is infinitesimal. Consequently, nonstandard analysis only asserted that $dy/dx \cong f'(x)$, or that the standard part $(dy/dy)^{\circ} = f'(x)$.

139 Robinson himself may have thought better of this, for he did not repeat mention of Gödel and the matter of consistency in *Nonstandard Analysis.* However, he did make the same point about Leibniz's lack of a formal language, as advanced in Robinson 1962a, p. 124, in virtually the same way, word for word, in Robinson 1966, p. 266.

140 Robinson 1966, p. 266.

141 Robinson 1966, p. 266; with slight variation bracketed from Robinson 1962a, p. 124.

142 Robinson 1962a, p. 124.

Although this might seem cumbersome notationally, the more complex formalism of nonstandard analysis was but "a small price to pay for the removal of an inconsistency."[143]

Moreover, Leibniz never specified exactly what rules governed infinitesimals and infinities (except for the vague assertion that they were presumed to behave exactly like familiar finite numbers). But if so, then why did Leibnizian infinitesimals fail to satisfy the axiom of Archimedes? Thanks to nonstandard analysis, Robinson could give a clear, and "indeed brilliant" answer (in the words of Martin Davis) to this question.[144]

Robinson's historical analysis went on to look carefully at Lagrange and d'Alembert, both of whom were concerned with the "evident lack of soundness" of the Leibnizian calculus. Lagrange tried to eliminate "any consideration of infinitely small or vanishing numbers, or of limits, or of fluxions." D'Alembert was of greater interest to Robinson because, rather than eliminate infinitesimals, he tried to show how the calculus might actually be based on the notion of limit.[145]

Limits were a central issue for Cauchy, a figure of surpassing interest to Robinson because Cauchy embraced actual infinitesimals (although for Cauchy, the infinitesimals he had in mind were variables, not numbers, i.e., variables with limit zero).

Robinson did not try to guess what "the precise picture of an infinitely small quantity may have been in Cauchy's mind." Instead, he posed a more reasonable question—what did Cauchy's mathematics look like when his infinitely small and infinitely large quantities were understood "in the sense of Nonstandard Analysis"?[146]

Clearly it is unfair to Cauchy (and indeed anachronistic) to judge him in light of the Weierstrass-Dedekind continuum that excludes infinitesimals. Robinson noted, instead, that Cauchy's theorem—that the sum of a convergent series of continuous functions was continuous—would be perfectly correct as stated if the convergence of the series were assumed to hold for nonstandard as well as for standard points of the interval (i.e., for points of the form $x + \eta$, where x is real and η is infinitesimal). This follows from Robinson's result that uniform convergence of an infinite series in an interval (the sufficient condition for the continuity of the sum) is in fact equivalent to pointwise convergence *if* nonstandard points are included.

[143] Robinson 1966, p. 266.

[144] Martin Davis to JWD, E-mail note of September 7, 1994.

[145] Robinson called attention in particular to a number of articles in the *Dictionnaire encyclopédique des mathématiques* (for example, on *limite* and *différentiel*) in which d'Alembert appealed to the notion of limit. See Robinson 1966, pp. 267–268.

[146] Robinson 1966, p. 270.

In other instances, Robinson also provided nonstandard counterparts for various definitions and theorems in which Cauchy freely used infinitesimals and infinitely large quantities. But whereas the conclusions of the nonstandard reformulations were nonproblematic, Robinson realized that the same could *not* be said of Cauchy's original versions. For example, passing from a ratio of infinitesimals to a ratio of standard numbers could be justified in nonstandard analysis by a rigorous proof, but in Cauchy's case, "it is a logical *non sequitur* and may be said to involve an unconscious use of the 'principle of continuity,' whose application had been criticized by Cauchy himself."[147]

Robinson went on to examine concepts introduced by Bolzano, Riemann, Weierstrass, and others, all in light of nonstandard analysis. As he explained, when the new techniques advocated by Weierstrass began to gain currency, references to infinitesimals:

> began to be taken automatically as a kind of shorthand for corresponding developments by means of the δ, ϵ-approach (or, later on, for some more sophisticated method). . . . However, it may well have been this attitude which gave rise to the impression that Cauchy's theory was already basically the same as Weierstrass', and that the former used infinitesimals really as a kind of shorthand. The question remains whether Cauchy himself also held this opinion. As we have seen he did in fact regard them (at least toward the end of his life) as in some way provisional but there seems to be no indication that he wished to, or would have been able to, eliminate them altogether.[148]

Especially remarkable at the end of the nineteenth century, to Robinson, was the revolutionary creation of transfinite set theory by Georg Cantor, which ultimately prompted an "upswing in the theory of infinitesimals." This, in its way, was ironic because Cantor adamantly opposed any notion of infinitesimals, and always condemned them as the *cholera bacillus* of mathematics. In fact, Cantor believed that the techniques of transfinite set theory could be used to prove the logical impossibility of infinitely small numbers.[149]

Cantor was so convincing about this that many mathematicians accepted his view, including Giuseppe Peano and Bertrand Russell. Even Robinson's mentor, Abraham Fraenkel, seems to have held a similar position, believing that infinitesimals could never be introduced into

[147] Robinson 1966, p. 275.

[148] Robinson 1966, p. 277

[149] Robinson 1966, p. 279. Cantor referred to infinitesimals as a *cholera bacillus* in a letter to the Italian mathematician Vivanti, December 13, 1893, cited in Dauben 1979, p. 233. Cantor also declared that infinitesimals were worthless, that they belonged "in the waste basket as *nothing but paper numbers*." See Dauben 1979, p. 131.

mathematics the way Cantor's transfinite numbers had been, although he did hold out the remote possibility that "someday a second Cantor may give an unobjectionable arithmetic foundation for new infinitely-small numbers which could prove to be mathematically useful and which might even provide a simple approach to the infinitesimal calculus."[150] Despite Cantor's rejection of infinitesimals, Robinson was apparently prepared to see himself as a second Cantor, having succeeded (as Cantor had with transfinite set theory) in creating a rigorous theory, albeit at the opposite extreme, of the infinitely small.

Like Cantor, Robinson also appreciated philosophical objections to the infinite–and to infinitesimals. Mentioning briefly general rejection of the actual infinite in antiquity and the Middle Ages, Robinson considered at greater length opposition to the infinite in the writings of Locke and Berkeley. As he said of the latter, "it is in fact not surprising that a philosopher in whose system perception plays the central role, should have been unwilling to accept infinitary entities."[151] It was in light of such opposition to the actual infinite, Robinson suggested, that "when Cauchy constructed his edifice in response to renewed doubts he quite probably chose the notion of a variable as basic since, by its very nature, the idea of a variable seems to express potentiality and not actuality."[152]

Robinson harbored similar doubts, at least on the subject of the reality of the actually infinite. As he once confided to William Lambert—one of his graduate students, who was surprised that despite Robinson's renown for nonstandard analysis (where he worked with uncountable languages, superstructures, etc.), he nevertheless mentioned one day, quite casually —"I am not sure I really BELIEVE in the set of all natural numbers!"[153]

Although to a logical positivist, any discussion of the infinite, ontologically, would be taken as meaningless, even a positivist would have to concede the *historical* importance (Robinson believed) of expressions involving the term "infinity" and of the (possibly subjective) ideas associated with such terms.[154]

Recalling the position he had already expressed on foundations in "Formalism 64," Robinson viewed nonstandard analysis from a formalist point of view *syntactically*. He had not introduced new entities, but "new deductive procedures":

> Whatever our outlook and in spite of Leibniz' position, it appears to us

[150] Quoted from Fraenkel 1928, in Robinson 1966, p. 279.
[151] Robinson 1966, p. 281. [152] Robinson 1966, pp. 280–81.
[153] W. M. Lambert, Jr., to JWD, letter of January 8, 1991.
[154] Robinson 1966, p. 281–282.

> today that the infinitely small and infinitely large numbers of a nonstandard model of Analysis are neither more nor less real than, for example, the standard irrational numbers. This is obvious if we introduce such numbers axiomatically, while in the genetic approach both standard irrational numbers and nonstandard numbers are introduced by certain infinitary processes.[155]

This was exactly the position Cantor had championed in claiming the acceptability of his own transfinite numbers a century earlier. For as Cantor insisted, once irrational numbers were accepted, there was no reason not to admit other consistent concepts of number that drew upon infinitary processes in a similar way.[156] It is interesting that both Cantor and Robinson should have felt compelled to examine the history of mathematics, in part to help justify their own attempts to deal with the infinite. In Robinson's case, nonstandard analysis ended with an historical survey of the calculus, not only to help vindicate many of his illustrious predecessors, but to show how even on questions that were purely historical, nonstandard analysis provided new insights. Not only would nonstandard analysis serve to rewrite the calculus, it would rewrite its history as well.

Doctoral Students at UCLA: 1965

Robinson was always a committed teacher, and genuinely delighted in the success of his students. The brightest ones were also a challenge—but he would work with the poorest as well, abandoning no one, as long as their interest was sincere:

> Robinson was always involved with his students. We had a student who had failed twice, and failed a third time, and yet Robinson kept trying, and wanted to give him a 50%, a barely passing mark for the course.[157]

Yiannis Moschovakis, who was responsible for grading this particular course, warned Robinson that "if we pass him, he'll want to do a thesis." Robinson nevertheless insisted on passing the student, whereupon indeed he wanted to do a thesis. But as Robinson said, "I have never said 'no' to a student."[158]

With his weaker students, Robinson's solution was to try and find a suitable problem or subject on which they could work. In every respect, Robinson was "always helpful to younger people," just as he was also

[155] Robinson 1966, p. 282.

[156] For discussion of this aspect of Cantor's justification of transfinite set theory, consult Dauben 1979, pp. 128–129.

[157] Moschovakis I-1991.

[158] Moschovakis I-1991.

the "ideal senior colleague" who sought to promote younger faculty in a positive way—in fact, in all respects he seems to have been "a great mentor."[159]

The first three dissertations that Robinson oversaw at UCLA were completed in 1965. One was on mathematics education, by E. Mark Gold, who wrote on "Models of Goal-seeking and Learning." Allen Bernstein submitted his dissertation on "Invariant Subspaces for Linear Operators," the main result of which Robinson had already announced (at the Bristol meeting), and which would become more widely known through their joint publication the following year.[160]

The third of Robinson's dissertation students that year was William Lambert, who defended his thesis on "Effectiveness, Elementary Definability, and Prime Polynomial Ideals" on May 27, 1965.[161] This was a subject close to Robinson's heart—and had been among his earliest interests in model theory—although it drew as well upon ideas he had worked on more recently with C. C. Elgot at IBM, especially the notion of "schematic definability."

Lambert had studied mathematics at the University of Wisconsin (B.A. 1958), and at UCLA began working with C.-C. Chang, "who tossed me in to sink or swim with infinitary logic. I sank!" In desperation, Lambert turned to Robinson:

> When Robinson asked me what I was interested in, I mentioned ideals, thinking of his "metamathematical ideals." Whether he misunderstood me or felt this might be a dead end I don't know, but instead he sent me to look at the work of G. Hermann, K. Hensel and E. Noether with the idea of formalizing (schematically) in first-order field language the ideas of irreducibility and absolute irreducibility.
>
> I was still at UCLA as an assistant when this started—indeed I was his research assistant and wound up catching typos in the rather unfortunate Model Theory book till he told me to "stop already!" He was very good at nudging me on without being too "heavy" about it. Then I went to Loyola of Los Angeles (now Loyola-Marymount) with a full-time job.
>
> When I felt I had the idea down right (I think this was after I had gone to Loyola) he threw me the curve: "OK, now what about making these notions EFFECTIVE?" He then explained the approach he had discussed with C. C. Elgot. As it turned out, this tail wound up wagging the dog; the only serious publication that came out of this (my fault, not his!) ignored

159 Moschovakis I-1991.

160 Bernstein and Robinson 1966. For discussion of the content and significance of this paper, see above, pp. 327–329.

161 Lambert 1965.

the first-order stuff in favor of the notion of decidability (the December 1968 *JSL* paper).

I remember one summer alternating between the office at Loyola and the pool and sauna literally sweating this out. The final draft was written at my family home in Wisconsin during the 1964 Christmas break (exposing family friends to just how unsociable a person doing the "minor details" of the final draft can be).[162]

There was worse to come! Finishing the thesis was not without its tribulations, but demonstrated what a "good sport" Robinson could be as every student's worst nightmare came to pass:

> An idea of how GOOD a person Abraham was: I left the draft in a locker at O'Hare airport while waiting for my California flight. When I went back to the locker . . . empty! In a state of total emotional annihilation I asked Abraham if he could possibly try to work with the carbon copy—this of course was long before everything would be on a floppy—THREE floppies for anything important!!!! Now my typing was, to put it mildly, bad to begin with—it was tolerable only to those who had had to endure my handwriting!—but he consented. Fortunately, in fact, I had not locked the locker properly and eventually the original got to me via O'Hare security and the airline.[163]

In his thesis, Lambert showed that, for ideals of polynomials over a perfect field, both primeness and absolute primeness could be characterized by a single predicate. The predicate was independent of the particular field, and depended only on the number of polynomials chosen for a basis of the ideal, the number of indeterminates, and a bound on the degrees of the basis polynomials. His approach, Lambert acknowledged, depended largely on methods pioneered by Kurt Hensel and Emmy Noether, as developed by Grete Hermann.[164]

Moreover, by using the idea of "schematic definability" which Robinson and Elgot had been considering in their joint work at IBM—basically a generalization of Herbrand-Gödel-Kleene recursion applied to arbitrary structures—Lambert showed that the characterizing predicates were weakly effective (closely related to Michael Rabin's notion of computability for structures, as Lambert pointed out). As for primeness:

> The characterizing predicates for primeness and for absolute primeness are universal quantifications of predicates which correspond directly, and uniformly in the numerical parameters involved, to a schematically defin-

[162] W. M. Lambert, Jr., to JWD, letter of January 8, 1991.
[163] W. M. Lambert, Jr., to JWD, January 8, 1991.
[164] Hermann 1926 and Lambert 1968.

> able function of polynomials having polynomial values. Here polynomials are taken as finite sequences from the ground field. Further, these characterizing predicates can be considered to correspond to a test for [absolute] primeness which is effective, in the sense of schematic definability, relative to the factorization of polynomials over the field.[165]

Lambert eventually published some of his results in an article for the *Journal of Symbolic Logic*. This was only after Robinson had already left UCLA for Yale, due to the slowness of both the refereeing process and subsequent revision of the paper.[166]

Robinson's Publications: 1965

In addition to the second printing of *Model Theory*—a sure indication of the interest nonstandard analysis was generating—Robinson's major mathematical publication that year was another contribution to the growing list of impressive contributions nonstandard analysis could make to mathematics, this time to the theory of complex variables. The opportunity was a *Festschrift* for the Finnish mathematician Rolf Nevanlinna, to which Robinson was invited to contribute.

In thinking about what to write for a *Festschrift* in honor of Nevanlinna, Robinson naturally thought of the prominent role Nevanlinna had played in the development of complex variable theory, especially Picard theory and the theory of univalent and multivalent functions.[167] This in turn had been inspired by qualitative theories associated with the "beautiful theory of normal families" developed largely by Paul Montel. And Robinson could in turn trace this earlier work back even further to Newton and Leibniz and the "first flirtation with infinitesimals." This, of course, brought Robinson right back to where he wanted to be—nonstandard analysis.

Robinson, of course, was interested in showing yet again how a nonstandard treatment might win another new, even spectacular result. The classic approach to the theory of normal families involved the extraction of sequences and subsequences of complex functions. This could be avoided by taking a nonstandard approach, whereby the theory of normal families was applied to complex analysis by considering the

165 Lambert 1965, pp. vii–viii.

166 As Lambert put it, Robinson "helped me deal with an appropriately demanding referee for the *JSL* article, but by the time this came out he had gone to Yale." Lambert to JWD, June 15, 1994. In fact, changes suggested by one of the reviewers "led to a radical reformulation of the presentation," along with related modifications in Lambert's original scheme. See Lambert 1968, p. 579.

167 Robinson 1965a.

"ideal" elements of a family of functions, namely the functions $f(z)$ in F^*.

Sketching briefly a nonstandard theory of functions of a complex variable required little more than introducing generalized functions which could take infinitely small or large values, both in their ranges and domains of definition. Robinson's aim was to show how such generalized functions could be "correlated with the classical theory of normal families" and thereby used "to obtain results in standard complex variable theory." Specifically, Robinson was in search of applications to univalent or multivalent functions, to algebraic or algebroid functions, and to lacunary polynomials.[168]

Robinson was now ready to make his point: a theorem that had not previously been stated in classical function theory, but which naturally suggested itself in nonstandard terms:

> THEOREM 5.1 (STANDARD): *Let a_0 be a fixed complex number, $a_0 \neq 0$, and let q be a fixed positive integer. Let F be the family of the functions $f(z) = a_0 + a_1 z + a_2 z^2 + \ldots$, which are holomorphic and (at most) q-valent in the unit circle and let r and δ be two positive numbers, $r > 1$. Then there exists a positive $\alpha = \alpha\,(a_0, r, \delta)$ such that for every $f(z) \in F$ there are q points $z_1, \ldots, z_q$, $|z_j| \leq r, j = 1, \ldots, q$, such that $|f(z)| > a$ provided $|z| \leq r$ and $|z - z_j| > \delta, j = 1, \ldots, q$.*[169]

As a second example, this time to lacunary polynomials (so-called because for a polynomial of rank $r > 1$ there are likely to be gaps in the degrees of the monomials constituting the polynomial), Robinson also proved:

> THEOREM 6.2 (STANDARD): *Let r be a positive integer and let F be the family of all lacunary polynomials of the form display $p(z) = z + a_2 z^{n_2} + \ldots + a_r z^{n_r}$. $a_j \neq 0$, $j = 2, \ldots, r$, $1 < n_2 < \ldots < n_r$ which have no zero $z \neq 0$ in the unit circle. Then there exists a positive ρ such that all elements of F are univalent for $|z| < \rho$.*[170]

At the end of his Nevanlinna paper, Robinson again took the opportunity to propagandize on behalf of nonstandard analysis. Directly confronting the question of what value there was to be found in nonstandard analysis, due to the fact that it was always possible to find conventional

[168] The only opportunity Robinson had had previously to outline the basic feature of a nonstandard theory of complex functions was in a little-known tract, Robinson 1962a. Although his *Introduction to Model Theory and to the Metamathematics of Algebra* had appeared, it did not go into detail about nonstandard complex analysis, a topic Robinson promised to present systematically in his forthcoming book, *Nonstandard Analysis*.

[169] Robinson 1965a, p. 177; Robinson 1979, 2, p. 80.

[170] Robinson 1965a, p. 181; Robinson 1979, 2, p. 84.

proofs for nonstandard arguments, he offered a pragmatic reply:

> The answer is that, though this may complicate matters, it is always possible to "translate" a proof of the kind given here into a standard proof by the use of ultrapowers, more particularly hyper-real fields. Nevertheless, we venture to suggest that our approach has a certain natural appeal, as shown by the fact that it was preceded in history by a long line of attempts to introduce infinitely small and infinitely large numbers into Analysis.[171]

Robinson's Last Summer with IBM: 1965

Robinson had spent three summers with IBM, working mostly with Calvin Elgot at the Thomas J. Watson Research Center in Yorktown Heights, New York. Together they were investigating a largely uncharted but potentially exciting new area, covered first in a paper they coauthored at IBM in 1964, "Random-Access Stored-Program Machines, an Approach to Programming Languages."[172] This was especially impressive because it "broke ground for new ideas."[173]

In order to characterize a digital computer more realistically than a Turing machine, they settled on a model based upon a general semantic notion of "instruction." Robinson was especially interested in two kinds of programming languages: those oriented to users and therefore "user friendly" (i.e., problem-oriented languages, POL) versus programming languages designed for the actual operation of the computer itself (i.e., machine languages, ML). What the Robinson-Elgot paper did was to set up a means of evaluating both kinds of languages, not only in terms of their functions but also the relations between them (including extensions or improvements).

Rather than investigate their syntactical form, Robinson and Elgot treated programming languages simply as instructions stored in a machine (real or ideal). Beginning with two schematizations for what basically happens in a standard digital computer, they showed that with an intermediate model for programming (IMP) and its more powerful companion, a random-access stored program (RASP), "any particular RASP capable of computing two simple functions and the equality relation is capable of computing all partial recursive functions."[174]

Robinson was well aware that given a RASP capable of computing recursive functions, the question of effective calculability naturally arose. Together he and Elgot showed that the operations carried out either by

[171] Robinson 1965a, p. 184; Robinson 1979, 2, p. 87.

[172] Elgot and Robinson 1964.

[173] Calvin C. Elgot, quoted in Keisler 1979, pp. xxxvii.

[174] Elgot and Robinson 1964, p. 373; Robinson 1979, 1, p. 635.

an IMP or a RASP were recursive.[175] They also confronted the most interesting problem facing the infinite model they had introduced, namely, the objection that infinite systems should never be used in any "realistic" theory of computers. On the contrary, Robinson and Elgot maintained that any restriction to finite systems rendered discussion of recursiveness *pointless.* Any assumptions, they argued, under which a given system was necessarily "very large," was better represented by an *infinite* model.

One of the most novel ideas developed in their 1964 paper was "instruction modification." Again insisting that no finite machine could compute all recursive sequential functions with fixed programs, the modification of programs turned out to be essential. Indeed, Robinson and Elgot were able to show that with programs whose instructions were modifiable, a finite machine could compute any partial recursive sequential function.

The following year (1965), he and Elgot extended their results on random-access machines to multiple control RASPs (i.e., to computers capable of parallel processing).[176] This again proved to be a pioneering effort which resulted in a mathematical model for parallel computation, at the time a "poorly developed field."[177] Working now with J. D. Rutledge, what they hoped to do was bring some of the "arsenal of mathematics" to bear on problems related to programming highly parallel computers, which in turn, they expected, would greatly accelerate the speed of processing time.[178]

The trick, of course, was to construct an operating system to direct the parallel processing—not an easy task because at the time most programming languages were not capable of expressing algorithms to be processed simultaneously.[179] Thus what Elgot, Robinson, and Rutledge faced in writing about parallel programming was very much a pioneering effort in 1965. Whatever the difficulties involved, their paper ended

[175] In the cases Robinson and Elgot considered, the RASPs in question were generated by a finite number of recursive instructions. Elgot and Robinson 1964, p. 385; Robinson 1979, 1, p. 647.

[176] Elgot, Rutledge, and Robinson 1965.

[177] Calvin C. Elgot, quoted in Keisler 1979, 1, pp. xxxvii.

[178] The first paper on parallel programming by Stanley Gill appeared in the *Computer Journal* in April 1958. Subsequent papers on the subject did not appear for another seven years—when several articles were published in 1965, including the one by Robinson and Elgot. See Bell 1983, p. 4. A decade later, interest in parallel programming had increased dramatically. For 1973, Bell lists fourteen papers; by 1982, no less than fifty were devoted to the subject (the bibliography consists of articles published in eighteen major computer science journals, including four ACM publications).

[179] Languages like Cobol, Fortran, Pascal, and Basic are all serial languages; not until the 1970s were new languages like ADA designed to facilitate parallel programming.

with a "plausibility" theorem showing that parallel programs could indeed be constructed in a reasonable way.

Metaphysics of the Calculus: London 1965

That summer a major international colloquium devoted to philosophy of science was held in London, at Bedford College in Regent's Park, from July 11–17. The meeting was divided into three main parts, and Robinson was invited to deliver a keynote lecture in the section on "Problems in the Philosophy of Mathematics."

On Sunday evening, June 11, W. C. Kneale opened the colloquium with a brief "Presidential Welcome" on behalf of the British Society for Philosophy of Science, which jointly sponsored the meeting along with the London School of Economics. The plenary meeting was then addressed at 8:30 P.M. by Sir Karl Popper, who spoke on "Rationality and the Search for Invariants."[180]

The next morning, Robinson was the first speaker; he presented a paper on "The Metaphysics of the Calculus," followed by some forty-five minutes of discussion. Despite the philosophical nature of his title, Robinson presented a fundamentally historical paper, meant to examine the history of foundations of the calculus. This covered largely similar territory already reconnoitered in the last chapter of Robinson's book on *Nonstandard Analysis*. But one issue that served to give the London paper its own focus was the emphasis it placed on philosophical issues—and the extent to which these were more critical to questions about foundations than were purely technical matters.

The title of Robinson's lecture, in fact, was borrowed from an historic phrase of d'Alembert, *La théorie des limites est la base de la vraie Métaphysique du calcul différentiel*.[181] Were limits the best foundation for the calculus? Robinson set out to explore the history of this question, from the earliest thoughts on the subject in the seventeenth century to the present, which due to nonstandard analysis, of course, made it possible "to give a more precise assessment of certain historical theories than has been possible hitherto."[182]

Beginning with Newton and Leibniz, Newton's views on foundations were dismissed as "somewhat ambiguous," having based his view of the calculus essentially on limits. Leibniz, who wanted to base the calculus "clearly and unambiguously" on a system which included infinitely small

[180] See the brief schedule of the colloquium given in Lakatos 1965, p. xx.

[181] Quoted by Robinson from the article by d'Alembert on "Limite" in the *Dictionnaire encyclopédique des mathématiques*. See Robinson 1965b, p. 28, and Robinson 1979, 2, p. 537.

[182] Robinson 1965b, p. 28; Robinson 1979, 2, p. 537.

quantities, was clearly Robinson's champion in this history. He again quoted the Marquis de l'Hôpital at length, whose *Analyse des infiniment petits* was the first to popularize the new calculus.[183] The Marquis, unambiguously, asserted "the reality of the infinitely small quantities," and it was "this robust belief in the reality of infinitely small quantities which held sway on the continent of Europe through most of the eighteenth century."[184] Leibniz, however, was critical of l'Hôpital's belief in the reality of infinitesimals, and took Fontenelle to task as well for his approval of l'Hôpital's views.

However different Leibniz and l'Hôpital's understanding of infinitesimals may have been ontologically, operationally, Robinson explained, they both believed that two quantities might be regarded as equal if they differed by only an infinitesimal amount (relative to the two quantities in question). On the other hand, they also believed that infinitesimals should behave arithmetically like ordinary numbers. However, as Robinson warned, "it is evident, and was evident at the time, that these two assumptions cannot be accommodated simultaneously within a consistent framework."[185] Nonstandard analysis, however, resolved this dilemma by emphasizing equivalence instead of equality, a point Robinson had already elaborated in his historical chapter at the end of *Nonstandard Analysis* (see above, pp. 349–355).

In similar fashion, Robinson also brought his paper for the London Colloquium to a close with a brief look at Cauchy (also discussed in more detail above, pp. 352–354). This time, however, Robinson stressed the apparent paradox of Cauchy's well-known success in bringing new rigor to the foundations of the calculus: "It is generally believed that it was Cauchy who finally put the Calculus on rigorous foundations. And it may therefore come as a surprise to learn that infinitesimals still played a vital role in his system."[186] But Cauchy's views on infinitesimals were, on balance Robinson believed, negative:

> Cauchy did not wish to regard the infinitesimals as numbers. And the assumption that they satisfy the same laws as the ordinary numbers, which had been stated explicitly by Leibniz, was rejected by Cauchy as unwarranted. Moreover, Cauchy stated, on a later occasion, that while infinitesimals might legitimately be used in an argument they had no place in the final conclusion.[187]

[183] For further discussion of the eighteenth-century debate over foundations of the calculus, especially in France, see Grabiner 1981, pp. 16–46, and Belhoste 1985, pp. 101–112; English translation, Belhoste 1991.

[184] Robinson 1965b, p. 33; Robinson 1979, 2, p. 542.

[185] Robinson 1965b, p. 34; Robinson 1979, 2, p. 543.

[186] Robinson 1965b, p. 35; Robinson 1979, 2, p. 544.

[187] Robinson 1965b, p. 36; Robinson 1979, 2, p. 545.

Despite such pronouncements, Cauchy often treated infinitesimals as if they were ordinary numbers, and in all but one famous case–that of his mistaken proof that the sum of a series of continuous functions is continuous–he usually reached correct results. Where did Cauchy go wrong in the case of the series of continuous functions? "Here again, Nonstandard Analysis, in spite of its different background, provides a remarkably appropriate tool for the discussion of Cauchy's successes and failures."[188]

In his London paper, Robinson was more explicit than he had been earlier in saying that Cauchy's "entire notion of a variable" was untenable, "as a mathematical entity, *sui generis*, [it] has no place." Similarly, Cauchy's infinitesimals in Robinson's opinion fared no better, "Cauchy's infinitesimals still are, to use Berkeley's famous phrase, the ghosts of departed quantities."[189]

Cauchy's importance, historically, lay elsewhere for Robinson, in his having established the concept of limit as central. Due to the long history of uncertainty over foundations of the calculus, "a grateful public was willing to overlook the fact that, from a strictly logical point of view, [Cauchy's] new method shared some of the weaknesses of its predecessors and, indeed, introduced new weaknesses of its own."[190]

It was not long, however, before Weierstrass brought further improvements to Cauchy's innovations, and Robinson recounted them, as well as Georg Cantor's remarkable condemnation of infinitesimals, just as he had in *Nonstandard Analysis* (see above, pp. 353–355). Robinson drew a very strict lesson from Cantor's metaphysics, however, at the end of his London talk:

> Cantor's belief in the actual existence of the infinities of Set Theory still predominates in the mathematical world today. His basic philosophy may be likened to that of de l'Hôpital and Fontenelle although their infinite quantities were thought to be concrete and geometrical while Cantor's infinities are abstract and divorced from the physical world. Similarly, the intuitionists and other constructivists of our time may be regarded as the heirs to the Aristotelian traditions of basing Mathematics on the potential infinite. Finally, Leibniz' approach is akin to Hilbert's original formalism, for Leibniz, like Hilbert, regarded infinitary entities as ideal, or fictitious, additions to concrete Mathematics. Thus, we may conclude this talk with the observation that although the very subject matter of foundational research has changed radically over the last two hundred years, there is a remarkable permanency in the concern with the infinite in Mathematics

[188] Robinson 1965b, p. 36; Robinson 1979, 2, p. 545.
[189] Robinson 1965b, p. 37; Robinson 1979, 2, p. 546.
[190] Robinson 1965b, p. 38; Robinson 1979, 2, p. 547.

and in the various philosophical attitudes which have been adopted towards this notion.[191]

Following Robinson's remarks, there was considerable discussion, some over historical details. Both Hans Freudenthal and Arend Heyting, however, were concerned about "Technique Versus Metaphysics in the Calculus," with Heyting viewing Robinson's lecture as largely about how initially metaphysical concerns "can later on be considered as merely technical questions." As he put it, "If you consider nonstandard analysis as technical, at the same time you consider it as an interpreted set theory, and from that point of view it contains some metaphysics."[192]

Yehoshua Bar-Hillel took up a very strong position, asserting "the irrelevance of ontology to mathematics," stressing what he took to be the "confusions created by using the material mode of speech on an inappropriate occasion."[193] Mario Bunge, on the other hand, was interested in "non-standard analysis and the conscience of the physicist." He thanked Robinson for having shown that reliance on infinitesimals was not misplaced after all, clearing the "bad conscience" of physicists who had continued to use them all along, despite "rumors of the Dedekind-Weierstrass revolution."[194]

Robinson's response to these comments was basically polite, and in one instance served to clarify his own philosophical position. As he responded specifically to what Bar-Hillel and Heyting had said:

> It will be evident that in my paper I have dealt with questions of reality in Mathematics from the detached point of view of a historian. However, I am willing to go further and commit myself to the point of stating that in my view these problems should not be discussed in the cavalier fashion advocated by Bar-Hillel. As to what is technical and what is essential, I certainly did not want to suggest that the very differences of opinion between constructivists and platonists are merely technical. But it seems to me likely that questions of detail within transfinite set theory such as the correctness of the continuum hypothesis or the existence of very large cardinals, will come to be regarded as philosophically irrelevant, although I yield to no one in my admiration for the ingenious methods which have been devised to cope with these problems.[195]

Following Robinson's lecture that morning, Andrzej Mostowski spoke on "Recent Results in Set Theory," after which there was again lively discussion. Among points raised by the audience was a remark from

[191] Robinson 1965b, pp. 39–40; Robinson 1979, 2, pp. 548–549.
[192] Robinson 1965b, p. 43; Robinson 1979, 2, p. 552.
[193] Robinson 1965b, p. 44; Robinson 1979, 2, p. 553.
[194] Robinson 1965b, pp. 45–46; Robinson 1979, 2, pp. 553–554.
[195] Robinson 1965b, p. 45; Robinson 1979, 2, p. 554.

Robinson raising again the relevance of recent results in axiomatic set theory for the foundations of mathematics:

> The wide acceptance achieved by the axioms of Zermelo-Fraenkel is merely a manifestation of their basic validity and that the same should apply to new axioms, yet to be discovered. On the other hand those who, like myself, have a deep distrust of ontological assertions concerning infinite totalities will find some support for their attitude in the current bifurcation (or multifurcation) of Set Theory and will even consider the possibility of a similar bifurcation in Number Theory. This is not to say that I would welcome the spreading of total anarchy in the foundations of Mathematics, nor even that I consider such a development likely. Indeed, as I have explained elsewhere (at the Jerusalem Congress, 1964), while I would describe myself as a formalist, I feel compelled to admit the existence of certain basic forms of thought–in Logic, in Arithmetic, and perhaps in Set Theory–which are prior to the arbitrary choice of mathematical axioms.[196]

Perhaps it was the fear of "anarchy" at the foundations of mathematics that led Robinson to place some faith in the existence of "certain basic forms of thought prior to the arbitrary choice of mathematical axioms." If nothing else, the intuitive naturalness of the Zermelo-Fraenkel axioms was the source of their "basic validity." But just as in geometry, where debates over non-Euclidean geometries in the nineteenth-century never led to anarchy, but indeed served to clarify many aspects of mathematics, Robinson found no reason to fear any real "anarchy" in mathematics stemming from the most recent developments in set theory. Later he would consider the examples of non-Euclidean geometries and non-Cantorian set theories as cases in point. In fact, he would come back to all of these themes again after he had settled at Yale, where he wrote several papers of historical and philosophical interest, and even offered an interdepartmental seminar on mathematics and philosophy with Stephan Körner in the fall of 1973 (see chapter 9).[197]

Robinson Returns to the Mathematics Department

When Robinson returned to UCLA from his very successful summer abroad, his teaching and administrative responsibilities were now entirely in mathematics. His joint appointment with philosophy had been demanding, a time-consuming four years of administrative responsibilities and students whose interests were not always up to the high standards of mathematical logic which Robinson would have liked to set.

[196] Robinson quoted in Lakatos 1965, pp. 103–104.

[197] See Seligman 1979, p. xxxi.

After his peripatetic year as chairman of the University-wide Educational Policy Committee, and the equally time-consuming monthly meetings of the Academic Council, Robinson decided that it was no longer wise to divide his time—and energy—between two departments. According to Yiannis Moschovakis, he was above all tired of the double administrative work. The Philosophy Department had too many meetings, discussions were sometimes lengthy over trivial matters, and Robinson came to feel that his time was being "eaten up."[198]

Because of his decision to leave philosophy, when the question of a salary increase for Robinson arose in the fall of 1965, Donald Kalish wrote to Lowell Paige in support of Robinson, but suggested that Paige initiate the promotion:

> In view of the fact that Professor Abraham Robinson will no longer be officially connected with the Department of Philosophy as of July 1, 1966, the Department considers it more appropriate to support than to initiate any action relating to his status in the academic year 1966–67. The Professors of our Department are unanimous in their support for an overscale merit raise for Professor Robinson, effective July 1, 1966.
>
> Professor Robinson has carried an important teaching load in philosophy at the upper division and graduate level; indeed, the breadth of his interests and abilities, and his competence to teach and direct research in a wide variety of subjects is truly extraordinary. Teaching was not his only service to our Department; although he presently holds a joint appointment in Mathematics and Philosophy, his participation in our departmental affairs (and the regularity of his attendance at our departmental meetings) equaled that of any staff members in amount and surpassed that of most, in quality.[199]

That spring (May 1966), when Robinson's "overscale merit increase" was approved, UCLA Chancellor Franklin D. Murphy wrote to explain:

> Advancement beyond the established salary scale has by University policy and tradition been reserved for those individuals who, in the judgment of their faculty peers and the administration, have demonstrated outstanding accomplishment and whose service is deemed exceptionally meritorious. Your advancement therefore represents more than a simple increase in monetary compensation; it represents an accolade in recognition of your many contributions to the University, your discipline, and community.[200]

[198] Moschovakis I-1991.

[199] Donald Kalish to Lowell Paige, September 28, 1965, Robinson's administrative file, Department of Philosophy, UCLA.

[200] Franklin D. Murphy to Abraham Robinson, May 20, 1966, Robinson's administrative file, Department of Philosophy, UCLA.

Robinson's Last Publication on Applied Mathematics

In October of 1965 Robinson submitted a paper written jointly with his UCLA colleague (and former student from Toronto), Albert Hurd, "On Flexural Wave Propagation in Nonhomogeneous Elastic Plates," to the editors of the *SIAM Journal of Applied Mathematics*.[201] This was to be Robinson's last publication in applied mathematics, and was devoted to the effects of variations in thickness and elastic parameters on the propagation of flexural waves in thin elastic plates. By using the method of characteristics to analyze systems of hyperbolic equations, Robinson and Hurd modified earlier results of Raymond Mindlin,[202] whereby they succeeded in describing the propagation of sharp discontinuities.[203]

Hurd wrote his masters thesis with Robinson at Toronto, where his early training had been in engineering. Although Robinson suggested the propagation problem to Hurd, he soon found that despite exhaustive calculations, he was getting nowhere. It was a tedious project, and Robinson was basically content to leave him alone, although from time to time Robinson offered his encouragement and insisted that Hurd keep at it, for he was sure the approach "has got to be right." But due to the massive amount of calculation involved, Hurd found that he was constantly making small errors, which later explained why the problem at first did not work out as they had expected.[204]

From Toronto, Hurd went to Stanford, then to MIT, where he thought about publishing the work he had done with Robinson. The editor of the *Journal of Applied Mathematics* at MIT, however, rejected it out of hand as the product of work done by "a logician who dabbles in applied mathematics." Consequently, the paper "sat around for another couple of years," after which Hurd went to UCLA. Coincidentally, finding Robinson there as well, he asked Robinson about trying to copublish the paper. Robinson objected, saying that this would be highly unusual—it was virtually all Hurd's work except for the problem itself (which had admittedly been Robinson's suggestion). Nevertheless, with both their names on the paper, there was no difficulty getting it into print.

St. Catherine's College, Oxford: Fall 1965

It was John Crossley who arranged for Robinson to visit St. Catherine's College, Oxford, in the fall of 1965:

[201] Hurd and Robinson 1968. [202] Mindlin 1951.

[203] Robinson had basically used this same approach successfully in his earlier study of the propagation of disturbances both in beams and in heterogeneous elastic mediums. See Robinson 1957 and 1957a.

[204] Hurd I-1991.

> Logic was booming in Oxford when I got Abraham Robinson invited to a Visiting Fellowship at St. Catherine's. As part of the process he submitted a handwritten C.V. with a publication list of 70 items or so including, much to my surprise, a book on "Wing Theory"–I couldn't decipher the handwriting at first and thought it must be "Ring Theory" but it was the result, he told me, of work with Sir James Lighthill. . . . He stayed with his wife in Rom Harré's house at Iffley and gave regular lectures–like most mathematicians, his seminars didn't provoke much discussion–at the Mathematical Institute–then in the old ex-maternity home at 10 Parks Road. Attendance was probably about 15 or so and of course the topic was Nonstandard Analysis.[205]

Crossley's offer could not have come at a better time, and Robinson was grateful for the opportunity to spend several quiet months in Oxford. Not only were Abby and Renée pleased to be back in England, they were delighted with Oxford, "that sweet city with her dreaming spires," as Matthew Arnold had called it. "So serene! . . . whispering from her towers the last enchantments of the Middle Age."[206] St. Catherine's College, however, was hardly of the Middle Ages. It was the newest college at Oxford, and seemed to have been designed with the eye of a mathematician in mind. Architecturally it was "a striking piece of geometry, a great rectangle, containing rectangles, with here and there a circle, symmetrical, well balanced, in a large frame of grass and trees.[207]

The design was the work of a Danish architect, Arne Jacobsen; the site was located on the east bank of the Cherwell River, in Holywell Great Meadow, surrounded by countryside. Jacobsen designed everything himself, from the gardens outside (with ponds of waterlilies graced with sculptures by Barbara Hepworth and Henry Moore on the lawns), to the furniture, tableware, and even the ashtrays. One visitor warned that the "implements" at St. Catherine's, thanks to their Danish design, "may cause initial alarm. What a stranger may take at first for a chop stick is in fact his soup spoon."[208] The Dining Hall, a gift of Esso Petroleum, was the largest in either Oxford or Cambridge, which prompted a quip from Sir Harold Macmillan, when dedicating the building: "Surely this must be Esso's largest filling station!"[209]

[205] John N. Crossley to JWD, letter of July 5, 1991. The topic was especially timely since Robinson was just then finishing his *Nonstandard Analysis,* Robinson 1966. As the page proofs began to arrive, he enlisted Moshé Machover to help with the proofreading (see below, pp. 373 and 393).

[206] Arnold 1865, "Preface," p. xv; quoted from Super 1962, p. 290.

[207] Wooley 1972.

[208] Balsdon 1970, p. 140.

[209] Quoted in Thackrah 1981, p. 141.

Robinson at St. Catherine's College, Oxford.

Robinson at Trinity College, Cambridge.

St. Catherine's, however, was an excellent place for mathematics and for science in general. As for logic, Oxford had a long tradition going back to such fourteenth-century figures as Walter Burley, Duns Scotus, William of Ockham, and Roger Swineshead (Swyneshed).[210] In the late twentieth century, logic was still doing very nicely.

Robinson's contribution was a series of lectures on nonstandard analysis. Colleagues and graduate students from near and far took advantage of the opportunity to hear Robinson lecture in person (Moshé Machover and Rohit Parikh made the trip together regularly to Oxford from Bristol). Among others in the audience was John Bell:

> As a graduate student in Oxford in 1965, I attended Robinson's lectures on nonstandard analysis. As I recall, the lecture hall was packed—the audience included Moshé Machover, Alan Slomson, Peter Aczel, John Wright, Frank Jellett, John Crossley, and Joel Friedman (his student who had accompanied him from UCLA). These lectures were very absorbing—it was obvious that Robinson was presenting something of fundamental importance—and delivered with what I can only describe as an endearing lack of slickness. For example, Robinson had a circuitous method of proving mathematical propositions at the blackboard which apparently proceeded

[210] For an overview of the earlier tradition in logic associated with Oxford and its later development, see Thomas 1964, pp. 297–311.

as follows. To prove a proposition P, he would start by assuming *not* P. He would then prove P completely independently of the assumption *not* P, deduce that the latter must be false, and finally infer the truth of P. This is not the familiar form of *reductio* argument:

$$\frac{\vdash \neg P \rightarrow P}{\vdash P}$$

but rather what I came to call the "Robinsonian" form:

$$\cfrac{\cfrac{\vdash P}{\vdash \neg\neg P}}{\vdash P}$$

At the end of the course Robinson held a party to which all the members of his audience were invited. I remember this as a very warm and enjoyable occasion.[211]

Overtures from Yale: Fall 1965

At the end of November, while at work in Oxford, Robinson received a letter that was soon to change his career–once again. Nathan Jacobson wrote from Yale, explaining that the administration had recently transferred a position in logic (vacated by Alan Anderson) from philosophy to mathematics.[212] An ad hoc committee (consisting of Church, Kleene, Wang, and Montague) had reviewed matters and recommended that the position be moved to mathematics. Moreover, everyone seemed to agree that Robinson was the best choice to direct mathematical logic at Yale:

> We have been deliberating in our Department on the choice of a mathematical logician for this position and we have consulted a number of eminent logicians. The conclusion of all of this has been that we feel very strongly that you would be the ideal person to help us to develop in this new direction for our Department. Also the Philosophy Department has backed us up with great enthusiasm in our decision. Accordingly, I am writing you to ask if you might be interested in an offer of a Professorship in mathematics at Yale.
>
> Our Department is small compared to that of U.C.L.A. However, I think it is a strong Department, particularly in algebra, functional analysis and

[211] John L. Bell to JWD, March 12, 1994.

[212] According to Jacobson, the Philosophy Department was not interested in making another appointment in logic. By transferring the position, mathematics, "which had never had anyone in logic before, acquired an incremental tenure slot in this area." See Jacobson 1989, 3, p. 2.

> topology. The tenured members of the Department are: Ore, Hedlund, Jacobson (Chairman), Kakutani, Rickart, Massey, Mostow, Tamagawa, Feit and Seligman as Professors and Hahn and Szczarba as Associate Professors on tenure. . . . I have already noted that Professor Fitch is in the Philosophy Department. Another member of this Department who is interested in logic is Professor Rulon Wells whose main interest is linguistics. In Statistics (which is a separate Department) we have Anscombe and Savage. Also Yale is in the process of building a Department of Applied Science.[213]

Remarkably, only a few days later, hard on the heels of Jacobson's letter, Robinson received a second offer. This one came from Queens University in Kingston, Ontario, where a senior position in mathematics was about to become available as Israel Halperin was leaving for the University of Toronto. Both Harold Lightstone and Hubert Ellis were at Queens, and very much wanted to add Robinson to the faculty in Kingston.[214]

There was never any real chance, however, that Robinson would opt for Queens when Yale was making a serious offer. On December 16 he wrote back to Nathan Jacobson to say that he would like to visit Yale on his way back to California from Europe. What he wanted to discuss, above all, were details including salary, fringe benefits, summer grants—and to see what the local housing market was like.[215]

Paris and Rome: 1966

Following the term at Oxford, Robinson was invited to the University of Paris where he had been appointed professeur associé at the Institut Henri Poincaré. The Robinsons stayed at the Hotel Bretagne on the rue Jacob, and while Renée spent most of her time going from one gallery to another, especially to the Louvre to see the Rembrandts, Abby spent most of his time working.

Again, he had no specific obligations but did give a series of lectures on *L'Analyse non-standard*, including one dealing with the metaphysics of nonstandard analysis. He shared an office with Gustave Choquet, who had many discussions with Robinson during the month or so he was in Paris:

> I did not attend his lectures, but he told me often about the influence of

[213] N. Jacobson to Abraham Robinson, November 29, 1965, RP/YUL.

[214] A.J. Coleman to Abraham Robinson, December 2, 1965, RP/YUL.

[215] Jacobson to Abraham Robinson, December 16, 1965, RP/YUL. See also Jacobson 1989, 3, p. 2.

Broadsheet announcing Robinson's lectures on nonstandard analysis at the University of Paris, spring 1966.

> Leibniz on the birth of his ideas; I don't remember any other source he ever mentioned to me. Leibniz was really his great man.[216]

Robinson spent a good portion of his time at the Institut Poincaré where he was also putting the finishing touches on his book, soon to appear from North-Holland, *Nonstandard Analysis*:

> Robinson was more interested in writing this book than in giving courses or meeting students. We didn't have many offices, so he shared mine.[217]

Actually, at that point Robinson had finished the book and was going over page proofs in Paris. Moshé Machover was also reading the page proofs for Robinson, and sent these along with some questions about various applications that he and Rohit Parikh had been thinking about:

> I sent you a proof (along with proofs of your book) of the fixed point theorem for Hilbert space, using Brouwer's fixed point theorem. Now Parikh has produced a nonstandard proof of the Brouwer theorem itself. . . .
>
> Parikh and I would like to know whether you had noticed the possibility of proving the Brouwer theorem and its generalisation to Hilbert space using your methods? Also, if the answer is negative, do you think it worthwhile and proper for us to publish a short note on this?
>
> Parikh also would like to know whether you have considered applying NSA to the following problems:
>
> 1) Topological invariance of homology groups
> 2) Pontrjagyn [Pontryagin] duality theorem for locally compact topological Abelian groups

[216] Gustave Choquet, equipe d'analyse, Université de Paris, to JWD, letter of May 29, 1984.

[217] Choquet I-1984.

3) Existence of Haar measure[218]

4) Roth's theorem on approximation to algebraic numbers by rationals

We consider NSA great fun and would like to apply it to one such subject.[219]

Indeed, there was growing enthusiasm for nonstandard analysis not only in the United States and England, but in Europe as well. Considerable interest was brewing in France, for example, especially in Strasbourg where an active school had begun to form around Georges Reeb. Reeb first became enthusiastic about the promise of nonstandard analysis as a tool, especially in topology and differential equations, after reading Robinson's *Nonstandard Analysis* in the early 1970s.[220] Throughout the Department of Mathematics in Strasbourg, Reeb proclaimed "in every corridor that Nonstandard Analysis was 'something really new, an actual revolution.'"[221] He was certain that it would prove especially useful in applications to perturbation problems where it would be possible, thanks to nonstandard analysis, to consider fixed objects instead of having to master functions. Soon a group had formed in Strasbourg, symposia and meetings on nonstandard analysis were encouraging its development, and active research groups were also emerging in Mulhouse and Oran (Algeria). After the appearance of Edward Nelson's "Internal Set Theory" in 1977, infinitesimals were all the more attractive for mathematicians in natural applications that Reeb and his school began to exploit with increasing success.[222] At Mulhouse, Robert Lutz and Michel Goze were inspired to write their book on *Nonstandard Analysis*, including typical algebraic applications, and in Oran, the Reeb school's best-known application of nonstandard analysis to perturbation theory resulted in the study of solutions to slow-fast systems in terms of Van Der Pol's equation with two parameters, $e\ddot{x} + (x^2 - 1)\dot{x} + x = a$, known as "canards."[223] As Reeb's students began to "pick up" on Robinson's

[218] Parikh did publish his results on a nonstandard Haar measure. By considering an infinitesimal neighborhood of the identity, Parikh showed how the Haar measure of a compact set could be defined in terms of the minimum number of translates needed to cover the set. See Parikh 1969, pp. 279–284.

[219] M. Machover to Abraham Robinson, January 17, 1966, RP/YUL.

[220] Lutz I-1994. See also G. Reeb 1977, p. 8–14. Reeb was famous for insisting that *Les Entiers Naïfs ne remplisent pas N.* A general appreciation of his role in popularizing nonstandard analysis in France is given by Renée Peiffer-Reuter in her preface to Claude Lobry's 1989, esp. p. 7. Lobry's book is an anecdotal history of the resistance to nonstandard analysis in France, the first chapter of which provides a brief account of Reeb and his contributions, pp. 13–21.

[221] As Robert Lutz and Michel Gose responded to Reeb's enthusiasm, "we believed it and thus had to write a book on it." See Lutz and Goze 1981, p. vi; Lutz I-1994.

[222] Nelson 1977.

[223] For a detailed discussion of "canards," the origin of the term and a photograph of

version of nonstandard analysis, France soon became one of the most active centers outside of the United States for its serious development.[224]

Following his lectures in Paris, Robinson was invited to spend a few weeks at the Castelnuovo Institute at the University of Rome. Again, he was called upon to offer several lectures on nonstandard analysis. These actually led to several publications, including a pair of notes for the *Proceedings* of the Accademia dei Lincei. Another paper, "On Some Applications of Model Theory to Algebra and Analysis," also appeared that year in the *Rendiconti di Matematica e delle sue Applicazioni*.[225]

Robinson's two "Notes" on *Logica Matematica* were communicated to the Academy by Beniamino Segre. Basically, both of these were devoted to a nonstandard theory of Dedekind domains. This actually went back to work on the theory of *idèles* and *adèles*, and to ideals of Dedekind domains in the theory of algebraic numbers (or functions) initiated by Prüfer, von Neumann, and Krull.[226]

Robinson's lectures at the University of Rome also resulted in a paper, "On Some Applications of Model Theory to Algebra and Analysis."[227] This work was conceived as another general introduction to show mathematicians the useful results one might obtain by applying model theory to algebra and analysis, and also included several results he believed to be entirely new.

From a very brief introduction to the basics of model theory, Robinson moved on to applications in algebra, where he began with theories of commutative and ordered fields, stressing the relation of the latter to Hilbert's seventeenth problem and Artin's theorem on positive definite functions, for which Robinson sketched a model theoretic proof.

Nonstandard analysis was then applied in a section devoted to lattice theory. Here the key, once again, was Robinson's use of compactness properties, replacing standard methods with nonstandard counterparts. For example, classical results usually proved using Mahler's compactness principle for lattices were now proved using Robinson's methods, and he even showed how Mahler's principle itself could be derived using the nonstandard approach. Whether this was really significant or not Robinson dismissed as a matter of individual preference:

---

the "nonstandard duck-hunters of Oran," see Diener 1984. See also Lutz and Goze 1981, chapter 4, "Nonstandard Analysis as a Tool in Perturbation Problems," especially "Lesson 8" on canards, pp. 173–177.

224 Choquet I-1984. For brief remarks about the group working in Strasbourg with Georges Reeb, see Reeb 1989, pp. 3–9.

225 Robinson 1966a and 1966b.

226 Prüfer 1925, von Neumann 1926, and Krull 1935.

227 Robinson 1966b, pp. 562–592; in Robinson 1979, 2, pp. 158–188.

> It is a matter of taste whether we wish to regard our present method as a remote reformulation of such arguments or whether we wish to assert rather that compactness arguments (e.g. selection principles) were introduced into Analysis in order to fill a gap due to the historical breakdown of the method of infinitesimals.[228]

Perhaps more to the point is how Robinson came to reach his results in the first place. There was a strong similarity in approach to his use of model theory in dealing with Hilbert's seventeenth problem based on Artin's use of positive definite functions, and his more recent conclusions on lattices. In both cases he began by establishing "ideal (weak) solutions" from which he then showed the existence of "real (strong) solutions." Inspired by the success of this approach in the theory of partial differential equations, Robinson in turn tried to do something similar for positive definite functions and lattices from a nonstandard point of view.[229]

Robinson's lectures made a strong impression in Rome, and afterwards, Segre wrote to thank him in particular for his contribution to the *Rendiconti*.[230] Another letter from Lucio Lombardo Radice was even more enthusiastic, and invited Robinson to return for a longer visit as visiting professor for two months in 1966–1967.[231]

New Haven: Cold Weather but a Warm Reception

On their return from Europe, the Robinsons stopped in New Haven, as planned, to take a look at Yale:

> The department gave a luncheon for Robinson to which a number of colleagues from other departments were invited. Simultaneously, Florie [Jacobson] hosted a luncheon for Mrs. Robinson and the department wives. At this luncheon, Mrs. Robinson made it clear that she was reluctant to leave the Los Angeles area, especially since they had recently purchased a beautiful house in a lovely part of Los Angeles.
>
> Our formal offer was sent to Robinson on February 4. On February 19, I received his reply that "with a heavy heart" he had decided to decline our offer. This was a serious blow, but, because of Mrs. Robinson's attitude, it was not totally unexpected.[232]

[228] Robinson 1966b, p. 580; Robinson 1979, 2, pp. 185.

[229] Robinson 1966b, pp. 589–590; Robinson 1979, 2, pp. 185–186.

[230] B. Segre to Abraham Robinson, April 15, 1966, RP/YUL.

[231] L. Lombardo Radice to Abraham Robinson, May 31, 1966. Robinson was offered a two-month appointment as visiting professor at 450,000 lire per month, RP/YUL.

[232] Jacobson 1989, 3, pp. 2–3. Jacobson was right about Renée's reluctance to leave Los Angeles, but their house was not a recent purchase; the Robinsons had lived in their

Back in Los Angeles, the administrative turmoil and constant department meetings were a sharp contrast to the very pleasant half-year Robinson had just spent abroad. The return to UCLA must have come as something of a shock, especially as conditions university-wide continued to deteriorate. Others felt the strain as well, and faculty were slowly beginning to go elsewhere in favor of better salaries, less teaching, less tension—and significantly less bureaucracy.

Indeed, financial woes made it all the more difficult for UCLA to retain its best faculty, and sunny skies alone were no longer sufficient to lure replacements. As the *Bruin* put it, "Warm climate, academic independence, other 'Fringe Benefits' have lost past effectiveness." For one department chairman, who left for the University of Michigan in 1963, a major enticement was reported simply as "big increase in pay."[233]

The mood of faculty was captured bluntly in an article the *Bruin* devoted to examining "Why Faculty Members Leave UCLA." As one departing professor asked rhetorically:

> How many graduate students do I have here? Thirty six! A veritable platoon. I can't remember thirty-six names, much less thirty-six thesis titles. . . .[234]

For Robinson, confronting the question of whether or not to stay at UCLA was only a matter of time. Taking advantage of a brief visit to New York that spring, he telephoned Nathan Jacobson to say that he wanted to come briefly to Yale to discuss the offer he had earlier declined. By then, however, the department had already made an offer to Dana Scott. Even so, Jacobson met Robinson at the New Haven train station the same day Robinson had called, and they had dinner together that evening at a local restaurant:

> [Robinson] told me that he would now accept the offer if this could be reinstated. I mentioned the complication of our negotiations with Scott that had gone quite far and, though I was confident that our offer could be reinstated, I needed to have some time to discuss this with the professors in the department and with the provost. All were enthusiastic about this turn of events and Provost Taylor agreed to making the offer again to Robinson, while at the same time continuing our effort to persuade Scott to come to Yale. In the end, Scott declined our offer and also an offer we made for him to try out Yale on a visiting basis. In order not to make too

---

home in Mandeville Canyon since shortly after their arrival at UCLA.

[233] *UCLA Daily Bruin*, Thursday, January 3, 1963. The article about faculty recruitment at UCLA, it should be added, was in response to UC President Clark Kerr's insistence that Governor Brown's 5 percent faculty pay increase was not enough.

[234] *UCLA Daily Bruin*, Wednesday, April 24, 1963, p. 8.

abrupt a break with UCLA, Robinson requested that his position at Yale begin in 1967.[235]

UCLA and the Vietnam War

In the spring of 1966, the American war in Vietnam began to escalate dramatically. Protests became larger and more frequent on campuses across the country, and UCLA was no exception.[236] When "giant" B-52 bombers began to assault North Vietnam for the first time, the reaction was immediate.[237] As the *Bruin* headlined on April 19: "Profs Hit War in LBJ Letter."[238] Among the 136 signatures, Robinson's was among them (along with many members of the Mathematics and Philosophy Departments).[239] In addition to calling for an end to offensive operations, the letter proposed "a settlement of the war by negotiations participated in by the Saigon and Hanoi governments and the National Liberation Front":

> Dear Mr. President: we believe as Walter Lippmann put it, "that a complete military victory in Vietnam, though theoretically attainable, can in fact be attained only at a cost far exceeding the requirements of our interest and our honor."[240]

The letter went on to affirm the right of the Vietnamese to determine their own political future, and called upon the U.S. military to cease the bombing of North Vietnam, continuing only defensive military operations in South Vietnam while committing no additional troops for a "reasonable but extended" period of time. Instead, to the dismay of many, by the end of the month more than four thousand new troops had landed "in a further buildup of American forces in the Southeast Asian nation."[241]

At UCLA, Donald Kalish was especially active in organizing teach-ins against the Vietnam War. The University of Michigan had pioneered the idea in April of 1965; six months later, UCLA's first teach-in was held on November 12.[242] Kalish, as treasurer of the University Commit-

[235] Jacobson 1989, 3, p. 3.

[236] *UCLA Daily Bruin*, Monday, February 7, 1966.

[237] *UCLA Daily Bruin*, Tuesday, April 12, 1966.

[238] *UCLA Daily Bruin*, April 19, 1966.

[239] The letter, in fact, was drafted in the Philosophy Department by Wade Savage and Charles Chastain. Savage was cocoordinator of the UCLA Committee on Vietnam; the letter to the president, however, was not sent in the name of the committee but from the signing faculty members as individuals.

[240] *UCLA Daily Bruin*, April 19, 1966.

[241] *UCLA Daily Bruin*, April 29, 1966.

[242] Pilisuk 1965, in Menashe and Radosh 1967, p. 13.

tee on Vietnam, found that Robinson was not only very supportive, but whenever money was needed for an advertisement in the *Bruin* or to rent a loud speaker, Robinson was always ready to help with a contribution. In return, during the Israel-Arab War when the Israelis occupied the West Bank in 1967, Robinson came to Kalish and said with a wide grin, "You've been at my door for the last two years, now I have an opportunity to knock on yours."[243]

Although Robinson always took social and political issues seriously, Kalish recalls that despite obvious concern, he was never distraught or upset, but approached even the most sensitive issue with warmth, a smile, and an appropriate sense of humor. Kalish appreciated this, and on one occasion was pleased to have waiting a "*pre*-partee"—a repartee he had been waiting to use. Kalish, who had taught at Swarthmore in the 1940s and thus had a close association with Quakers, responded to their suggestion to have a silent vigil in protest of the Vietnam War and organized one at UCLA in 1966. These became a regular noontime feature on Wednesdays, and on the appointed first day, Robinson was among the earliest to arrive and stand with Kalish. As they stood together, Kalish turned to Robinson with his "*pre*-partee" and said, with a smile, "Some of our best Jews are friends."[244]

Three More Ph.D.s: 1966

As soon as Robinson was back from his term at Oxford, he found that three of his graduate students were in the final throes of writing their dissertations. As usual, they were working on a wide variety of subjects.

L. E. Travis wrote his dissertation under Robinson's supervision on "A Logical Analysis of the Concept of Stored Program: A Step Toward a Possible Theory of Rational Learning." Travis was interested in stored programs, which offered "a powerful kind of self-modification by machine." He hoped that applications might also prove useful in a theory of rational learning, "a theory concerned with how people learn by appraising and then deliberately changing themselves."[245]

Robinson oversaw a second thesis that year in philosophy, "A Set Theory of Proper Classes," written by Joel Friedman. The main object of Friedman's dissertation was the development of a general set theory of proper classes (STC). As he explained:

243 Kalish I-1991.

244 Kalish I-1991, with further details in a letter to JWD of December 1, 1992.

245 Travis 1966, p. xii. In addition to thanking Robinson for his help, Travis singled out Allen Newell of the Carnegie Institute of Technology, and J. C. Shaw of the Rand Corporation, for their work which had inspired his own on stored programs.

> Here, proper classes are regarded as elements. Traditionally, they have been defined as those classes which are not elements. But here, the notion of set is extended so that sets may contain proper classes. This theory is regarded as a generalization of ordinary set theory.[246]

Indeed, in the course of his thesis Friedman not only considered various models of STC, but considered alternative theories as well, many of which were shown to be interpretable in STC. Friedman also considered Oberschelp's theory of proper classes, and the set theory of properties, both extensional and intentional, in comparison with STC.

Lawrence D. Kugler was among Robinson's students at UCLA interested in nonstandard analysis. He had done his undergraduate work at Caltech (B.S. in mathematics, 1962), and gone on to UCLA where he took his first course with Robinson on set theory in 1964–1965:

> [Robinson] was a patient, methodical teacher, and I felt very comfortable in the course, until the oral final exam on which I performed poorly. (I had studied hard on Gödel's proof of the consistency of the continuum hypothesis, and was unable to do simple ordinal number computations he asked for). In that year, encouraged by other more senior graduate students who reported that Robinson had a very high success rate in producing Ph.D's and fascinated by nonstandard analysis, I summoned the courage to ask him to supervise my dissertation. He immediately agreed, gave me two possible topics: almost periodic functions or uniform spaces, and suggested I read up on both and select one. It was only a short time before I picked almost periodic functions. He started me off with a basic idea about the distribution of "near-periods" in $*R$ (that turned out, of course, to be correct and pivotal) and I went to work.[247]

The subject of almost periodic functions was, in many ways, peculiarly amenable to the techniques of nonstandard analysis. As Kugler pointed out, the formal statement of the main definition in the theory was reminiscent of the basic definitions of elementary calculus for which nonstandard analysis had already proven "very successful":

> Informally, this definition of almost periodicity states that certain translates of the graph of an almost periodic function are within an arbitrary preassigned distance of the graph of the function itself. Beginning with this definition, [Harald] Bohr showed that a generalized Fourier series can be obtained for any almost periodic functions, and that many basic theorems in the classical theory of Fourier series for purely periodic func-

[246] Friedman 1966, p. v. Friedman was generous in his thanks to Robinson, noting that "Each chapter of this dissertation is largely the result of his fruitful suggestions and general insights. Also, I am indebted to his biting criticism."

[247] Lawrence D. Kugler to JWD, letter of February 7, 1991.

tions carry over to the almost periodic case. In particular, generalized versions of the Parseval equality and the trigonometric polynomial approximation theorem hold, and these theorems constitute the focal points of this dissertation.[248]

Kugler gave three characterizations of "almost periodicity" in terms of infinite and infinitesimal numbers, upon which his nonstandard interpretation was based. Having shown how a Fourier series could be generated for an almost periodic function, he established the Parseval equality, and after investigating some of its applications, went on to investigate Bohr compactifications and the approximation theorem, all from a nonstandard point of view.[249]

When it came time to approve Kugler's dissertation, Robinson was clearly pleased:

> He asked me a few questions on the trickiest arguments, and declared himself satisfied. In discussing a committee for dissertation defense, he gave me several names of faculty outside the mathematics department, including an engineer (UCLA required one or two nondepartmental committee members). I must have looked surprised at the engineering suggestion and he good-naturedly said, "I've served on a lot of his students' doctoral committees; he owes me![250]

From Tübingen to Moscow: Summer 1966

Despite numerous offers to lecture and teach during the summer, Robinson could not refuse the chance to return to Tübingen when Peter Roquette wrote to say that he had arranged a guest professorship for Robinson that summer.[251] He flew directly to Zürich on July 1, and then went by train to Tübingen, where Roquette had found him a room for the month. Robinson had requested that his lectures be scheduled midweek (Tuesday-Wednesday-Thursday), but promised to be available

[248] Kugler 1966. Bohr's work went back to a generalization of the theory of Fourier series developed in H. Bohr 1925. An abridged version of the major results of Kugler's thesis was presented at the Pasadena Symposium on Applications of Model Theory to Algebra, Analysis, and Probability held at Caltech in 1967 (see below, pp. 390–393). This was subsequently published as Kugler 1969. See also Kugler 1969a.

[249] A few years later, Robinson reported some of the results from Kugler's dissertation in one of his own papers. At the time Robinson was interested in Kugler's representations of Bohr compactifications, and was pleased to add that the nonstandard approach made it possible to discuss—and sometimes identify—compactifications which solved the universal mapping problems for particular groups and rings. See Robinson 1969. Further discussion of this paper is given in chapter 9.

[250] Lawrence D. Kugler to JWD, letter of February 7, 1991.

[251] See Robinson's letters to Roquette of May 23 and June 2, 1966, RP/YUL.

Robinson in Tübingen.

for additional colloquium lectures as well, if so desired. He also asked Roquette for some background information about the level and preparation of the students he would be teaching: "Do they know some logic? Or should I begin at the beginning?"[252] Basically, he started at the beginning, using two of his own books, *Model Theory* and *Nonstandard Analysis*. These constituted the primary reading for the course, which was designed to cover the fundamentals of model theory (including applications to algebra and nonstandard analysis).

The summer seminar proved to be a great success—not only was Robinson invited regularly thereafter to offer such seminars in the future, but the following December he received a gratifying letter from H. H. Schaefer, reporting that a study group they had formed at Tübingen had finished reading chapter 2 of Robinson's *Nonstandard Analysis*, declaring, "It's a lot of fun, even if it gave us some difficulties."[253]

International Congress of Mathematicians: Moscow 1966

Thanks to the International Congress of Mathematicians held in Moscow late in August of 1966, Robinson was invited to spend a few days in Finland, where he was called upon to lecture at Turku University. He and Renée also took this opportunity to explore Helsinki as well. One afternoon Wim Luxemburg joined them for lunch in an historic castle where they all had their picture taken together against a background of tapestries and waitresses in local costume.

The Congress opened on August 16 in Moscow with greetings from M. B. Keldysh, president of the USSR Academy of Sciences. For logicians, a memorable highlight of the first plenary session was the Fields Medal, awarded that year to Paul Cohen for his work on the continuum problem.[254] The opening ceremonies were also enlivened by a perfor-

[252] Robinson to Roquette, June 2, 1966, RP/YUL.

[253] H. H. Schaefer to Abraham Robinson, December 29, 1966, RP/YUL. Not long thereafter, when Roquette left Tübingen for a position at the University of Heidelberg, he arranged for Robinson to continue his summer visits as a guest professor there.

[254] Congress 1966, pp. 15–20. Other Fields medalists that year were Michael Atiyah,

Robinson lecturing at the University of Helsinki.

mance of Dmitry Shostakovich's one-act ballet, *The Young Lady and the Hooligan.*

The Congress itself met in the auditoriums and classrooms of Moscow University on the Lenin Hills.[255] More than 4,300 attended from 54 countries, including delegations from Cuba, North Korea, and North Vietnam. More than a dozen other Latin American, African and Asian countries also made their first appearance at the Moscow Congress. Conspicuously absent, however, were both Mainland China and Taiwan (although Taiwan had sent representatives to the 1962 Congress in Stockholm). Even so, as Lee Lorch later reported, "This [Congress] represented the first large-scale contact between the mathematical communities of the USSR and nonsocialist countries, undoubtedly the most valuable contribution of the Congress."[256]

Not only was a special stamp issued to commemorate the Congress, but the Soviet press carried extensive reports on the Congress and those in attendance. Many Moscow mathematicians had social gatherings in their homes, to which foreign mathematicians were invited, and the young Soviet mathematicians "threw a huge party and entertainment for all the young mathematicians at the Congress; some of the older ones crashed the affair."[257]

As for the scientific part of the program, there were over two thou-

---

Alexander Grothendieck, and Stephen Smale.

[255] A brief notice of the Congress was published in *Notices of the American Mathematical Society*, 13 (1966), pp. 312–313. Few details about the Congress are included in the volume of *Proceedings* published in 1968.

[256] Lorch 1967. [257] Lorch 1967, p. 161.

Renée and Abby at the Moscow Congress, August 1966.

sand individual papers presented, in addition to special symposia and plenary sessions. A. I. Mal'cev, naturally, was called upon to give one of the special invited hour lectures, which he devoted to "Some Questions Bordering on Algebra and Mathematical Logic."[258] Although there is no record of Robinson having given any paper of his own during the Congress, he was engaged with a meeting of the editorial board for North-Holland's series, Studies in Logic. The agenda included updates from authors whose manuscripts were expected, and decisions on who should be offered contracts for new books. The board was disappointed to learn, however, that Paul Cohen had declined an invitation to write a volume, as the board had hoped, for the Studies series.[259]

In addition to visiting major tourist attractions in Moscow, including the Kremlin and renowned GUM department store, the Robinsons took advantage of being close to Klin to visit Tchaikovsky's home there, with its spacious gardens full of flowers. The house itself, the last of several in which the composer had lived in Klin, was a two-story brick and wood structure, where they saw the music room with Tchaikovsky's grand piano and all of the portraits and photographs of family and friends lining the walls.

From Moscow the Robinsons ventured on to Leningrad, and spent several days seeing the city. They enjoyed many hours at the Hermitage,

[258] Mal'cev's lecture was published, in Russian, in Congress 1966, pp. 217–231. An English translation of Mal'cev's lecture is available as Mal'cev 1968.

[259] See Abraham Robinson to M.D. Frank, October 31, 1966, RP/YUL.

where Renée took Abby's picture in front of the museum. When they were not visiting museums, they enjoyed browsing in sidewalk bookstalls, strolling in the parks, or catching a glimpse of the canals for which Leningrad is famous. And periodically they stopped to relax in one or another of the local cafés. They even saw *Die Fledermaus* at the Kirov Theater, and left Leningrad having found it an especially charming city.[260]

Back at UCLA: Fall 1966

When Robinson returned to UCLA in the fall of 1966, the university had since instituted the quarter system. His old course on logic and foundations was divided into three separate, one-quarter courses, while model theory, decidability and undecidability, and recursive functions were all compressed into one-quarter courses. Set theory was also a three-quarter course, although distributive lattices and Boolean algebras was covered as a two-quarter course.[261]

If the new system seemed to overly fragment the academic year, UCLA also seemed on the verge of overly expanding in every direction. As Robinson began his last academic year in California, the campus had reached its presumed maximal enrollment. The *Bruin* called it the "UCLA Numbers Game: 27,581 Students."[262]

But there was more than size to disturb the tranquillity of the Westwood campus, as the Vietnam war continued to preoccupy faculty and students alike. At UCLA, Donald Kalish was steadfastly on hand, continuing to oversee the weekly silent protest he had begun earlier that year. As described by the *Daily Bruin*:

> Approximately 120 people lined both sides of Election Walk yesterday in a silent protest of the war in Vietnam. Silent vigils on campus have been taking place every Wednesday since last June. . . . According to Kalish, the protest is independent from any particular organization.[263]

Robinson, in his quiet way, was throughout a mainstay of UCLA's Vietnam Day Committee:

> He was equally adamant in his opposition to the Vietnam war and he was one of those who organized with some of the other professors, mainly from the philosophy and mathematics departments, a committee against the Vietnam war. He was the spirit of this committee, who tried to keep the action of it rational and dignified to prevent the protest from becom-

[260] For a concise yet historical and evocative description of Leningrad, "Venice of the North," see chapter 3 of Robinson (no relation to AR) 1982, pp. 83–113.

[261] UCLA 1966.

[262] *UCLA Daily Bruin*, Tuesday, September 27, 1966.

[263] *UCLA Daily Bruin*, Thursday, October 6, 1966.

> ing something that tried to shock people rather than to convert by persuasion and by dignity, which unfortunately in the long run did not last.[264]

Indeed, real trouble was brewing once more at Berkeley. In November 1966, there was another student sit-in, and police were once again called onto the campus, whereupon the students voted to strike against the university (on December 2).[265] This time, however, the mood was noticeably different from what Robinson had experienced two years earlier when the Academic Council was forced to deal with the "Free Speech" crisis:

> Of all the differences, the most striking was the difference in mood. In 1964 the campus had a wealth of idealism and hope; the FSM had been good-natured, ironical, and humorous. In the months before the present crisis (1966), the campus was tired, humorless, and disillusioned. . . . In 1964 the students claimed "constitutional rights," . . . in 1966 they demanded "student power."[266]

Robinson, in fact, was not impressed by the new militancy, and found the stridency of student activism increasingly at odds with what he took to be traditional academic dignity–and courtesy. As the University of California became larger, more politicized, increasingly impersonal, he must have felt the frustration of trying to be an effective teacher in the midst of constant turmoil for both students and faculty.

UCLA and the Reagan Years

In the spring of 1967, as U.S. war planes continued to bomb North Vietnam with increasing intensity, the University of California was again under siege, this time politically from Republicans and the conservative right. The State of California, now under the leadership of Ronald Reagan (former actor turned politician, elected governor in November of 1966), sought a 10 percent cut in the university's requested $278 million budget, and imposed a student "educational fee" for the first time at both the state colleges and universities. One area singled out for "trimming" was research funded by the university. And then, in an even more extraordinary move, the Regents fired Clark Kerr, by a vote of 14–8. According to the *Daily Bruin*:

> Kerr has blamed partisan politics for his ouster. "The University should serve truth and not political partisanship," he said. Governor Reagan . . .

[264] Straus R-1976.
[265] Wolin and Schaar 1970. See chapter 2: "The University Revolution," pp. 43–72.
[266] Wolin and Schaar 1970, pp. 45, 47.

> called the action "very reasonable." Mrs. Hearst said that Kerr was dismissed due to "lack of administrative ability."[267]

The faculty lost no time in responding, and Robinson, as a personal friend, was among those who supported several resolutions adopted at a special session of the UCLA Academic Senate, one expressing its "profound appreciation and gratitude for the leadership" of former UC President Kerr, the second deploring the "summary dismissal" of Kerr "in a climate of political pressure on the University."[268]

Meanwhile, the new fiscal realities were reflected in gloomy headlines in the *Daily Bruin*: "UCLA cutbacks begun."[269] Retaining faculty became all the more difficult, according to the *Bruin*, as "UC drops to 42nd spot on professor pay scale (in the past it was always in the top 10)."[270] As the term progressed, the news did not improve: "Budget crisis causes UCLA standstill";[271] "Reagan, Legislature Make Extensive Cuts."[272]

Apart from less attractive salaries, unwieldy size, and near legendary impersonality, among prominent reasons for many who left the University of California in the late 1960s was the Republican governor himself:

> Faculty members rarely say they are leaving because of Reagan, but include his influence on the University as one of the several factors prompting them to accept offers from other schools.[273]

All too soon, Robinson would also be gone. But if nothing else, his last term at UCLA was an extraordinarily rich and rewarding one mathematically.

Nonstandard Analysis: A Rising Star

Robinson was enjoying the celebrity that came with growing interest in his work—especially nonstandard analysis. Above all, Luxemburg at Caltech had decided to take advantage of the fact that he was nearby to schedule a special seminar that spring on nonstandard analysis. Robinson was to be an integral part of the working group in Pasadena, and Luxemburg consequently arranged his appointment as "Visiting Professor

[267] *UCLA Daily Bruin*, Monday, January 23, 1967.
[268] *UCLA Daily Bruin*, Tuesday, January 24, 1967, p. 12.
[269] *UCLA Daily Bruin*, Monday, February 27, 1967.
[270] *UCLA Daily Bruin*, Monday, March 27, 1967, p. 2.
[271] *UCLA Daily Bruin*, Monday, May 8, 1967.
[272] *UCLA Daily Bruin*, Thursday, July 7, 1967.
[273] *UCLA Daily Bruin*, Friday, May 5, 1967.

of Mathematics" for three months, from January 1 through March 31, 1967.[274]

One paper Robinson wrote in conjunction with the Caltech seminar, which drew on ideas he had begun to develop in Rome during his lectures at the Castelnuovo Institute, was devoted to a "Nonstandard Theory of Dedekind Rings."[275] After covering the basic information related in his earlier two notes published by the Lincean Academy, Robinson recalled that for a Dedekind ring D and its nonstandard enlargement $^*D$, there was an ideal μ in $^*D$ such that $\mu \cap D = \{0\}$ and such that the ideals of D correspond to principal ideals in the ring $D = {}^*D/\mu$. As Robinson had shown earlier, this in turn led to a theory of unique decomposition of the elements of D in D. Robinson was now prepared to go further, relating D to the standard theory of P-adic numbers and *adèles*.

After considering a Dedekind ring D and a prime ideal P in D, along with the P-adic valuation of D and its completion, the theory of enlargements enabled Robinson to obtain the following result:

> THEOREM 4.2: *Let D be a countable Dedekind ring which is not a field such that the quotient rings D/P are finite for all non-trivial prime ideals P in D. Let *D be a comprehensive enlargement of D and let μ be the intersection in all extensions *Q to *D of non-zero standard ideals Q in D. Then the ring* $\Delta = {}^*D/\mu$ *is isomorphic to the strong direct sum of the P-adic completions of D.*[276]

The "most interesting" special case Robinson then considered in connection with theorem 4.2 was the ring of integers in the field of rational numbers (or in a finite algebraic extension of the field of rational numbers). In either case, D is isomorphic to the ring of *adèles* with entire p-adic or P-adic components but without components (or with zero components) in the Archimedean completions.

[274] When initially offered the appointment as a "Research Associate," Robinson balked and asked that the title be changed to "Visiting Professor." This was done, with a stipend of two thousand dollars for the three-month period. The appointment was supported by a grant for "Research in Nonstandard Analysis" under a contract from the Office of Naval Research. Robinson was expected to "actively contribute to the seminar on Nonstandard Analysis, which will be conducted during the academic year." All of this was duly noted in a memorandum from Marshall Hall, executive officer for mathematics, to Abraham Robinson, June 9, 1966, RP/YUL. L. A. DuBridge, president of Caltech, wrote to confirm Robinson's appointment as visiting professor on November 22, 1966.

[275] Robinson 1967.

[276] Robinson 1967, p. 451; in Robinson 1979, 2, p. 129.

More Results on Nonstandard Arithmetic

Among the special lectures Robinson was slated to give that spring was an hour invited address at the American Mathematical Society's Far-Western Section meeting on April 22, in San José, where he decided to talk about "Nonstandard Arithmetic."[277] This was a perfect occasion to develop further his recent ideas on nonstandard algebra and arithmetic, which he had presented to a limited audience at the University of Rome the year before, and on which he had been working during the Caltech seminar that semester.

In San José, Robinson was determined to win over the general audience he knew would be on hand to the value of nonstandard methods. His paper was consequently designed to shed interesting light on such familiar topics as the theory of ideals, the theory of p-adic numbers and valuation theory, and above all, class field theory. In particular, he showed how the nonstandard approach offered new foundations for infinite Galois theory and the theory of *idèles*.

The most remarkable aspect of Robinson's San José paper was its sudden departure, as Robinson drew his remarks to an end, from his usual insistence that mathematical logic–rather than the method of ultrapowers–was preferable in dealing with nonstandard analysis. Instead, perhaps in light of his audience of mathematicians, Robinson suggested that the use of ultrapowers might not only be advantageous, but even necessary for certain applications. Acknowledging the "mathematician's desire for obtaining an intuitive picture of his universe of discourse," Robinson concluded that:

> The use of ultrapowers as representations of enlargements is entirely appropriate, although even this does not lead to a categorical (unique) enlargement except by means of artificial restrictions. Beyond that, the use of ultrapowers or of other special models may actually be required in order to prove particular propositions.[278]

Even so, Robinson still defended the generality that mathematical logic provided, and which he unmistakably preferred:

> On the other hand, it would seem wasteful to give up the logical basis of our method altogether, for it alone provides the setting within which we may deduce the validity of statements in M^* quite generally from their validity in M and vice versa. Without this setting, any property which is

277 Robinson 1967a.

278 Here Robinson had in mind one of Luxemburg's results on enlargements of topological spaces that were saturated models. Luxemburg presented this result (after Robinson's San José lecture) at the Caltech Symposium on model theory in May. See Luxemburg 1969a.

> known to apply to M has to be established for M^* separately in each case, and while this is certainly possible it has been contrary to good mathematical practice ever since the days of Theaetetus.[279]

Whatever choice one might make about how to approach nonstandard analysis, Robinson stressed that he found it remarkable that the subject had "enough vitality to make a meaningful contribution to a subject as far removed from its origin as the theory of algebraic number fields."[280] Indeed, this was to become a topic of increasing interest to Robinson in the years immediately following his departure for Yale from UCLA. In fact, at the beginning of his San José lecture, Robinson was pleased to acknowledge "several stimulating conversations" with W.A.J. Luxemburg, A. M. MacBeath, O. Todd, P. Roquette, E. G. Straus, and H. Zassenhaus. It was with Peter Roquette, in the years ahead, that Robinson would begin to develop the serious possibilities of nonstandard number theory in unexpected detail.

The Caltech International Symposium on Model Theory: May 1967

As a timely follow-up to the nonstandard analysis seminar that had been meeting all semester with Robinson at Caltech, Luxemburg organized an international symposium on applications of model theory to algebra, analysis, and probability, which was held in Pasadena from May 23 to 26. Although Thoralf Skolem had pioneered the subject of model theory in the 1920s, producing nonstandard models for arithmetic in his paper of 1934, Luxemburg was interested in more recent events:

> A whole new development began in 1960 when Abraham Robinson applied similar ideas to analysis. This development led to the establishment of new structures which are proper extensions of the real number system. This enabled Robinson to give the complete solution for the century-old problem of introducing infinitely small and infinitely large numbers in the Differential and Integral Calculus which was so strongly advocated by Leibniz and later discarded as unsound and replaced by Weierstrass by the δ, ϵ-method. Going beyond this, Robinson showed that this new method, which has become known as nonstandard analysis, can be applied fruitfully to other mathematical structures.[281]

The Caltech Symposium, indeed, was a turning point for nonstandard analysis. Whereas Robinson had been working methodically, pub-

[279] Robinson 1967a, p. 842; Robinson 1979, 2, p. 156.
[280] Robinson 1967a, p. 842; Robinson 1979, 2, p. 156.
[281] Luxemburg 1969, pp. v–vi.

lishing article after article, along with several books, demonstrating as best he could (along with the work of his students) how valuable model theory could be over a wide spectrum of classical mathematics, the Caltech Symposium for the first time brought together a large number of mathematicians, all of whom were doing their part to show what nonstandard analysis in particular could do. Among those on hand for the Pasadena meeting, in addition to Luxemburg and Robinson, were Allen Bernstein, J.L.B. Cooper, Georg Kreisel, Jerome Keisler, R. D. Kopperman, Lawrence Kugler, D.W. Müller, Rohit Parikh, Dana Scott, Arthur Stone, R. F. Taylor, Frank Wattenberg, and Elias Zakon.

In his introductory presentation, which offered "A General Theory of Monads," Luxemburg provided a broad survey of the basics of nonstandard analysis in language accessible to mathematicians. Despite the similarity of terminology, Luxemburg could not help but add that "any resemblance between the mathematical concept of a monad introduced in the present paper and the philosophical concept of a monad in the monadology of Leibniz is purely coincidental!"[282]

Beginning with logical foundations for nonstandard analysis based upon ultrapowers, including enlargements and saturated models, Luxemburg introduced monads and various results, including one very basic theorem: in an enlargement, the monad of a filter is internal if and only if it is a principal filter.[283] Above all, saturated models were used to show how the general theory of monads could contribute specifically to topology in terms of the properties of near-standard points. Here questions related to compactness that Robinson had not himself treated earlier were discussed, including Luxemburg's proof that the standard part of an internal set is closed, which he was able to express for a uniform space "very elegantly," as Luxemburg noted, as an intersection. He also gave a new, nonstandard proof of the Ascoli theorem for compact families of continuous functions.[284] He also investigated precompactness, along with the completion and nonstandard hull of a uniform space, which Luxemburg showed to be complete. The paper, a veritable *tour de force* of the range of conclusions nonstandard methods could help to establish, ended with various applications to the theory of normed spaces.

More on Nonstandard Number Theory

As for Robinson's own contributions to the Caltech Symposium, his

[282] Luxemburg 1969a, p. 19.

[283] Independently, Joram Hirschfeld also proved that the monad of a nonprincipal filter is always external. See Hirschfeld 1975.

[284] Luxemburg 1969a, pp. 70–71.

most ambitious was a broad overview of a new area to which he would devote considerable attention in the next few years, "Topics in Nonstandard Algebraic Number Theory."[285] This was yet another in his ongoing series of papers devoted to nonstandard number theory, and could be considered as a direct extension of his San José lecture for the American Mathematical Society.

Beginning with an overview of the theory of entire ideals in an infinite algebraic number field (which constituted an infinite algebraic extension of the field of rational numbers), failure of the cancellation rule in the classical multiplicative theory of such ideals was a major obstacle. A nonstandard approach, however, made it possible to avoid this problem. In turn, Robinson was led to the further development of class field theory for infinite algebraic number fields, which he had already outlined a month earlier in his San José lecture. He also tied his earlier results more closely to the classical theories of Chevalley and Weil.[286]

Robinson, however, had more up his sleeve for the Caltech Symposium, and also presented another short paper on "Germs," as well as a contribution written jointly with Elias Zakon. Briefly, the paper on "Germs" was yet another exercise in nonstandard topology. Set germs, equivalence classes of subsets of a topological space, and function germs, related to equivalence classes of functions, were investigated from a nonstandard point of view. Germs, Robinson noted, had become popular due to their convenience in discussing the local behavior of functions.[287]

There were drawbacks, however, to their standard treatment: set germs were not actual sets of points, and function germs were not actual functions (instead, they were related to equivalence classes defined for subsets and equivalence relations). Robinson's nonstandard approach remedied this, replacing set germs and function germs by actual sets and functions. This in turn led to a number of interesting applications, including new proofs of *Nullstellen* theorems of Hilbert and Rückert. One advantage Robinson stressed of the nonstandard approach was that it did not require the rings under consideration to be Noetherian. For example, by showing the existence of a generic point in the prime ideal in the ring of nonstandard analytic function germs (of n variables at the origin in the complex field), Rückert's theorem for analytic germs followed as a direct consequence.

As for his joint paper written with Elias Zakon on "A Set-Theoretical Characterization of Enlargements," here interest focused on a reinter-

[285] Robinson 1969a.

[286] See, for example, Chevalley 1937.

[287] Robinson 1969b.

pretation of nonstandard analysis that abandoned Robinson's preferred use of type theory in favor of set theory. Such basic concepts as enlargements and both standard and internal entities were defined using injections (monomorphisms) of one model of set theory into another.[288] As an alternative to either the type-theoretic or ultrapower approach to nonstandard analysis, this gave a more concrete picture of what was involved, especially some of the essential characteristics of internal elements that emerged from the set-theoretic characterization of enlargements. Among other results they established using this approach, Robinson and Zakon proved that an element was internal if and only if it was an element of a standard set.[289]

W.A.J. Luxemburg: Nonstandard Analysis for Mathematicians

It was hard to believe, as Moshé Machover warned Robinson as he read the page proofs for *Nonstandard Analysis*, that mathematicians would readily respond to his new ideas if they were presented in the unfamiliar framework of type theory. The complexities of type theory that Robinson had initially used to introduce the subject would probably prevent most readers from ever getting to the heart of the matter. Machover eventually resolved this problem with his own approach: "Nonstandard Analysis Without Tears (An Easy Introduction to A. Robinson's Theory of Infinitesimals)."[290]

Not only was Robinson's approach foreign to the intuitive, set-theoretic way in which most mathematicians thought, but it was also unnecessarily complicated as far as Machover was concerned. The only theorem one needed from logic for nonstandard analysis was the compactness theorem. As Machover explained in lectures he gave first in Bristol in 1966, and later at Jerusalem in 1966–1967, the compactness theorem "becomes a bit complicated to state for a theory formulated in terms of types, and its intuitive meaning is then somewhat difficult to explain. For a type-less language, on the other hand, the theorem can easily be explained with adequate rigor, even without going into a great deal of formal and technical detail."[291]

This proved to be the essence of Machover's presentation of nonstandard analysis "without tears," and aimed to "explain Robinson's theory in a way which departs as little as possible from the usage and

[288] Zakon and Robinson 1969.

[289] Zakon and Robinson 1969, p. 114; Robinson 1979, 2, p. 211.

[290] Machover 1967.

[291] Machover 1967, p. 2.

conventions of ordinary mathematics."[292] But as Machover later acknowledged, it was Wim Luxemburg who was the first to facilitate the spread of Robinson's ideas, by presenting "an elementary treatment (using an ultrapower of the first-order structure of real or complex numbers) without explicit heavy use of logic."[293]

Although Luxemburg and Robinson had known each other since the early 1950s, when they were both at the beginnings of their careers, teaching at Toronto, it was Arthur Erdélyi at Caltech who persuaded Luxemburg to give nonstandard analysis a serious look in 1961. Erdélyi had long been interested in nonarchimedean systems of infinitesimals, and his own work on delta functions provided a natural context within which he had already considered infinitesimals in some detail.[294] He was also familiar with the approach Carl Schmieden and Detlef Laugwitz had taken, whereby they too had successfully introduced infinitesimals in a series of publications, but in a very different way from what Robinson had done.[295] Erdélyi was interested in finding out what made both theories tick. In particular, he wanted to know what accounted for their differences. Luxemburg agreed to examine both Robinson's nonstandard analysis and the papers of Schmieden and Laugwitz, and presented his overview in a special seminar at Caltech, where he succeeded in shedding considerable new light on the recent interest in infinitesimals.

One major drawback of the Schmieden-Laugwitz extension of the real numbers was that it proved to be too large, resulting in a partially ordered ring, but not in an ordered field. Robinson's R^*, however, thanks to the model-theoretic approach he had taken, was designed specifically to insure that his nonstandard elements, infinitely large and small, behaved arithmetically just like ordinary numbers.

Erdélyi believed that Robinson's methods were not only entirely new, but also more fundamental than what Schmieden and Laugwitz had done. This was in part due to the fact that Robinson had succeeded in giving a *proper* extension of the real numbers.[296] But it was Luxemburg who discovered exactly how the two theories were related.[297] By formulating Robinson's theory in terms of ultrapowers, Luxemburg then represented the elements of R^* as equivalence classes of sequences of real

[292] Machover 1967, p. 2. Although the duplicated version of Machover 1967 received only a limited circulation, the approach was also used in the more widely available *Lecture Notes on Nonstandard Analysis,* Machover and Hirschfeld 1969.

[293] Bell and Machover 1977, p. 575. The last chapter was devoted to "Nonstandard Analysis," and adopted the approach taken in Machover 1967, and Machover and Hirschfeld 1969.

[294] Erdélyi 1961.

[295] Schmieden and Laugwitz 1958; Laugwitz 1959, 1961 and 1961a.

[296] Erdélyi 1967, pp. 764–766. [297] Luxemburg 1962.

numbers. This in turn made it possible to compare what Schmieden and Laugwitz had done with Robinson's nonstandard analysis, which at the same time revealed the source of difficulties in the Schmieden-Laugwitz approach.

In terms of the set of all mappings of the natural numbers *N* into the real numbers *R*, R^N is a ring when addition and multiplication are defined as pointwise operations. By choosing the ultrapower in this way, R^N is in fact the ordered ring used by Schmieden and Laugwitz. However, R^N is not totally ordered, but is a nonarchimedean ordered ring with divisors of zero.

Luxemburg managed to remove these "unpleasant features," as he called them, by introducing free ultrafilters (Schmieden and Laugwitz had used a Fréchet filter on *N* to determine equality between the elements of R^N, rather than a free ultrafilter).[298] Using the ultrafilter to establish the equivalence relation on R^N needed to give Robinson's nonstandard R^*, Luxemburg went on to show that R^* was not only a field, but in fact a nonarchimedean totally ordered field. In contrasting the properties of Robinson's R^* with the Schmieden-Laugwitz approach to introducing infinitesimals, Luxemburg stressed that:

> These two theories are very much different from each other. Schmieden and Laugwitz arrived at their theory by means of a new approach to Cantor's definition of the system of real numbers. The generalized system of numbers they obtained is an ordered (= partially ordered) ring with divisors of zero and which contains infinitesimals and infinitely large numbers.
>
> Robinson's approach is based on the fact that the system of real numbers *R* permits proper extensions which possess all the properties of *R* that are formulated in a formal language such as a first order language. . . . A nonstandard model for the system of real numbers has the feature of being a non-archimedean totally ordered field which contains a copy of the real number system.[299]

Thus, at the very beginning of its own history, the advantages of Robinson's nonstandard analysis were made very plain over the drawbacks of its closest rival. But the ultimate value of what Robinson had accomplished would be reflected in results, results that an increasing number of mathematicians had begun to achieve. Thanks to the Caltech Symposium, a number of mathematicians using nonstandard analysis in a variety of ways clearly helped to demonstrate its versatility and promise. And thanks to Luxemburg, mathematicians as early as 1962

[298] Luxemburg 1962, pp. 15–16.

[299] Luxemburg 1962, pp. 1–2.; W.A.J. Luxemburg to JWD, June 15, 1994.

were able to make use of nonstandard analysis using familiar tools of analysis and set theory, rather than having to struggle with the metamathematical context Robinson himself preferred.

Robinson, Zakon, and Enlargements

Among the old friends Robinson enjoyed seeing during the Caltech Symposium was Elias Zakon. Together they had been working on enlargements, and following presentation of their joint paper on enlargements during the Pasadena meeting, Zakon sent Robinson a revised version early in June, along with a note:

> I send on a new note of mine, and a revised copy of our paper, "A Set Theoretical Characterization of Enlargements." Unfortunately, I was not able to decipher some of the text written by you and tried to fill the resulting "Dedekind cuts" by some "reasonable" guesswork of mine [Robinson's handwriting was often undecipherable!]. If you wish to change some of it, please give your corrections to Dr. Luxemburg directly. I shall give him a similar copy when he visits Kingston later this month.[300]

Zakon, back in Canada, was about to attend the forthcoming meeting of the Canadian Mathematical Congress (on June 9). Again, he planned to give a talk, and hoped it would help "contribute to the popularization of Nonstandard Analysis in Canada."[301] But he had also begun to worry about nonstandard terminology. In reflecting on all of the papers that had been presented during the Pasadena meeting, he had some serious concerns:

> It seems to me that the term "standard element" or "standard set" is being used . . . in several incompatible meanings. Firstly, an entity of a nonstandard model $*M$ is called standard if it has the form $*x$ with $x \in M$ (i.e. x is an entity of the original full structure M). Secondly, x *itself* is often referred to as a "standard" element or set, (even though it may even not be internal) only because $x \in M$ (though not $x \in *M$). Thirdly, a set A which is an entity in M is being confused with $\{*x \mid x \in A\}$ . . . and likewise erroneously called a "standard" set. The resulting confusion is so great that sometimes it is difficult to understand what actually is meant, and one gets the impression that the author has simply overlooked the possibility that, for entities of higher type, A need not be a subset of $*A$ and may even be disjoint from $*A$. . . .
>
> I think that something must be done to avoid this confusion before it gets out of hand. I would suggest that only entities of the form $*x$ be called

[300] Zakon to Abraham Robinson, June 8, 1967, RP/YUL.
[301] Zakon to Robinson, June 8, 1967, RP/YUL.

standard elements. Individuals in M then are standard entities, but *sets* in M are, in general, not standard, and should not be so called; this also applies to sets of individuals, like N or R, which are *not* standard (being external). Possibly, it might be useful to circulate a letter on this question, or raise it at the next symposium. I would be glad to have your reaction to it.

Possibly, I am too pedantical in this respect and still too "green" to make suggestions of this kind. However, I am very much afraid that, unless a precise and consistent terminology is adopted right from the start, confusion will prevail in the future.[302]

Following the Canadian Congress, Zakon attended the annual Summer Research Institute at Queens University in Kingston, Ontario. There he and Luxemburg discussed Zakon's worries about nonstandard terminology. Shortly thereafter, Zakon wrote to Robinson that something must be done:

I met Luxemburg at Kingston, and he shares my opinion that something must be done to avoid ambiguity (in fact, he seems to have arrived at this conclusion even before I did). . . . He intends to talk over this matter with you.

On my part, I am all in favour of occasional "*abus de language* [*sic*]" where no complicated mathematical notions are involved and where it is harmless. However, the very phrase "*abus de language* [*sic*]" presupposes that there is some fixed *precise* usage (not an "*abus*"), and this seems to be lacking in our case. I think that it would be of great help to agree on some precise terminology, to be used at least in the *formulation* of theorems and definitions, so that one would not have to lose hours (as I did) to decipher the arising multiple ambiguities. I sincerely hope that something can be done in this respect.[303]

UCLA Institute on Axiomatic Set Theory: Summer 1967

The highlight of Robinson's final summer in Los Angeles was the Institute on Axiomatic Set Theory, which he co-organized with Paul Cohen and Dana Scott. This was the Fourteenth Annual Summer Research Institute sponsored by the American Mathematical Society, jointly with the Association for Symbolic Logic, with financial support from the National Science Foundation. It was a fitting academic farewell to the years Robinson had spent at UCLA.[304]

302 Zakon to Robinson, June 14, 1967, RP/YUL.
303 Zakon to Robinson, June 29, 1967, RP/YUL.
304 The summer institute was actually the kick off event for an entire "Logic Year" at

Although Kurt Gödel had been invited to attend the meeting as guest of honor, he had declined the invitation. He demurred, as he usually did, citing poor health (but primarily because his interests, mathematically and philosophically, had moved in very different directions, he said, away from axiomatic set theory).[305] Robinson, in a letter to Gödel that May, explained that the Organizing Committee (as approved by the councils of both the American Mathematical Society and the Association for Symbolic Logic) wanted permission to dedicate the meeting to Gödel, to which Gödel responded positively:

> I accept gladly, with one reservation however, namely that the wrong impression should be avoided as if these highly interesting recent developments were only an elaboration of my ideas. I really took only the first step. Moreover, I perhaps stimulated work in set theory by my epistemological attitude toward it, and by giving some indications as to the further developments, in my opinion, to be expected and to be aimed at. I did not, however, give any clues as to how these aims were to be attained. This had become possible only due to the entirely new ideas, primarily of Paul J. Cohen and, in the area of axioms of infinity, of the Tarski school and of Azriel Levy. I request that this should be mentioned in the introduction to the book.[306]

When the institute convened on July 10, Robinson read a part of Gödel's letter to everyone, and indeed mentioned Cohen and Levy as Gödel had requested.[307] The meeting was a gratifying success, with more than 125 participants who attended the four-week program from July 10 through August 5. In addition to two ten-lecture series, one offered by Dana Scott, the other by Joseph Shoenfield, individual one-hour lec-

---

UCLA. Plans had been made to bring a panoply of logicians to Los Angeles in 1967–1968, with year-long courses to be offered on logic and set theory, including recent advances in recursion theory and model theory as well. Graduate students were encouraged to apply: "The coming academic year will afford an unusually favorable environment in which to begin, continue, or complete graduate studies in logic and to become acquainted with a broad range of problems, methods, and latest developments." Teaching assistantships, research assistantships, traineeships, and basic fellowships were all available. See "Logic Year at UCLA," *Notices of the American Mathematical Society*, 14 (1967), p. 36.

[305] See Gödel's letter to Paul J. Cohen, April 27, 1967, Gödel Papers no. 010417, Princeton University Library.

[306] Gödel to Abraham Robinson, July 7, 1967, Gödel Papers no. 011944, Princeton University Library.

[307] See Robinson's letter to Gödel, May 4, 1967, Gödel Papers no. 011941, Princeton University Library. The *Proceedings* from the meeting were eventually published, in two parts, after various delays. See Scott 1971 and Jech 1974. No dedication to Gödel was made in either volume, nor did Robinson present a paper during the Institute.

Participants at the UCLA Summer Institute on Axiomatic Set Theory, July 1967.

tures were also given, usually four each day during the first three weeks. "By the last week this was reduced to three per day, as the strength of the participants had noticeably weakened. Nevertheless, most of the success of the institute was due to the fact that nearly everyone attended all of the sessions."[308]

Robinson's Last Students at UCLA

Shortly after Robinson left UCLA for Yale, Peter Tripodes finished his dissertation on "Structural Properties of Certain Classes of Sentences," which he wrote under Robinson's supervision.[309] Among the last of Robinson's students at UCLA to work with him on nonstandard analysis was Robert Phillips, whose thesis was devoted to "Some Contributions to Non-Standard Analysis."[310] Phillips had already received a master's degree in mathematics from the University of California at Santa Barbara, and then went to Berkeley in hopes of a Ph.D. Trying to support his family there proved difficult, but he was worried that the offer of an assistant professorship from California State College at Los Angeles, though financially welcome, might interfere with his plans for a doctorate:

[308] Dana Scott, in the "Foreword" to Scott 1971.

[309] Tripodes 1968. Yiannis N. Moschovakis was cochairman with Robinson of the committee in charge of Tripodes' thesis.

[310] Phillips 1968.

> Moving to Los Angeles, I thought, jeopardized my chances to obtain a Ph.D, but I was already interested in Robinson's work and, very optimistically, I thought I might be able to work with him. I had attended an MAA talk he gave entitled "In Praise of Leibniz," and this led me to read his Model Theory book which, I felt, greatly clarified the theory I had spent a year of study on at Berkeley.
>
> It was almost two years before I could find the courage to ask Robinson if I could work with him. I had friends at Berkeley who had trouble getting professors they had studied with to agree to direct their dissertations. Therefore I had nearly convinced myself that Robinson, whom I had never met, would not give me the time of day. Hence, when I was convinced that I had nothing to lose, I went to him. I did it this way: I waited outside his office, without an appointment, until he returned from his class. When he appeared, I asked if I could speak to him for a few minutes about working with him on nonstandard Analysis and he agreed to talk with me. Therefore, my first impression of Robinson was of a kind considerate man who would at least hear one out. Throughout the next couple of years while I worked on my dissertation, and after that, he always treated me in a helpful, kind, and courteous manner.
>
> During this first meeting, I explained that I was familiar with his Model Theory book and that I understood nonstandard analysis at least to the extent that it was presented in his Model Theory book. He then took me to the library and helped me find some papers on a theorem of Liouville's about conformal mappings and he told me to try to reduce the hypotheses of the theorem using nonstandard methods. This I finally did, with Robinson's help, and this work became the central part of my dissertation.[311]

Phillips surmised that Robinson was willing to work with him primarily because he was anxious to promote awareness of—and interest in—nonstandard analysis any way he could. The best way to do this was with results—as many as possible. Phillips' thesis was more ammunition in the nonstandard arsenal.[312]

The last of Robinson's students from UCLA to finish her dissertation was Diana Dubrovsky, who wrote on "Computability in *p*-adically Closed Fields and Nonstandard Arithmetic," which she successfully defended

[311] Robert Phillips, now a member of the Department of Mathematics, Computer Science, and Engineering, University of South Carolina at Aiken, to JWD, letter of January 10, 1991.

[312] "Looking back, I doubt if I could have found a better thesis advisor and I am thankful now that I moved to Los Angeles." Phillips was also pleased that in letters of recommendation, Robinson always wrote about Phillips' work "in the kindest terms possible. He said that it was truly original, which it was and which is probably the best that could be said for it." R. Phillips to JWD, January 10, 1991.

in 1971.[313] Several years later she published an account of some of the results she had obtained in her thesis, and was pleased to acknowledge the help Robinson had given her. Primarily, what she emphasized from her thesis was the completion Q_p of the rationals with respect to the *p*-adic valuation, and showed that recursive *p*-adic numbers also form a *p*-valued subfield $Q_p(R)$ of Q_p, which was itself a *p*-adically closed field. She also considered computable and arithmetically definable fields, from which a nonstandard analog for Q_H, the Henselization of Q inside Q_p was introduced, namely Q_H*. Dubrovsky showed the existence and uniqueness of Q_H*, which she then extended to all arithmetically definable *p*-valued subfields of Q_p,. Ultimately, she was able to conclude that all arithmetically definable *p*-valued subfields of Q_p, are elementarily closed relative to the standard natural numbers *N*.[314]

Robinson and Yale: Looking Ahead

Robinson's decision to leave UCLA for Yale after the spring term, 1967, was in part influenced by questions of prestige–and style. Robinson was very favorably impressed by the way he was received at Yale, compared with the more casual attitude that he felt was shown him at UCLA. He was particularly struck by the difference between having lunch at the UCLA Faculty Club with Foster Sherwood, then vice chancellor of UCLA, whose attitude and demeanor Robinson regarded as perfunctory, compared with dinner at Yale with its president, Kingman Brewster, who clearly appreciated Robinson's status as an internationally renowned scholar. There was no doubt in his mind: the prestige and personal treatment would be better at Yale.[315]

There was also a deeper, almost imperative sense, however, to Robinson's decision to go to Yale. This was something he discussed at length with David Kaplan, for there was a "cosmic" element to the offer that Robinson could not dismiss. "Yale was a great university, with a great mathematics department. As Robinson said, 'You can't say "no." The institution demands it.'"[316]

> Moreover, there was an important academic side Robinson had to consider:
>
> At UCLA it was hard even for Robinson to attract consistently good graduate students, but this would not be a problem at Yale. Robinson was an ambitious man. It was not a transparent, burning ambition as in some—he was too refined for that—but he knew his value, his importance, and he

[313] Dubrovsky 1971. [314] Dubrovsky 1974, pp. 473–491.
[315] Hales I-1991; A. Hales to JWD, June 22, 1994.
[316] Kaplan I-1991.

> not only respected this, but expected the respect that this also entailed. Robinson was also ambitious to push his field, and was ambitious in the sense that he wanted very much to advance the fortunes of symbolic logic.[317]

Ultimately, Yale held out a number of attractive possibilities:

> A smaller private university offered him close contact with all members of a distinguished senior faculty in mathematics, a regular flow of talented postdoctoral researchers and graduate students, and the opportunity for easy contact with scholars in fields other than his own. No doubt his awareness of tradition lent appeal to the case of a university nearly twice as old as the oldest of his previous affiliations.[318]

Yale, in the all too brief period that Robinson would be there, witnessed the extraordinary evolution of his full mathematical powers. In the Biblical span of the last seven years of his life, Robinson not only began to pass on his abilities to a strong group of graduate students, but gained international recognition and respect for both nonstandard analysis and the essential value of model theory for mathematics in general.

[317] Kaplan I-1991.
[318] Seligman 1979, 1, p. xxvii–xxviii.

CHAPTER NINE

Robinson Joins the Ivy League: Yale University 1967–1974

> Mathematics may be likened to a Promethean labor, full of life, energy and great wonder, yet containing the seed of an overwhelming self-doubt. . . . This is our fate, to live with doubts, to pursue a subject whose absoluteness we are not certain of, in short to realize that the only "true" science is itself of the same mortal, perhaps empirical, nature as all other human undertakings.
>
> –*Paul J. Cohen*[1]

NEW HAVEN was a town "whose one distinction, apart from Yale, was its reputation as the birthplace of vulcanized rubber, sulfur matches and the hamburger."[2] Yale, by contrast, was one of America's oldest and most outstanding universities, an oasis of learning. Architecturally, it reflected a mishmash of styles running from American colonial and collegiate gothic to sleek alabaster walls of the Beinecke Rare Book Library. The heart of the university was the historic old campus:

> The feel of that old part of the campus was Victorian. . . . Grand, dark brownstone dormitories surrounded it, elm trees shaded the grass, the wooden Yale fence lined flagstone paths. Battell Chapel, superbly hidden, was in one corner. At the other end stood the two remaining Colonial brick buildings. In front of one a statue of Yale alumnus Nathan Hale, clean cut, handsome in his long hair and Colonial dress. . . .[3]

LOGIC: A NEW ERA BEGINS AT YALE

Robinson arrived in New Haven knowing that the future of mathematical logic at Yale depended upon his success in bringing the subject into focus, giving it a more prominent place within the Department of Mathematics, and attracting a coterie of colleagues and graduate students to build a substantial program. The way had been prepared by Michael Rabin, who the semester before had been at Yale teaching logic. Much of what he had to say came as a revelation to his students:

> Prior to Michael Rabin's course, none of the students had heard of model

[1] Cohen 1971, p. 15.
[2] Kakutani 1982, p. 280.
[3] Moore 1982, p. 197.

theory, or the Completeness Theorem, for that matter. In the philosophy department courses were available on the foundations of mathematics, modal logic and model set theory, natural deduction systems, and similar topics, primarily under the direction of Professor Fitch.

At this time Paul Cohen's work in set theory was considered very exciting and mysterious. Professor Fitch taught an undergraduate seminar on the Gödel monograph, and presided over an informal seminar on Cohen's work, starting with a report I gave on Scott's lectures at Rockefeller University on the Boolean valued interpretation of forcing. For the most part we saw the formal validity of the results of Gödel and Cohen without real understanding; this applies to all concerned.

Both Rabin and Robinson impressed us deeply with the force of the Compactness Theorem. In fact I doubt that I could distinguish clearly between "Model Theory" and "The Compactness Theorem" at this time.[4]

Yale's Department of Mathematics was located on Hillhouse Avenue, once described by Charles Dickens as "the most beautiful street in America."[5] Distinguished homes built in the nineteenth century for the finest of New Haven's society dotted the avenue. Over the years they had been bought or bequeathed, and now (for the most part) were departmental and administrative buildings of the university.

Robinson's office was in Daniel Leet Oliver Hall, built at the turn of the century for the Sheffield Scientific School as a gift from Mrs. James Brown Oliver, in 1908. Located at 12 Hillhouse Avenue, this is a large three-story building with offices and classrooms. Robinson's office was on the second floor, and was famous for its traces of cigar smoke—a telltale sign that he was in his office.

Mathematics: A Student Perspective

Mathematics had a strong reputation at Yale, but as the *Yale Course Critique* in 1968 warned:

> Freshmen coming to Yale and seeking a major will probably not choose mathematics. With so many exciting professors in other departments, only about two percent of each year's class decides to take more mathematics than the poorly taught elementary calculus courses, Math 10 and 15. This is too bad because Yale has one of the finest pure mathematics departments in the country.[6]

[4] Cherlin R-1974. Rabin taught Math 160, model theory, in the spring of 1967, on Mondays, 3:00–5:00 P.M.

[5] Holden 1967, pp. 123–124.

[6] *Yale Course Critique* (1968), p. 103.

Part of the problem was the subject itself, one regarded as overly abstract and inherently difficult:

> Understanding of the mathematics major at Yale follows from two postulates. First, mathematics is taught at Yale as the theoretical development of an orderly system; it is presented more as a philosophy or an art form learned for its own sake rather than as a tool for the physicist. Second, the major is designed only for the brilliant or the masochistic.
>
> Two professors whose stars the major should follow in particular include Mr. Kakutani and Mr. Robinson. Mr. Kakutani teaches the 31a, 32b sequence. The department recently brought Mr. Robinson to Yale to teach its leading offerings in logic, one of mathematics' most popular "new" fields.[7]

Among the faculty already at Yale when Robinson arrived in 1967 were Walter Feit, Nathan Jacobson, Shizuo Kakutani, William S. Massey, George Daniel Mostow, Øystein Ore, Charles Rickart, and George Seligman. It was an active, friendly department. Distinguished faculty attracted eager and promising graduate students, and now—thanks to Robinson—not only did bright young Ph.D.s begin to accept postdoctoral positions in mathematical logic, but established logicians began coming to Yale for periods of a semester or more:

He was soon the mentor of a lively group of younger logicians. Promising recent Ph.D.'s, among them K. J. Barwise, P. Eklof, E. Fisher, P. Kelemen, M. Lerman, D. Saracino, J. Schmerl, S. Simpson, and V. Weispfenning took postdoctoral positions at Yale and profited from his keen questioning and the generous sharing of his time and wisdom. Established logicians, among them A. Levy, G. Sacks, and G. Sabbagh, were happy to make visits of a semester or longer. Graduate students were charmed and excited by the promise held out for model theory and nonstandard methods in analysis and arithmetic by his courses, and many sought him out as adviser. He refused none, and gave generously of his attention to all.[8]

Getting Settled

In the fall of 1967 Robinson taught a graduate seminar on model theory for which, as he explained to Nathan Jacobson, "...the idea is to make the course suitable both for those who will have attended Michael Rabin's lectures and for those who won't."[9] To help integrate him into campus life as quickly as possible, Robinson was given an affiliation with Daven-

[7] *Yale Course Critique* (1968), p. 112–113.

[8] Seligman 1979, p. xxviii.

[9] A. Robinson to N. Jacobson, October 31, 1966. In Robinson's administrative file, Department of Mathematics, Yale University.

Renée and Abby at the annual department picnic, in Sleeping Giant State Park, Hamden, Connecticut.

port College, one of Yale's undergraduate residences.[10] Almost immediately, he was a devoted member of its Senior Common Room. On Wednesdays, when the fellows of Davenport would meet for dinner and wide-ranging discussion, Robinson was almost always on hand.

Since 1966, Horace Dwight Taft had been master of Davenport. Taft had done his graduate work at Chicago in physics, joining the Yale faculty in 1956. Intellectually, this was a good match for Robinson, but Taft and his family were also very social, and one of the highlights every year, as the Robinson's soon discovered, was their annual Christmas party. The evening would begin with an invitation for drinks at the master's house, after which everyone proceeded to the college dining hall:

> It was beautiful, with seasonal decorations and a big table with a glistening ice sculpture and a buffet all around. There were mostly small tables set with white table cloths, all very elegant. After dinner, we all went back again to the Master's house for after-dinner drinks and more lively conversation.[11]

Meanwhile, the Robinsons had been living in a suburb of New Haven, renting temporarily in Orange, Connecticut. Renée spent much of her time that fall house hunting, and eventually found a spacious, modern home which they bought early in 1968. The setting was idyllic, in

[10] Holden 1967, p. 207. The college was named after John Davenport, founder of the New Haven Colony, a distinguished preacher and the first to suggest a college in New Haven.

[11] Renée Robinson I-1992.

Sleeping Giant State Park, just off of Blue Trail—reminiscent of the Vienna woods—with the house set well back, hidden from the road. Leisurely walks along the hillside trails were especially enjoyable, and if any place in America could have given them a feeling of the old Europe of their childhoods, it was the wooded seclusion offered by Sleeping Giant State Park.

The Association for Symbolic Logic

The esteem Robinson's colleagues felt for him, as well as their trust in his administrative abilities, were reflected in his election that fall to the presidency of the Association for Symbolic Logic. Robinson was due to take over the reins from William Craig after the association's annual meeting that December in Boston (officially, Robinson's term began on January 1, 1968, and ran through the end of 1970).[12]

In preparation for the Boston meeting, Robinson wrote to Craig about the agenda. Robinson wanted to discuss the question of setting limited terms for consulting editors of the *Journal of Symbolic Logic*: "I know that we do not quite agree on this, but I am anxious to see the reaction of the Council to my suggestion that—unlike Roman senators—consulting editors should not necessarily stay for life."[13]

Robinson had good reason for concern. Within months of his assuming the ASL presidency, the touchy issue of editorial reappointments for the *JSL* arose in earnest. This began with a "Strictly Confidential" letter from Robinson to Hartley Rogers in Cambridge (England), with copies to Paul Benacerraf and Alonzo Church. Rogers had agreed to chair a Nominating Committee for positions on the journal's editorial board. Robinson warned Rogers that disagreement over one of the JSL editors was already brewing due to what some felt was an uncritical acceptance of manuscripts.[14] In considering editorial reappointments, Robinson reiterated his earlier opinion expressed only to Craig: "I do not think that an editor should be asked to serve for a second term automatically," and this message was now relayed to all of the members he had appointed to the Nominating Committee.[15]

By mid-June dozens of letters had been exchanged, most marked "Confidential." The appointment of an overall, central managing editor had been suggested, various names had been circulated, and finally Robinson

[12] For a report of the meeting, see the *Journal of Symbolic Logic*, 33 (1968), pp. 636–644.

[13] Abraham Robinson to William Craig, November 9, 1967, RP/YUL.

[14] Abraham Robinson to Hartley Rogers, May 21, 1968, RP/YUL.

[15] Abraham Robinson to members of the Nominations Committee, May 31, 1968, RP/YUL.

told everyone that he would handle all contacts with prospective editors himself.[16] Among the changes made was a division of responsibility between mathematics and philosophy. Robinson wrote to Arthur Prior, who was invited to continue for three more years as a consulting editor, but with the proviso that he handle only papers in philosophy. Papers of a more mathematical character were to go to Azriel Levy.[17]

Prior wrote back, accepting the reappointment but wondering whether the division of labor Robinson had suggested was necessary or even appropriate. Prior had just returned from a "Philosophy thing in Vienna," and had mislaid Robinson's letter but remembered "the general drift of it." He accepted reappointment with reservation, and stressed that on his own experience, mathematics *versus* philosophy was not "the most significant division nowadays." He was willing, however, to exchange papers with Azriel Levy, but hoped that the process would not become "too much regimented."[18]

The protracted discussion over Prior's reappointment convinced Robinson that there were a number of fundamental issues that the Association for Symbolic Logic should face at its next annual meeting. Above all, he wanted to discuss general editorial policy more fully. Consequently, he eventually wrote to all members of the ASL Advisory Committee, making it clear that this time he would set the agenda for the next annual meeting in New Orleans, at which he definitely intended to discuss the *JSL*.[19]

Meanwhile, Leon Henkin had raised another problem, not unrelated to the growing tensions between philosophers and mathematicians among logicians. Was it, he asked, still a good idea to reserve certain places just for mathematicians, others for philosophers—since more and more members of the ASL now had to be classified as "computer scientists"? Should the "unwritten rule" for disciplinary parity be scrapped, as well as the pattern of alternating fields which traditionally had been followed in the list of presidents and vice presidents of the Association?[20]

[16] Abraham Robinson to members of the Nominations Committee June 13, 1968, RP/YUL.

[17] Abraham Robinson to A. N. Prior at Balliol College, Oxford, August 16, 1968, RP/YUL. Prior had been the subject of much discussion, and was criticized strongly by some for being out of touch with the mathematical side of things, such as accepting material for the *JSL* too uncritically. The question, in retrospect, was sadly irrelevant when Prior died just a year later, on October 6, 1969.

[18] Arthur Prior to Abraham Robinson, from Balliol, September 13, 1968, RP/YUL.

[19] Abraham Robinson to ASL Advisory Committee, September 24, 1968, RP/YUL.

[20] Leon Henkin, writing to Abraham Robinson from the Mathematical Institute in Oxford, July 1, 1969, RP/YUL. Also reflecting differences in interest and emphasis among its members was a decision to launch a new series of publications with North-Holland.

Yale: Spring 1968

The spring of 1968 was a tense and difficult one, not only on American college campuses but for the nation as a whole. In January Yale's university chaplain and antiwar activist William Sloan Coffin was indicted by a federal grand jury for leading a conspiracy "calling for a nationwide program of resistance to the Selective Service System."[21] On February 23, Robinson joined more than five hundred other concerned faculty at Yale who signed an open letter protesting the Coffin case as "a tragic heightening of the domestic costs of the Vietnam War."[22]

The assassination of Martin Luther King on April 4 in Memphis, Tennessee, prompted demonstrations across the nation. In New Haven, as the *Yale Daily* reported, "Militants Gather On Green; Yale Closed in King Tribute."[23] Days later, Yale President Kingman Brewster released a plan for greater urban commitment on Yale's part. Many felt this was too little too late. As the *Yale Daily* reported only days later, "Blazes Break Out in Tense Elm City."[24]

Heidelberg, Jerusalem, and Warsaw: Summer 1968

Having agreed to attend meetings in Poland and Italy that summer, Robinson felt obliged to decline an invitation from Peter Roquette in Heidelberg. Robinson was reluctant to spend "the entire summer" in Europe, but Roquette sent a formal invitation anyway, offering a month's salary and an additional travel stipend. This tipped the scales, and Robinson agreed to go "not only for the honor of being a Visiting Professor at Heidelberg," but because the invitation came from Roquette himself. Robinson planned to stay less than four weeks, but promised to "cram in a month's worth of lectures." As a general theme he suggested "Model Theory and its Applications," and so that he and Gert Müller would not be too bored, he also promised to introduce some of his latest results.[25]

In May Robinson wrote to M. D. Frank about a decision the Association for Symbolic Logic had made to establish the *Annals of Mathematical Logic*. See Abraham Robinson to M. D. Frank, President, NHPC, May 14, 1968, RP/YUL.

21 *Yale Daily News*, Monday, January 8, 1968, p. 1.

22 *Yale Daily News*, Friday, February 23, 1968, p. 3.

23 *Yale Daily News*, Monday, April 8, 1968, p. 1. On April 5 the headlines read, "King Slain in Memphis, Shocked Reaction Here."

24 *Yale Daily News*, Wednesday, April 10, 1968, p. 1.

25 Abraham Robinson to P. Roquette, December 19, 1967, and January 31, 1968, RP/YUL. Later, Gert Müller also offered to pay for a full-fare ticket (Robinson had opted for an inexpensive charter flight) if this would mean that he would come for an entire month. See G. Müller to Robinson, April 9, 1968. Müller added that they would use May and

Renée did not go with Abby to Heidelberg, but they did spend most of August back in Israel. Owing to the Six Day War (June 1967), Jerusalem was newly reunited, and once again it was possible to enter the Old City:

> It was for both of us an exciting event and I shall never forget what I saw. . . . We went through Jaffa Gate. The streets were quite empty in comparison to what it became in later years. You saw mainly Arabs and some tourists. The shops were kept by Arabs. The houses on both sides of the street were occupied by Arabs. Often their washing was out, hanging from their windows. Old men sat next to each other along the streets, smoking from long water pipes.
>
> There were not many religious Jews along the Wailing Wall except on Friday evening and Saturday morning when they came from their homes, quite a long way with their garments, to pray there. . . . We went to the Omar Mosque, which was quite empty. There were a few men washing their feet in the special area outside, before they entered the mosque. We had to leave our shoes outside too. . . . Of course we walked along the Via Dolorosa and to the Church of the Holy Sepulcher. . . . Outside the walled city we went to the Rockefeller Museum and of course one walked to [Mt. Scopus] to see the University, where Abby showed me the small house which was the Math Department. Standing at the edge of this hill you had a wonderful view of the area below, all quite deserted. . . .[26]

Above all, Abby and Renée enjoyed walking along the walls of the Old City, although they traveled further afield as well, going to Safad and Qumran. They also spent some time at kibbutz Nezer, where they stayed with friends, and also visited Abby's uncle who lived in a small house with his wife and together raised chickens.[27] As the summer drew to a close, the Robinsons were next on their way back to Europe where Abby had a two-day meeting of the Association for Symbolic Logic in Warsaw. The meeting was organized around another week-long conference, this time devoted to "Construction of Models for Axiomatic Systems."[28]

June to prepare for the special seminar on model theory. Robinson suggested that the Heidelberg group concentrate on Chapters 1–5 (along with the last chapter) of his *Introduction to Model Theory* (1963), as well as Kreisel and Krivine 1967. See Robinson to Müller, March 16, 1968, RP/YUL.

[26] Renée Robinson to JWD, December 11, 1992.

[27] Renée Robinson to JWD, December 11, 1992.

[28] See the announcement of the meeting in *The Journal of Symbolic Logic*, 33 (1968), p. 320, and the final report of the meeting in 34 (1969), pp. 533–544. There is no indication that Robinson presented a paper at the Warsaw meeting.

Model Theory in Varenna

Leaving Poland, Renée and Abby drove south to Italy, where Robinson had also been invited to a week-long international meeting in Varenna. This was the third summer institute organized by the Centro Internazionale Matematico Estivo. Renée and Abby were assigned an apartment in the Villa Monastero, a palatial residence where formal gardens offered pleasant venues for leisurely afternoon strolls. They especially enjoyed seeing old friends again, like Andrzej Mostowski.

Of the four courses given that September, Robinson's was a compact, one-week introduction to model theory, which he organized around one "famous result yielding the solution of a problem of E. Artin's." This focused on the work Ax and Kochen had done with respect to Artin's conjecture on the existence of p-adic zeros of forms over p-adic fields.[29] But where Ax and Kochen had relied upon a combination of ideas from logic and valuation theory to show the model-theoretic equivalence of the field of p-adic numbers and formal power series, Robinson used nonstandard analysis to avoid many of the complexities from algebraic field theory that Ax and Kochen had required.[30] This he regarded as "one of the most striking applications of model theory to date." In the terminology of nonstandard analysis, this depended upon the introduction of an infinite prime p, the p-adic fields Q_p and the fields of Laurent series $R_p((t))$, where R_p is the prime field of characteristic $p = 2, 3, 5, \ldots$. Basically, Robinson's proof of the Ax-Kochen theorem reduced to showing that for infinite p, the Q_p and $R_p((t))$ are elementarily equivalent.[31]

A *Festschrift* for Arend Heyting

One of the first courses Robinson offered at Yale was devoted to "chapters in history of mathematics," which he taught in the spring of 1968. This undoubtedly helps to explain why later that year he chose an historical theme for a *Festschrift* in honor of Arend Heyting. Heyting had made his own contributions to foundations, and this was the subject, considered historically, upon which Robinson decided to focus. Although mathematics seemed to reflect eternally certain truths, Robinson wondered why its foundations remained so elusive:

> There are but few mathematicians who feel impelled to reject any of the

[29] Robinson 1968b. This paper was not reproduced in Robinson 1979.

[30] Ax and Kochen 1965a; Ax and Kochen 1965b showed that Q_p is a decidable field.

[31] Robinson 1968b. Robinson showed the equivalence in question by using a test (see his theorem 5.1). Ax published his proof of the decidability of the theories of finite fields and p-adic fields in Ax 1968.

major results of Algebra, or of Analysis, or of Geometry and it seems likely that this will remain true also in future. Yet, paradoxically, this iron-clad edifice is built on shifting sands.[32]

Robinson took Euclid's axiomatization of geometry as a case in point: "The cautious formulations of the second and fifth postulates seems (sic) to show a trace of the distaste for infinity that we find already in Aristotle." An even more important step, in terms of its consequences for foundations, was the discovery of non-Euclidean geometry. As Robinson explained, "by contradicting the axiom of parallels it denied the uniqueness of geometrical concepts and hence, their reality."[33]

Ultimately, this served to displace geometry as the "ultimate foundation" for mathematics, just as rigorous theories of real numbers and set theory were beginning to play essential roles as underpinnings for analysis:

> An ironic fate decreed that only after Geometry had lost its standing as the basis of all Mathematics its axiomatic foundations finally reached the degree of perfection which in the public estimation they had possessed ever since Euclid. Soon after, the codification of the laws of deductive thinking advanced to a point which, for the first time, permitted the satisfactory formulation of axiomatic theories.[34]

Just as Euclidean geometry had once claimed to model "reality"—so too set theory in the twentiethth century was axiomatized under the assumption that "the axioms were still supposed to describe 'reality,' albeit the reality of an ideal, or Platonic, world." Thanks to Paul Cohen's results, however, by the mid-sixties "the belief that Set Theory describes an objective reality was dropped by many mathematicians"—Robinson among them.[35]

Once again, Robinson had found it convenient to use the history of mathematics to make an important philosophical point about the foundations of mathematics. He closed his article for the Heyting *Festschrift* by stressing yet again the intuitive value of nonstandard analysis: "there is every reason to believe," he noted, "that the codification of intuitive concepts and the reinterpretation of accepted principles will continue also in future and will bring new advances, into territory still uncharted."[36]

[32] Robinson 1968.

[33] Robinson 1968, pp. 191–192; Robinson 1979, 2, pp. 571-572.

[34] Robinson 1968, p. 192; Robinson 1979, 2, p. 572.

[35] Robinson 1968, p. 192; Robinson 1979, 2, p. 572. Robinson acknowledged that after he had finished writing his paper, Cohen and Hersh 1967 made many of the same points he had stressed about the parallels between the respective histories of geometry and set theory.

[36] Robinson 1968, p. 193; Robinson 1979, 2, p. 573.

Model Theory and Contemporary Philosophy

Shortly after Robinson was settled at Yale, offprints of an article he had written primarily in Los Angeles (but signed Yale University) arrived in the mail. Raymond Klibansky (president of the International Institute of Philosophy), had requested a very compressed account of model theory for his latest series of volumes on contemporary philosophy.[37] In focusing this time on the decade between 1956 and 1966, Klibansky was especially conscious of the need to cover the extraordinary changes that had occurred in mathematical logic:

> In the previous series, Logic and the Philosophy of Science were surveyed in one volume. The vigorous activity in these fields . . . the interesting developments which have taken place during the last decade make it necessary to accord them a more detailed treatment in two volumes, of which the first covers Logic and the Foundations of Mathematics, the second the Philosophy of Science.[38]

Beginning with a brief history of model theory, Robinson credited Tarski as responsible for the first detailed analysis of the interpretation of formal sentences in a "universe of discourse," and Tarski's "school" at Berkeley as the most active center for the study of model theory. Among the major methods and results developed in model theory since the mid-fifties, Robinson included quantifier prefix problems, the connection between completeness and elementary equivalence, Beth's theorem, and ultraproducts. He also emphasized applications to algebra and analysis, including his own success in determining numerical bounds for Hilbert's seventeenth problem, the Ax-Kochen theorem, and of course, nonstandard analysis. In concluding this sweeping overview of the subject, Robinson stressed connections between model theory and set theory, so recently invigorated by the exciting new method of forcing Paul Cohen had introduced in his work on the continuum hypothesis in 1966. Robinson would soon make his own impressive contributions to methods of forcing in model theory.

Promoting Mathematical Logic in Japan

In his capacity as President of the Association for Symbolic Logic, Robinson was determined to advance the subject, especially where he felt there was significant untapped potential (for his efforts to encour-

[37] Earlier, Klibansky had edited a collection of essays devoted to philosophy in the postwar period 1949–1955, published in four volumes on the occasion of the Twelfth International Congress of Philosophy held in Venice, Italy, in 1958. See Klibansky 1959.

[38] Klibansky 1968, vol. 1, p. ix.

age mathematical logic in Latin America, for example, see below, pp. 432–433 and 452–453). In the fall of 1968, with Japan in mind, he submitted a proposal to the National Science Foundation for a joint seminar in mathematical logic:

> Mathematical Logic has made great strides in recent years. In Japan, the emphasis has been placed largely on the development of Proof Theory while less attention has been paid to other fields. At the same time much progress has been made, in the United States and elsewhere, in Model Theory, Set Theory, and Recursion Theory. An exchange of information in the fields mentioned would thus serve a very useful purpose. The seminar is to be held in Japan in order to enable also a number of younger people, who are interested in the subject, to attend. In this connection we note that no meeting in this field has taken place in Japan for some years whereas in the United States, events of this kind take place at frequent intervals. . . .

Donald Kalish, reviewing the proposal for NSF, rated it "Excellent." In arguing the potential significance of the seminar, he emphasized the success of similar ventures elsewhere, chiefly in Poland and the Soviet Union:

> The proposal made by Professor Abraham Robinson strikes me as extremely important and timely. Professor Robinson's own status in his fields of study, and his broad perspective evidenced by his positions in England and Israel, as well as a joint position in philosophy and mathematics at UCLA which preceded his present position at Yale, enable him to have a thorough knowledge of what is being studied where and what types of communication would be useful. Frequent exchange of scholars from Russia and Poland with those from the United States has had a profoundly beneficial effect on the development of mathematical logic in the countries mentioned. A similar exchange between Japanese and American scholars would be equally beneficial.[39]

Robinson doubtless would have liked to count himself as among possible participants in the Japan Symposium, but he was too busy teaching at Yale when the U.S.-Japan Logic Seminar actually met in Tokyo the following October, 1969.[40]

[39] Donald Kalish, NSF review, in Robinson's administrative file, Department of Philosophy, UCLA. The proposal itself, quoted above, accompanies the review by Kalish in Robinson's file.

[40] For a report of the meeting, which ran from October 16–21, see the *Journal of Symbolic Logic*, 36 (1971), pp. 357-359.

The Beauty of Mathematics

In January of 1969 Robinson submitted a proposal to offer a two-semester course at Davenport College on the very general topic: "Mathematics is Beautiful."[41] College seminars at Yale were meant to satisfy "the intellectual curiosity and diversity of interest of the present undergraduate generation," and were offered for credit "in local settings." Conceived as experimental courses, they were usually taught with limited enrollments and were expected to offer "some leavening of the curriculum."[42]

Although Robinson's course was eventually approved, the Martz Committee overseeing college seminars asked for a slight but significant change in title, from the assertive "Mathematics is Beautiful" to the less judgmental (and more prosaic) "The Beauty of Mathematics." The course was designed to introduce students to "selected chapters" of higher mathematics:

> No previous training in University Mathematics will be required. The course will contain a good deal of historical background material. No single textbook will be used but a list of books suggested for consultation is appended. The students will be asked to carry out home assignments. In addition a simple term paper on a selected topic will be required each semester.[43]

Robinson began with an introduction to the natural numbers, rings and fields, prime numbers, and prime power factorizations, with ideals forming the algebraic core of the first part of the course. Coordinate and projective geometry led to discussion of the classic theorems due to Pappus and Desargues. Groups were studied in terms of permutations and permutation groups, continuous transformation groups and matrices. Vectors in two, three, and higher dimensions led students up to the concept of Hilbert space. Set theory provoked discussion of transcendental numbers and Russell's paradox. There was even time for the propositional calculus, truth tables, Boolean algebra, and a brief introduction to probability theory (in terms of tossing coins and throwing dice). Finally, the seminar concluded, after a very rich two semesters, with some topological ideas including graphs, Möbius strips, the Klein bottle, and the four-color problem.[44]

[41] This was submitted to the Martz Committee, which was responsible for discussing and approving hundreds of courses offered in the colleges each semester.

[42] Datlin 1982, p. 193.

[43] Robinson to the Martz Committee, January 9, 1969, RP/YUL.

[44] Robinson to the Martz Committee, January 9, 1969, RP/YUL. For basic reading Robinson used his own *Numbers and Ideals*, Robinson 1965, as well as Courant and Robbins 1941 and Ore 1963.

This was certainly an ambitious undertaking, but perfectly in keeping with Robinson's "missionary zeal—to say something about mathematics to nonmathematicians. But he was disappointed [with the course at Davenport College] because he felt the students did not get as much out of the course as he had hoped."[45] More satisfying was a course he offered jointly with Stephan Körner several years later on the philosophy of mathematics, but this was a seminar designed more for specialists which made a significant difference.

Nonstandard Analysis and Quantum Physics

In the spring of 1969, Robinson received a letter from a graduate student at Indiana University, Peter Kelemen, who had just been introduced to nonstandard analysis in a course taught by Andrew Adler. Wanting to know more about possible applications in physics, Kelemen wrote to Robinson, but expected nothing more than a short note or a reprint in reply:

> To my surprise I received a long letter. My ego was truly inflated. It is unusual for a student at a lesser known university to receive a letter from one of the world's best mathematicians. The letter raised the spirits of all the graduate students in our department, because we felt less isolated. Later I learned that Professor Robinson had the policy of visiting small four-year colleges that could not afford to invite visiting speakers. . . . The excitement those visits generated lasted for weeks, as I was told by one of the colleagues who experienced it.[46]

As a result of this correspondence, Kelemen began to consider various nonstandard constructions that might provide new insights for quantum physics. Robinson responded enthusiastically, thanking Kelemen for his letter (of April 14):

> which impressed me greatly. Using non-standard analysis to eliminate the divergences of quantum field theory has been suggested before, but no one has acted on this suggestion. However, your remarks on parity were completely new to me and I would very much like to see the details.[47]

Robinson offered Kelemen one hundred dollars for a lecture, and wondered if he would like to come to Yale for the year in 1970–1971? Robinson was also prepared to help with a suitable fellowship. "At any rate," he added, "I should very much like to meet you sometime in the

[45] Mostow I-1993.

[46] Kelemen R-1974.

[47] Robinson to Kelemen, April 18, 1969, RP/YUL.

near future."[48] Kelemen responded with a manuscript, to which Robinson replied in early June:

> I do not think it should be too difficult to produce a suitable "Delta-like" function for your problem 3. On the other hand, I do not have enough of the physical background to be able to say anything authoritative on the parity breeding question. . . .[49]

A few weeks later, however, Robinson had a few reservations:

> Permit me to disagree with you. Although the treatment of infinities and divergences would seem a "natural" for nonstandard analysis, the treatment of Quantum Mechanics within **H* would seem to me more basic, and might even lead to an understanding of more divergences. As you may know, some people are in fact talking about the use of extended Hilbert spaces in Quantum Mechanics. Among them is a young German physicist whose dissertation was sent to me a few months ago. At that time I wrote back suggesting the use of **H* but I have not heard from him since. If we could get together we might be able to sort this out, but I appreciate that at the moment you are probably quite busy with your dissertation. However, I am fairly certain that the structure used by that man is just a quotient space of a subspace of some **H*. So it may be better than you think.[50]

Robinson and Kelemen soon had a chance to discuss nonstandard physics together when Kelemen made a three-day trip to Yale that August:

> I arrived at his office punctually on a Monday morning. Again to my surprise he was there waiting. Through the years of knowing him, I know of no incidence [*sic*] when he kept anyone waiting. He always considered my time as valuable as his own. . . .[51]

After they met, Robinson made good his offer to find support for Kelemen at Yale, and in 1970 he accepted a position as an instructor in physics. Working with Robinson was a special pleasure, above all because Robinson let Kelemen work at his own pace:

> I never felt pressured. To reassure me at times he said, "My accomplishments are the fruits of one percent inspiration and ninety-nine percent perspiration." He advised me to set an attainable goal, as he did while

[48] Robinson to Kelemen, April 18, 1969, RP/YUL.

[49] Robinson to Kelemen, June 6, 1969, RP/YUL.

[50] Robinson to Kelemen, June 17, 1969, RP/YUL. The physicist was Sohn. As Robinson explained, Sohn's "approach seemed more congenial than a paper on. . . . Hilbert spaces I worked at some time ago." Quoted from Robinson to Kelemen, July 18, 1969, RP/YUL.

[51] Kelemen R-1974.

> writing a book, of producing three pages each day, and achieve it under all circumstances. . . .[52]

Eventually, Robinson and Kelemen jointly authored two papers on nonstandard quantum theory. As they explained:

> We feel that nonstandard analysis can be used advantageously in physics. Thus, many calculations can be simplified by its use through the avoidance of passages to the limit at certain stages. Also, using infinitely large numbers one can give a rigorous meaning to self-energies and renormalization. Finally, one may treat certain nonseparable Hilbert spaces with the same ease as separable ones.[53]

More On Ax-Kochen: Robinson's Mind at Work

During the summer, Robinson was back in Heidelberg working with Peter Roquette on nonstandard number theory. Robinson had drafted a paper to which Roquette had offered comments, which Robinson was pleased to accept *in toto*. On the other hand, Robinson confessed:

> I am not optimistic regarding the "characteristic 0" covers. The Ax-Kochen result in question relies heavily on the fact that a Hensel field *K of characteristic 0* contains a field which is isomorphic to the residue class field of *K*. I am slightly more hopeful that something could be done about the other question mentioned by you. I know that Greenberg's statement . . . is somewhat more precise than mine. Has he said anything more about this in a paper that I do not know? Do you think that the "sufficiently large" in question should or should not depend on the particular field under consideration? However, I would prefer to leave all these questions for later.[54]

It was in his collaboration with Roquette over the last half-dozen years of his life that Robinson revealed something of how he operated as a mathematician, how he posed and thought about mathematical prob-

[52] After two years at NYU, Kelemen left the university for a position with Xerox. See Kelemen R-1974. His collaboration with Robinson did result in two papers, Robinson and Kelemen 1972 and 1972a. He also published a third article of his own, Kelemen 1972.

[53] Robinson and Kelemen 1972, p. 1873; in Robinson 1979, 2, p. 273. The $\lambda{:}\phi_2^4(x){:}$ model of quantum field theory was studied earlier in Glimm and Jaffe 1968, 1970, and 1970a.

[54] Abraham Robinson to Peter Roquette, August 14, 1969, RP/YUL. Michael Artin had called Robinson's attention to a result of Greenberg's on rational points in Henselian discrete valuation rings, Greenberg 1966. Robinson approached the problem using infinite subscripts and a nonstandard embedding of the field of Laurent series in a field of characteristic 0. See Robinson 1970b, which Roquette communicated to the *Journal of Number Theory* in 1969.

lems as he worked on them. Robinson's brainstorming approach to research was basically experimental—try *X* or *Y*—whatever inspiration will work. Try to suggest something that might fit—hope for inspiration, see what one idea might suggest in terms of others. This, it seems, is exactly how Robinson worked with Roquette in their joint research, applying nonstandard analysis to number theory. In fact, Robinson even wrote somewhat apologetically to Roquette, hoping that Roquette did not mind if he sometimes thought aloud, letting his mind "fantasize" as new ideas occurred to him. "Even if that doesn't always lead immediately to results, I find that your comments are always very stimulating."[55]

Indeed, Robinson and Roquette were beginning to make exciting progress with their nonstandard approach to number theory, especially when they found that "it simplified that part of Siegel's work that looked most complicated relative to inductions." This was because "Siegel's work is very intricate. To improve on Siegel is no mean feat. A mathematician who is not a logician will find the standard proof turgid and opaque, but the nonstandard approach makes it manageable."[56] What made the Robinson-Roquette collaboration so productive was the interaction of their very deep understanding of two different areas, logic and number theory.

Meanwhile, Robinson's colleagues at Yale also provided stimulation of the highest caliber and, in 1969, he published a paper thanking Professors Feit, Jacobson, Kakutani, Massey—and Luxemburg (at Caltech) for "relevant discussions."[57] They had all contributed to his thoughts on nonstandard enlargements of separated Hausdorff topological groups (and similar results applied to topological rings). His use of compactifications of groups and rings in a nonstandard setting drew in part on his work on nonstandard arithmetic (and nonstandard Dedekind rings) developed earlier in similar contexts.[58]

One of the main conclusions of the paper (theorem 2.1) was that for every compactification of *G* there exists a homomorphic mapping which (in his theorem 3.2) Robinson used to give a "concrete" description of the universal compactification of *G*. In certain cases, Robinson was able to give an explicit description of universal compactifications (as in theorem 6.5) for the ring of integers *G* in an algebraic number field and the

[55] Robinson to Roquette, December 3, 1969, RP/YUL.

[56] Mostow I-1993.

[57] Robinson 1969. Although he was only concerned in this paper with compactification of discrete groups and rings, he hoped compactifications of nondiscrete groups or rings might yield interesting results as well.

[58] Robinson 1967a. Robinson also mentioned Larry Kugler's success in dealing with Bohr compactifications, and the possibility, under certain conditions, of even identifying compactifications which gave the universal mapping problems for particular groups and rings.

universal compactification isomorphic to the direct product of all *p*-adic integers over *G*.[59]

Saul Visits: September 1969

The highlight that fall socially for the Robinsons was a short visit from Abby's brother Saul. Saul had been invited to Brown University, where he spent nearly two weeks giving an entire series of lectures on international comparative education, after which he spent a few days with Abby and Renée. The two brothers talked for hours on end in the Robinson's living room overlooking the magnificent panorama of Sleeping Giant National Park, where they also liked to take long walks together through the woods. Saul's failing heart, however, was an increasing problem, and they would have to stop frequently to let Saul catch his breath. But Robinson was grateful to see his brother, for it was one of the few opportunities they had taken in recent years to spend any substantial time together.[60]

Saul visits: at home with Abby in Hamden, Connecticut.

The Shearman Memorial Lectures: Spring 1970

That spring Robinson was on leave from Yale. Following a series of lectures in York, England, at the invitation of the British Mathematical Colloquium, he presented the first of three Shearman lectures (devoted to "Logic as the Science of Mathematical Reasoning") at the University of London beginning April 22.[61] Two days later he launched a second series of "University Lectures in Mathematics," all of which were devoted to nonstandard analysis. In these he introduced the analytic theory of polynomials, covered compactness arguments in function theory, and concluded with a lecture on germs.[62]

[59] Robinson 1969, p. 586; Robinson 1979, 2, p. 253.

[60] Renée Robinson I-1992.

[61] Seligman 1979, p. xxix. For details about Shearman (1866–1937), who taught logic at University College London, see Grattan-Guinness 1985, pp. 109–110 and 119–120.

[62] Further details about the lectures may be found in a letter from the University of London dated April 24, 1970, and in a letter Robinson wrote to W.A.J. Luxemburg, May

From London the Robinsons flew to Israel, returned briefly to see their friends on the kibbutz at Nezer, and went to Herodion the day before it opened as an official Israeli archaeological site. Like Masada, it was one of the palace fortresses built by Herod the Great to protect Judaea. As Renée and Abby walked through the hilltop ruins, a little boy ran after them, selling ancient coins—of which Abby bought three.

Especially memorable that spring was a lecture they heard by David Ben Gurion at the Weizmann Institute, where again efforts were renewed to entice Robinson back to Israel with the offer of a position. When Renée objected emphatically that she did not like the houses, that they were all too close together and the institute itself was too near Tel Aviv, they were assured that a new home could be built to their specifications! Even so, Renée told Abby that in any case she did not like the locale.[63]

Turmoil at Yale: Bobby Seale and May Day, 1970

The Robinsons had picked a good time to be away from Yale, for the spring of 1970 was divisive and tense. Bobby Seale, prominent dissident and cofounder (with Huey P. Newton) of the Black Panther Party, was on trial in New Haven for having kidnapped and then murdered a fellow Panther. Black militant groups had aroused passions—and fears—throughout all of New Haven, and a rally scheduled for May 1 on the New Haven green in support of the Black Panthers (and calling for a moratorium on the war in Vietnam as well) threatened trouble.

Yale prepared for the weekend-long rally by opening its doors to any who needed food or shelter, while edgy administrators hoped for the best. Indeed, when it was all over, damage was minimal, and John Hersey, master of Yale's Pierson College, was even self-congratulatory for a city that had not gone up in flames as some had predicted.[64]

May Day elsewhere proved more tragic. In the wake of President Nixon's televised speech on April 30 announcing the large-scale military invasion and bombing of Cambodia, the Ohio National Guard responded at Kent State University, where "rumors of impending violence by outside agitators, communists, Maoists, were spreading like wildfire. . . ."[65] After a night of minor confrontations, as classes were breaking for lunch, frightened, ill-trained guardsmen suddenly fired on a gathering of students, killing four in a period of thirteen seconds.[66] The

---

4, 1970, RP/YUL.

[63] Renée Robinson to JWD, interviewed October 26, 1992.

[64] Hersey 1970, pp. 105, 111.

[65] Banks 1989, p. 72.

[66] Banks 1989, p. 79.

community at Yale, having survived its own May Day, joined the protests that swept the entire country in the aftermath of the Kent State killings.

Nonstandard Analysis Triumphs: 1970

1970 was an impressive year for Robinson and for nonstandard analysis. That spring the Mathematical Association of America released a film it had made with Robinson providing an hour's introduction to the subject.[67] Shot in black and white at Harvard University's Carpenter Center, Robinson appeared in a classroom setting that was not a classroom but a stage set, without benefit of students or any sort of lecture-audience interaction.[68] Facing the unresponsive eye of the camera, Robinson tried to explain nonstandard analysis as if he were addressing a bright group of undergraduates.

The film opens with Robinson in a dark blue suit, his left hand almost always in his trousers pocket. His appearance conveys a sense of casual elegance, with a no-nonsense, methodical quality to his presentation. Thick-rimmed glasses lend an air of professorial seriousness, even heaviness. His English is fluent, but accented, with occasional traces of British pronunciation. Throughout the film Robinson's delivery is slow and methodical, although there is a sense of rapport with his unseen audience.

True to Robinson's interest in history, he begins with a brief comment about Leibniz—who Robinson explains sought foundations for the calculus in the realm of infinitely small and large numbers, but whose insights were never developed. Concerned with the apparent paradoxes of the infinite and nagged by doubts as to why the calculus worked, nineteenth-century mathematicians took what seemed a safer approach, the now familiar finite delta-epsilon methods of Cauchy and Weierstrass. At this point, leaning casually on the lectern, Robinson rather nonchalantly notes that, due to the advent of nonstandard analysis, Leibniz can now be justified—actually vindicated. This is possible thanks to the formal language of mathematical logic.

[67] *Non-Standard Analysis*—a filmed one-hour lecture presented by the Mathematical Association of America, 1970. See Robinson 1969c.

[68] Carol Wood, one of Robinson's first graduate students at Yale, felt that it was an uncomfortable film to watch when she first saw it, screened as a tribute to Robinson during a meeting in Robinson's memory held at Yale in the summer of 1975, shortly after his death. She found it captured Robinson in a rather "two dimensional" way, and that there was something artificial about it. "The film did not seem to me to portray his lecturing style adequately, probably due to the poor state of technology of such filming in 1970 and the resulting awkwardness for him." Carol Wood to JWD, E-mail note of June 24, 1994, and Wood I-1992.

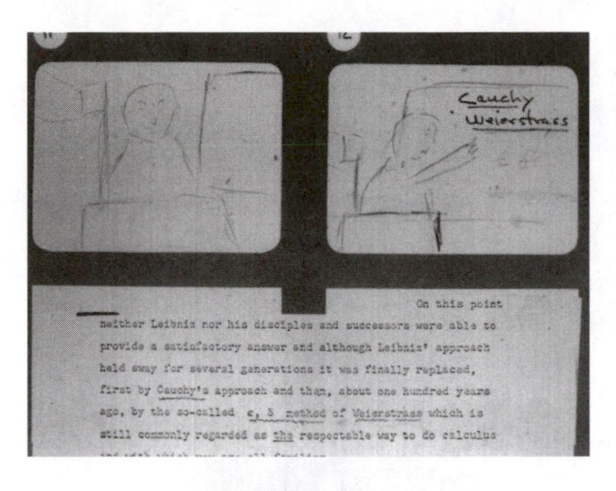

Storyboard illustration used in designing the sequence of filming for *Nonstandard Analysis.*

Robinson filming *Nonstandard Analysis.*

Throughout, Robinson's style is laconic, sometimes witty. After a very brief introduction to symbolic logic, some basic mathematical concepts are introduced. Commenting on the limit of a sequence, he is disarmingly candid! In the case of

$$\lim_{n \to \infty} s_n = a$$

"A delicate conscience," he concedes, "compels me to admit there exists a problem to interpret the meaning."

To be precise, with a formal language L, a basic theorem in logic permits extensions of the real numbers R. What do such extended structures look like? There are many such extensions, but Robinson is

Robinson at the blackboard, explaining a nonstandard theorem.

only interested, he explains, in *one*–R^*–an ordered field that is *non*-archimedean. Uppermost is the fact that while R is archimedean (a fact not expressible in the first-order language), the extended structure R^* is not. There are many properties, however, of R that carry over directly to R^*, and Robinson indicates how every notion described for R by a sentence in L will carry over directly to R^*, that is, if $N(x)$ is satisfied in R by N, then $N(x)$ is satisfied in R^* by N^*.

With this general introduction in mind, part 2 of the film goes on to express derivatives, integrals, and limits of sequences, all in terms of R^*. The presentation of nonstandard integration is particularly graphic, with Robinson drawing little rectangles on the blackboard. "We divide the interval from a to b into an infinite number of subintervals," he says, adding with a jolly smile, "I'm afraid I can't draw that very well":

> Some of you may feel that I've been leading you up the garden path–since no one has ever divided an interval into an infinite number of subintervals, you (may have noticed) the trouble I was having. We may treat these infinite sums formally just as if they were finite sums for finite n; but in a pure formalism it doesn't matter. So far as the formalism is concerned, it doesn't matter if the sums are finite or infinite.[69]

At this point, Robinson seems to shift gears, changing the direction in which the film is headed. "Can we develop all calculus with these methods?" he asks. "Yes, it is possible, but I now talk of wider possibilities." This requires a higher-order language $\mathcal{L}$ capable of dealing with sets and relations, which leads to an example of the power nonstandard

[69] Transcript of Robinson film, Robinson 1969c, p. xxx.

analysis had to offer specifically in the case of infinite dimensional real separable Hilbert spaces H. A nonstandard approach, Robinson tells his viewers, led to the (justly) famous resolution of a previously unresolved invariant subspace problem for linear operators.

With this powerful example of the usefulness of nonstandard analysis in hand, Robinson turns to his audience with a basic question: "You might well believe that no such structure exists. Does R^* exist?" Appealing to the formal language $\mathcal{L}$, the existence of R^* may be proved without difficulty, he notes, by a finiteness principle (i.e., the compactness principle) of mathematical logic.

Admitting there are several ways to establish the existence of R^* (he mentions the ultraproduct method for one), Robinson offers two alternatives. Given $0 < a < 1$, the Archimedean axiom may be expressed in various ways, for example, A_1 or A_2:

$$A_1: \underbrace{a + a + a + \ldots + a}_{(n \text{ times})} > 1$$

A_2: There exists n such that $na > 1$.

Since A_2 can be formulated as a sentence in $\mathcal{L}$, it follows by the transfer principle that it must also hold in R^*. Typical of Robinson's ironic sense of humor, he takes advantage of the apparent terminological contradiction to explain exactly what he has been driving at all along:

> There is only one point to be added here, that in this case there is indeed a natural number n, only that this natural n is *un*-natural.[70]

By unnatural Robinson means, of course, that n must be a nonstandard (or infinite) element of N^*. The film comes to an end with Robinson looking into the camera, a wry smile on his face, followed by a slow fade to the credits. . . .

Although film-making may not have been so natural to Robinson, promoting model theory was. As soon as *Nonstandard Analysis* was a "final wrap," he was off to Norway to present his most recent interest, a paper on "Infinite Forcing in Model Theory."

Forcing in Model Theory: Oslo 1970

Early in June the Robinsons flew to Norway where they boarded a coastal steamer in Bergen to see the famous Fjords. Bad weather, however, made it difficult to see anything of the islands and inland waterways. June 18 they arrived in Oslo for the Second Scandinavian Logic Symposium, a

[70] Robinson 1969c.

three-day meeting sponsored by the Association for Symbolic Logic. Robinson was among thirty participants, and had decided to talk about his most recent results on "Infinite Forcing in Model Theory."[71]

One of the model-theoretic concepts that had long interested Robinson was model-completion. Unfortunately, many theories do not have model-completions. To get around this, model-companions offer some help as a broader class, but there are still important theories that lack both model-completions or model-companions. Robinson-forcing was one way he hoped to get beyond both of these limitations.

Robinson had first presented his new ideas on forcing at a colloquium on model theory held in Rome during November of 1969. Subsequently, Robinson-forcing proved to be a useful tool for constructing models, and along with his students, Robinson began to exploit the technique in a number of directions. The Rome paper, which was confined to finite forcing conditions, considered the set of sentences which were forced weakly from a set of sentences K as a natural generalization of model-completions in cases where "true" model-companions did not exist.[72]

It was Paul Cohen's method of forcing in set theory that inspired Robinson to try a similar approach in model theory. While Cohen had used forcing with extraordinary facility in his work on the independence of the continuum hypothesis and the axiom of choice, the basic idea had similarly powerful consequences when applied in the context of model theory. Robinson followed Cohen's idea of forcing closely, formulating his version relative to a fixed nonempty, consistent set of sentences K of a first order language having no function symbols but an infinite number of constants.

For a countable language, the forcing companion K^f is the set of sentences true in all K-generic models (K^f may or may not include K). In terms of forcing and model-completion, in cases where there was no model-completion (for a given set of sentences K), the set of sentences K' forced weakly from K seemed to Robinson a natural generalization of model-completion. But as he cautioned, "it remains to be seen to what extent the models of K' which are obtained in these cases possess properties which make them analogous to the well-known structures obtained for certain ordinary model-completions."[73]

On the subject of infinite forcing, Robinson worked out some of the earliest implications of his ideas for infinite languages and forcing in model theory with Jon Barwise while he was an assistant professor at Yale (1968–1970). Together they took the ideas Robinson had outlined

[71] Robinson 1970. [72] Robinson 1971.

[73] Robinson 1971, p. 80; Robinson 1979, 1, p. 216.

in his Rome paper, and found they led in a natural way to the definition of so-called *K*-generic structures.

For uncountable languages, there was a more general version of K^f. The focus of the Robinson-Barwise paper, in fact, was *K*-generic models and K^f, for which the main result was the following characterization:

> COROLLARY 3.5. *K^f is model-complete if and only if every model of K^f is K-generic.*[74]

Separate sections of the Robinson-Barwise paper were devoted to generic structures, examples of forcing companions and generic models. They also gave a Löwenheim-Skolem theorem for forcing, which was of real interest since, given a theory *K*, Robinson and Barwise had as yet only shown the existence of a *K*-generic structure in the case where *K* was countable, or where every model of K^f was *K*-generic. But the fact that there was very little difference between countable and uncountable theories when it came to results which did not mention *K*-generic structures was explained, to some extent at least, by a result which allowed passage from an uncountable theory *K* to a countable subtheory K^*. Applications (in a section on forcing companions and generic models) included necessary and sufficient conditions for an inductive theory *K* to have a model-completion, a model-companion, or equivalently, for K^f to be model-complete.[75]

The paper closed with several open questions. Robinson and Barwise acknowledged, for example, that "we have not been able to determine K^f for the case where *K* is the set of axioms for Peano arithmetic."[76] Among the questions raised in section 8: Are there any uncountable theories *K* for which no *K*-generic models exist? What is the forcing companion of Peano arithmetic? Under what model theoretic conditions on *K* is it possible to assert that every model, or every countable model, of *K* can be embedded in a *K*-generic model?

For the Second Scandinavian Logic Symposium in Oslo, Robinson further developed his ideas on "Infinite Forcing in Model Theory."[77] Unlike his earlier work with Barwise, this paper did not rely on sets of sentences but identified infinite forcing conditions with *structures*. The results were analogous with those Robinson had already established for finite forcing, but he was aware of some significant differences as well.

What he had found was a new connection between the forcing relation and model theory:

[74] Barwise and Robinson 1970, p. 130; in Robinson 1979, 1, p. 230.
[75] Barwise and Robinson 1970, pp. 140–141; Robinson 1979, 1, pp. 240–241.
[76] Barwise and Robinson 1970, p. 120; Robinson 1979, 1, p. 220.
[77] Robinson 1970.

this leads us to a kind of compactness theorem for forcing and to the axiomatization of classes of generic structures by infinitary sentences. Some of these results are developed also for finite forcing.[78]

Working within the lower predicate calculus (with equality), Robinson began by ranking WFF, beginning with atomic formulae at rank one. Higher ranks were defined in terms of the relative number of connectives and quantifiers. For a given class of similar first-order structures Σ, he defined the forcing relation $M \Vdash X$ (M "forces" X) in terms of satisfaction, $M \models X$ (M "satisfies" X), as follows:

For an atomic sentence X, $M \Vdash X$ if and only if $M \models X$, M satisfies X; $M \Vdash X$ where $X = Y \wedge Z$ if and only if $M \Vdash Y$ and $M \Vdash Z$; $M \Vdash X$ where $X = Y \vee Z$ if and only if M forces at least one of Y and Z; $M \Vdash X$ where $X = (\exists\, y)\ \sigma(y)$ if and only if $M \Vdash Q(a)$ for some constant a. Finally (as the only but crucial departure from the satisfaction relation) $M \Vdash X$ where $X = \neg Y$ if and only if there is no $M' \in \Sigma$ which is an extension of M such that $M' \Vdash Y$.[79]

Here Robinson was interested in the very close relation between forcing and satisfaction. If Σ contains only a single proper structure M, then for a sentence X in M, M forces X if and only if M satisfies X. This connection naturally sparked Robinson's attention, for, as he put it, "we may therefore regard the forcing relation more generally as a generalization of the relation of satisfaction."[80]

Now, for a nonempty and consistent set of sentences K, Robinson defined $\Sigma = \Sigma_K$ as the class of all structures consistent with K (i.e., substructures of models of K, together with $\varnothing$). An element M of Σ_K was said to be K-generic if all sentences of K are defined in M (these do not necessarily need to hold in M), and if for every sentence X which is defined in M, either $M \Vdash X$ or $M \Vdash \neg X$. Two theorems then followed:

THEOREM 2.4: *Let $M \in \Sigma$. Then M is contained in a K-generic structure $M^* \in \Sigma$.*

THEOREM 2.5: *In order that a structure M of Σ be K-generic it is necessary and sufficient that the sentences of K be defined in M and that for every sentence X defined in M, $M \models X$ if and only if $M \Vdash X$.*[81]

Robinson pointed out that theorem 2.5 could also be used as an alternative definition for a K-generic structure. Basically, this means that a

[78] Robinson 1970, p. 317; in Robinson 1979, 1, p. 243.
[79] Robinson 1970, p. 318; in Robinson 1979, 1, p. 244.
[80] Robinson 1970, p. 318; Robinson 1979, 1, p. 244.
[81] Robinson 1970, p. 320; Robinson 1979, 1, p. 246.

proper structure M is K-generic if and only if K is defined in M and the relations of satisfaction and forcing on M coincide. Thus it was possible to replace "M satisfies X if and only if M forces X" with "M satisfies X entails M forces X".[82]

Next, Robinson devoted considerable attention to weak forcing. M is said to force X "weakly"–$M \Vdash {}^*X$–if and only if there is no $M' \in \Sigma$, $M' \supset M$ such that $M' \Vdash \neg X$. If X is defined in M, then $M \Vdash {}^*X$ if and only if $M \Vdash \neg\neg X$. This also follows if $M \Vdash X$, and if X is defined in M and $M \Vdash {}^*\neg X$, then $M \Vdash \neg X$.

Moreover, if $M \Vdash {}^*X$ and $M' \supset M$ for $M' \in \Sigma$, then $M' \Vdash {}^*X$. Also, no $M \in \Sigma$ can force weakly both X and $\neg X$, for if $M \Vdash {}^*\neg X$ and $M' \supset M$ is an element of Σ_K in which X is defined, then M' cannot force $\neg\neg X$. Therefore, $M'' \Vdash \neg X$ for some $M'' \supset M'$ and so M cannot force X weakly. It follows from his earlier theorems 2.4 and 2.5 that $M \models {}^*X$ if and only if X holds in all K-generic structures $M' \supset M$ in which it is defined.[83]

For any $M \in \Sigma_K$, Robinson defined $K^F(M)$, namely the set of all sentences defined in K and forced weakly in M. By putting K^F=$K^F(\emptyset)$, K^F is the *forcing companion* of K, and he called F the *forcing operator*. Thus K^F is the set of all sentences defined in K which hold in all K-generic structures of Σ_K. It was one of Robinson's basic aims to investigate the properties of K^F and its models. He did so in the Oslo paper with twelve theorems and related corollaries.

Robinson was especially interested in the joint embedding property. Given K and two structures M_1 and M_2 whose relations and functions are denoted by symbols of K, when is there a model M of K which is an extension of both M_1 and M_2? This came down to whether or not there were injections $M_1 \to M$ and $M_2 \to M$ such that

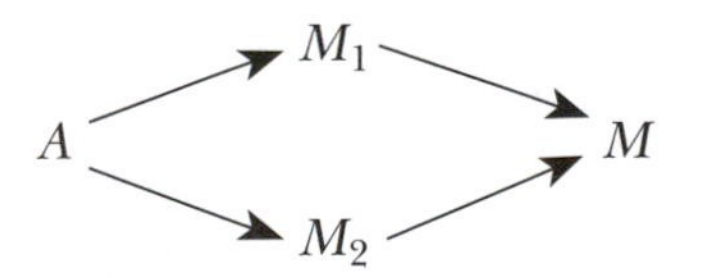

is commutative. Whenever this was the case, then the set K had the *joint embedding property*:

THEOREM 4.1: *For given K, A, M_1 and M_2, there exists a joint embedding of M_1 and M_2 in a model M of K, as detailed above, if and only if for any existential sentences X_1 and X_2 whose individual constants belong to A or to K and such that $M_1 \models X_1$, $M_2 \models X_2$, the set $K \cup \{X_1, X_2\}$ is consistent.*[84]

[82] Robinson 1970, p. 320; Robinson 1979, 1, p. 246.
[83] Robinson 1970, p. 320; Robinson 1979, 1, p. 246.
[84] Robinson 1970, p. 324; Robinson 1979, 1, p. 250.

Robinson also gave additional theorems showing that in order for K to have the joint embedding property, it was necessary and sufficient that for any two existential sentences X_1 and X_2 defined in and consistent with K, $\{X_1, X_2\}$ is also consistent with K. Analogous to a theorem he had already established for finite forcing, he was able to show that for K^F to be complete, it was necessary and sufficient that K possess the joint embedding property. Here the forcing operator selected one set K^F from every class of mutually model-consistent sets K. The set K^F is complete if and only if K again possesses the joint embedding property.

Robinson then went on to consider forcing companions and model-companions. K^* was called a model-companion of K if K^* had the same vocabulary as K, was mutually model-consistent with K, and was model-complete. If K^* is an extension of K (and both model-consistent and model-complete relative to K), then it was called a model-completion of K, ideas he had already developed in his earlier papers on forcing. He then showed that K could have only one model-companion (and therefore, no more than one model-completion, up to logical equivalence).

If K is model-complete, then the class of K-generic structures is just the class of models of K. A corollary: if K is model-complete, then K^F is the deductive closure of K. Moreover, if K possesses a model-companion, K^*, then K^F is the deductive closure of K^*. In particular, whenever K possesses a model-companion, $K^F = K^f$. In particular, if $K = N$ is the set of all true sentences of Arithmetic, then $N^f = N$ (as Robinson had shown in his paper with Jon Barwise). However, N^F N.[85]

Indeed, Robinson regarded N^F as a "remarkable" subject of study. Although different from N, N^F included many theorems of standard arithmetic, in particular all $\forall\exists$ sentences which are true for the natural numbers.[86] Moreover, N^F is complete, since N is complete and, therefore, has the joint embedding property. It follows that N^F is not even recursively enumerable.[87]

Considering the case of the axioms P for Peano arithmetic P_A, then P^F N^F (since by a theorem Yuri V. Matiyasevich had proven, P_A is a proper subset of the set of all true universal sentences of arithmetic). Moreover, as Michael Rabin had suggested to Robinson, by an argument which involves an application of Matiyasevich's theorem, P does

[85] This Robinson showed using a result Michael Rabin had obtained, which would have ensured that there does exist an existential sentence X which is defined in a model of N^F, N_1, and holds in some N_2 (a model of N), N_2 . . . N_1, and yet does not hold in N_1 itself. Consequently, N^F N. See Rabin 1962.

[86] It also includes Raphael Robinson's system of axioms Z, and since Q is essentially undecidable it follows that N^F is also undecidable.

[87] For details, see Robinson 1970, p. 328; Robinson 1979, 1, p. 254.

not possess the joint embedding property, and so P^F cannot be complete. However, P^F still includes Raphael Robinson's set of axioms *Q*, and thus is essentially undecidable.[88]

Robinson then considered subclasses of Σ_K the most obvious of which was the subclass G_K given by *K*-generic structures. He also gave a compactness theorem for infinite forcing. Robinson ended with some comments on how infinite forcing might be useful for finite forcing. The latter could be interpreted in terms of basic sentences which are diagrams of finite structures of models of *K*, whereby he was also able to produce analogs for theorems previously established, but with significant simplifications.[89]

Robinson's use of forcing in model theory made an especially strong impression on Kurt Gödel, who assessed it as follows:

> Since 1969 Robinson has developed a method of forcing in model theory which is an analogue of Paul Cohen's method in set theory. This has proved to be an extremely fruitful idea which is being developed by Robinson and his students. Robinson's paramount influence in model theory is evidenced by the large number of students and younger logicians who have been attracted to him.[90]

Similarly, Jerome Keisler has succinctly summarized the value of the method as follows:

> Robinson showed that the generic models constructed by his forcing are closely related to model-completions. If a theory has a model-completion, then it must be the set of sentences true in all generic models of the theory. Whether the theory has a model-completion or not, all generic models are existentially closed in the theory. These results created new interest in model-completeness and suggested many questions in particular areas of

[88] See Matiyasevich 1970. R. M. Robinson's axioms *Q* were based on a weaker (and simpler) system than Tarski's undecidable subtheory of arithmetic $\bar{Q}$. See section 2.3 of Tarski 1953, pp. 51–55.

[89] Robinson 1970, p. 338; Robinson 1979, 1, p. 264. Robinson also worked on forcing with Edward R. Fisher, and together they too coauthored a paper on "Inductive Theories and Their Forcing Companions." Here Fisher and Robinson showed how a decomposition theorem for ideals of a distributive lattice was related to a classification of the generic models of an arbitrary inductive theory. This made it possible to generalize, for example, the classification of algebraically closed fields according to their characteristics. For details, see Fisher and Robinson 1972.

[90] "Data for Nomination [to membership in the Royal Society of London]. Abraham Robinson," Gödel papers no. 011983, p. 3, Princeton University Archives, Firestone Library (hereafter, PUA/FL). Robinson acknowledged in a letter to Gödel that Mitsuru Yasuhara had given him some interesting observations concerning his recent work on forcing in model theory. See Robinson to Gödel, January 29, 1971, Gödel papers no. 011953, PUA/FL.

algebra. As a result there has been a substantial increase recently in research activity in the whole area of model-theoretic algebra.[91]

Logic in Latin America: July 1970

No sooner than the meeting in Oslo was over, Robinson was almost immediately on his way to Chile to participate in a summer school for logic which was already in progress. Robinson had discussed the possibilities of doing something to promote the subject in Latin America with Rolando Chuaqui during the "Logic Year" at UCLA in 1967. At the time, Chuaqui was in Los Angeles, visiting from Chile, and already had an interest in mathematical logic. Successful meetings of the ASL in Europe suggested similar possibilities elsewhere, and William Craig, who preceded Robinson as president of the association, had also raised the idea. The first "concrete steps" were not taken, however, until Robinson appointed an advisory committee comprised of himself, David Kaplan (UCLA), Antonio Monteiro (Universidad Nacionál del Sur, Bahía Blanca, Argentina), and Rolando Chuaqui (who served as committee chairman, from the University of Chile, in Santiago), to suggest ways to promote logic in Latin America.

Robinson subsequently submitted a grant proposal to help fund a meeting in 1969. Three-week seminars were to be offered in model theory, set theory, and recursion theory (all at intermediate levels), after which a three-day colloquium was expected to draw upon both invited lectures and contributed papers.[92]

Due to internal problems at the University of Chile, the proposed meeting did not take place until the following year. In the meantime, Chuaqui had moved to the Catholic University in Santiago, where the rector, Fernando Castillo Velasco, offered enthusiastic support for the logic meeting. Consequently, the First Latin American Symposium on Mathematical Logic (SLALM I) was finally held at the Catholic University of Chile, in Santiago, in the summer of 1970.[93]

The seminars met during the first three weeks of June. Short courses

[91] Keisler 1979, p. xxxv. See the presentation of various forcing techniques, including the relation of Robinson-forcing to others in Keisler 1973, as well as Hirschfeld and Wheeler 1975.

[92] Robinson to Craig, January 24, 1968, RP/YUL.

[93] Robinson was an "indefatigable worker for the success of this meeting." Indeed, he sent letters to dozens of individuals, foundations, universities, and governmental agencies in the United States and Latin America, requesting financial support. See "A Short History of the Latin American Logic Symposia," in Arruda 1977, pp. ix–xviii, esp. p. ix. This report of the Santiago meeting includes Robinson's "A Proposal for the Organization of a Seminar and Colloquium in Mathematical Logic, to be Held in Chile (Revised Version, June 3, 1969)," pp. x–xi.

were given by Azriel Levy on foundations of set theory, Joseph Shoenfield on degrees of unsolvability, and Roman Sikorski on Boolean algebras. The colloquium, which followed as an official meeting of the Association for Symbolic Logic (jointly sponsored by the Catholic University of Chile, the Organization of American States, and the International Union for the History and Philosophy of Science) consisted of thirteen one-hour lectures and as many shorter communications.[94] As a report of the meeting later remarked, Robinson participated actively—even to the point of giving one lecture in his hotel because he had fallen ill.[95] In fact, Robinson gave two invited one-hour lectures, the first on "Model-theoretic Aspects of Algebra," and the second on "Finite and Infinite Forcing in Model Theory."[96]

Despite the apparent success of the first Latin American Logic Symposium, only Argentina, Brazil, and Chile were represented with actual participants. From the beginning, Robinson and his advisory committee had found it difficult to locate individuals in Latin America interested in logic. Although they had tried to contact members of the ASL living in Latin America, this produced few responses; the only viable contacts were made through friends.

Even so, SLALM I, which was a great success for all who participated (especially in terms of publication) encouraged Robinson to discuss future possibilities with many of the logicians present. As a result, it was decided that the next symposium would be held two years later in Brazil.[97]

Nonstandard Analysis and the First Oberwolfach Meeting: 1970

The next meeting on Robinson's schedule that summer was in Germany. But first he wanted to spend a few days in Heidelberg, primarily to see his old friends at the Institute of Mathematics. He had taken special care to arrange a pleasant evening with Peter Roquette and Gert Müller, both of whom he had invited to dinner at Restaurant Rheinback

[94] In fact, the OAS generously treated the symposium as two meetings—the seminar and the colloquium—and therefore gave double its usual amount. See Arruda 1977, p. xiii, as well as the report of this meeting in Chuaqui 1970.

[95] See "A Short History of the Latin American Logic Symposia," in Arruda 1977, pp. xii–xiii.

[96] For a report of the meeting, see the *Journal of Symbolic Logic*, 36 (1971), pp. 576–580.

[97] A new ASL Advisory Committee for Latin America was agreed upon, consisting of the ASC president (*ex officio*), namely Robinson, as well as Newton da Costa, Antonio Monteiro and Rolando Chuaqui.

Abraham Robinson and Wim Luxemburg at Oberwolfach, 1970. (Photo courtesy of Manfred Wolff, Tübingen)

on July 17.[98] Two days later, he was in the Black Forest for the first international meeting on nonstandard analysis, organized by W.A.J. Luxemburg and Detlef Laugwitz at the Mathematical Research Institute in Oberwolfach, from July 19–25. Virtually all of the early pioneers were present, including Larry Kugler, Joram Hirschfeld, Moshé Machover, Keith Stroyan, Peter Loeb, and of course, Abraham Robinson.[99] Peter Loeb found the meeting very stimulating, and Robinson "charismatic."[100]

Loeb had just recently come to appreciate the power of nonstandard analysis. Having learned about its application to ideal boundaries when Robinson gave a visiting lecture at UCLA in 1969, Loeb gave it serious study a year prior to the Oberwolfach meeting when he took a copy of Robinson's book on nonstandard analysis with him to a potential theory meeting in Stresa (Italy). There he read the first chapter which introduced type theory and the ultrapower construction. Outside, a fierce thunderstorm was raging. Loeb kept reading the chapter over and over again. "The seventh time it suddenly dawned on me that our mathematical world really consisted of models and our only hold on those models was with the use of language."[101] By the time Robinson gave the opening lecture at the Oberwolfach meeting, Loeb was fully prepared to appreciate its significance.

Robinson spoke on "Nichtarchimedische Körper" (in German). This drew on research by Detlef Laugwitz on the field L of general power series $\Sigma a_k t^{\nu_k}$ (a_k, ν_k real; $\nu_0 < \nu_1 < \nu_2 \to \infty$), which followed research done earlier by Levi-Cività and Ostrowski. Laugwitz extended ordinary infinitely differentiable functions to this field and asked whether or not

98 Abraham Robinson to Peter Roquette, July 1, 1970, RP/YUL.

99 Luxemburg I-1991. Details are provided in the report of the "Erste Oberwolfach Tagung über Nonstandard Analysis," *Vortragsbuch* for 1970. For details of the participants and papers presented at this meeting, see pp. 175–187.

100 Peter Loeb, interviewed in Frankfurt, Germany, June 15, 1991.

101 Loeb I-1991.

these satisfied the intermediate- and the mid-value theorems? Robinson was able to show that such was not always the case, and generalized his results by embedding L in a remainder field of a subring of a nonstandard model (e.g., an ultraproduct) R^* of the real numbers.[102]

The last of Robinson's obligations that summer was the 1970 International Congress of Mathematicians convened in Nice on September 1. Yale was well represented. Walter Feit gave one of the general, plenary lectures ("The Current Situation in the Theory of Finite Simple Groups"), and Robinson presented a fifty-minute invited lecture on "Forcing in Model Theory." In fact, the Congress broke with former practice by eliminating entirely all shorter presentations, and instead distributed a booklet of 265 "individual communications."[103]

Director of Graduate Studies: 1970

Earlier that May, W. S. Massey, then chairman of the Yale Mathematics Department, had asked Robinson to accept a three-year appointment as director of graduate studies. Massey explained that he had already served three years in the position himself, and that "although it was time-consuming during certain weeks of the year, I always had a feeling that what I was doing was necessary for the Department to function effectively. This is *not* true of many other academic jobs I have held."[104]

Robinson agreed to accept the position which began in the fall and included a reduction in his teaching load (by one semester course a year). As usual, he managed to find time for everything, including an expanding group interested in logic at Yale:

> Robinson ran the entire operation in logic. With just one appointment, he seemed to do it effortlessly. He wrote very fast, and composed letters very quickly. Even as President of ASL, this was no hardship. At Yale, he

[102] Robinson, in Oberwolfach *Vortragsbüch* 1970, p. 175. See also *Tagungsberichte 1970*, no. 25, "Nonstandard Analysis," Oberwolfach: Mathematisches Forschungsinstitut, p. 2. Later, Robinson and Luxemburg edited a collection of papers from the Oberwolfach meeting. See Robinson and Luxemburg 1972. Robinson wrote a brief introduction to this volume, and explained in a letter to Luxemburg that "It is very prosaic.... I did not feel that this was the right place to sing a song in praise of NSA." See Robinson to Luxemburg, August 4, 1971, RP/YUL. The lecture Robinson delivered at Oberwolfach is *not* the paper published in Luxemburg and Robinson 1972.

[103] At the closing ceremonies, Jean Dieudonné asked for a show of hands of those present, and more than two to one favored continued elimination of any short ten-minute communications. See Congress 1970, 1, p. xiii.

[104] W. S. Massey to Abraham Robinson, May 18, 1970, RP/YUL. Massey added that "Charles Rickart has been working hard to reduce the amount of work involved in the job; in particular, he has been training one of our secretaries, Mrs. Slowen, to help him with many aspects of the job. I'm sure she would be of great assistance to you next year."

> sat on the Appointments Committee in the Sciences. Whenever he was asked to serve, he never said "no"—and it was the same for lectureships. If invited, he felt it was an obligation to "spread the light."[105]

Spreading the light also meant his teaching, which included his undergraduate course on naive set theory (Math 70a). One of his students that year captured the flavor of what a course with Robinson was like:

> Robinson's lectures were central to an understanding of the course. For the most part they were quite clear and, by the standards of the math department, they were remarkably easy to follow. Although there was an assigned text, it was referred to only rarely. Problem sets were infrequent and relatively short. There was no problem section, but occasionally a class period would be devoted to answering questions about problems.
>
> Math 70's worst problem was that there was not sufficient application of the theory covered. There were not enough interesting problems, and the exam questions were best answered by reproducing one's class notes verbatim. Consequently, few students could become excited by the subject's intrinsic beauty. In sum, Math 70 is a worthwhile course for those who care to learn set theory. Those seeking excitement or inspiration in their math classes, however, had best look elsewhere unless the course can be redirected toward more practical aspects of set theory.[106]

Apart from his teaching, Robinson spent most of the fall term in New Haven. When the weather was especially pleasant he and Renée would go for long walks in the Sleeping Giant woods, and when necessary on weekends, Abby would rake up massive piles of autumn leaves, the only yard work Renée would permit him to do.

Meanwhile, Renée was busy developing her artistic talents in new directions. She had become increasingly involved in printmaking, and soon was attending workshops and being invited to show her work in local galleries in New Haven. She discovered a new interest in silk screening, and later even gave lessons during the term she and Abby spent at the Institute for Advanced Study (1963, see below, pp. 457–459), where their apartment displayed the results of new pieces she did in Princeton, as well as some favorites they had brought with them.[107] From seriography she went on to colographs, and consistently won prizes when invited to exhibit. Many of her prints found their way onto the walls of their home in Hamden, into the homes of friends, or were bought by admirers.

Socially, Renée and Abby continued the tradition of party-giving they had enjoyed wherever they lived. Typically, a party at the Robinson's

[105] Mostow I-1993.
[106] *Yale Course Critique* (1970), p. 102.
[107] Renée Robinson I-1993.

home in Hamden would include a congenial mix of artists and mathematicians.[108] Among the most important of their shared interests, however, was travel, and in the fall of 1970 they made time for another brief trip to Mexico, thanks to the annual meeting of the AMS/MAA that year in San Antonio.

K-Generic Structures and Algebraic Closedness

In November Robinson sent a paper he had just finished, "On the Notion of Algebraic Closedness for Noncommutative Groups and Fields," to the *Journal of Symbolic Logic*. The results he described were the outcome "of some ongoing research on the forcing method in model theory" (although the paper itself did not require a knowledge of forcing and did not use it explicitly). Instead, this was a paper meant to prompt interest in *K*-generic structures as the metamathematical counterpart of algebraic closedness.[109] Robinson offered no proofs in this paper, preferring to emphasize how the major metamathematical properties of algebraically closed commutative fields could be exploited to achieve results for a wide class of first-order theories (including the theories of commutative and skew fields, and of commutative and general groups).

Just before Christmas Renée and Abby visited their old friends the Kaydens in New York, where they had been invited over the years to spend the Thanksgiving or Christmas holidays. Mrs. Kayden had grown up in Palestine, where she and Abby had first met in the 1930s (he helped to teach her Hebrew when she first arrived from Europe).

The real highlight that winter was the Robinson's year-end trip to St. Lucia in the Caribbean. It was warm, relaxing, and the volcanic landscape of the island itself was unusual and dramatic. Abby was at work on another paper, and wrote his legendary three pages a day, but he also made time with Renée for mule racing on the hotel lawn in Castries.[110] Renée did not find St. Lucia as attractive as Martinique, but they were both grateful to be relaxing in the Caribbean, and they especially enjoyed dancing in the evenings. Both were surprised, however, when Santa Claus appeared on Christmas Day in red flannel and white beard, despite the heat!

[108] Renée Robinson I-1993.

[109] Robinson called attention to the title of his paper, which referred to "closedness" rather than "closure." As he stressed, "our analysis will not touch upon the question of the existence of a 'closed' extension which is in some sense minimal as is the algebraic closure of a commutative field," Robinson 1971b, p. 441; Robinson 1979, 1, p. 478.

[110] Robinson I-1982.

Cherlin's Dissertation: Infinite Forcing

Gregory Cherlin studied with Robinson from the fall of 1969 through the spring of 1971:

> In my case Robinson's supervision was almost completely nondirective. He suggested that I think about whatever I liked for around a year. We then met formally twice a month to talk things over (our contact was of course in no way limited to these scheduled meetings). I don't recall much serious mathematics being discussed apart from my own brief reports on certain ideas that Joram Hirschfeld and I were pursuing in nonstandard analysis.[111]

Robinson did make a pointed remark, once, that made a definite impression: "Since you are specializing in Logic, perhaps you should take a course in it." Robinson had in mind a basic course, since Cherlin had only taken courses in model theory. On the subject of Cherlin's dissertation, Robinson offered the same advice he gave to all of his graduate students—"take the idea and push it as far as it will go." But first Cherlin had to decide exactly what he wanted to work on:

> In August or September 1970 Robinson suggested that I should perhaps do something definite, either in forcing or in nonstandard analysis. . . . One afternoon a few weeks later I came up with the first generic hierarchy (Σ^n), wrote out the theory of infinite forcing from this point of view, and then told Robinson about it with great enthusiasm. Not surprisingly, I recall that conversation rather well. After I finished my outline, Robinson's first remark was, "Where would you like to work next year?" I replied by reminding him of a former student who had spent a year in Massachusetts perfecting a betting system for horse racing, and suggested I might do something similar, e.g. make movies in Hawaii. With his customary severity, Robinson suggested I might be interested in a job at the University of Hawaii. . . .[112]

Instead of horse racing or surfing, Cherlin won a visiting fellowship at Princeton, where (according to Robinson) he wanted time "to think things over." Whatever fears he may have had, Robinson confessed in a subsequent letter of recommendation that Cherlin "has now decided to stay in the field for which Fate quite obviously predestined him."[113]

Cherlin's thesis was devoted to "A New Approach to the Theory of Infinitely Generic Structures," and was based on Robinson's extension

[111] Cherlin R-1974. See also the paper by Cherlin and Hirschfeld in Luxemburg and Robinson 1972.

[112] Cherlin R-1974.

[113] Robinson in a letter of recommendation for Cherlin of May 23, 1972, RP/YUL.

of forcing in model theory. Robinson had already shown that every class of models of a first-order inductive theory had a unique model-completion, and Cherlin built upon this work giving two constructions (one of which avoided the assumption that the class was elementary). After examining classes intermediate between the class and its model-completion, Cherlin went on to apply model-completion methods to more powerful, higher-order languages.[114] Here Cherlin drew upon results Carol Wood had obtained for her thesis, which considered specifically higher-order languages. As Cherlin acknowledged, Wood had shown that "a good deal of forcing theory may be developed in languages more powerful than ordinary first order logic."[115]

Carol Wood: Forcing for Infinitary Languages

As an aphorism at the beginning of her thesis, Wood chose a Latin passage from Virgil's *Aeneid.* Having survived the rigors of Scylla, Charybdis, and the Cyclops, Aeneas now looked forward to better days:

> We are going on
> Through whatsoever chance and change, until
> We come to Latium, where the fates point out
> A quiet dwelling-place, and Troy recovered.
> Endure, and keep yourself for better days.[116]

Wood's dissertation was devoted to "Forcing for Infinitary Languages." When it came to recommending her results for publication, Robinson praised her work in a letter to Henry Bauer, an editor of *Inventiones Mathematicae.* Robinson stressed in particular her treatment of differentially closed fields of characteristic p 0, for which she established the existence of a model-companion.[117]

The third thesis written that year under Robinson's direction was by David Randolph Johnson, Jr., and covered "The Construction of Elementary Extensions and the Stone-Cech Compactification."[118] This was devoted to aspects of model-theoretic topology, for which Johnson

[114] Cherlin 1971. [115] Cherlin 1971, p. 2.

[116] Virgil, *Aeneid*, I, 1, 203–207. English from the verse translation, Humphries 1951, p. 10. Wood quoted the original Latin, to which Robinson responded after reading the thesis by writing *imprimatur* at the top near the passage in Latin, "and seemed quite pleased at this small joke." Carol Wood to JWD, E-mail note of June 24, 1994.

[117] A. Robinson to H. Bauer, RP/YUL. For publication of her results on differential fields, see Wood 1973. Wood 1972 also contains material from her thesis related to forcing and infinitary languages.

[118] Johnson 1971. Although Johnson never published his dissertation, he did publish a joint paper with D. A. Mattson on nonstandard proximity spaces. See Johnson and Mattson 1975.

provided two model-theoretic constructions, one the *z*-ultrapower construction (a continuous analog of the ultrapower construction), and the other hyper-models of densely ordered sets.

Tarski Symposium: Berkeley 1971

Tarski's seventieth birthday was celebrated in grand style at Berkeley, from June 23–29, 1971, thanks to a generous grant from the National Science Foundation. The meeting included virtually every branch of mathematics to which Tarski had contributed–set theory, model theory, decidability, general algebra, lattice theory, algebraic logic, foundations of geometry, nonclassical logic, proof theory, the concept of truth in both natural and formalized languages, as well as methodology of science.[119]

Robinson, for his part, decided to focus on the subject of model theory in relation to Tarski's classic paper, "A Decision Method for Elementary Algebra and Geometry," which Tarski first published in 1948, although he had the main results as much as a decade earlier.[120] Robinson, in light of his own work, wanted to investigate just how "the successive widening of our model theoretic point of view has shed new light on Tarski's result."[121]

Considering the theory of real-closed ordered fields *K*, it followed from Tarski's work that *K* is complete. Thanks to his own results in model theory, Robinson was able to show that any true statement made in the language of *K* for the field of real numbers must then be true as well for any other real-closed ordered field. As Robinson explained:

> This, of course, is not exciting if the proof of the result under consideration was carried out in that language from the outset. However, there are several important results of this kind which concern real numbers and whose proof involves topological or other higher order notions, and for these the principle just stated is effective and striking. By contrast, although Tarski's method provides an effective procedure for describing whether or not a given sentence of the lower predicate calculus is a theorem of the theory of real-closed ordered fields, no substantial problem has been settled *directly* by it so far. Although a considerable effort has been invested in the attempt to design computer programs for this purpose, this is still an open issue.[122]

[119] For a brief announcement of the meeting, see the *Journal of Symbolic Logic*, 36 (1971), p. 190.

[120] Tarski 1939 and 1948.

[121] Robinson 1971a, p. 140; Robinson 1979, 1, p. 491.

[122] Robinson 1971a. Material on the Tarski Symposium in Tarski papers, box 11, folder 1, The Bancroft Library, University of California, Berkeley.

Gregory Cherlin, in reviewing Robinson's contribution to the Tarski symposium, later praised it as an "excellent brief introduction to the modern theory of existentially complete and generic structures." Robinson was in the best position to do this, he added, "having personally developed most of the fundamental notions and many of the applications in this area."[123]

The Hedrick Lectures, Then Bucharest: 1971

At the end of the summer, Robinson delivered the twentieth series of Hedrick lectures at the summer meeting of the Mathematical Association of America, held at Pennsylvania State University.[124] He surprised no one in talking about "Nonstandard Analysis and Nonstandard Arithmetic."[125] As soon as the last lecture was finished, he was immediately on a plane, headed for Romania. The Fourth International Congress for Logic, Methodology, and Philosophy of Science had opened on August 29, but ran until September 4. Although Robinson missed the first few days of the Congress, he was in Bucharest by the end of the week, in time to deliver his paper on Friday morning, a half-hour lecture on "Nonstandard Arithmetic and Generic Arithmetic."[126]

The basic point of Robinson's paper was inspired by recent thoughts on generic arithmetic and the new work he and his graduate students at Yale had been doing on model-companions and related topics.[127] Robinson began with the fact that the theory of algebraically closed fields is the model-companion of the theory of fields. Similarly, the theory of real-closed fields is the model-companion of the theory of formally real fields. But there was an important difference between the two examples, namely, the fact that while the theory of algebraically closed fields is model-complete relative to the theory of fields, the theory of real-closed fields is not model-complete relative to the theory of formally real fields. His paper then focused on existentially complete and generic structures for arithmetic.[128]

The generic arithmetic that Robinson considered in the Bucharest paper focused on three examples: the natural numbers, the ring of

[123] Gregory Cherlin, *Mathematical Reviews*, 51 (1976), no. 2902. Further developments of this material may be found in Hirschfeld and Wheeler 1975.

[124] The meeting (held jointly with the AMS) ran from Monday, August 30, to Wednesday, September 1, 1971. See Alder 1971, pp. 1047–1061.

[125] Alder 1971, p. 1048.

[126] Robinson 1971c. For more complete details and subsequent developments, see Hirshfeld and Wheeler 1975.

[127] Wood I-1992.

[128] Existentially complete structures had also been studied previously by Eklof and Sabbagh 1970, by Robinson himself in his Oslo paper 1970, and by Simmons 1972.

rational integers, and the field of rational numbers. For the natural numbers, Robinson dealt primarily with the extent to which the corresponding forcing companion coincided with ordinary arithmetic–and the extent to which it did not.[129]

For example, considering the theory N of natural numbers and its forcing companion N^F, Robinson showed that the set N^F is not arithmetical (theorem 2.22). This answered very nicely a question Martin Davis had posed at the Nice Congress, and added further detail to Robinson's own earlier result that N^F is not recursively enumerable.[130]

Robinson was particularly interested in existentially complete and generic structures of arithmetic because they served as canonical examples for other cases, which in turn raised interesting possibilities for important structures that had long concerned Robinson, like algebraically closed fields and real-closed fields. To stress this fact, Robinson showed, "quite generally, the generic structures of a class Σ possess an analog of the theory of algebraic varieties for algebraically closed fields which does not exist for Σ itself."[131]

There was more to the meeting in Romania, of course, than the technical details of new mathematical ideas. Among old friends he saw in Bucharest was Julia Robinson. Abby took advantage of the opportunity to invite her to give the next Frank J. Hahn lectures at Yale on Hilbert's tenth problem–and although she had resolved not to do any more teaching or lecturing, she told Abby that for him she would break her resolve and accept his invitation. Not only was it an honor to do so, she said, but it was, after all, the subject of her talk in Bucharest.[132]

Sterling Professorship: 1971

The idea of naming Robinson to one of Yale's Sterling Professorships arose in the fall of 1971. As G. D. Mostow (at the time chairman of the Mathematics Department) wrote to Charles Taylor, acting president of the university:

> I am writing to propose Abraham Robinson for the Sterling Professorship of Mathematics. This proposal has the strong support of my colleagues in the Mathematics Department.

[129] Robinson had already studied the corresponding class of generic structures for the natural numbers in Robinson 1970 and 1971b. Robinson also pursued this case in further detail in section 2 of the Bucharest paper.

[130] See Robinson 1970 or 1971b; Robinson 1971c, p. 146; in Robinson 1979, 1, p. 289.

[131] Robinson 1971c, p. 139; in Robinson 1979, 1, p. 282.

[132] See Julia Robinson to AR, September 28, 1971, RP/YUL. Julia Robinson added that she thought Matiyasevich should have been the one invited to give the lecture, but for political reasons, this was unfortunately not feasible. As she added, "I think mathematics and hence mathematicians would benefit greatly if there weren't these artificial barriers."

> It is our understanding that a Sterling Professor should ideally be a researcher of preeminent achievement in his field and at the same time a scholar of impressive breadth. We feel that Robinson meets these two criteria with much to spare. He is a significant figure, on a historic scale, in the theory of logic: his work on non-standard analysis has succeeded in putting Leibniz's hitherto intuitive "infinitesimals" on as firm a footing as ordinary number theory.
>
> I have had frank conversations, particularly with other worthy candidates in our department for this chair, about the choice of Robinson, and I am pleased to report that their wish coincides with mine.[133]

Robinson's official photograph as Sterling Professor, Yale University.

In due course, Robinson was named to a Sterling Professorship. At the departmental Christmas party that year, the graduate students presented him with a chair covered with aluminum foil. A special feature of these annual parties was a skit, traditionally put together by the second-year students. Larry Manevitz, who was one of Robinson's students, took the part of official presenter and gave the "sterling" chair to Robinson at the beginning of the skit, which meant that Robinson got, literally, a front-row seat and was thus properly positioned for later jokes. There was also a parody set to the words of "Jesus Christ Superstar," a popular hit song at the time.[134] The words devised just for the occasion were a nonstandard spoof, the chorus running:

> Abra-ham! RRRR-Star!
> What kind of fools do you think we are?
> Del-ta, Ep-si-lon,
> They're what we used to rely upon!

[133] G. D. Mostow to Charles Taylor (acting president, Yale University), October 4, 1971, Robinson's administrative file, Department of Mathematics, Yale University.

[134] That Robinson figured so prominently in the skit that year reflected the high esteem in which he was held; he was the graduate student advisor and a central figure in the department. The skit was written almoost entirely by a second-year student mathematical commune which called itself "Nicholas Bourbaki, Jr." Members included Georgia Benkart, Carl Bulmiller, Mark Kidwell, Larry Manevitz, James Mulflur, and Jeffrey Nunemacher. Bourbaki Jr. even had an entry in the New Haven telephone directory. Larry Manevitz, in an E-mail note to JWD, September 11, 1994.

Nonstandard Economics

Among the applications of nonstandard analysis that began to interest Robinson greatly in the early 1970s were results he and his Yale colleague, Donald J. Brown, were working on together. Brown had actually written his dissertation in logic, but turned to economics "to escape logic, because I thought that logic was really an area of economics that did nothing other than really talk to itself." Brown was on a postdoctoral fellowship at Yale when he first met Robinson, and from time to time they would see each other over coffee. Later, after Brown had joined the Economics Department, they were both walking down Hillhouse Avenue from one of the faculty parking lots:

> [Robinson asked] "What are you doing now?" And I said I'm teaching this seminar which is primarily concerned with Aumann's paper on continuing traders, and I was astounded, he knew about that paper. I said how could you know about that paper?[135]

It was doubtless due to the unique combination of their talents, the fact that Brown was well versed in logic and Robinson widely interested in everything, including the promotion of nonstandard analysis wherever possible, that their collaboration was so natural. As Robinson began to think about economics, there were a number of obvious ways in which a nonstandard perspective might find potential applications. In particular, as he and Brown soon discovered, the nonstandard approach fit nicely in the context of "nonstandard" economies in which the effect of any individual in an infinite economy of traders might be taken as negligible, or infinitesimal. Considered in such terms, Robinson saw a clear chance for nonstandard applications.

The first paper to appear in print was actually the last to be written, and was devoted to "A Limit Theorem on the Cores of Large Standard Exchange Economies."[136] Basically, what they did was to translate a conjecture of F. Y. Edgeworth's about the core–that it approaches the set of competitive allocations as the number of traders increases–into a theorem about the core of an infinite economy, or in terms of a limit theorem about the cores of very large yet finite economies.

[135] D. Brown, transcript of an interview by Lynn Steen, RP/YUL. In the record of Brown's transcript in RP/YUL, R. J. Aumann's name is mistranscribed as Almon, but corrected above. Aumann's work on "Markets with a Continuum of Traders" had already come to Robinson's attention when the two were together on the faculty at the Hebrew University, where they had apparently discussed Aumann's research in some detail, much to Brown's surprise.

[136] Brown and Robinson 1972. This paper was intended primarily to "announce" results that appeared much later elsewhere. It was T. C. Koopmans who sent their paper to the National Academy of Sciences for publication in March of 1972.

Using nonstandard analysis, Brown and Robinson gave a precise formulation of the concept of perfect competition underlying Edgeworth's theorem. They were then able to show (in the nonstandard setting) that the two concepts of the core and competitive equilibrium are the same.[137] As they put it concisely in a paper published later in *Econometrica*:

> The idealized notion of perfect competition is that the action of any finite set of traders in terms of their willingness to buy or sell at the competitive prices should have a negligible effect on these prices. Clearly this cannot be the case for any finite economy, but perfect competition is a meaningful concept for an economy having ω traders, where ω is an infinite integer, and where the endowment for each trader relative to that of the whole economy is infinitesimal.[138]

This was the essence of a theorem which they proved in a separate paper on "Nonstandard Exchange Economies." Here Brown and Robinson also proved a "less intuitive," equivalent theorem dealing with sequences of economies. This in turn led them to establish several asymptotic results dealing with unbounded families of standard exchange economies. In particular they proved a "limit theorem" which showed that for core allocations in large economies there exist prices such that the average number of exceptional traders is small, a result they proved in "The Cores of Large Standard Exchange Economies."[139] This paper, in the words of M. Ali Khan, "is important particularly in that it shows that results proved for the abstract nonstandard limit economy translate, essentially without any further work, into results for sequences of large but finite economies."[140]

When plans to hold a special colloquium on mathematical economics were being made at Berkeley for two weeks in August of 1974, Robinson was invited to participate.[141] In turn, he was delighted by the recognition nonstandard economics had begun to receive:

> I am genuinely pleased to think that the joint impact of Don Brown and

[137] Brown and Robinson defined the "core" as "the set of all allocations *X* that are not dominated by any allocation *Y* via any non-negligible coalition." See Brown and Robinson 1972, p. 1259; Robinson 1979, 2, p. 280.

[138] Brown and Robinson 1975, p. 42; in Robinson 1979, 2, p. 356. This paper summarizes virtually all of the results reported in the earlier papers Brown and Robinson had published; it is the one to read, although its date of publication is very misleading–the manuscript was actually submitted to *Econometrica* in June of 1971!

[139] Brown and Robinson 1974; in Robinson 1979, 2, pp. 345–354. This paper was submitted in January of 1972.

[140] M. Ali Khan, *Mathematical Reviews*, 57 no. 15318. For a recent evaluation of nonstandard methods applied in economic analysis, see Anderson 1990.

[141] Gerard Debreu to Abraham Robinson, January 4, 1974, RP/YUL.

> myself may have added something to the theoretical tools of the economists. Accordingly, I greatly appreciate your invitation to the August meeting.[142]

Unfortunately, Robinson felt obliged to decline:

> I have just recently come up against a medical problem which will mean that I shall have to be rather more economical with my energies than I used to be. I still hope to be in Vancouver where I am due to give a one-hour talk [at the International Congress of Mathematicians], but my activities for the rest of the summer may have to be severely curtailed.[143]

In 1972, however, Robinson was still in excellent health, and when Charles Ryavic invited him to give the DeLong lectures later that semester at the University of Colorado in Boulder, he was pleased to accept (joining a distinguished list of previous speakers, including Paul Cohen, Irving Kaplansky, and Mark Kac).[144] Robinson decided to lecture on "A Framework for Algebraic Extension Theory," and in notes on his copy of the letter he sent to Ryavic accepting the invitation, he listed several possible topics for his own reference:

1. What is an algebraically closed field?
2. Generic structures in Algebra.
3. Generic structures for mathematics.[145]

Saul B. Robinsohn (1916–1972)

When Abby's brother Saul died suddenly from a heart attack on April 9, "His untimely death at the early age of 55 came as a shock."[146] Indeed, his importance as an international expert on comparative education had just been recognized the year before, in June of 1971, when he was elected President of the Comparative Education Society in Europe. As one of his colleagues, who worked closely with Robinsohn, recalled:

> In a relatively short period of time his work made him one of the world's foremost comparative educationists. His research, with colleagues, into the reform of school systems, curriculum developments, teacher education and the politics of educational change added greatly to our understanding of these processes in an international context.[147]

[142] Robinson to Debreu, January 24, 1974, RP/YUL.
[143] Robinson to Debreu, January 24, 1974, RP/YUL.
[144] Charles Ryavic to Abraham Robinson, February 10, 1972, RP/YUL.
[145] Robinson to Ryavic, February 25, 1972, RP/YUL.
[146] Becker 1972.
[147] Holmes 1972.

Renée with Hilde and Saul at their home in Berlin (Grunewald).

But Holmes was also candid about what it was like to work with Robinsohn:

> He was not always an easy person but I quickly came to respect his integrity, energy and intellectual abilities. He was able to develop the point he wanted to make systematically and thoroughly. As a man he had many qualities which enabled him to gain the affection and win the loyalty of many of those with whom he worked. I was one of them.[148]

Despite certain striking similarities between the two brothers, Saul was not the easy going, even-tempered person that Abby was. Abby could be short-tempered, but not often. Usually when offended or upset he would simply fall silent, but he would never argue.

When Saul and Abby were together, they often spoke of their childhood, and especially of Jerusalem. Israel, in fact, was always a strong bond that both brothers shared in a very visceral way.[149] And in one very important respect, the two were very much alike; both were totally devoted to their work. And yet Saul was also driven in a way that Abby was not. Although Abby worked steadily, he always had time for his students, time for coffee, time for a cigar, the paper, and a chat about the latest news. He was never the stickler for detail or perfection that his brother was. Saul was much more intense:

> He was a glutton for work. I well remember working with him on the expert meeting he called in Hamburg to prepare for the CESE Amsterdam Conference. With short breaks for coffee we spent hours going through, word by word, our joint introductory paper. Others will doubtless testify how in Hamburg and Berlin he drove himself and them on to achieve

[148] Holmes 1972, p. 283.

[149] Renée Robinson I-1980.

> success in the professional tasks he set them. It may not always have been easy to accept his demand for perfection but, for my own part, I enjoyed the intellectual stimulus of his company, shared many of his professional aims, and benefited from his attention to detail.[150]

In Hamburg, Robinsohn had been responsible for an "impressive series" of publications for UNESCO based on "carefully prepared expert meetings and conferences." When he moved to Berlin in 1964, to head the Max-Planck-Institut für Bildungsforschung, and to teach at the Freie Universität (where he held an appointment as professor extraordinarius), he was in charge of a talented group of young research assistants who were also interested in many aspects of educational reform.

The ideal of education as a means of self-improvement, something that should be available to all, was a commitment both Abby and Saul felt deeply—and practiced. Robinsohn also had forward-looking ideas about a comprehensive, systematic "scientific utopia" of modern teacher education. This actually had some influence in the German Council on Education and was a factor in discussions at the time when actual reforms were being made.[151] In the end, however, it was the personal side of Saul Robinsohn that was most impressive:

> Yet finally I shall remember Saul as a man—with his weaknesses and strengths. He was sensitive and kind, with wide interests and deep feelings. I shall remember him best not as a colleague at professional meetings but when he talked to me about the art of Rembrandt when we were together in Amsterdam, or for his relaxed friendship when we walked together in Kew Gardens, and finally for his hospitality in the home he and his wife Hilde built on the edge of the woods in Dahlem. There I found another Saul Robinsohn and it is he I shall miss most.[152]

A Call from Berlin

Renée was the first to hear the bad news:

> We were in our home in Hamden. I think it was a weekend. Abby went in the morning into the woods for a walk. I did not. When a call came from Berlin, it was the Director of the Max Planck Institute, who told me about Shaul's [Saul's] death.[153]

When Abby returned from his morning walk, Renée told him there

[150] Holmes 1972, pp. 283–284.

[151] For a full list of Robinsohn's publications, see the bibliography of his works in Robinsohn 1973, pp. 453–462.

[152] Holmes 1972, p. 284.

[153] Renée Robinson to JWD, December 11, 1992.

had been a call from Berlin. It was as if he already knew, for all he said was "Shaul is dead."[154] Suddenly, with a single telephone call, he had lost his brother—his lifelong mentor, surrogate father, and best friend.

Abby made immediate arrangements to fly to Berlin, to be with Hilde and help her as best he could with the details and legal formalities following Saul's death. Not long thereafter, he received copies of a *Gedächtnisrede für Schaul* by Hellmut Becker, which Abby acknowledged had touched him deeply.[155] He also thanked Detlef Glowka (another of Saul's colleagues in Berlin) for a *Nachruf* he had written, and which Robinson also confessed was "especially moving."[156]

Saul's death was a special shock for Abby, who now began to fear that he, too, would fall victim to an inherited genetic flaw. His fifty-fifth birthday loomed on the horizon, his heart now, more than ever, seemed an unpredictable time bomb.[157] Renée was aware of this and urged Abby to cut back his lecture commitments and irrepressible interest in foreign places. Abby, however, refused to dwell on such things, although he confided to Gert Müller that "my wife is convinced that I travel more than is good for me."[158]

Perhaps it was the sudden realization of his own mortality that prompted Robinson to write an uncharacteristic letter that June to Gerald Sacks. Robinson had recently organized a logic meeting at Yale, and when Sacks arrived, he gave Robinson a copy of his just published book on Model Theory. Sacks had inscribed it "to the *première* model theorist," and upon seeing this, Robinson was understandably delighted. But later, after he had read the introduction in which Sacks emphasized

[154] Renée Robinson I-1980.

[155] As Robinson wrote, *hat Sie mich tief beeindruckt.* See Abraham Robinson to Hellmut Becker, July 2, 1972, RP/YUL.

[156] Robinson to Becker, July 2, 1972, RP/YUL.

[157] When Luxemburg went to New Haven to see Robinson for the last time at Yale, "he looked like a ghost, so thin, easily tired. 'I always thought I would die from a heart attack like my father and brother,' he said," Luxemburg I-1991. But Luxemburg was also impressed by the extent to which Robinson remained attentive, quite serious about professional matters, especially where logic was concerned: "Despite the advanced state of the disease and the narcotics, he also talked to me about mathematics. He told me that he was happy that I would take care of the book with Lightstone. Although he would have loved to do some more work on it, he felt that the manuscript was ready for publication and that North-Holland's Library Series was the appropriate place for it. He also mentioned that he had some new ideas about his work with Roquette, but since he tired quickly, he could not elaborate on this. He was also aware that we had made an appointment at the junior level in logic [at Caltech], and congratulated me, saying 'You found an excellent person.' I mention this to show that even then he was still very busy in his mind with his work during the few periods that he was awake." W.A.J. Luxemburg to JWD, June 15, 1994.

[158] Abraham Robinson to Gert Müller, May 24, 1972, RP/YUL.

the pioneering contributions of Alfred Tarski, Robinson's mood changed entirely. "He grabbed me that night, and was clearly displeased. . . . He sat next to me at dinner, and was so upset that this in turn upset me, and I couldn't eat anything."[159]

Looking back, although Sacks does not blame him, Robinson could not have read more than the introduction. "Had he gone on to look at the book itself in greater detail, he would have discovered that his own work was mentioned over and over again, while Tarski was only mentioned twice. . . . In fact, Robinson succeeded in developing a much better treatment of Model Theory than Tarski; above all, his theory of model completeness is more important than Tarski's elimination of quantifiers. Nevertheless, Tarski did, after all, start the field in its modern form."[160]

Robinson, however, was concerned about the record of his own accomplishments—and contributions—to mathematical logic. In a letter to Sacks the following June, Robinson focused on what he took to be Tarski's self-conscious, "Whiggish" history, attributing to himself (or his school) virtually all of the great achievements of modern logic. As Robinson complained:

> Thank you for your letter. First of all, to quote the Aramaic version of a well-known proverb, I do not wish to do unto others what is objectionable to me. Although I know that my earlier work had some influence on Simon Kochen, I still admire him and Ax for their achievements. Nor do I dislike Tarski. His nervous, nagging temperament has been a major factor in the development of logic. I also believe he has tried manfully, though unsuccessfully, to overcome his own jealous and ungenerous nature. Tarski certainly was the man who put the truth concept on the map, and this was an important phase in the pre-history of model theory. But so was the work of Skolem, Gödel and—really no longer *pre*-history—of Mal'cev. The term was indeed coined by Tarski in the early fifties and this is where Mostowski in his "Thirty years of Foundational Studies" (according to which I apparently started my career in 1963) places the beginning of the subject.
>
> However, if you were to look at my "On the Metamathematics of Algebra" you will find that it contains not only algebraic applications but also the general framework of model theory (e.g. the general scheme of classes of sentences *versus* classes of models). At the same time I do not wish to belittle Henkin's influence on later developments.
>
> In any case, I am not surprised to observe, again and again, that Tarski has trained his students (and that includes Mostowski) to see history in

159 Gerald Sacks in an E-mail note to JWD, September 28, 1994.
160 Sacks I-1993.

> the way he wants them to. But it does upset me that *your* sense of fairness has not prevented you from perpetuating this myth. If you really want to know my feelings, they are that in this case a private letter or a complimentary dedication are not enough. A misleading statement of this kind should be balanced by another statement that will also form part of the literature, not necessarily today or in a month's time but not, I hope, only as part of my obituary either.[161]

Another NSA Symposium and Two More Ph.D.s: Spring 1972

The Victoria symposium on nonstandard analysis, held at the University of Victoria, Canada, from May 8–11, 1972, was organized "in the same spirit" as the first Caltech Symposium held in Pasadena in 1967 and the Oberwolfach meeting in 1970. The Victoria Symposium aimed "to have everyone working in nonstandard analysis in North America attend the conference."[162] Robinson was pleased to return to Canada. Although the paper he gave reflected his most recent work with Don Brown on "Nonstandard Exchange Economies," he submitted a different paper for the published proceedings (a brief account of "Enlarged Sheaves").[163]

Robinson had written on sheaves as early as 1969.[164] His work on "enlarged" sheaves was inspired by various standard results (on complex functions) considered in a nonstandard setting.[165] Enlarged sheaves involved holomorphic functions of several complex variables and the critical role of monads: "as might be expected by now, the question of the coherence of a given sheaf is determined entirely by the behavior of its monadic sections." Indeed, in a nonstandard way, Robinson showed that every locally finitely generated subsheaf of the sheaf of germs of holomorphic functions on a given domain is coherent (Oka's theorem).[166]

Doubtless the most pleasant responsibility Robinson fulfilled that spring was graduating two more of his Ph.D. students: Joram Hirschfeld, who wrote his dissertation on "Existentially Complete and Generic Structures in Arithmetic," and William H. Wheeler, who also worked on aspects of generic structures, specifically "Generic and Existentially Complete Algebraic Structures." Sadly, Hirschfeld and Wheeler would

[161] Abraham Robinson to Gerald [Sacks], June 8, 1972, RP/YUL.

[162] And this goal was, indeed, "almost completely realized." See Hurd and Loeb 1974, p. iii.

[163] Robinson 1972a.

[164] Robinson 1969b; in Luxemburg 1969, pp. 138–149; Robinson 1979, 2, pp. 220–231.

[165] Among the standard results Robinson established was the "closure of modules" theorem. See the theorem as proved in section D, chapter 2, of Gunning and Rossi 1965.

[166] Robinson 1972a, p. 259; Robinson 1979, 2, p. 343.

Participants in the SLALM II, Brasília.

be Robinson's last graduate students to finish under his direct supervision. Ethan E. Kra, although he began his research with Robinson on "Infinitary Forcing for Languages with the *Q*-Quantifier: Feferman-Vaught Preservation Theorems for Infinitary Forcing," did not finish until 1974, after Robinson's death in April. The following year, Larry M. Manevitz finished his dissertation on "Model Theoretic Forcing and Absoluteness," as did Peter M. Winkler, who wrote on "Assignment of Skolem Functions for Model-Complete Theories."

SLALM II: July 1972

After considering various locales, the Organizing Committee for the Second Latin American Logic Seminar, to be held in July of 1972, settled on the University of Brasília. Robinson's last major undertaking, in fact, as outgoing president of the ASL, had been a letter to the Rector of the University of Brasília, offering the support (and cosponsorship) of the

Association for Symbolic Logic. SLALM II emphasized courses of about ten lectures each. One of these was given by Robinson himself on nonstandard analysis and forcing in model theory.

Thanks to the success of the first two symposia, logic went on to establish a firm hold in Latin America.[167] Robinson was particularly supportive, not only because he was devoted to the idea of mathematical logic, but because he always sought to "spread the word" wherever and whenever he could. There is no doubt, however, that among his most receptive audiences were the mathematicians and logicians of Latin America, especially those who participated in the first Latin American symposia on logic in the early 1970s. After Robinson's death in 1974, they went on to establish symbolic logic there as a major professional subdiscipline among mathematicians and philosophers in various centers throughout the region.[168]

What Robinson sought to do for logic in Latin America was to educate and inspire a first generation that could then, on its own, begin to further the development of mathematical logic locally. Subsequently, with more than two decades of hindsight and experience since SLALM I first met in 1970, Robinson's best hopes for logic in Latin America have been realized with notable success.

Late Summer Lectures: 1972

At the end of August Robinson was off again, this time to France for a meeting at the University of Orléans where he presented a paper on "Generic Categories" at the Congrès 1972 de logique d'Orléans. In addition to his invited hour lecture in the section on model theory, Robinson chaired a special panel discussion (with Alistair Lachlan, Angus Macintyre, and Gerald Sacks) on unsolved problems and possible future directions of research in model theory.[169]

Robinson's lecture on "Generic Categories" explored the connections between mathematical logic and category theory. Primarily, he was interested in the significance of basic ideas in model theory for abelian categories, and his own recent work in forcing in model theory provided the context. Given a system of first-order axioms for abelian cat-

[167] When the third symposium was held in São Paulo in 1976, it consisted of a substantial number of contributions from South America, fifty-eight in all, including participants from Brazil, Chile, Argentina, Peru, and Colombia. Moreover, the number and quality of papers presented were deemed of sufficient importance that they were published in 1977 as a volume of the Studies in Logic and the Foundations of Mathematics series. See Arruda 1977.

[168] Arruda 1977, p. xviii.

[169] See the report of the meeting in the *Journal of Symbolic Logic*, 39 (1974), pp. 371–389.

egories, with the definitions designed to insure that the theory would be inductive (including axioms for monomorphism, epimorphism and isomorphism, kernels and cokernels, and exact category), the paper led up to a final section in which he considered generic abelian categories and posed a number of open questions for future consideration.[170]

A New Generation for Logic at Yale

Robinson invited Gabriel Sabbagh to spend the fall of 1972 at Yale on a one-year visiting professorship. Although Robinson was planning to be away for part of the year (at the Institute for Advanced Study in the Spring), he played this down by noting that Jon Barwise, "an expert in the thriving subject of infinitary logic," would also be in residence.[171] Attempts to get Jerome Keisler to leave the University of Wisconsin on a permanent basis unfortunately failed, although he wrote back to say that he was nevertheless grateful for Robinson's interest in bringing him to Yale.[172]

Robinson did succeed, however, in persuading the young Angus Macintyre to accept a three-to-five year tenure-track position that year as associate professor.[173] Despite the short time they were to work together, Yale was soon to become, for a brief moment, one of the most intellectually stimulating centers for mathematical logic in the world.

Robinson's Retiring ASL Presidential Address

The 1972–1973 Annual Meeting of the Association for Symbolic Logic was held at the Fairmont Hotel in Dallas, Texas, on January 25–26, 1973 (in conjunction with the American Mathematical Society's annual meeting). Robinson had been asked to give an hour talk at the AMS meeting, where he decided to talk about "Sheaves."[174] The major reason he was

[170] Robinson 1979a, in Robinson 1979, 1, pp. 298–306. This paper, in a twenty-page typescript version, was found among Robinson's papers at the time of his death, and was previously unpublished.

[171] Abraham Robinson to Gabriel Sabbagh, April 25, 1972, RP/YUL.

[172] Jerome Keisler to Abraham Robinson, May 23, 1972, RP/YUL. Keisler I-1993.

[173] Abraham Robinson to Angus Macintyre, who was then at Aberdeen, June 16, 1972, RP/YUL.

[174] Later, Robinson told Dana Scott that the title was carefully chosen—pun intended! As he explained, the idea "grew naturally out of my paper on Germs in the Luxemburg volume (pun). I would classify it as nothing more than a non-trivial exercise. However, it does point up the place of 'internal' (as opposed to 'standard') functions in nonstandard SCV [several complex variables]," Abraham Robinson to Dana Scott, August 23, 1973, RP/YUL. Although prior to the meeting, H. F. Weinberger wrote to Robinson on behalf of the editors of the AMS *Bulletin*, asking to publish Robinson's lecture on "Sheaves," it

in Dallas, however, was not to present his AMS paper, but rather to deliver his retiring address as past-president of the ASL.

Robinson had been working carefully on the lecture for some time, and a preprint had already circulated the department at Yale during the fall of 1972. This was an important occasion for Robinson, and he took it very seriously. His aim, clearly, was to imitate Hilbert's famous "23 Problems" lecture at the 1900 International Congress of Mathematicians in Paris. Robinson's address as retiring ASL president was similarly intended to highlight a wide range of significant, open "Metamathematical Problems."[175]

As usual, Robinson wanted to offer something new, and this time, using his experience in model theory, he took it upon himself to pose twelve open problems or areas which were not only of intrinsic interest, but whose solution he believed would require "weapons whose introduction would close definite gaps in our armory." Although some of the questions Robinson raised were well known, even classical, others he was pleased to say, were "put on the map" for the first time in his Dallas lecture.

The problems Robinson outlined ranged from broad questions–like providing a metamathematical framework for Baire's category theory with obvious parallels to the compactness theorems in topology in mind–to very specific questions, like the decidability of the theory of real numbers constructible by straightedge and compass alone. Other topics concerned differential algebra, fields of power series, Hilbert's thirteenth problem, complex function theory, and the comparison of number fields with algebraic function fields. It all reflected Robinson's impressive command of a wide range of mathematical interests.

Robinson also offered insights of a philosophical nature about what he took to be the essence of mathematical logic. What did logicians face when using their own "characteristic tools," he asked? What was the point of formalized languages, explicit relations between symbols and objects, rigidly expressed and controlled rules of deduction? The aim was to understand structures, methods, theories, and theorems of mathematics in general, as best one could. This was Robinson the model theorist:

> We may then expect him [the logician] to adopt the attitude of the physicist or psychologist who (whatever his professed philosophy) feels that he deals with phenomena of the external world, whose rules cannot be im-

---

was never printed. See H. F. Weinberger to Abraham Robinson, December 5, 1972, RP/YUL.

[175] Robinson 1973. Robinson was pleased to thank Simon Kochen and Kurt Gödel for "valuable conversations" on some of the topics he discussed in his paper.

> posed by him arbitrarily. He, or those that come after him, may indeed use the understanding thus gained in order to modify these phenomena, but as a scientist he would not regard this possibility as his only justification.[176]

Robinson ended with some comments on foundations of mathematics, an area "fraught with difficulties," he said, if one expected to frame an absolutely meaningful problem. Here Robinson had his eye on just such an absolute question: "Is there," he asked, "a well-formed first-order assertion about the natural numbers which can be neither proved nor refuted by a formal or by a generally acceptable informal argument?"[177]

Answers to such questions, Robinson knew, could divide mathematicians into opposing camps depending upon the sort of assumptions they were willing to make:

> One cannot predict for certain that some argument in arithmetic will not one day split the classical mathematicians down the middle, perhaps precisely because it leads to the kind of result adumbrated in my question. . . .
>
> While others are still trying to buttress the shaky edifice of set theory, the cracks that have opened up in it have strengthened my disbelief in the reality, categoricity or objectivity, not only of set theory but also of all other infinite mathematical structures, including arithmetic. I am thus taking sides in an ancient controversy that has appeared and reappeared in different forms over thousands of years. In our time no less a man than Paul J. Cohen has indicated his agreement with my point of view. However, it is equally true that the array ranged against it is formidable. Among the platonic realists, it includes the greatest of all mathematical logicians, Gödel, as well as Bernays, whom we revere as the Nestor of our subject, and the dynamic and influential Professor Kreisel. Not long ago, Bernays, in an article published in *Dialectica*, criticized my attitude in his usual gentle manner, while Kreisel had stated his disagreement with me previously, also in his usual manner. From another direction comes the voice of the constructivists who deplore the fact that the formalists, whose philosophical attitude towards the actual infinite in mathematics is akin to their own, do not join with them also in their mathematical practice. However, I must confess that neither invective nor moral indignation have (sic) induced me to change my opinion in this matter.[178]

Despite disagreements over foundations, Robinson believed that "all mathematicians have a strong feeling for the beauty and fascination of

[176] Robinson 1973, p. 500; in Robinson 1979, 1, p. 43.
[177] Robinson 1973, p. 514; Robinson 1979, 1, p. 57.
[178] Robinson 1973, pp. 514–515; Robinson 1979, 1, p. 57–58.

Robinson in front of Fuld Hall, Institute for Advanced Study.

their subject," and it was this that he had hoped to convey, he said, in his retiring presidential lecture.

Institute for Advanced Study: Spring 1973

When Kurt Gödel invited Robinson to the Institute for Advanced Study in 1971, he had to decline. A year later, however, when Gödel asked again, Robinson was pleased to accept and agreed to come as a visiting member that spring.[179]

As it turned out, this proved to be an especially busy term for Robinson, and it was fortunate that he was otherwise free of his usual teaching and administrative responsibilities at Yale. Invitations to give visiting lectures, some of them very distinguished, now came regularly. One from David Ray, chairman of mathematics at Bucknell University, asked Robinson to visit as distinguished professor for one to three weeks.[180] In early March, he also went to North Carolina to present a paper in the Mathematics Colloquium at Duke.[181]

Meanwhile, there were many opportunities for Robinson and Gödel to interact at the institute. Gödel had been urging Robinson to spend as much time in Princeton as he liked, and had come increasingly to admire Robinson's work. Gödel was especially optimistic about applications of nonstandard analysis in both analysis and number theory, and wanted to know specifically what the technical advantages of nonstandard analysis might be. Robinson replied at length:

> There are indeed many "standard" theorems whose proof can be—and has been simplified by the use of NA. This includes elementary theorems like Bolzano-Weierstrass, Heine-Borel and other basic facts of the differential

[179] Abraham Robinson to Kurt Gödel, January 29, 1971, Gödel papers no. 011953, PUA/FL.

[180] David Ray to Abraham Robinson, March 1, 1973, RP/YUL.

[181] Jack Lees to Abraham Robinson, January 11, 1973, RP/YUL.

> and integral calculus. But it also includes Riemann's mapping theorem and Picard's theorems (as done by myself), Parseval's and Herglotz's theorems in Harmonic Analysis (Luxemburg), the basic theorem on almost periodic functions (Kugler), measure theory (Bernstein-Wattenberg and Loeb). Moreover, sometimes the greater flexibility of Nonstandard Analysis compared with the available "classical methods" has led to new results in several areas. I mentioned one of them, concerning an invariant subspace problem of Smith and Halmos, in my survey. Others are due to Luxemburg, Behrens and Stroyan (algebra of bounded analytic functions), Hurd (dynamical systems) Larson (algebraic ring theory) Marcus and Juhasz (topology of hyperspaces).
>
> In many areas Nonstandard Analysis adds a special intuition which leads naturally to concepts and constructs that, by other methods, have been or would have to be introduced *ad hoc*. Thus, profinite groups (used, for example, in algebraic number theory) are homomorphic images of "starfinite" groups (groups whose order is an infinite natural number) and this determines many of their properties.
>
> However, as you know, Nonstandard Analysis is conservative so that ultimately its use is optional. As it is, the fact that a rigorous method of infinitesimals, though a latecomer to the mathematical scene, provides a viable alternative to established methods is at least of some interest from a historical or philosophical or methodological point of view.[182]

Gödel was asking Robinson for details about his work for very specific reasons. He had long been impressed with Robinson and was now building a case for bringing him to the institute for an extended period of time, with the aim in mind that Robinson would eventually be Gödel's successor. Gödel had already asked Robinson to come to the Institute for a five-year period. This might mean having to resign his position at Yale, but Robinson had agreed to give Gödel's proposal serious consideration.[183]

In one very fundamental respect Gödel and Robinson, despite their differences on foundations, were kindred spirits in agreeing upon the importance—and efficacy—of using the methods and results of mathematical logic for the rest of mathematics. In this, Gödel recognized in Robinson a peer of exceptional ability. Logic was not merely a peripheral subject for the esoteric—those with obsessive interests in foundations or logically detailed, rigorous proofs. Logic, for both Gödel and

[182] Robinson to Gödel, January 3, 1972, Gödel papers no. 011960, PUA/FL; copy in RP/YUL.

[183] Robinson to Gödel, April 14, 1971, Gödel papers no. 011957, PUA/FL; copy in RP/YUL.

Robinson, embodied a creative force that could invigorate all of mathematics. As Gödel seems firmly to have believed:

> In my opinion Nonstandard Analysis (perhaps in some non-conservative version) will become increasingly important in the future development of Analysis *and* Number Theory. The same seems likely to me, with regard to all of mathematics, for the idea of constructing "complete models" in various senses, depending on the nature of the problem under discussion.[184]

Gödel had even asked Robinson to list what he regarded as the major recent landmarks in mathematical logic. According to Robinson:

> As the outstanding individual achievements in logic since 1963 I would list (in chronological order):
>
> 1) The work of Ax-Kochen on *p*-adic fields.
> 2) Solovay on measurable sets, and Tenenbaum-Solovay on Souslin's conjecture.
> 3) Matiyasevich's solution of Hilbert's 10th problem.
>
> However, perhaps this list does not give a true picture of the development of logic in the past ten years. There has been substantial progress on broad fronts elsewhere in set theory (e.g., the theory of constructible sets, Boolean valued logic), in model theory (infinitary model theory, model theoretic forcing and generic structures, nonstandard analysis), recursion theory (generalized recursion theory, e.g., in ordinals, the structure of the system of r.e. degrees), decision procedures and finally, proof theory. (I must admit that I do not appreciate exactly what went on in the last-mentioned area).[185]

Mostowski's Sixtieth Birthday: Topological Model Theory

One of the papers Robinson wrote at the Institute for Advanced Study that spring was a note dedicated to Andrzej Mostowski for his sixtieth birthday. Of the "Metamathematical Problems" Robinson had posed in his ASL presidential address concerned the "emerging field" of topo-

[184] Gödel to Robinson, December 29, 1972, Gödel papers no. 011962 (draft), PUA/FL.

[185] Robinson to Gödel, April 6, 1973, Gödel papers no. 011966, PUA/FL; copy in RP/YUL. It should be noted that Matiyasevich's theorem (3 above) is also known as the MRDP theorem for the work on Hilbert's tenth problem done earlier by Julia Robinson, Martin Davis, and Hiliary Putnam. While Matiyasevich solved the problem by showing how to construct an equation whose solutions grow in a certain manner, Robinson (Julia), Davis, and Putnam had shown that it was just such an equation that was needed in order to establish the main result. For details, see Davis, Matiyasevich, and J. Robinson 1976.

Presentation of the Brouwer Medal.

Leiden, April 26, 1973.

logical model theory, he now began to make good on some of the questions he had been considering on this subject for some time.[186] Robinson also completed another commemorative article that appeared in 1973, this one in memory of his Russian colleague, A. I. Mal'cev, "On Bounds in the Theory of Polynomial Ideals."[187]

The Brouwer Medal: 1973

The highlight of Robinson's career as a mathematician came unexpectedly, when he was named the second recipient of the L.E.J. Brouwer Medal in 1973. This was awarded triennially, by the Dutch Mathematical Society, and in addition to the medal (which required delivery of a "Brouwer Memorial Lecture"), included a certificate and payment of all expenses. In return, he was also expected to spend a week in Holland.

Robinson was following in excellent but not unrelated footsteps. The

[186] Robinson 1973.   [187] Robinson 1973e.

first Brouwer medalist had been René Thom in 1970, who spoke on topology. It was already decided that the 1973 award was to be given for "foundations," and Robinson was delighted to accept.[188] After the medallion was presented in Leiden (on April 26, 1973), Robinson gave his Brouwer lecture and afterwards made a point of asking Renée what she thought. "It couldn't have been better," she assured him.[189]

Robinson devoted his lecture to "Standard and Nonstandard Number Systems."[190] Nonstandard analysis had made its debut in the *Proceedings of the Royal Dutch Academy of Sciences*, and he now offered a grand overview of the subject in a lecture that made clear why he had won the Brouwer Medal. It was, in its way, a *tour-de-force*, and a tour indeed of the major highlights of his best-known work among mathematicians at large.

Robinson specifically intended his lecture "to honor BROUWER as a philosopher of mathematics and as the founder of intuitionism." Furthermore:

> BROUWER's intuitionism is closely related to his conception of mathematics as a dynamic activity of the human intellect rather than the discovery of an immutable abstract universe. This is a conception for which I have some sympathy and which, I believe, is acceptable to many mathematicians who are not intuitionists. It shall be the underlying theme of this lecture that the dynamic evolution of mathematics is an ongoing process not only at the summit, where intricate new results are piling up with great rapidity but also at the more basic level of our number systems, which may seem as eternal to a new generation as yesterday's technological innovations are in the eyes of a child of today. I shall carry on my discussion within the framework of classical mathematics although I may recall that the nature of our numbers was also one of BROUWER's main preoccupations.[191]

Immediately Robinson launched into one of his favorite topics—how the complex numbers "after several hundred years of marginal existence," finally won respectability in the nineteenth century. Other systems were also introduced and debated in the decades since, and as Robinson noted, "It is, inevitably, a matter of convention, which algebraic or arithmetical structures are to be honored by the name of *number systems*." As Robinson agreed:

[188] A. C. Zaanen, President of the Dutch Mathematical Society, to AR, March 13, 1972. As Robinson told Wim Luxemburg, "I have been asked ... to be the next Brouwer lecturer. There are several reasons—you are one of them—why I appreciate this honor particularly." Robinson to Luxemburg, April 18, 1972, AR/YUL.

[189] Renée Robinson I-1980.

[190] Robinson 1973b.

[191] Robinson 1973b, p. 115; in Robinson 1979, 2, p. 426.

Generally speaking, the more naturally a new system seems to grow out of a *number system* of acknowledged standing, and the more similar to such a system its features, the more likely it is that the new entities also will be regarded as numbers.[192]

Before demonstrating the power of nonstandard analysis, Robinson first sketched out the basic elements of the theory, then discussed the by-now familiar example of how a nonstandard theory of Hilbert space first led to the standard result of the existence of invariant subspaces for linear operators whose squares are compact. He then presented the Ax-Kochen result for infinite primes, and gave as a "last example in the nonstandard way of thought"–his newest research on the theory of rational points on curves.

In closing his Brouwer memorial lecture, Robinson offered some sobering thoughts:

[Nonstandard analysis] does not present us with a single number system which extends the real numbers, but with many related systems. Indeed, there seems to be no natural way to give preference to just one among them. This contrasts with the classical approach to the real numbers, which are supposed to constitute a unique or, more precisely, categorical totality. However, as I have stated elsewhere, I belong to those who consider that it is in the realm of possibility that at some stage even the established number systems will, perhaps under the influence of developments in set theory, *bifurcate* so that, for example, future generations will be faced with several coequal systems of *real numbers* in place of just one.

This, in turn, led Robinson to reflect on the character and novelty of his best-known creation, nonstandard analysis:

Nonstandard analysis is now over twelve years old. While the question of its use is and may remain a matter that is up to the choice of the individual mathematician one can, by now, hardly doubt its effectiveness. I shall now describe yet another class of systems, which have emerged only within the last three years. While these systems have not yet been applied to classical mathematics, I shall endeavor to convince you, by the natural way in which they present themselves, that they also deserve our attention.[193]

The new class of systems was the class of generic structures G, with many properties in common with the ring of integers satisfying, above all, the axiom of induction (at least to a specified extent). Moreover, the

[192] Robinson 1973b, p. 116; in Robinson 1979, 2, p. 427.

[193] Robinson 1973b, p. 130; in Robinson 1979, 2, p. 441. According to Carol Wood, Robinson "regarded time as a highly valuable ingredient in testing the accuracy and the importance of a new result." Carol Wood to JWD, E-mail note of June 24, 1994.

Robinson enjoying a glass of champagne with Arend Heyting.

elements of G also shared some of the most desirable properties of algebraically closed fields in the class of fields. "Thus, we may regard the structures of G and of E [the class of all existentially complete structures in Σ] as new number systems which are closely related to arithmetic and whose further investigation would be of obvious interest."[194]

Leiden—and the Rest of the Netherlands

Following the Brouwer lecture there was a lively reception, and Robinson mingled warmly with colleagues and friends.

That evening, there was a festive gala dinner in Robinson's honor at the Vergulden Turk, with appropriate wines and a series of toasts. The menu reflected Robinson's exotic tastes as a world traveler:

> Cocktail Florida; Toast et Beurre
> Consommé Queue de Kangourou en tasse
> Asperges en Branche
> Steak de Veau à la Turc doré
> Pêche Melba – Moka
> Vins: Bernkastler Riesling 1970
> Château Vernon 1961[195]

From Leiden Abby and Renée returned to Amsterdam, but visited Utrecht where Dirk van Dalen was their resident host. At one of the local museums, the Kröller-Müller Museum in the impressive Hoge Veluwa National Park in Otterloo, he was especially impressed by Robinson's knowledge of Dutch painting.[196] Robinson also gave several additional lectures while he was in Holland, at the University of Utrecht, where he spoke on "Logic as a Framework for Algebraic Theories," and

[194] Robinson 1973b, pp. 131–132; in Robinson 1979, 2, pp. 442–443.
[195] April 26, 1973, RP/YUL.
[196] van Dalen I-1986.

Robinson with the van Dalens, visiting the Kröller-Müller Museum, Otterloo.

(a few days later, on May 4) at the University of Amsterdam, "Some Remarks on Topological Model Theory."

Following the excitement of the Brouwer Medal and Robinson's lectures in Holland, Abby and Renée were looking forward to spending a few weeks in Israel. On the way they enjoyed some mountain hiking in Switzerland, but Abby, as always, was happiest when they were back in Jerusalem. After all of his travels, this was the one city to which he was continuously drawn.

A year earlier, Samuel Karlin had asked Robinson to give a talk at the Weizmann Institute, and at last Robinson was able to oblige.[197] Although he and Renée were only in Israel for two weeks (May 15–June 2), this gave Abby sufficient time to renew old acquaintances and to ease back into the familiar surroundings of home-away-from-home—the Department of Mathematics at the Hebrew University.[198]

Upon his return to Yale in early June, Robinson had a letter from G. W. Pierson in history, congratulating him on the Brouwer Medal and asking, "What public largesse does a medalist have to dispense?"[199] Robinson offered a beer: "a Brouwer medalist owes his friends a glass of beer (Brouwer = Brewer, I believe]. I shall be happy to oblige."[200]

Robinson also made a point of writing to Heyting, confessing that he was extraordinarily moved by the award:

> I shall cherish it as a token of remembrance of a great moment in my life. The honor of having received the Brouwer medal is even dearer to me

[197] S. Karlin to Abraham Robinson, March 15, 1972, RP/YUL.
[198] Abraham Robinson to Hellmut Becker (in German), April 22, 1973, RP/YUL.
[199] G. W. Pierson to Abraham Robinson, May 31, 1973, RP/YUL.
[200] Robinson to Pierson, June 21, 1973, RP/YUL.

Robinson, on a wall of the Old City with the Dome of the Rock in the distance, Jerusalem.

Robinson at Masada with Joram Hirshfeld, summer 1973.

because it was conferred upon me by the people of the Netherlands who have on so many occasions, and again recently, shown their friendship for my people, even at high cost to themselves.[201]

Heidelberg and Bristol: Summer 1973

Gert Müller had invited Robinson back to Heidelberg, and after discussion of various possibilities, it was agreed that he would come for about a week, July 8–14. Although Müller was planning to be in Japan, he looked forward to seeing Robinson at a meeting they were both to attend in Bristol. Müller also reassured Robinson that as for Heidelberg, "The entire group of model theoreticians in my seminar and from Roquette's circle will be in Heidelberg for your visit."[202]

From Heidelberg Robinson flew immediately to Bristol, where a European meeting of the Association for Symbolic Logic met from July

[201] Abraham Robinson to Arend Heyting, November 27, 1963, RP/YUL.
[202] G. Müller to Abraham Robinson, February 22, 1973, RP/YUL.

15–21. Emphasis was given to philosophy of mathematics, metamathematics of algebra, proof theory, category theory and theory of computation.[203] Robinson indeed saw Gert Müller, who arrived from Japan, and he was pleased to see other old friends as well, including Stephan Körner, Bill Boone, Paul Bernays, Michael Dummett, Saunders Mac Lane, Dana Scott, and Calvin Elgot, among others.[204]

Among the fourteen invited papers, Robinson's was unique in its concern with developments in philosophy of mathematics. Basically, he was interested in two very straightforward questions: "What is the kind of progress that has been achieved, over the years, in the philosophy of mathematics and what further progress may we expect of it in the future?"[205]

Characterizing recent technical developments in mathematical logic and the foundations of mathematics as "spectacular," Robinson was struck by the extent to which "the evolution of our understanding of the essential nature of mathematics has been hesitant and ambiguous." While the results of one school of thought might be rejected by others, he confessed that the one philosophical problem "that has the greatest fascination for me—and I maintain that it is a genuine problem—is that of the existence, or reality, or intelligibility, or objectivity, of infinite totalities."[206]

Robinson was convinced, based upon the most recent developments in mathematics, that Platonism was not the answer: "The incompleteness of Peano arithmetic and the undecidability of the continuum conjecture have not led to a general abandonment of Platonism although it would seem to some, including myself, that they provide evidence against it."[207] For Robinson, the point at issue was the reality of infinite totalities. Remarkably, and perhaps something of a surprise, Robinson thought part of the answer might well depend upon neurophysiology:

> the possibility of analyzing in detail the neurophysiological processes in the brain which correspond to its mathematical activity. . . . In particular, proof theory, in the wider sense of a general theory of (mathematical) proofs, and the problems of pure and applied mathematics, are two of the areas that are likely to be affected.[208]

[203] Abstracts were printed in the *Journal of Symbolic Logic*, 39 (1974), pp. 406–432. Later, the proceedings offered "expanded and developed versions of the corresponding lectures given at the conference." See Rose and Shepherdson 1973.

[204] It was the last time he saw many of these old friends, including Müller, who recalls that they went to a good restaurant, had a substantial steak—they were both hungry—and a brandy afterwards. "He seemed full of health, body and soul," Gert Müller I-1992.

[205] Robinson 1973a.

[206] Robinson 1973a, p. 42; Robinson 1979, 2, p. 556.

[207] Robinson 1973a, p. 45; Robinson 1979, 2, p. 560.

[208] Robinson 1973a, pp. 48–49; Robinson 1979, 2, pp. 563–564. On the question of

But this was far in the future. Back to reality, Robinson still had to draw some conclusions about the "essential nature of mathematics." Believing that purely technical advances were unlikely to resolve the differences between one school of thought and another, Robinson tried instead to make his position clear by imagining a conversation between a Platonist (or "Fregean objectivist") *P*, and a formalist *R* (for Robinson himself?):

> *P* believes that the universe of sets and the universe of natural numbers are in some sense real or objective and hence, that their structure and properties are unique. *R* does not believe in the reality or objectivity of these infinite universes. He thinks that it is not absolutely meaningful to talk of the structure of these universes and expects that sooner or later the mathematicians will accept several co-equal theories of sets and even of arithmetic. They agree to test their views against the well-known incompleteness results of arithmetic and set theory.
>
> *R* points out that Gödel's incompleteness theorem applies not only to first order Peano arithmetic but also (indeed, in the original version) to the higher order theory of *Principia Mathematica*. *P* replies that the Gödel sentence (or any similar sentence) though unprovable and irrefutable within the given formal framework, can nevertheless be shown to be true by informal but rigorous arguments (or by a generally acceptable widening of the formal framework).
>
> *R* concedes that this is correct and suggests that they move on to set theory. *P* is agreeable and *R* brings up the continuum hypothesis, pointing out that it is neither provable nor refutable in accepted axiomatic set theory and that, in this case, the question cannot be settled either by an intuitive argument.
>
> *P* retorts that he is aware of this fact but that the discovery of a generally acceptable axiom of intuitive argument may be expected, sooner or later, to decide this problem also. *R* rejoins that he cannot exclude this but insists that if *P* wishes to enter the realm of hypothetical developments, he should also include the possibility that, sooner or later, someone will discover an absolutely undecidable sentence of elementary arithmetic.
>
> Both realize that the conversation has arrived at an impasse and they consider the possibility of meeting again one thousand years hence, in order to continue the debate in the light of developments that will occur in the meantime.[209]

---

infinite totalities, Robinson admitted to accepting largely Körner's views on this subject. See Körner 1960.

[209] Robinson 1973a, pp. 49–50; Robinson 1979, 2, pp. 564565.

One cannot help but wonder if Robinson expected much progress to have been made by then. But he was convinced of a certain "asymmetry" in the positions adopted by *P* and *R*. If *P* were ever persuaded that there were absolutely undecidable sentences in either set theory or arithmetic, he would either have to give up Platonism or accept the existence of "unknowable truths." On the other hand, even if the formalist were persuaded that any possible statement of set theory or arithmetic were ultimately decidable, *R* could still maintain his basic position. "Indeed," Robinson argued, "even now the infinite structures whose truth sets are recursive are no more real to him than all the others."[210]

Robinson concluded that future developments might give greater prominence to a "dialectical" approach to the ideas of logical positivism or even nominalism. But as a formalist, he ended on a strong note that clearly betrayed his interest in balancing pure versus applied mathematics:

> I expect that future work on formalism may well include general epistemological and even ontological considerations. Indeed, I think that there is a real need, in formalism and elsewhere, to link our understanding of mathematics with our understanding of the physical world. The notions of objectivity, existence, infinity, are all relevant to the latter as they are to the former (although this again may be contested by a logical positivist) and a discussion of these notions in a purely mathematical context is, for that reason, incomplete.[211]

Back at the Institute for Advanced Study

Robinson finished his summer that year with what was to be his last, all too brief visit with Gödel at the Institute for Advanced Study (August 15–18). Gödel had written to Robinson in July:

> I am looking forward to your visit in August and hope very much that you will stay for more than a few days. I am sure payment of your expenses and housing in the Institute can be provided for any period of time you wish to spend here. . . . I hope you had a pleasant and interesting trip abroad and that I shall hear from you about your experiences. Our discussions last spring interested me very much and I am looking forward to seeing you again soon.[212]

Robinson wrote back directly, and on July 6 offered Gödel a sketch of

[210] Robinson 1973a, p. 50; Robinson 1979, 2, p. 565.
[211] Robinson 1973a, p. 51; Robinson 1979, 2, p. 566.
[212] K. Gödel to Abraham Robinson, July 2, 1973, RP/YUL.

the Bristol meeting, along with a copy of his paper.[213] Philosophically, he knew Gödel would disagree. After they had discussed the matter at greater length in August, Robinson found himself trying to explain his position on the subject of their philosophical differences:

> I am distressed to think that you consider my emphasis on the model theoretic aspects of Nonstandard Analysis wrongheaded. As a "good formalist" I am quite willing to write a book on the assumption that the reals are unique, and in fact have done so. However, even on this basis the present evidence for the uniqueness of a nonstandard ω-ultrapower of the reals is not strong. In addition it would not make any difference to my arguments, although I realize its foundational significance.
>
> Concerning the Bristol paper . . . as far as the main thrust of my paper is concerned, with which you are bound to disagree, I can only quote Martin Luther: "Hier stehe ich, ich kann nicht anders."[214]

Yale: Fall 1973

The activism of the sixties had finally given way to what Yale President Kingman Brewster referred to as the "grim professionalism" of the seventies.[215] The difference in student attitudes was dramatic:

> The Yale of the early seventies was an academic pressure cooker. . . . The first campus-wide controversy I remember, one month into freshman year [September 1970] wasn't about Vietnam but rather a plan to cut back on the hours of Sterling Library. There followed such an uproar that it was withdrawn. May Day and the Panther Trial and the strike and the days of Abbie Hoffman were over (thank God), and it was once again important to get into law school, med school or Harvard Biz. A full-page ad taken out by Kodak in the *Yale Daily* read: *Maybe the way to change the world is to join a large corporation.*[216]

Students, once again, were serious about their education. According to Christopher Buckley, "The bursars' office announced that tuition, room and board for '73–'74 would probably go to the unheard of $5,000 a year; it was calculated that every class cut or slept through cost nine-

213 Robinson to Gödel, July 6, 1973, RP/YUL.

214 "Here I stand–I can not do otherwise." Luther quoted by Robinson to Gödel, August 23, 1973, RP/YUL.

215 Michiko Kakutani felt this as well. As she notes, "A visiting reporter characterized the student mood as 'disappointed, disillusioned, drained and exhausted,' and Kingman Brewster began to decry 'grim professionalism' instead of the inability of black revolutionaries to get a fair trial," Kakutani 1982, p. 282.

216 Buckley 1982, p. 266. As for "Grim Professionalism," as Buckley added, the phrase "replaced 'Eat my shorts' as the epithet of choice," p. 267.

teen dollars. . . ."[217] For nineteen dollars, an undergraduate in 1973 could attend one meeting of Robinson's course that fall, Math 70a, on set theory, or the joint seminar he taught with Stephan Körner on "Philosophical Foundations of Mathematics."

Before the fall semester actually began, H. D. Taft, dean of the faculty, asked Robinson to serve on the Senior Appointments Committee in the social sciences as one of the two "outside" members for 1973–1974.[218] As usual, he agreed, and so was immediately back to business as usual at Yale after what had been a very full and productive semester away.

Similarly, when Andrzej Mostowski, on behalf of the Association of Symbolic Logic, asked if Robinson would please represent the Division of Logic, Methodology, and Philosophy of Science at IUHPS meetings to decide the distribution of funds for various projects, he could not refuse. As Mostowski argued, "we ought to ask our most distinguished colleagues to represent us in this work . . . one of the rare opportunities to make ourselves useful to the whole scientific community."[219]

What Robinson looked forward to most that fall, however, was returning to his graduate students, to his joint seminar with Körner on the foundations of mathematics, and to the pending arrival of Peter Roquette. It was going to be a very full semester.

Philosophy of Mathematics: The Seminar with Stephan Körner

When Stephan Körner accepted an appointment in the Philosophy Department at Yale, it was doubtless only a matter of time before he and Robinson would decide, given their mutual interests, to offer a joint seminar on some aspect of the philosophy of mathematics. This they finally did in the fall of 1973. After discussing the matter, they agreed to divide the course into eight parts, following a scheme Körner had worked out as follows:

Outline of Proposed Seminar

Philosophical Foundations of Mathematics

I. Logical Preliminaries (AR)
II. Historical Preliminaries (Plato-Leibniz-Kant) (SK)
III. Relevant Developments in Mathematics (AR)

[217] Buckley 1982, p. 269.

[218] Horace D. Taft to Abraham Robinson, September 5, 1973, RP/YUL.

[219] Andrzej Mostowski to Abraham Robinson, October 19, 1973, RP/YUL. Robinson sent his acceptance of the appointment to F. Greenaway, chairman of the ICSU Committee in question, on October 30, 1973.

IV. Realism versus Nominalism (with special reference to Cantor) (AR)
V. Logicism (SK)
VI. Formalism (AR)
VII. Intuitionism (SK)
VIII. The Application of Mathematics (SK)[220]

Robinson later drew up a course description for the catalogue:

Philosophical Foundations of Mathematics

Logical preliminaries. Historical preliminaries (Plato-Leibniz-Kant). Relevant developments in mathematics. Realism versus nominalism. Logicism. Formalism. Intuitionism. Other trends. The application of mathematics (RP).

Texts: S. Körner, *Philosophy of Mathematics*;
Benacerraf and Putnam, *Philosophy of Mathematics, Selected Readings*.[221]

By all accounts the seminar was a great success, not only for the students, but especially for Körner and Robinson. As Körner remembers, "He once expressed his doubts that the students enjoyed the seminar. When asked for his reason, his answer was: 'Because we are enjoying it too much.'"[222]

Robinson's Philosophy of Mathematics

As the last forum in which Robinson had an opportunity to express his most mature views on the foundations of mathematics, his intensive interaction with Körner is of special interest, both historically and philosophically. Körner sees Robinson as a follower of—or at least working in the same spirit as—Leibniz and Hilbert. Like Leibniz, Robinson rejected any empirical basis for knowledge about the infinite. Leibniz nevertheless adopted both the infinitely large and the infinitely small in mathematics for pragmatic reasons, because they permitted an economy of expression and an intuitive, suggestive, heuristic picture. Ultimately, Robinson agreed with Leibniz on the ontological status of infinities and infinitesimals as well-founded fictions—*fictiones bene fundatae*—in the

220 Stephan Körner to Abraham Robinson, November 29, 1972, RP/YUL.

221 RP/YUL. In a brief opening paragraph to the Committee on the Course of Study, Robinson explained: "On behalf of Professor Körner and myself, I am submitting this proposal for a course on the above subject, to be given in the Departments of Philosophy and of Mathematics in the fall semester of 1973-74 and to be listed as undergraduate (senior) and as graduate courses in both departments."

222 S. Körner in a letter to Renée Robinson, RP/YUL; and Körner I-1982.

sense that their applications served to penetrate the complexity of natural phenomena and helped to reveal relationships in nature that purely empirical investigations would never have produced.

On the subject of infinitesimals, however, Robinson succeeded where Leibniz and his successors had not. Leibniz, for example, never demonstrated the consistent foundations of his calculus, for which his work was sharply criticized by Nieuwentijt, among others.[223] The consequences, historically, lasted for centuries. Because "Leibniz's approach was considered irremediably inconsistent, hardly any efforts were made to improve this delimitation."[224]

At issue, in part, were philosophical questions about the nature of number, rigor, and above all, the infinite and infinitesimals. In 1888 the German mathematician Richard Dedekind had posed similar questions in a famous monograph, *Was sind und was sollen die Zahlen?* (What are numbers and what should they be?). Robinson saluted Dedekind's important work in an article of his own meant to raise similar foundational issues about numbers in a clever, informal account for the *Yale Scientific Magazine*: "Numbers: What are They and What are They Good For?"[225]

This was another piece in Robinson's continuing effort to popularize—and justify—nonstandard analysis. Although it might represent an unfamiliar, even controversial new system including infinitesimals and infinitely large numbers, Robinson assured his readers that number systems were constantly evolving. Indeed, from time to time new ones were always being added, often facing initial, if not long periods of doubt and uncertainty before finally being welcomed into the community of acceptable mathematical concepts. Robinson had no doubt that the same would prove true of nonstandard analysis. As he quipped:

> Number systems, like hair styles, go in and out of fashion—it's what's underneath that counts.[226]

This might well be taken as the leitmotif of much of Robinson's entire mathematical career, for his surpassing interest since the days of his dissertation written at the University of London in the late 1940s was

[223] For a recent survey of the controversies surrounding the early development of the calculus, including Nieuwentijt's critique of Leibniz and the better-known opposition of Bishop Berkeley to Newton's calculus, see Hall 1980.

[224] Körner 1979, p. xlii. Körner notes, however, that an exception to this generalization is to be found in Vaihinger's general theory of fictions. Vaihinger tried to justify infinitesimals by "a method of opposite mistakes," a solution that was too imprecise, Körner suggests, to have impressed mathematicians. See Vaihinger 1911, pp. 511 ff.

[225] Robinson 1973c.

[226] Robinson 1973c, p. 14.

model theory, and especially the ways in which mathematical logic could not only illuminate mathematics, but have very real and useful applications within virtually all of its branches. In discussing number systems, he wanted to demonstrate, as he put it, that:

> The collection of all number systems is not a finished totality whose discovery was complete around 1600, or 1700, or 1800, but that it has been and still is a growing and changing area, sometimes absorbing new systems and sometimes discarding old ones, or relegating them to the attic.[227]

To make his point, Robinson offered the example of W. R. Hamilton, who had been the first to demonstrate a larger arithmetical system than the complex numbers, namely, his quaternions. These were soon supplanted by the system of vectors developed by Josiah Willard Gibbs of Yale and eventually transformed into a vector calculus. This was a more useful system, much better than quaternions for the sorts of applications for which it had been invented.

Similarly, Georg Cantor vastly enlarged the mathematician's vocabulary with a revolutionary theory of transfinite numbers. For Cantor a cardinal number was the power of a set, or the concept that represented all sets (regardless of their elements or their order) that were equivalent to the set in question. The advantage of this view of the nature of numbers, of course, was that it could be applied to infinite sets, producing transfinite numbers and eventually leading to an entire system of transfinite arithmetic. Its major disadvantage, however, was that it led Cantor to reject adamantly any mathematical concept of infinitesimal.[228] Cantor had based his theory of transfinite numbers on abstractions from sets, and he could not conceive of any set from which, as an abstraction, infinitesimals could result. Robinson, with the tools of mathematical logic, again succeeded where one of his historic predecessors had not.

There was another important lesson to be learned, Robinson believed, in the eventual acceptance of new ideas of number, despite their novelty or the controversies they might provoke. Ultimately, utilitarian realities could not be overlooked or ignored forever.

With an eye on the future of nonstandard analysis, Robinson was also impressed by the fate of another theory devised late in the nineteenth century which also attempted, like those of Hamilton, Cantor, and Robinson, to develop and expand the frontiers of number. In the 1890s Kurt Hensel introduced a whole series of new number systems, his now familiar p-adic numbers. Hensel realized that he could use his p-adic

[227] Robinson 1973c, p. 14.

[228] For details, see Dauben 1979.

numbers in order to investigate properties of the integers and other numbers. He also realized (as did others) that the same results could be obtained in other ways. Consequently, many mathematicians came to regard Hensel's work as a pleasant game, but as Robinson himself observed, "many of Hensel's contemporaries were reluctant to acquire the techniques involved in handling the new numbers and thought they constituted an unnecessary burden."[229]

The same might be said of nonstandard analysis, particularly in light of the transfer principle which demonstrates that theorems true in R^* can also be proven for R by standard methods. Moreover, many mathematicians are clearly reluctant to master the logical machinery of model theory with which Robinson developed his original version of nonstandard analysis.[230] But for the doubts of those who regarded nonstandard analysis as a fad, little more than a "pleasant game," like Hensel's p-adic numbers, Robinson believed the subsequent history of Hensel's ideas should give skeptics an example to ponder. For today, p-adic numbers are regarded as coequal with the reals, and have proven a fertile area of mathematical research.

The Omega Group: Summer 1973

Robinson had long been a consultant to the Dutch mathematics publisher, North-Holland, and as one of the editors of its Studies in Logic series, was always on the lookout for possible new books and authors. But in July, 1973, Einar Fredricksson wrote about North-Holland's concern that recently several important authors had been signed by Springer-Verlag for new books in logic. He was worried, he said, not so much by the threat of competition, as he was about the possibility of overlap. What did Robinson think they should do?

As Robinson explained, much of the problem was related to the "Ω" group, a new "Working Group on Mathematical Logic" headed by Gert Müller in Heidelberg, with close ties to the Heidelberg Academy, Springer-Verlag and the Mathematical Research Institute in Oberwolfach.[231] Robinson, of course, had known Müller for years (they had

[229] Robinson 1973c, p. 16.

[230] This problem has been resolved by Keisler, Luxemburg, and Machover, among others, who have presented nonstandard analysis in ways accessible to mathematicians without their having to take up the difficulties of mathematical logic as a prerequisite. See Luxemburg 1962 and 1976, Machover 1967, and Keisler 1971.

[231] The name of the group had been discussed at length, and was carefully chosen. In trying to characterize what was special about the group's common interests, Müller felt it was the characteristic approach mathematical logicians took to the infinite. He offered a silver dollar, "an old, pure, silver dollar," to whoever came up with the best name

met first in 1954, in Zürich, when Müller was Bernays' assistant at the ETH).[232] Over the years they had become good friends, especially since Robinson had been spending summers in Heidelberg. Robinson was certainly familiar with the challenge the Ω-group represented to North-Holland:

> One becomes more and more aware of the existence of the Ω-group. Müller's conception of an international Bourbaki group is very effective, especially as it includes periodic meetings in the Black Forest. Since it seems unlikely that North Holland/Elsevier will offer its authors free vacations in the Swiss Alps, we can only continue in our own small way, relying on our reputation. However, I agree more and more with your idea of a one or two volume collective enterprise and am willing to have a hand in it.[233]

In one very important way, however, the Ω-group was not at all like Bourbaki, a group which focused on a very specific view of mathematics, namely structure, and sought to build up all of modern mathematics largely in terms of modern algebra.[234] The Ω-group, Müller insists, "never followed any philosophical dictum."[235]

Fredricksson was hopeful that North-Holland might find a more direct way to meet the challenge posed by the Ω-group:

> In approaching some potential authors recently I have heard the argument that they are interested in publishing their work with the Ω-group because "it has the advantage that you can have your manuscript thoroughly discussed with the group." (This particular quote is from Dag Prawitz.) As I see it, every project submitted to the STUDIES is being discussed among the Editors. True, by letter it has not always been an easy and efficient way.[236]

Fredricksson suggested the editors of the Studies series might consider meeting at least once a year during international logic gatherings,

---

for the group, and it was eventually won by Dirk Siefkes. The Ω was taken from Georg Cantor, an "omega" not only in the sense that the infinite was not an indefinite approximation but something well-defined, but also in the sense that the Ω should stand for "everything," while averting any paradoxes. The group was pleased to note that there was also the Swiss "Ω" watch—a timely connection that seemed equally auspicious. Müller I-1992.

[232] See a letter of recommendation Robinson once wrote for Müller, dated February 27, 1972, RP/YUL.

[233] Robinson to Fredricksson, July 30, 1973, RP/YUL. Officially, the Ω-group initially consisted of R. O. Gandy, H. Hermes, A. Levy, G. H. Müller, G. E. Sacks, and D. S. Scott.

[234] See Dieudonné 1970 and 1977.

[235] Müller I-1992.

[236] Fredricksson to Robinson, October 30, 1973, RP/YUL.

or perhaps assigning "guiding editors" to each work in the series to work more closely with authors. Robinson was not convinced:

> I really do not think you should take Prawitz' remarks literally, but should interpret it [*sic*] psychologically or sociologically. The authors of the Ω-group do not care any more than any other authors for interference with their ideas as such. What they love is the feeling of togetherness in an important group, with a joint purpose, and meeting in agreeable places. This, and other support, has been made possible by a large grant from a foundation. With due respect, I do not believe that the interpolation of more people in our own procedures would match that in any way, it would only complicate matters.[237]

One alternative that North-Holland pursued immediately was to produce a collaborative *Handbook of Mathematical Logic*, to be out in time for the twenty-fifth anniversary of the Studies series in 1976.[238]

Meanwhile, Müller's solution, as he once explained it to another of North-Holland's senior executives, was to set different goals. If North-Holland was really concerned to avoid competition, perhaps Studies in Logic should concentrate on special topics at a high level, since the Ω-series was intended to take a broader view, aiming at a more general audience.[239]

As for any potential rivalry between publishers, Müller himself was not "married" to either Springer or to North-Holland, although admittedly Springer did provide some of the initial funding for the group's semiannual meetings, and served as publisher of the official Ω-series.[240]

[237] Robinson to Fredricksson, November 7, 1973, RP/YUL.

[238] Fredricksson to Robinson, December 21, 1973, RP/YUL. Jon Barwise had already expressed his interest to Fredricksson in editing the *Handbook*, and indeed, it did appear under his editorship in 1977. Fredricksson also expressed his concern at preliminary talks Joseph Shoenfield was having with Springer about its taking over publication of the *Journal of Symbolic Logic*. This, he warned Robinson, would be "an unhappy turn of events and I hope that a decision will not be taken unless we have a fair chance to bid for publishing this journal."

[239] The Ω-group sought to produce texts that would help to orient mathematicians quickly and easily to the main body of knowledge, along with its versatility and many strengths, of mathematical logic. With this in mind, one of the most ambitious undertakings was Müller's *Ω-Bibliography of Mathematical Logic*. As noted in his preface, "work on the Bibliography started at the same time as the Ω-group came into being, early in 1969. The financial support was provided by the Heidelberger Akademie der Wissenschaften in the framework of the Ω-group project," Müller 1987, pp. xii–xiii.

[240] Müller I-1992. Müller, in fact, edited a book for North-Holland that appeared in 1976, *Sets and Classes*, and Robinson was about to sign a contract with Springer to do a book with V. B. Weispfenning on *Algebra–A Model Theoretic Viewpoint*. The latter, unfortunately, has never appeared, although the signed contract is among Robinson's papers, YUL. See also Robinson to Müller, February 3, 1973, RP/YUL.

Nonstandard Number Theory: Summer 1973

Robinson only had a brief opportunity to see Peter Roquette that summer in Heidelberg, but he knew Roquette would be at Yale in the fall. As planned, Roquette was in New Haven by the middle of September. What they hoped to do was push their long-standing mutual interest in nonstandard number theory as far as they could."[241] In fact, their collaboration nicely fulfilled Gödel's best hopes that nonstandard analysis might well make a substantive contribution to number theory. As Robinson was pleased to report:

> . . . in recent weeks, Peter Roquette and I have produced a nonstandard proof of Siegel's theorem on integer points on curves. This theorem is regarded as one of the high points of 20th-century number theory. We also believe that in one direction we can go slightly beyond Siegel's result. As you may know, Roquette, of Heidelberg University, is a leading "standard" expert in this area. He was visiting here for a month and will come back later to continue our collaboration.[242]

When Roquette returned to Heidelberg, he realized what a wonderful opportunity the month with Robinson had been:

> My all too short trip to you seems today to me as a visit in another world, and I assure you, that the peace and freedom to live exclusively mathematics again, is something I cherish. I'm glad something came of our conversations—although since then I've not had time to work on the Siegel Theorem.[243]

This classic result on diophantine equations, proved in nonstandard terms by Robinson and Roquette, was originally due to C. I. Siegel, namely, "Any irreducible algebraic equation $f(x, y) = 0$ of genus $g > 0$ admits only finitely many integral solutions."[244] Kurt Mahler generalized the theorem, allowing not only integer solutions but rational solutions as well (as long as their denominators were divisible by primes belonging to a given finite set).[245] It was Robinson's inspiration to ex-

[241] Roquette I-1991. Correspondence about Roquette's visit to Yale began as early as January 23, 1973. On March 15, Robinson wrote about various administrative formalities, saying that the NSF had promised to support Roquette, but "I always work by the old saying, that one can only rely on written *Verpflichtungen*." In mid-July, Roquette wrote to thank Robinson for the "interesting lectures" Abby had just given in Heidelberg, and to say that he could only come to Yale for a month beginning September 15, with possibly a second month the following spring. See Roquette to Robinson, July 17, 1973, RP/YUL.

[242] Robinson to Gödel, October 16, 1973, RP/YUL.

[243] Roquette to Robinson, November 3, 1973, RP/YUL.

[244] Siegel 1929.

[245] Mahler 1934, pp. 168–178.

ploit the basic idea of enlargements, specifically of an algebraic number field in a nonstandard setting, which gave Robinson and Roquette a means for comparing the arithmetic of such an enlargement with the functional arithmetic in the function field defined by $f(x, y) = 0$.[246]

Earlier, Robinson had already achieved some interesting results with nonstandard number theory, and had been lecturing on various related topics for some time. In 1970, at Oberwolfach, he had presented some early work on "Algebraic Function Fields and Non-standard Arithmetic."[247] Initially, Robinson's efforts to develop nonstandard number theory had been inspired by André Weil's decomposition theorem and Siegel's result that the number of integers on a curve of positive genus is finite.[248] Later, he continued to push his early results still further, which led to a paper for the *Journal of Number Theory*, "Nonstandard Points on Algebraic Curves."[249] Eventually this all crystallized in his mind as a viable means of approaching the Siegel-Mahler theorem. As Peter Roquette reflected on the summer of 1973, when Robinson spent a week in Heidelberg lecturing on algebraic function fields and nonstandard arithmetic:

> He expounded his ideas on embedding function fields into nonstandard models of their ground fields, or into finite extensions of such models. If the ground field is an algebraic number field, then this yields a representation of algebraic functions by (nonstandard) algebraic numbers. In this way it should be possible to explain the arithmetic structure of function fields directly, using well-established nonstandard principles only, by means of the arithmetic structure carried by the ground field.[250]

It was this idea that he and Roquette began to develop in their joint attack on the Siegel-Mahler theorem. According to Roquette:

[246] Roquette I-1991. [247] Robinson 1972.

[248] Starting from the field of rational numbers Q, Robinson considered the nonstandard extension Q^*, and showed that for an algebraic curve Γ defined over Γ, Γ had an infinite number of rational points if and only if the function field of G over Q, i.e., $Q(\Gamma)$, had an embedding in Q^*. Robinson considered internal valuations induced from a nonstandard prime number in Q^*, by a standard prime number in Q, or by the Archimedean valuation of Q^*. His primary result showed that every valuation of a function field $Q(\Gamma)$ embedded in Q^* is induced from an internal valuation of Q^*. The concept of distribution used by Weil to prove the Mordell-Weil theorem is equivalent to Robinson's restriction of internal valuations to $Q(\Gamma)$, and Robinson's paper, in fact, closely parallels Weil's. For details, see Weil 1928.

[249] Robinson 1973d. This paper established one of Siegel's basic inequalities for rational points on algebraic curves. In fact, Robinson was pleased that his theory led "almost immediately" to a number of the basic inequalities of the theory of rational points on algebraic curves.

[250] Roquette and Robinson 1975, p. 424.

> In my opinion, these ideas of Abraham Robinson are of far-reaching importance, providing us with a new viewpoint and guideline towards our understanding of diophantine problems. It seems worthwhile to put these ideas to a test in order to verify their usefulness and applicability in connection with explicit diophantine problems. Perhaps a good test in this sense would be the explanation, in nonstandard terms, of Weil's theory of distributions and, closely connected with it, the theorems of Mordell and Weil and of Siegel and Mahler.[251]

The Robinson-Roquette paper offered just such a test, and in Roquette's opinion, "a successful one." As he added, "Although we deal explicitly with the Siegel-Mahler theorem only, it will be clear to anyone familiar with the subject that Weil's theory of distributions also can be explained in this context."

Unfortunately, Robinson did not live to see the final results that he and Roquette had achieved actually appear in print. Roquette's description of how closely they had worked together on the paper's outline and results is, in itself, a moving tribute:

> The paper was written as a result of several weeks of close collaboration with Abraham Robinson at Yale. He completed the first draft by his own hand in November 1973. His severe illness and tragic death prevented him from participating in the discussion of the following versions. Hence, although I want to make it clear that the basic ideas are Robinson's, I have to take full responsibility for the form of presentation of the subject. . . .[252]

Robinson and Roquette dedicated their paper, "On the Finiteness Theorem of Siegel and Mahler Concerning Diophantine Equations," to Kurt Mahler on the occasion of his seventieth birthday, which had been celebrated earlier that year in Hamburg. Basically, what nonstandard analysis did for number theory in this case was to exploit, yet again, the power of the compactness theorem. The basic idea was as follows: if Γ is a curve with infinitely many integral points on it in some number field K, then the extension of Γ to a nonstandard enlargement K^* of K must have some nonstandard integer point (x, y) on it. The nonstandard Siegel-Mahler theorem asserted that if F is an algebraic function field of genus $g > 0$ in one variable over K such that $K \subset F \subset K^*$, then every nonconstant element x of F has at least one nonstandard prime divisor of K^*.

[251] Roquette and Robinson 1975, p. 424.

[252] Peter Roquette, in Robinson and Roquette 1975, p. 175; in Robinson 1979, 2, p. 424. Originally, Robinson (in his first draft) had included a section on effectiveness relative to the bounds given in Roth's theorem. This was not published in the paper with Roquette, however, but appeared separately later, edited by Takeuti. See Robinson 1988, "On a Relatively Effective Procedure Getting All Quasi-integer Solutions of Diophantine

The importance of the nonstandard argument lies in the way *F* can be regarded as both a field of functions over *K* and as a subfield of K^*. Basically, what Robinson and Roquette did was to provide a transfer principle translating arithmetical properties of *F* (as a subfield of K^*) into equivalent functional properties of *F* (as a function field). The beauty of this approach was the way it expressed the equivalence between the arithmetical properties and the functional properties of the field *F*.

Robinson was exceptionally pleased with the new results he and Roquette had achieved. As he told Roquette, "I've been asked to give a general hour lecture at Vancouver next August at the International Congress. I intend to include our joint effort in the presentation."[253]

Robinson, Lightstone, and Asymptotic Expansions

In contrast to Robinson's satisfaction with the work he had done with Roquette, he was becoming increasingly concerned about a collaborative effort he had agreed to undertake with his old friend Harold Lightstone. Lightstone had spent 1971–1972 at Yale, on leave from Queens University in Kingston, Ontario, thanks to a fellowship from the Canada Council. He and Robinson had begun work then on a joint venture, *Nonarchimedean Fields and Asymptotic Expansions.*[254]

Shortly before Robinson's cancer was diagnosed late in 1973, he wrote to Lightstone to voice certain doubts about their manuscript, urging they not hurry with revisions. He suggested a motto to be used at the beginning of the book, which in retrospect seems uncannily prophetic: "You may not finish the job. But you are not free to shirk it."[255]

Did Robinson already realize that he would not live to see the end of the book with Lightstone? Did he fear that trying to hurry it to completion, or cutting it back to a size he could perhaps manage, would result

---

Equations with Positive Genus." (The paper was actually "written by Gaisi Takeuti based on A. Robinson's notes.")

[253] Robinson to Roquette, November 7, 1973, RP/YUL. When Robinson sent his official acceptance to talk at the Vancouver Congress (August 21–29, 1974), he proposed as his topic, "Applications of Model Theory to Algebra, Analysis and Number Theory." See his letter to the Congress Program Committee, November 20, 1973, RP/YUL.

[254] As Lightstone later noted, "this monograph is based on a draft by the second author (A.R.), while the final text is due to the first author (A.H.L.)," Lightstone and Robinson 1975, p. vii.

[255] The phrase, Robinson told Lightstone, was from the *Pirqei Avoth*, "Chapters of the Fathers," part of the Mishnah consisting of sayings or pearls of wisdom from the Sages. I am grateful to Moshé Machover for identifying the passage as chapter 2, verse 21, attributed to Rabbi Tarphon (first century A.D.). Machover translates the Hebrew as "It is not up to you to finish the job, but you are not at liberty to skive off it." Machover to JWD in an E-mail message of December 7, 1993. See also Robinson to Lightstone, RP/YUL.

in a book in which he could take no pleasure? He already had major doubts about large parts of it, and suggested "two minor and two major pieces of surgery which are, in my view, essential." He was especially concerned that towards the end of their manuscript, "the connection with Nonstandard Analysis becomes extremely weak." Robinson hoped to add another forty pages to emphasize infinitesimals, which he reminded Lightstone, were "crucial for proper understanding of the method."[256]

Without telling Lightstone, Robinson was determined that the book *not* appear in North-Holland's Studies series. As he explained to Einar Fredricksson:

> I have given a good deal of thought to the form of publication and have again come to the conclusion that it would not be suitable for the *Studies in Logic*. In fact, although you know my preference for "glamour," it seems to me that the manuscript in question, in spite of its length, is suitable for the North Holland *Mathematics Studies* (in which Shoenfield's notes were published).[257]

Lightstone was equally clear to Robinson about his own preferences. He did not like the motto. Nor did he want to lengthen the book at all, and instead preferred to keep to their original deadline of January 1974. He felt it was not a good idea to refer to linguistics in the section of the manuscript devoted to the languages of R and R^*–"many mathematicians are turned off by any reference to linguistics," he noted–and rather than giving weight to the ultrapower theorem, he felt that it was the transfer theorem that should be stressed.[258]

On November 27, 1973, Robinson broke the news to Lightstone that he was going to hospital. As if in resignation, Robinson said that he was willing to go along with most of Lightstone's views, but was adamant about the need to strengthen the link between nonstandard analysis and asymptotic expansions. Later, Robinson confided to A. J. Coleman:

> We all have our cross to bear and in your case and mine this includes Harold. I have always known that he is obstinate and not too prone to accept other peoples' judgments. On the other hand, he *is* a serious mathematician who is willing to tackle problems that others in their mundane wisdom would rather leave alone. With regard to our manuscript, it had been my intention to bring it out only in photostatic form, but the referee to whom the editors sent it insisted on something better.[259]

[256] Robinson to Lightstone, November 6, 1973, RP/YUL.
[257] Robinson to Fredricksson, October 17, 1973, RP/YUL.
[258] Lightstone to Robinson, November 13, 1973, RP/YUL.
[259] Abraham Robinson to A. J. Coleman, February 18, 1974, RP/YUL.

When it was published, Lightstone dedicated *Nonarchimedean Fields and Asymptotic Expansions* to Robinson's memory, adding, "The manuscript of this book was sent to the publisher in December, 1973, four months before the sudden death of Abraham Robinson. His mastery of mathematics and his warm human qualities will be sorely missed."[260]

Renewed Overtures: Canada And Israel

It was exactly Robinson's mastery of mathematics and his human qualities that kept him in great demand. In November of 1973, he received word that Toronto wanted him back—this time as a visiting distinguished professor. George Duff wrote to explain that there was a newly founded visiting professorship (thanks to the Samuel Beatty Fund—subscribed by alumni) and hoped Robinson would agree to come for either a full year or a term in 1974–1975. Could he teach an undergraduate course on nonstandard analysis, perhaps, along with a graduate course of his choice? Toronto intended to match his salary at Yale, and Duff asked Abby to call him collect to discuss the offer.[261]

At the same time, the Weizmann Institute also hoped to lure Robinson back to Israel. Samuel Karlin wrote with an offer that played on Robinson's own sentiments and read more like an imperative:

> I am extremely anxious to persuade you to join us during the next academic year so that the two of us could do our best to develop a first-class mathematics center here. Let me emphasize how much I need strong senior people here who are highly respected.[262]

Although Robinson declined the offer to return to Toronto, the idea of returning to Israel was increasingly attractive to him. As he made clear to G. D. Mostow, "If it had not been for his illness, he would have gone to the Weizmann Institute. The arrangement with the Institute for Advanced Study would have made it all the more attractive."[263]

Robinson on Tour: Late 1973

That fall, Robinson received a record number of invitations to lecture, including a request that he address the Mathematics Colloquium at Rice University, and another that he give both a general lecture and a more

[260] Lightstone and Robinson 1975, p. viii. As Robinson had requested, the book did not appear in the Studies in Logic series, but was published in the Mathematical Library series.

[261] George Duff to Abraham Robinson, November 29, 1973, RP/YUL.

[262] Samuel Karlin to Abraham Robinson, December 3, 1973, RP/YUL.

[263] Mostow I-1993.

specialized seminar for the logic group at the University of Michigan in Ann Arbor.[264] For the Rice lectures, Robinson again told Roquette that "I am going to talk on nonstandard points on curves . . . and I would like to present at least part of our joint work there."[265]

There was also the Fifth Balkan Mathematical Congress, scheduled for the end of June (1974) in Belgrade, about which Kurepa wrote to say that he hoped Robinson would agree to take part.[266] Arnold Oberschelp also wanted Robinson to present an invited lecture for a summer institute and logic colloquium in Kiel, being organized by Gert Müller, Klaus Potthoff, and Oberschelp himself.[267]

By December, however, Robinson had begun to decline all such offers. When Gordon Keller wrote from the University of Virginia, for example, inviting him to give a special expository lecture, Robinson wrote back explaining that he was in no position to do so: "I am now in the hospital after surgery, and it would not be wise for me to accept an invitation at this stage."[268]

Robinson Faces Surgery: December 1973

When the Robinsons returned to New Haven in the fall of 1973, Abby was still in good health, although he had apparently begun to experience stomach pains:

> We do not know how long he bore these without complaining, but by November he had to acknowledge them to Renée. He continued his regular schedule until Thanksgiving recess at the end of November, when he entered the hospital for a series of tests. These gave abundant grounds for suspecting cancer of the pancreas, and an exploratory operation revealed that the disease was beyond surgical remedy.[269]

Robinson recuperated in the Yale Infirmary. Despite his usual optimism, he understood that the future was uncertain. As he reported to his friend and colleague at Caltech, Wim Luxemburg:

[264] J. Jacobowitz (Rice) to Abraham Robinson, September 28, 1973; Peter Hinman (Michigan) to Robinson, October 1, 1973, RP/YUL.

[265] Robinson to Roquette, November 14, 1973, RP/YUL. Robinson noted that the simplest proof of the Siegel-Mahler theorem used Riemann surfaces, but required a lemma that Roquette had proved but that Robinson had forgotten. He asked Roquette to send him the details as soon as possible.

[266] "The International Symposium on the Non Real Ecart and Related Topics, with Applications." D. Kurepa to Abraham Robinson, January 26, 1974, RP/YUL.

[267] A. Oberschelp to Abraham Robinson, November 6, 1973, RP/YUL.

[268] G. Keller to Abraham Robinson, December 6, 1973; Robinson to Keller, December 13, 1973, RP/YUL.

[269] Seligman 1979, p. xxxi.

I am writing this letter from the hospital, but at least this part of the problem seems to have gone well. As for the future, consult my horoscope in the *LA Times*.[270]

As soon as he was able, Robinson wrote to Joram Hirschfeld, who was in Israel. Robinson confessed that he was relieved to know that Hirschfeld was safe in the aftermath of the unexpected war that had begun on Yom Kippur in 1973:

I and, I am sure, all your friends here are pleased to learn that you are safe and sound. . . . I am dictating this letter because I am fighting a little battle of my own, i.e., . . . I am in the hospital after surgery. However, I still hope that I shall have some time left for Israel.[271]

As soon as Robinson was sufficiently recovered, George Seligman drove him home from hospital:

Along the way we had to stop at a pharmacy to pick up a prescription. When she brought it out after some time, Renée commented on the elaborate bookkeeping that had had to be done. Abby replied that this was only to be expected in the case of such medicines, with an aside to me that showed he was being given strong narcotics.[272]

News of Robinson's illness spread quickly, and soon letters expressing good wishes, or registering dismay, began to pour in. Most were from individuals, but others registered collective regards. On January 9, a postcard was mailed from Oxford, and simply read: "Dear Abraham: Concern, respect and affection from the Ω-Group."[273]

Systematically, Robinson now began to cancel one commitment after another. Among the first to whom he wrote was Stephan Körner, who was planning another philosophy of science meeting for Bristol in July 1974. Robinson had agreed to speak "On the Boundary Between Mathematical Theory and Mathematical Practice," but felt that prudence now required him to withdraw:

Dear Steven!

I regret to have to report that neither your Leibnizian optimism nor the impression given to me by my doctors have so far been confirmed. Although I give my course this term, I do not feel that in my present situation I shall be able to prepare for Bristol a paper which is good enough for the occasion. As my good friend, please do not contradict me.

I still hope that in the time available we shall be able to meet some-

[270] Abraham Robinson to W.A.J. Luxemburg, December 13, 1973, RP/YUL.
[271] Abraham Robinson to Joram Hirschfeld, December 14, 1973, RP/YUL.
[272] Seligman 1979, p. xxxi.
[273] Postcard of January 9, 1974, RP/YUL.

where privately, perhaps in London or elsewhere in England, immediately before or after your conference. At the beginning of July I shall take part in a conference on non-standard analysis in Oberwolfach. Since this is a case where they want me as a statue rather than a living person, I do intend to go to that one.

So, to avoid any misunderstanding, I regretfully withdraw from the Bristol conference. But we shall be very happy to meet you and Ditti a couple of days somewhere in Europe before or after that time.[274]

Similarly, he wrote to Peter Hinman at the University of Michigan:

If your grapevine is good enough, you may have heard that the state of my health leaves something to be desired. Accordingly, I feel obliged to cancel my visit to Ann Arbor next March. For the time being I shall have to confine my outside activities to what is absolutely necessary.[275]

Indeed, Robinson faced his fate with a sense of grim reality:

Renée was constantly at hand, doing her best to help him conserve time and energy for the tasks he felt were most important. He was under no illusions. He insisted that no extraordinary measures to prolong life be undertaken. He would do the best he could for as long as he could. Carol Wood, his former student, asked him in early 1974 how she should answer inquiries about his health. He instructed her to tell them frankly that he was dying.[276]

Kurt Gödel, when he heard the news that Robinson was seriously ill, sent strangely comforting words of his own. Gödel, well known for his hypochondria, was no friend of doctors or modern medicine:

In view of what I said in our discussion last year [about Robinson coming to the Institute for an extended period of time] you can imagine how very sorry I am about your illness, not only from a personal point of view, but also as far as logic and the Institute for Advanced Study are concerned.

As you know, I have unorthodox views about many things. Two of them would apply here:

1. I don't believe that any medical prognosis is 100% certain.
2. The assertion that our ego consists of protein molecules seems to me one of the most ridiculous ever made.

[274] Robinson to Körner, January 31, 1974. Apparently, Robinson hoped that he and Renée would get to Switzerland for a brief visit sometime "around Easter," and Körner suggested that if so, they also plan to stay with him and his wife. See S. Körner to AR, January 24, 1974, RP/YUL.

[275] Robinson to Hinman, January 31, 1974, RP/YUL.

[276] Seligman 1979, p. xxxi.

I hope you are sharing at least the second opinion with me. I am glad to hear that, in spite of your illness, you are able to spend some time in the mathematics department. I am sure this will provide some welcome diversion for you.[277]

Fighting On: Spring 1974

When the new term began at Yale, Robinson insisted on continuing business as usual. This meant offering at least his graduate course on model theory:

> It was soon evident that he could not stand to lecture, and the classroom was awkward for a seated lecture. The class was removed to his modest office, where a dozen or so hearers crowded in. The disease and the drugs forced him to struggle to concentrate, but his wit still could flash out, and his listeners' laughter would then fill the narrow corridor outside his office. He was still able to discuss the genesis of some of the central ideas, and he gave insights into the psychology of his own mathematical invention that he might have been too self-effacing for under other circumstances.[278]

What came as most welcome news arrived at the end of March, when Robinson received word from H. C. Becker at the Max-Planck-Gesellschaft in Berlin that his brother Saul's papers were being reprinted. This information arrived along with condolences for Abby's own predicament. He replied, fully aware of the irony: "I feel touched by the letter concerning my illness," he wrote. "It is indeed remarkable that my own turn has come so soon after my brother's."[279]

In trying to get his affairs into order as best he could, Robinson began to sort through his papers, saving very little. Phyllis Stevens, the department secretary, did her best to make the strain as light as possible:

> Robinson began to dismantle his office. Phyllis was helping him clear out his files. Before he went to the hospital he emptied his desk, preparing for his death. Phyllis was in tears. As the illness progressed, Robinson could be found sitting in his office, exhausted. Sometimes he would come into the Department with Renée, she would be holding him up, to give a lecture or a seminar. He was trying to carry out his duties as best he could.[280]

[277] Kurt Gödel to AR, March 20, 1974, Gödel papers no. 011998, PUA/FL.
[278] Seligman 1979, p. xxxi.
[279] Abraham Robinson to H. C. Becker, March 28, 1974, RP/YUL.
[280] Mostow I-1993.

A (Brief) Reprieve: March 1974

When March came, however, there seems to have been a short respite, when Robinson felt momentarily that his health was improving. Piero Mangani wrote from Florence, inviting him to offer a CIME summer course in model theory in Italy in 1975, and suggested "Nonstandard Analysis and its Applications" as a possible topic.[281]

Robinson replied, tentatively, but was optimistic:

> At the moment the state of my health has improved to such an extent that I am willing to accept provisionally. However, if this statement is too vague and you want a definite commitment now, then I must decline directly.[282]

Indeed, Robinson had already made arrangements to pay a brief visit to Gödel and spend a few days sometime in March at the Institute for Advanced Study. Unfortunately, this trip was not to be.[283]

Other meetings, however, were still on the horizon. Detlef Laugwitz was beginning to make final plans for the next nonstandard analysis meeting at Oberwolfach that summer, and asked diplomatically if Abby and Renée would like to stay in one of the institute's bungalow apartments? Laugwitz also invited Robinson to give a lecture in Darmstadt, if he had the time and interest.[284]

Meanwhile Roquette, who had only recently heard through Weispfenning that Robinson had just undergone a serious operation, was encouraged about plans for a *Kuraufenthalt* in Europe later that spring. Hoping that Robinson was already on his way to recovery, Roquette said that he looked forward to seeing Abby—in Heidelberg or elsewhere.[285] A few weeks later, Roquette wrote again, asking rhetorically: "Where should one look for solace in the face of bitter fate?"[286]

Increasingly, Robinson found solace in thinking about mathematics—and about Israel:

> As the pain wore on, he knew he only had a few months to live, and there were only a few things he wanted to hear about—what was going on in the Department, in mathematics, and in Israel. These were the only topics that commanded his interest. He was especially troubled that he couldn't take the offer to go to the Weizmann Institute.[287]

281 Piero Mangani to Abraham Robinson, December 13, 1973, RP/YUL.

282 Robinson to Mangani, March 8, 1974, RP/YUL.

283 See the confirmation Robinson received on January 22, 1974, RP/YUL. The institute agreed to cover all expenses for both of the Robinsons at the institute.

284 Laugwitz to Robinson, December 29, 1973, RP/YUL.

285 Roquette to Robinson, January 8, 1974, RP/YUL.

286 Roquette to Robinson, January 23, 1974, RP/YUL.

287 Mostow I-1993.

Robinson's grave on Givat Shaul, Jerusalem.

April 1974

April proved itself the cruelest month. All too soon, in fact, Robinson began to lose his strength. He was forced to cancel his one class and return to the Yale Infirmary. Conveniently, it was just across the street from the Mathematics Building at 17 Hillhouse Avenue. Colleagues and friends could easily stop by to see him, and there were even occasional moments of hope when his strength would improve. But he had lost considerable weight, and as Wim Luxemburg recalled after visiting Abby just days before he died, there was only a weak shell left of his former self.[288] It was a strain on everyone, but especially on Renée:

> His periods of alertness grew rarer and shorter. Nevertheless, he could brighten at the visits of old friends and former students, who came to see him one more time. Out of his very limited store of energy, he could still

[288] Luxemburg I-1991. But as Luxemburg also vividly recalls, "despite the advanced state of the disease and the narcotics, he also talked to me about mathematics." They discussed the book Robinson had been writing with Lightstone, and Robinson said he had some new ideas about nonstandard number theory that he had been working on with Peter Roquette, although he did not have the energy to say anything in detail. He was interested in a new junior appointment they had just made in logic at Caltech, and indicated his approval of their choice. "Even then, Abby was still very busy in his mind with his work during the few periods that he was awake." W.A.J. Luxemburg to JWD, June 15, 1994, quoted *in extenso* on p. 449, n. 157.

draw together enough to concentrate a few minutes as a graduate student (in this case, Peter Winkler) reported the latest progress of his thesis. In a very few brief questions he was able to show what kinds of problems lay beyond the result and to encourage Winkler to get on with them.

By the beginning of April, it seemed the end must be very near. Visitors often found Abby asleep under the influence of the drugs and Renée in a chair asleep with fatigue.[289]

Robinson died quietly in hospital on Thursday afternoon, April 11, 1974. Only a few days before, he had been voted a member of the U.S. National Academy of Sciences, and Nathan Jacobson, chairman of the section of mathematics, "went out on a limb," as he put it, and told Robinson he had been elected before the results were actually official.[290] It was a fitting if final accolade honoring the life and work of Abraham Robinson.

[289] Seligman 1979, pp. xxxi–xxxii.

[290] Jacobson I-1985. G.D. Mostow, who was visiting Robinson when Jacobson stopped by, was asked to step out of the room briefly. Later, Abby told him that Jacobson had brought very good news. In retrospect, Mostow speculates that Robinson must have been very pleased, since among mathematicians, few logicians had been elected to the academy. Mostow I-1993. Jacobson sent a memorandum about the election to all members of the section of mathematics: "As you may know, Robinson died on April 11 shortly before the April meeting of the Academy. He had received enough votes to assure election. In such cases it has been the practice that a deceased nominee whose standing in the Preferential Balloting appears to assure election has been elected posthumously. This procedure was followed in Robinson's case." Copy of memorandum in Tarski papers, folio 11.9 (National Academy of Sciences), The Bancroft Library, University of California, Berkeley.

Abraham Robinson, lecturing.

Abraham Robinson: The Man and His Mathematics

> Playfulness is an important element in the makeup of a good mathematician.
>
> —*Abraham Robinson*[1]

> He never separated the mathematical craftsman from the human being.
>
> —*Jürgen Schmidt*[2]

ABRAHAM ROBINSON published more articles during the seven years he was at Yale than he did during any other period of his career. Yale, of course, placed a premium on research, offered reduced teaching loads and attractive leaves for concentrated study and writing. This, coupled with the constant stimulation of the brightest graduate students, prominent colleagues, and the constant stream of visitors to a renowned academic center like New Haven, all help to account for this extraordinary burst of productivity.

Abraham Robinson, smiling.

But there is another ingredient as well—Robinson had matured as a mathematician, and his efforts peaked during the time he spent at Yale. He was always pleased to dispel the myth that the best mathematicians were under thirty and that a mathematician did her or his best work early, at the very start of one's career. As a striking counterexample, Robinson's best mathematics was only beginning to reap the benefits of his wide experience when, suddenly, at the age of fifty-five he died—at exactly the same age as his brother—and just as Robinson had feared.

It might be argued that Robinson's success as a mature mathemati-

[1] Robinson 1973c, p. 15.

[2] J. Schmidt to George Seligman, letter of May 29, 1977, RP/YUL.

The first China Symposium on Nonstandard Analysis (Xinxiang), 1978.

cian was in part a function of the kind of mathematics he did. Robinson's contribution was especially significant in showing the power of model theory for mathematics, and he demonstrated clearly the benefits of bringing techniques from one area of mathematics to the service of others. In doing so he achieved startling results.

Nonstandard analysis, of course, was the most prominent—and controversial—of Robinson's contributions. Whatever its merits, it was a continuing but often tangential interest at Yale. Robinson appreciated the fact that as a tool it was powerful in certain contexts, and historically far more interesting, even revolutionary, than most new discoveries. But as a tool it required finesse, and he was one of the few mathematicians who knew enough about other branches of mathematics to exploit its possibilities—at least in the early days. As he once told Gregory Cherlin, "at first it was easy to get results—now you have to do more."[3]

Important results have continued to come from nonstandard methods in mathematics, but perhaps of greater significance in the course of history will be Robinson's steadfast faith and promotion of mathematical logic in general, and of model theory in particular, as a subject not

[3] Cherlin R-1974.

just of local interest to logicians, but of global interest to mathematicians.

From Robinson's grave site in Jerusalem, looking out from Givat Shaul to the hills surrounding the city–the city which he loved above all others–one cannot help but think of the extraordinary life he had led, and the final place to which his own personal journey had brought him to rest. But through it all, he was the consummate mathematician–a mathematician's mathematician–who lived the scientific life to its fullest. In the course of his fifty-five years, he accomplished more than most can claim to have accomplished in far longer lifetimes. Indeed, he was a man who made mathematics a thing of beauty, and equally important, he had the remarkable ability to reveal that beauty to all who wished to learn from his example.

BIBLIOGRAPHY

Abbreviations

AR	Abraham Robinson
I	Interview with JWD unless otherwise noted
JSL	The Journal of Symbolic Logic
JWD	Joseph W. Dauben
PUA/FL	Princeton University Archives, Firestone Library, Princeton, New Jersey, 08540.
R	Reminiscences about AR
RP/YUL	Robinson Papers, Department of Manuscripts and Archives, Yale University Library, Box 1603A, Yale Station, New Haven, Connecticut 06520.

Interviews

Birkhoff I-1989. Garrett Birkhoff, interviewed at his home in Cambridge, Massachusetts, June 8.

Brown I-1982. Donald Brown, interviewed in his office in the Department of Economics, Yale University, March 5.

Chang I-1991. C.-C. Chang, interviewed by telephone at home in Los Angeles, January 30.

Choquet I-1984. Gustave Choquet, interviewed in his office at the University of Paris VII, June 28.

Coxeter I-1990. H.S.M. Coxeter, interviewed in his office at the University of Toronto, November 12.

Duff I-1990. George F.D. Duff, interviewed in his office at the University of Toronto, November 13.

Feit I-1992. Walter Feit, interviewed in his office at Yale University, June 22.

Fraenkel I-1982. Malka (Mrs. Abraham) Fraenkel, interviewed at her apartment in Jerusalem, January 11.

Hales I-1991. Alfred Hales, Chairman of the UCLA Department of Mathematics, interviewed in his office, January 29.

Henkin I-1989. Leon Henkin, interviewed in his office at the University of California, Berkeley, November 22.

Hurd I-1991. A. E. Hurd, interviewed by telephone, May 8.

Jacobson I-1985. Nathan Jacobson, interviewed in his office at Yale University, April 27.

Kakutani I-1992. Shizuo Kakutani, interviewed in his office at Yale University, June 22.

Kalish I-1991. Donald Kalish, interviewed in his office, UCLA, in the Department of Philosophy, January 30.

Kaplan I-1991. David Kaplan, interviewed in his office in Murphy Hall, UCLA Academic Senate, UCLA, January 29.

Katzenstein, Hannah

I-1993. Interviewed in her home at 14 Hanassi Street, Jerusalem, January 17.

I-1994. Interviewed by telephone from Jerusalem, June 18 and 19.

Keisler I-1993. H. Jerome Keisler, interviewed during the Symposium on "History of Model Theory" during the Nineteenth International Congress of History of Science, Zaragoza, Spain, August 25.

Körner I-1982. S. Körner, interviewed in his office at Yale University, November 12.

Kuhlmann I-1994. Salma Kuhlmann, interviewed during the meeting on *Nichtstandard Analysis und Anwendungen*, Mathematisches Forschungsinstitut, Oberwolfach/Walke, Germany, February 3–4.

Levy I-1993. Azriel Levy, interviewed in Jerusalem, January 15.

Loeb I-1991. Peter Loeb, interviewed in Frankfurt, Germany, June 15.

Lutz I-1994. Robert Lutz, interviewed during the meeting *Nichtstandard Analysis und Anwendungen*, Mathematisches Forschungsinstitut, Oberwolfach-Walke, Germany, February 3.

Luxemburg, W.A.J.

I-1981. Interviewed in his office at the California Institute of Technology, January 23.

I-1991. Interviewed in his office, January 28.

Machover I-1991. Moshé Machover, interviewed at his home, Queens Park, London, May 1.

Macintyre I-1982. Angus Macintyre, interviewed at Silliman College, Yale University, March 5.

Moschovakis I-1991. Y. N. Moschovakis, interviewed in his office at UCLA, January 31.

Mostow I-1993. George Daniel Mostow, interviewed in his office at Yale University, February 23.

Müller I-1992. Gert H. Müller, interviewed in his office at the University of Heidelberg, May 22.

Pinczower, Eliezer

I-1982. Interviewed in his home at 22 Hatibbonim Street, Jerusalem, January 11. The interview was conducted in German.

I-1993. Interviewed in his home, January 17.

Rabin, Michael

I-1993a. Interviewed in his office at Harvard University, October 7.

I-1993b. Interviewed in his office, December 8.

Rickart I-1992. Charles E. Rickart, interviewed in his office, Yale University, June 23.

Robinson, Renée

I-1980. Interviewed at Yale University, June 23.

I-1982. Interviewed at her home, Hamden, Connecticut, April.

I-1983. Interviewed at her home, May 10.

I-1989. Interviewed at Yale University, October 9.

I-1990. Interviewed at her home, October 10.
I-1991. Interviewed at the Metropolitan Museum of Art, New York City, February 9.
I-1992. Interviewed at the Yale Law School Commons, October 26.
I-1993. Interviewed at Yale University, February 23.
Rooney I-1990. P. G. (Tim) Rooney, interviewed in his office at the University of Toronto, November 12–13.
Roquette I-1991. Peter Roquette, interviewed in Heidelberg, May 24.
Ross I-1990. A. R. Ross, interviewed in his office at the University of Toronto, November 14.
Sacks I-1993. Gerald Sacks, interviewed in his office at Harvard University, December 7.
Seligman I-1992. George B. Seligman, interviewed in his office at Yale University, June 23.
Taussky I-1981. Olga Taussky Todd, interviewed in her office at the California Institute of Technology, January 23.
van Dalen I-1986. Dirk van Dalen, interviewed at the Harvard Club of New York City, April 11.
Wellman I-1982. Henry Wellman, interviewed in his office at Rockefeller Center, New York City, Spring 1982.
Wood I-1992. Carol S. Wood, interviewed at Wesleyan University, Middletown, Connecticut, November 24.

MANUSCRIPTS FROM THE ROBINSON PAPERS, *Department of Manuscripts and Archives, Yale University Library*

Many of the following manuscripts and typescripts are undated, but most were solicited by George Seligman in 1976. Undated manuscripts have been dated in brackets based on direct or other best internal evidence available in the Robinson archive. In quoting from the recollections of Robinson listed below that were based on interviews, in some cases I have made minor editorial changes to facilitate the reading of oral remarks.

Abramsky R-1976. Schimon (Chimen) Abramsky, 4 pp. typescript [1976].
Cherlin R-1974. Gregory Cherlin, "Reminiscences of Abraham Robinson (1974)," 8 pp.
Hermann R-1975. Miriam Hermann, transcript of an interview conducted by Mrs. Abraham Robinson, in Jerusalem.
Jahn R-1976. Hermann Jahn, 2 pp. typescript [1976].
Kelemen R-1974. Peter J. Kelemen, "My Memories of Professor Abraham Robinson," 2 pp. typescript.
Lipschitz R-1975. Typescript description, in German, of Robinson's responsibilities and accomplishments as an administrator at the Hebrew University during 1959–1961, by Mrs. Lipschitz, secretary of the Natural Sciences faculty, the Hebrew University.
Pinczower R-1977. Eliezer Pinczower, 6 pp. typescript, in German, dated June 22.

Rabin R-1976. Michael Rabin, 4 pp. typescript [1976].

Robinsohn R-1912. Autobiographical note by Abraham Robinsohn, May 12.

Robinsohn R-1917. An account, in German, of the Robinsohn family history, written in Weidling, Austria, by Sara Robinsohn.

Robinson R-1933. A diary, in German, kept by Abraham Robinson in the spring of 1933 as he and his family were fleeing Germany for Palestine.

Robinson R-1940. An account, in Hebrew, written by Abraham Robinson early in the war, about his experiences as a foreign student in Paris from January through June of 1940, including his escape from France that June with Jacob Talmon and their subsequent arrival in London at the beginning of World War II. All passages cited here are based upon the English translation by Dr. Sheila Rabin, Ph.D. Program in History, The Graduate Center, City University of New York.

Simonson I-1975. Mrs. Max Simonson, transcript of an interview conducted by Renée Robinson, in Jerusalem (in German).

Steketee R-1976. J. A. Steketee, "Notes on A. Robinson in Toronto," sent by Steketee to George Seligman, August 17.

Straus R-1976. Ernst Straus, "Recollections of Abraham Robinson," 2 pp. typescript [1976].

Talmon R-1940. Jacob L. Talmon, "Our escape from France in 1940, and early days in England," 13 pp. manuscript in his own hand.

Talmon R-1976. Jacob L. Talmon, 3 pp. manuscript [1976].

Taussky R-1976. Olga Taussky Todd, letter to George Seligman, July 29, 2 pp. typescript.

Wood R-1976. Carol Wood, reminiscences, November 15.

Young R-1976. Alec Young, 3 pp. typescript [1976].

Works by Abraham Robinson

Asterisked entries indicate years in which Robinson published coauthored works. These are listed separately in the following section of the Bibliography.

1939 "On the Independence of the Axioms of Definiteness," *Journal of Symbolic Logic*, 4, pp. 69–72.

1939a "On Nil-ideals in General Rings"; in Robinson 1979, 1, p. 523.

1941 "On a Certain Variation of the Distributive Law for a Commutative Algebraic Field," *Proceedings of the Royal Society of Edinburgh*, 61, pp. 93–101; Robinson 1979, 1, pp. 525–533.

1943* See Robinson and Whitby 1943.

1945* See also Fagg, Montagnon, and Robinson 1945.

1945 "Shock Transmission in Beams," *Aeronautical Research Council, Reports and Memoranda No. 2265 (8769, 9306, and 9344)*. London: Ministry of Supply, 1945 (1950).

1945a "A Minimum Energy Theorem in Aerodynamics," *Technical Note No. S.M.E. 298*.

1945b "Shock Transmission in Beams of Variable Characteristics," *Report No. S.M.E. 3340*. Farnborough: Royal Aircraft Establishment.

1947* See Davies and Robinson 1947; Hunter-Tod and Robinson 1947; Motzkin and Robinson 1947.

1948 "On Source and Vortex Distributions in the Linearized Theory of Steady Supersonic Flow," *Quarterly Journal of Mechanics and Applied Mathematics*, 1, pp. 408–432.

1948a "On Some Problems of Unsteady Supersonic Aerofoil Theory," *Proceedings of the 7th International Congress for Applied Mechanics,* London, 2, pp. 500–514.

1948b "On the Integration of Hyperbolic Differential Equations," *College of Aeronautics preprint.* See also Robinson 1950b.

1949 "On Non-Associative Systems," *Proceedings of the Edinburgh Mathematical Society*, 8 (2), pp. 111–118; in Robinson 1979, 1, pp. 551–558.

1949a "On the Metamathematics of Algebra," *Journal of Symbolic Logic*, 14, p. 74.

1950 "On the Application of Symbolic Logic to Algebra," *Proceedings of the International Congress of Mathematicians.* Cambridge, Massachusetts, 1950. Providence: American Mathematical Society, 1951. Volume 1, pp. 686–694. In Robinson 1979, 1, pp. 3–11.

1950a "On the Integration of Hyperbolic Differential Equations," *Journal of the London Mathematical Society*, 25, pp. 209–217.

1950b "On Functional Transformations and Summability," *Proceedings of the London Mathematical Society*, 52, pp. 132–160; in Robinson 1979, 2, pp. 454–482.

1950c "Les rapports entre le calcul déductif et l'interprétation sémantique d'un système axiomatique." *Colloques internationaux du C.N.R.S.*, No. 36, Paris: Centre National de la Recherche Scientifique, 1953, pp. 35–52.

1951 *On the Metamathematics of Algebra*. Amsterdam: North–Holland.

1951a "On the Foundations of Dimensional Analysis," dated "RMS Franconia, August/September 1951." This is a 25-page manuscript cited as Robinson MS 1951; RP/YUL.

1951b "On Axiomatic Systems which Possess Finite Models," *Methodos*, 3, pp. 140–149; in Robinson 1979, 1, pp. 322–331.

1951c "Stochastic Processes," course lecture notes; RP/YUL.

1952 "L'application de la logique formelle aux mathématiques," *Applications scientifiques de la logique mathématique. Actes du 2. Colloque international de la logique mathématique,* Paris, Institut Henri Poincaré, 25–30 août, 1952. Paris: Gauthier-Villars, 1954, pp. 51–63.

1953 "Flow Around Compound Lifting Units," *Symposium on High Speed Aerodynamics,* Ottawa, Canada 1953, Ottawa: High Speed Aerodynamics Laboratory, National Aeronautics Establishment, pp. 26–29.

1953a "Non-Uniform Supersonic Flow," *Quarterly of Applied Mathematics*, 10, pp. 307–319; in Robinson 1979, 3, pp. 183–195.

1954* See also Lorentz and Robinson 1954.

1954 "On Some Problems of Unsteady Wing Theory," *Second Canadian Symposium on Aerodynamics,* Toronto, Canada, (publication details), pp. 106–122; in Robinson 1979, 3, pp. 196–211.

1954a "On Predicates and Algebraically Closed Fields," *Journal of Symbolic Logic*,

19, pp. 103–114; in Robinson 1979, 1, pp. 332–343.

1955* See also Gilmore and Robinson 1955.

1955 *Théorie métamathématique des Idéaux.* Paris: Gauthier-Villars.

1955a "Ordered Structures and Related Concepts," *Mathematical Interpretation of Formal Systems.* Amsterdam: North-Holland, pp. 51–56; in Robinson 1979, 1, pp. 99–104.

1955b "Note on an Embedding Theorem for Algebraic Systems," *Journal of the London Mathematical Society*, 30, pp. 249-252; in Robinson 1979, 1, pp. 344-347.

1955c "On Ordered Fields and Definite Functions," *Mathematische Annalen*, 130, pp. 257–271; in Robinson 1979, 1, pp. 355–369.

1956* See also Laurmann and Robinson 1956.

1956 "On the Motion of Small Particles in a Potential Field of Flow," *Communications of Pure and Applied Mathematics*, 9, pp. 69–84; in Robinson 1979, 3, pp. 212–227.

1956a "Further Remarks on Ordered Fields and Definite Functions," *Mathematische Annalen*, 130, pp. 405–409; in Robinson 1979, 1, pp. 370–374.

1956b "A Result on Consistency and Its Application to the Theory of Definition," *Koninklijke Nederlandse Akademie van Wetenschappen. Proceedings*, 59, and *Indagationes Mathematicae*, 18, pp. 47–58; in Robinson 1979, 1, pp. 87–98.

1956c "Note on a Problem of Leon Henkin," *Journal of Symbolic Logic*, 21, pp. 33–35; in Robinson 1979, 1, pp. 105–107.

1956d "Completeness and Persistence in the Theory of Models," *Zeitschrift für mathematische Logik und Grundlagen der Mathematik*, 2, pp. 15–26; in Robinson 1979, 1, pp. 108–119.

1956e *Complete Theories.* Amsterdam: North-Holland; second edition 1977.

1956f "Solution of a Problem by Erdös-Gillman-Henriksen," *Proceedings of the American Mathematical Society*, 7, pp. 908–909; in Robinson 1979, 1, pp. 559–560.

1957* See also Lightstone and Robinson 1957; Lightstone and Robinson 1957a.

1957 "Wave Propagation in a Heterogeneous Elastic Medium," *Journal of Mathematics and Physics*, 36, pp. 210–222; in Robinson 1979, 3, pp. 228–240.

1957a "Transient Stresses in Beams of Variable Characteristics," *Quarterly Journal of Mechanics and Applied Mathematics*, 10, pp. 148–159; in Robinson 1979, 3, pp. 241–252.

1957b "Some Problems of Definability in the Lower Predicate Calculus," *Fundamenta Mathematica*, 44, pp. 309–329; in Robinson 1979, 1, pp. 375–395.

1957c "Relative Model-Completeness and the Elimination of Quantifiers," published in typescript as part of *Summaries of Talks, Summer Institute for Symbolic Logic,* Cornell University, 1957. Princeton: Institute for Defense Analysis, Communications Research Division, 1958; 2d ed. 1960, vol. 1, pp. 155–158. Published under the same title in a more extensive version as Robinson 1958.

1957d "Applications to Field Theory," published in typescript as part of *Summaries of Talks, Summer Institute for Symbolic Logic,* Cornell University, 1957. Princeton: Institute for Defense Analysis, Communications Research Division, 1958; 2d ed. 1960, vol. 3, pp. 326–331. Not published in Robinson 1979.

1957e "Proving a Theorem (As Done by Man, Logician, or Machine)," published in typescript as part of *Summaries of Talks, Summer Institute for Symbolic Logic,* Cornell University, 1957. Princeton: Institute for Defense Analysis, Communications Research Division, 1958; 2d ed. 1960, vol. 3, pp. 350–352. Not published in Robinson 1979.

1958 "Relative Model-Completeness and the Elimination of Quantifiers," *Dialectica,* 12, pp. 394–407; Robinson 1979, 1, pp. 146–159.

1958a "Outline of an Introduction to Mathematical Logic," *Canadian Mathematical Bulletin,* 1 (1958), pp. 41–54, 113–127, 193–208; and 2 (1959), pp. 33–42.

1958b "On the Concept of a Differentially Closed Field," *Bulletin of the Research Council of Israel,* 8, pp. 113–128.

1959 "Solution of a Problem of Tarski," *Fundamenta Mathematica,* 47, pp. 179–204; in Robinson 1979, 1, pp. 414–439.

1959a "Obstructions to Arithmetical Extension and the Theorem of Łoś and Suszko," *Koninklijke Nederlandse Akademie van Wetenschappen. Proceedings,* 62, and *Indagationes Mathematicae,* 21, pp. 439–446; in Robinson 1979, 1, pp. 160–166.

1959b "Algèbre différentielle à valeurs locales," *Atti del VI Congresso dell'Unione Matematica Italiana.* Napoli: Edizioni Cremonese Roma (a one-page abstract).

1959c "Model Theory and Non-Standard Arithmetic," *Infinitistic Methods. Proceedings of the Symposium on Foundations of Mathematics. Warsaw, 2–9 September 1959.* Oxford: Pergamon Press, 1961, pp. 265–302; in Robinson 1979, 1, pp. 167–204.

1959d "On the Concept of a Differentially Closed Field," *Bulletin of the Research Council of Israel,* 8F, pp. 113–128, in Robinson 1979, 1, pp. 440–455.

1960* See also Halfin and Robinson 1960; Zakon and Robinson 1960.

1960 "Local Differential Algebra," *Transactions of the American Mathematical Society,* 97, pp. 427–456; in Robinson 1979, 1, pp. 561–590.

1960a "Recent Developments in Model Theory," in E. Nagel, P. Suppes, and A. Tarski, eds., *Logic, Methodology and Philosophy of Science. Proceedings of the 1960 International Congress for Logic, Methodology and Philosophy of Science.* Stanford, California: Stanford University Press, 1962, pp. 60–79; in Robinson 1979, 1, pp. 12–31.

1961 "Non-Standard Analysis." *Koninklijke Nederlandse Akademie van Wetenschappen. Proceedings,* 64, and *Indagationes Mathematicae,* 23, pp. 432–440; in Robinson 1979, 2, pp. 3–11.

1961a "A Note on Embedding Problems," *Fundamenta Mathematica,* 50, pp. 455–461, in Robinson 1979, 1, pp. 471–477.

1961b "On the Construction of Models," *Essays on the Foundations of Mathemat-*

ics. Jerusalem: The Magnes Press of The Hebrew University, 1961, pp. 207–217; in Robinson 1979, 1, pp. 32–42.
1962 "Airfoil Theory," in Wilhelm Flügge, ed., *Handbook of Engineering Mechanics.* New York: McGraw-Hill, section 72, pp. 1–24.
1962a *Complex Function Theory over Non-Archimedean Fields.* Arlington, Virginia: Armed Services Technical Information Agency, Arlington Hall Station, ASTIA Document No. 282416. Also issued in Jerusalem: The Hebrew University, June 1962.
1963* See also Griesmer, Hoffman, and Robinson 1963; Halfin and Robinson 1963.
1963 *Introduction to Model Theory and to the Metamathematics of Algebra.* Amsterdam: North-Holland, 1963; second printing, 1965. Translated into Russian 1967; Italian 1974; second English edition 1974.
1963a "On Languages which Are Based on Standard Arithmetic," *Nagoya Mathematics Journal,* 22, pp. 83–117; in Robinson 1979, 2, pp. 12–46.
1963b "Topics in Non-Archimedean Mathematics." In Berkeley 1963, pp. 285–298; in Robinson 1979, 2, pp. 99–112.
1963c "Some Remarks on Threshold Functions," *IBM Research Note NC-291.* Yorktown Heights, New York: Thomas J. Watson Research Center.
1964* See also Elgot and Robinson 1964.
1964 "On Generalized Limits and Linear Functionals," *Pacific Journal of Mathematics,* 14, pp. 269–283; in Robinson 1979, 2, pp. 47–61.
1964a "Formalism 64," *Proceedings of the International Congress for Logic, Methodology and Philosophy of Science, Jerusalem, 1964.* Amsterdam: North-Holland, pp. 228–246; in Robinson 1979, 2, pp. 505–523.
1965* See also Elgot, Rutledge, and Robinson 1965.
1965 *Numbers and Ideals. An Introduction to Some Basic Concepts of Algebra and Number Theory.* San Francisco: Holden-Day.
1965a "On the Theory of Normal Families," *Studia logico-mathematica et philosophica, in honorem Rolf Nevanlinna die natali eius septuagesimo 22.X.1965. Acta Philosophica Fennica,* 18, pp. 159–184; in Robinson 1979, 2, pp. 62–87.
1965b "The Metaphysics of the Calculus." In Lakatos 1965, pp. 27–46; in Robinson 1979, 2, pp. 537–555.
1966* See also Bernstein and Robinson 1966.
1966 *Nonstandard Analysis.* Amsterdam: North-Holland; reprinted 1970; revised edition 1974.
1966a "Logica matematica: A New Approach to the Theory of Algebraic Numbers. I and II," *Atti della Accademia Nazionale dei Lincei. Rendiconti della Classe di Scienze fisiche, matematiche e naturali,* 40, pp. 222–225 and pp. 770–774, respectively. In Robinson 1979, 2, pp. 113–116 and 117–121.
1966b "On Some Applications of Model Theory to Algebra and Analysis," *Rendiconti Matematica e delle sue Applicazioni,* 25, pp. 562–592; in Robinson 1979, 2, pp. 158–188.
1967* See also Zakon and Robinson 1967.
1967 "Nonstandard Theory of Dedekind Rings," *Koninklijke Nederlandse Akademie van Wetenschappen. Proceedings,* Series A, 70, and *Indagationes Mathe-*

maticae, 29, pp. 444–452; in Robinson 1979, 2, pp. 122–130.
1967a "Nonstandard Arithmetic," *Bulletin of the American Mathematical Society*, 73, pp. 818–843; in Robinson 1979, 2, pp. 132–157.
1968* See also Hurd and Robinson 1968.
1968 "Some Thoughts on the History of Mathematics," *Compositio Mathematica*, 20, pp. 188–193. Also in *Logic and the Foundations of Mathematics* (dedicated to A. Heyting on his 70th birthday). Groningen: Walters-Noordhoff, 1968, pp. 188–193; in Robinson 1979, 2, pp. 568–573.
1968a "Model Theory." In Klibansky 1968, vol. 1, pp. 61–73; in Robinson 1979, 2, pp. 524–536.
1968b "Problems and Methods of Model Theory," *Aspects of Mathematical Logic* (C.I.M.E. Corsi 3° Ciclo, Varenna). Roma: Edizioni Cremonese, 1969, pp. 181–266. Not published in Robinson 1979.
1968/1969 "Problems and Methods of Model Theory," *Aspects of Mathematical Logic. C.I.M.E. Corsi 3° Ciclo, Varenna, 1968.* Rome: Edizioni Cremonese, 1969, pp. 181–266.
1969* See also Zakon and Robinson 1969.
1969 "Compactification of Groups and Rings and Nonstandard Analysis," *Journal of Symbolic Logic*, 34, pp. 576–588; Robinson 1979, 2, pp. 243–255.
1969a "Topics in Nonstandard Algebraic Number Theory." In Luxemburg 1969, pp. 1–17; in Robinson 1979, 2, pp. 189–205.
1969b "Germs." In Luxemburg 1969, pp. 138–149; in Robinson 1979, 2, pp. 220–231.
1969c "Non-Standard Analysis," typed filmscript of a filmed one-hour lecture produced by the Mathematical Association of America; RP/YUL.
1970* See also Barwise and Robinson 1970.
1970 "Infinite Forcing in Model Theory," *Proceedings of the Second Scandinavian Logic Symposium, Oslo, 1970.* Amsterdam: North-Holland, 1971, pp. 317–340; in Robinson 1979, 1, pp. 243–266.
1970a "Forcing in Model Theory," *Proceedings of the International Congress of Mathematicians, Nice, 1970.* Paris: Gauthier-Villars, 1971, 1, pp. 245–250.
1970b "Elementary Embeddings of Fields of Power Series," *Journal of Number Theory*, 2, pp. 237–247; in Robinson 1979, 2, pp. 232–242.
1971 "Forcing in Model Theory," *Symposia Mathematica*, 5. Rome: Istituto Nazionale di Alta Matematica, 1969/70; repr. London: Academic Press, pp. 69–82; in Robinson 1979, 1, pp. 205–218.
1971a "A Decision Method for Elementary Algebra and Geometry—Revisited," *Proceedings of the Tarski Symposium,* Berkeley, California, 1971. Proceedings of Symposia in Pure Mathematics, 25. Providence, R.I.: American Mathematical Society, 1974, pp. 139–152; in Robinson 1979, 1, pp. 490–503.
1971b "On the Notion of Algebraic Closedness for Noncommutative Groups and Fields," *Journal of Symbolic Logic*, 36, pp. 441–444; in Robinson 1979, 1, pp. 478–481.
1971c "Nonstandard Arithmetic and Generic Arithmetic," *Proceedings of the Fourth International Congress for Logic, Methodology and Philosophy of Sci-*

ence, Bucharest, 1971, P. Suppes et al., eds. Amsterdam: North-Holland, 1973, pp. 137–154; in Robinson 1979, 1, pp. 280–297.

1972* See also Brown and Robinson 1972; Fisher and Robinson 1972; Kelemen and Robinson 1972; Luxemburg and Robinson 1972.

1972 "Algebraic Function Fields and Non-standard Arithmetic." In Luxemburg and Robinson 1972, pp. 1–14; Robinson 1979, 2, pp. 256–269.

1972a "Enlarged Sheaves," *Proceedings of the Victoria Symposium on Nonstandard Analysis, 1972, Lecture Notes in Mathematics,* 369. New York: Springer-Verlag, pp. 249–260; in Robinson 1979, 2, pp. 333–344.

1973 "Metamathematical Problems," *Journal of Symbolic Logic,* 38, pp. 500–516; in Robinson 1979, 1, pp. 43–59.

1973a "Concerning Progress in the Philosophy of Mathematics," *Proceedings of the Logic Colloquium at Bristol, 1973.* Amsterdam: North-Holland, 1975, pp. 41–52; in Robinson 1979, 2, pp. 556–567.

1973b "Standard and Nonstandard Number Systems," *Nieuw Archief voor Wiskunde,* 21, pp. 115–133; in Robinson 1979, 2, pp. 426–444.

1973c "Numbers—What Are They and What Are They Good For?" *Yale Scientific Magazine,* 47, pp. 14–16.

1973d "Nonstandard Points on Algebraic Curves," *Journal of Number Theory,* 5, pp. 301–327; in Robinson 1979, 2, pp. 306–332.

1973e "On Bounds in the Theory of Polynomial Ideals," *Izdat. "Nauka" Sibirsk. Otdel.,* pp. 245–252.

1973f "A Note on Topological Model Theory," *Fundamenta Mathematica,* 81, pp. 159–171; in Robinson 1979, 1, pp. 307–319.

1974* See also Brown and Robinson 1974.

1975* See also Brown and Robinson 1975; Lightstone and Robinson 1975; Roquette and Robinson 1975.

1975 "Otto Toeplitz." In *Dictionary of Scientific Biography,* ed. C.C. Gillispie. New York: Charles Scribner's Sons, vol. 13, p. 428.

1979 *Selected Papers of Abraham Robinson.* H. J. Keisler, S. Körner, W.A.J. Luxemburg and A. D. Young, eds. Volume 1: Model Theory and Algebra. Volume 2: Nonstandard Analysis and Philosophy. Volume 3: Aeronautics. New Haven: Yale University Press.

1979a "Generic Categories." In Robinson 1979, 1, pp. 298–306.

1988 G. Takeuti, ed., "On a Relatively Effective Procedure Getting All Quasi-Integer Solutions of Diophantine Equations with Positive Genus," *Annals of the Japan Association for Philosophy of Science,* 7, pp. 111–115.

Works Coauthored by Robinson

Barwise, Jon, and Abraham Robinson

1970 "Completing Theories by Forcing," *Annals of Mathematical Logic,* 2, pp. 119–142; in Robinson 1979, 1, pp. 219–242.

Bernstein, Allen R., and Abraham Robinson

1966 "Solution of an Invariant Subspace Problem of K. T. Smith and P. R. Halmos," *Pacific Journal of Mathematics,* 16, pp. 421–431; in Robinson 1979, 2, pp. 88–98.

Brown, Donald J., and Abraham Robinson

1972 "A Limit Theorem on the Cores of Large Standard Exchange Economies," *Proceedings of the National Academy of Sciences*, 69, pp. 1258–1260; in Robinson 1979, 2, pp. 279–281.

1974 "The Cores of Large Standard Exchange Economies," *Journal of Economic Theory*, 9, pp. 245–254; in Robinson 1979, 2, pp. 345–354.

1975 "Nonstandard Exchange Economies," *Econometrica*, 43, pp. 41–45; in Robinson 1979, 2, pp. 355–369.

Davies, F. T., and Abraham Robinson

1947 "The Effect of the Sweepback of Delta Wings on the Performance of an Aircraft at Supersonic Speeds," *Aeronautical Research Council, Reports and Memoranda No. 2476,* March 1947. London: Ministry of Supply (Her Majesty's Stationery Office), 1951. This paper is not reproduced in Robinson 1979.

Elgot, Calvin C., and Abraham Robinson

1964 "Random-Access Stored-Program Machines, an Approach to Programming Languages," *Journal of the Association for Computing Machinery,* 11, pp. 365–399; in Robinson 1979, 1, pp. 627–661.

Elgot, Calvin C., J. D. Rutledge, and Abraham Robinson

1965 "Multiple Control Computer Models," in *Systems and Computer Science* (Proceedings of a Conference held in London, Ontario, 1965). Toronto: University of Toronto Press, pp. 60–76; in Robinson 1979, 1, pp. 662–678.

Fagg, S. V., with P. E. Montagnon and Abraham Robinson

1945 "A Note on the Interpretation of V-g Records," *Aeronautical Research Council, Reports and Memoranda No. 2097 (9460).* London: Ministry of Supply.

Fisher, Edward R., and Abraham Robinson

1972 "Inductive Theories and Their Forcing Companions," *Israel Journal of Mathematics,* 12, pp. 95–107.

Gilmore, Paul C., and Abraham Robinson

1955 "Metamathematical Considerations on the Relative Irreducibility of Polynomials," *Canadian Journal of Mathematics,* 7, pp. 483–489; in Robinson 1979, 1, pp. 348–354.

Griesmer, J. H., A. J. Hoffman and Abraham Robinson

1963 "On Symmetric Bimatirx Games," *IBM Research Paper RC-959.* Yorktown Heights, New York: Thomas J. Watson Research Center.

Halfin, S., and Abraham Robinson

1960 "Local Differential Algebra—the Analytic Case," *Technical (Scientific) Note No. 9,* Contract No. AF61(052)-187. Brussels, Belgium: U.S. Air Force, Air Research and Development Command, European Office.

1963 "Local Partial Differential Algebra," *Transactions of the American Mathematical Society,* 109, pp. 165–180; in Robinson 1979, 1, pp. 591–606.

Hunter-Tod, J. H., and Abraham Robinson

1947 "Bound and Trailing Vortices in the Linearized Theory of Supersonic Flow, and the Downwash in the Wake of a Delta Wing," *College of Aeronautics Report No. 10.*

Hurd, A. E., and Abraham Robinson

1968 "On Flexural Wave Propagation in Nonhomogeneous Elastic Plates," *SIAM Journal of Applied Mathematics*, 16, pp. 1081–1089; in Robinson 1979, 3, pp. 253–261.

Kelemen, Peter, and Abraham Robinson

1972 "The Nonstandard $\lambda{:}\phi_2^4(x)$: Model. I. The Technique of Nonstandard Analysis in Theoretical Physics," *Journal of Mathematical Physics,* 13, pp. 1870–1974; in Robinson 1979, 2, pp. 270–274.

1972a "The Nonstandard $\lambda{:}\phi_2^4(x)$: Model. II. The Standard Model from a Nonstandard Point of View," *Journal of Mathematical Physics,* 13, pp. 1875–1878; in Robinson 1979, 2, pp. 275–278.

Laurmann, John A., and Abraham Robinson

1956 *Wing Theory*. Cambridge: Cambridge University Press.

Lightstone, A. Harold, and Abraham Robinson

1957 "Syntactical Transforms," *Transactions of the American Mathematical Society*, 86, pp. 220–245; in Robinson 1979, 1, pp. 120–145.

1957a "On the Representation of Herbrand Functions in Algebraically Closed Fields, *Journal of Symbolic Logic*, 22, pp. 187–204; in Robinson 1977, 1, pp. 396–413.

1975 *Nonarchimedean Fields and Asymptotic Expansions*. Amsterdam: North-Holland.

Lorentz, G. G., and Abraham Robinson

1954 "Core-Consistency and Total Inclusion for Methods of Summability," *Canadian Journal of Mathematics*, 6, pp. 27–34; in Robinson 1979, 2, pp. 483–490.

Luxemburg, W.A.J., and Abraham Robinson

1972 *Contributions to Nonstandard Analysis*. Amsterdam: North-Holland.

Motzkin, Theodore, and Abraham Robinson

1947 "The Characterization of Algebraic Plane Curves," *Duke Mathematical Journal*, 14, pp. 837–853; in Robinson 1979, 1, pp. 534–550.

Roquette, Peter, and Abraham Robinson

1975 "On the Finiteness Theorem of Siegel and Mahler Concerning Diophantine Equations," *Journal of Number Theory*, 7, pp. 121–176; in Robinson 1979, 2, pp. 370–425.

Whitby, R. H., and Abraham Robinson

1943 "Note on the General Design of T.B.R. Aircraft," (July 1943). Although this paper is not cited in the Bibliography of Robinson's works in Robinson 1979, a copy is preserved among Robinson's files in the archives of Yale University Library.

Zakon, Elias, and Abraham Robinson

1960 "Elementary Properties of Ordered Abelian Groups," *Transactions of the American Mathematical Society*, 96, pp. 222–236; in Robinson 1979, 1, pp. 456–470.

1969 "A Set-Theoretical Characterization of Enlargements," in Luxemburg 1969, pp. 109–122; Robinson 1979, 2, pp. 206–219.

Works by Other Authors

Addison, J. W.
1984 "Alfred Tarski: 1901–1983," *Annals of the History of Computing*, 6, pp. 335–336.

Addison, Paul
1951 *Now the War Is Over: A Social History of Britain, 1945–1951*. London: British Broadcasting Corp.; repr. J. Cape, 1985.

Agmon, Shmuel
1956 "Professor Fekete on his 70th Birthday," *Riveon Lematematika*, 10, pp. 1–8. A portrait accompanies the article.

Albers, Donald J.
1987 et al., *International Mathematical Congresses: An Illustrated History, 1893–1986* (rev. ed.). New York: Springer-Verlag.
1991 "Paul Halmos by Parts," in Ewing and Gehring 1991, pp. 3–32.

Albeverio, Sergio, Jens Erik Fenstad, Raphael Høegh-Krohn, Tom Lindstrøm, eds.
1986 *Nonstandard Methods in Stochastic Analysis and Mathematical Physics*. New York: Academic Press.

Alder, Henry L.
1971 "Report of the ASL Annual Meeting for 1971," *American Mathematical Monthly*, 78, pp. 1047–1061.

Alder, Henry L., and Ivan Niven
1990 "Yueh-Gin Gung and Dr. Charles Y. Hu Award for Distinguished Service to Leon Henkin," *The American Mathematical Monthly*, 97 (1) (January), pp. 3–4.

[Amirà, Binyamin (Benjamin)]
1970 "In Memoriam," *Journal d'analyse mathématique*, 23, pp. xii–xvi (with photograph).

Amitsur, Shimshon A.
1957 "The Scientific Work of Professor Jakob Levitzki," *Riveon Lematematika*, 11, pp. 1–6 (in Hebrew).
1974 "Jakob Levitzki," *Israel Journal of Mathematics*, 19, pp. 1–3 (in Hebrew).

Amouroux, Henri
1961 *La vie des Français sous l'occupation. Le peuple du désastre. 1939–1940*. Paris: A. Fayard.

Anderson, Robert M.
1990 *Nonstandard Methods in Mathematical Economics*, Berkeley, Calif.: Department of Economics.

Arieli, Yehoshua
1982 "Jacob Talmon—An Intellectual Portrait," in *Totalitarian Democracy and After. International Colloquium in Memory of Jacob L. Talmon*. Jerusalem, 21-24 June, 1982. Jerusalem: The Magnes Press of the Hebrew University, 1984, pp. 1–34.

Arnold, Matthew
1865 *Essays in Criticism*. Boston: Ticknor and Fields.

Aron, Robert, and G. Elgey
1958 *The Vichy Regime, 1940–44.* Trans. H. Hare. New York: Macmillan.
Arruda, A. I., N.C.A. da Costa and R. Chuaqui, eds.
1977 *Non-Classical Logics, Model Theory and Computability*. Amsterdam: North-Holland.
Ashby, Eric
1967 *Report of the Academic Advisory Committee on Birkbeck College*. London: Birkbeck College, University of London.
Aumann, Robert J.
1964 "Markets with a Continuum of Traders," *Econometrica*, 32, pp. 39–50.
Avermaete, Roger
1975 *Frans Masereel, 1889–1972.* Amsterdam: Arbeiderspers; repr. London: Thames and Hudson, 1977.
Ax, James
1968 "The Elementary Theory of Finite Fields," *Annals of Mathematics*, 88, pp. 239–271.
Ax, James, and Simon Kochen
1965 "Diophantine Problems over Local Fields. I," *American Journal of Mathematics*, 87, pp. 605–630.
1965a "Diophantine Problems over Local Fields. II. A Complete Set of Axioms for *P*-adic Number Theory," *American Journal of Mathematics*, 87, pp. 631–648.
Bachelard, Gaston
1945 "La Philosophie Scientifique de Léon Brunschvicg," in Brunschvicg 1945, pp. 77–84.
Baehr, Harry W.
1987 "Notes on a Century: The Paris Trib in the Mid-Forties: Starting to Think Internationally," *International Herald Tribune*, August 19, 1987, p. 5.
Balaban, Majer
1916 *Dzieje Zydow w Galicyi w Rzeczypospolitej Krakowskiej. 1772–1868.* Lwów: B. Poloniecki.
Balázs, János
1958 "The Scientific Work of the Late Michael Fekete," *Matematikai lapok*, 9, pp. 197–224.
Balsdon, Dacre
1970 *Oxford Now and Then*. London: Duckworth.
Banks, James and Paula
1989 "Kent State: How the War in Vietnam Became a War at Home," in Dumbrell 1989, pp. 68–81.
Bardoux, Jacques
1957 *Journal d'un témoin de la troisième: Paris, Bordeaux, Vichy.* 1 September 1939—15 July 1940. Paris: A. Fayard.
Barreau, Henri, ed.
1989 *La Mathématique Non-Standard*. Paris: Editions du CNRS.
Barwise, Jon
1970 See Barwise and Robinson, above.
1977 ed., *Handbook of Mathematical Logic*. Amsterdam: North-Holland.

Barwise, Jon, and S. Feferman, eds.
1985 *Model-Theoretic Logics.* New York: Springer-Verlag.
Batchelor, G. K.
1990 "Kolmogorov's Work on Turbulence," in Kendall 1990, pp. 47–51.
Baudouin, Paul
1948 *The Private Diaries (March 1940 to January 1941) of Paul Baudouin.* Trans. C. Petrie. London: Eyre and Spottiswoode.
de Beauvoir, Simone
1960 *La Force de l'âge.* Paris: Gallimard.
Becker, Hellmut
1972 "Saul B. Robinsohn," typescript eulogy dated Berlin, April 1972, and read on the occasion of the memorial service for Saul Robinsohn held at the Jewish cemetery in Berlin on April 13, 1972. A copy is included among the Robinson papers, archives of Yale University.
1973 "Einleitung," in Robinsohn 1973, pp. 7–14.
Behnke, H.
1949 "Otto Toeplitz zum Gedächtnis," *Mathematisch-Physikalische Semesterberichte,* 1, pp. 89–96.
Behnke, H., and G. Köthe
1963 "Otto Toeplitz zum Gedächtnis," *Jahresbericht der Deutschen MathematikeR-Vereinigung,* 66, pp. 1–16.
Belhoste, Bruno
1985 *Cauchy, 1789–1857. Un mathématicien légitimiste au XIXe siècle.* Paris: Belin.
1991 *Augustin-Louis Cauchy. A Biography.* Trans. F. Ragland. New York: Springer-Verlag.
Bell, D. H., et al.
1983 *Parallel Programming–a Bibliography.* New York: Wiley and Sons.
Bell, John L., and Moshé Machover
1977 *A Course in Mathematical Logic.* Amsterdam: North-Holland.
Bell, John L., and A. B. Slomson
1969 *Models and Ultraproducts: An Introduction.* Amsterdam: North-Holland.
Benacerraf, Paul, and Hilary Putnam, eds.
1964 *Philosophy of Mathematics. Selected Readings.* Englewood Cliffs, N.J.: Prentice-Hall; second ed., New York: Cambridge University Press, 1983.
Benis-Sinaceur, Hourya: See Sinaceur, Hourya.
Bentwich, Norman
1936 *The Refugees from Germany, April 1933 to December 1935.* London: George Allen and Unwin.
1953 *The Rescue and Achievement of Refugee Scholars. The Story of Displaced Scholars and Scientists. 1933–1952.* The Hague: Martinum Nijhoff.
1960 *The Jews in Our Time. The Development of Jewish Life in the Modern World.* Harmondsworth: Penguin Books.
1961 *The Hebrew University of Jerusalem, 1918–1960.* London: Weidenfeld and Nicolson.
Berkeley
1963 Addison, J. W., L. Henkin, and A. Tarski, eds. *The Theory of Models.*

Proceedings of the 1963 International Symposium at Berkeley. Amsterdam: North-Holland.

Beth, Evert W.

1953 "On Padoa's Method in the Theory of Definition," *Koninklijke Nederlandse Akademie van Wetenschappen, Proceedings*, 56, and *Indagationes Mathematicae*, 15, pp. 330–339.

1959 *The Foundations of Mathematics. A Study in the Philosophy of Science*. Amsterdam: North-Holland.

Birkhoff, Garrett

1960 *Hydrodynamics. A Study in Logic, Fact, and Similitude*. Princeton: Princeton University Press, 1950; repr. New York: Dover, 1955; second rev. ed., Princeton: Princeton University Press, 1960.

Bissell, Claude

1974 *Halfway Up Parnassus. A Personal Account of the University of Toronto, 1932–1971*. Toronto: University of Toronto Press.

Blum, Lenore

1968 "Generalized Algebraic Structures: A Model Theoretic Approach," *Ph.D. Dissertation*. Cambridge, Mass.: Massachusetts Institute of Technology.

1977 "Differentially Closed Fields: A Model-Theoretic Tour," *Contributions to Algebra*. New York: Academic Press, pp. 37–61.

Bohr, Harald

1925 "Zur Theorie der fastperiodischen Funktionen," *Acta Mathematica*, 45, pp. 29–127, and 46, pp. 101–214.

Boothe, Clare

1940 *Europe in the Spring*. New York: A.A. Knopf.

Born, Max

1940 "Obituary. Prof. Otto Toeplitz," *Nature*, 145, p. 617.

Bos, Hendrik J. M.

1974 "Differentials, Higher-Order Differentials and the Derivative in the Leibnizian Calculus," *Archive for History of Exact Sciences*, 14 (1974–1975), pp. 1–90.

Bothwell, Robert, Ian Drummond and John English

1989 *Canada since 1945: Power, Politics, and Provincialism*. Toronto: University of Toronto Press, 1981; rev. ed. 1989.

Bourbaki, N.

1949 "Foundations of Mathematics for the Working Mathematician," *Journal of Symbolic Logic*, 14 (1) (March), pp. 1–8.

Bovis, H. Eugene

1971 *The Jerusalem Question, 1917–1968*. Stanford, Calif.: Hoover Institution Press.

Boyer, Carl

1959 *The History of the Calculus and Its Conceptual Development*. New York: Dover, 1959 (reprinted from the earlier edition, *The Concepts of the Calculus*, New York: Columbia University Press).

Braun, Otto

1940 *Von Weimar zu Hitler*. New York: Europa Verlag.

Brecht, Arnold
1967 *Mit der Kraft des Geistes. Lebenserinnerungen. Part II, 1927–1967*. Stuttgart: Deutsche Verlags-Anstalt.
Brewster, Dorothy
1960 *Virginia Woolf's London*. New York: New York University Press, p. 317.
Bridgman, Percy W.
1931 *Dimensional Analysis*. New Haven: Yale University Press, 1922; rev. ed., 1931.
Bristol
1964 *Proceedings of the Logic Colloquium at Bristol, 1973*. Amsterdam: North-Holland, 1975.
de Broglie, Louis
1945 "Léon Brunschvicg et l'évolution des sciences," in Brunschvicg 1945, pp. 73–74.
Browder, Felix E., ed.
1976 *Mathematical Developments Arising from Hilbert Problems*, Proceedings of Symposia in Pure Mathematics, vol. 28. Providence, R.I.: American Mathematical Society, 1976.
Brown, Donald
1972 See Brown and Robinson 1972, above.
1974 See Brown and Robinson 1974, above.
1975 See Brown and Robinson 1975, above.
Brunschvicg, Léon
1912 Les Étapes de la philosophie mathématique. Paris: Alcan; repr. Paris: A. Blanchard, 1972.
1945 "Léon Brunschvicg: L'Oeuvre et l'homme," *Revue de métaphysique et de morale,* 50 (1–2) (January–April), pp. 1–140.
Büchi, J. Richard
1962 "On a Decision Method in Restricted Second Order Arithmetic," in *Logic, Methodology and Philosophy of Science. Proceedings of the 1960 International Congress*, E. Nagel, P. Suppes, and A. Tarski, eds. Stanford: Stanford University Press, 1962, pp. 1–13.
Buckingham, Edward
1914 "Physically Similar Systems," *Physical Review*, 4, pp. 345–376.
Buckley, Christopher
1982 "A Keening of Weenies," in Dubois 1982, pp. 260–278.
Campbell, Louis Lorne
1955 "Mixed Boundary Value Problems for Hyperbolic Differential Equations," *Ph.D. Thesis,* University of Toronto.
Carmichael, Joel, and M. W. Weisgal, eds.
1963 *A Biography by Several Hands. Chaim Weizmann*. New York: Atheneum: 1963.
Cayley, Arthur
1857 "On the theory of the Analytical Forms Called Trees," *Philosophical Magazine*, 13, pp. 172–176.
Chang, Chin-Liang, and R. C.-T. Lee
1973 *Symbolic Logic and Mechanical Theorem Proving.* New York: Academic Press.

Chapman, Guy

1968 *Why France Fell. The Defeat of the French Army in 1940*. New York: Holt, Rinehart and Winston.

Cherlin, Gregory

1971 "A New Approach to the Theory of Infinitely Generic Structures," *Ph.D. Dissertation*, Yale University.

1972 and Joram Hirschfeld, "Ultrafilters and Ultraproducts in Non-standard Analysis," in Robinson and Luxemburg 1972, pp. 261–279.

Chevalley, Claude

1937 "Généralisation de la théorie du corps de classes pour les extensiones infinis," *Journal de mathématiques pures et appliquées*, 15, pp. 359–371.

Chuaqui, R.

1970 "Meeting of the Association for Symbolic Logic, Santiago, Chile, 1970," *The Journal of Symoblic Logic*, 36 (1971), pp. 576–580.

Church, Alonzo

1956 *Introduction to Mathematical Logic*. Princeton: Princeton University Press.

Churchill, Winston S.

1959 *Memoirs of the Second World War*. Boston: Houghton Mifflin Company.

Clarke, Arthur C.

1949 "The Dynamics of Space Flight," *Journal of the British Interplanetary Society*, 8 (2), pp. 71–84.

Cleator, P. E.

1934 "Retrospect and Prospect," *Journal of the British Interplanetary Society*, 1 (1) (January), pp. 2–5.

Cohen, Paul J.

1971 "Comments on the Foundations of Set Theory," in *Axiomatic Set Theory. Proceedings of Symposia in Pure Mathematics. Part I*, D. S. Scott, ed., 13, Providence, R.I.: American Mathematical Society, pp. 9–15.

1967 with Reuben Hersh, "Non-Cantorian Set Theory," *Scientific American*, 217 (December), pp. 104–116.

Cohn, Emile B.

1944 *David Wolffsohn. Herzl's Successor.* Washington: The Zionist Organization of America.

Cohn, Ernst J.

1950 "Die Jüdische Gemeinde zu Breslau," chap. 1, n. 36 in *Breslau. Ein Buch der Erinnerung*, Niels von Holst, ed. Hamlen: F. Seifert.

Colette

1932 "Mannequins," chap. 4 in *My Paris*, Arthur K. Griggs, ed. New York: The Dial Press, pp. 32–33.

Colloquium

1950 *Les méthodes formelles en axiomatique. Colloque international de logique mathématique. Paris, Décembre 1950.* Paris: Centre national de la recherche scientifique, 1953.

1952 *Applications scientifiques de la logique mathématique. Actes du 2. Colloque international de logique mathématique. Paris, Institut Henri Poincaré, 25–30 août, 1952*. Paris: Gauthier-Villars, 1954.

Congress

1950 *Proceedings of the International Congress of Mathematicians. Cambridge 1950.* Providence, R.I.: American Mathematical Society, 1952.

1954 *Proceedings of the International Congress of Mathematicians. Amsterdam 1954.* Groningen: E. P. Noordhoff; Amsterdam: North-Holland, 1957.

1958 *Proceedings of the International Congress of Mathematicians. Edinburgh 1958.* Cambridge: Cambridge University Press, 1960.

1962 *Proceedings of the International Congress of Mathematicians. Stockholm 1962.* Djursholm: Institut Mittag-Leffler; printed in Uppsala: Auqvist & Wiksells, 1963.

1966 *Proceedings of the International Congress of Mathematicians. Moscow 1966.* Moscow: MIR, 1968.

1970 *Actes du congrès international des mathématiciens. Nice 1970.* Paris: Gauthier-Villars, 1971.

Cook, Marilyn

1961 "Foreign Student's Haven," *The Varsity* (student newspaper, University of Toronto), December 13, 1961, p. 5.

Cooke, Richard G.

1950 *Infinite Matrices and Sequence Spaces.* London: Macmillan.

1953 *Linear Operators. Spectral Theory and Some Other Applications.* London: Macmillan.

1960 "Paul Dienes," *Journal of the London Mathematical Society*, 35 (1960), pp. 251–256.

Cornell

1957 *Summaries of Talks, Summer Institute for Symbolic Logic. Cornell University, 1957.* Princeton: Institute for Defense Analysis, Communications Research Division, 1958; second ed. 1960, in three volumes.

Courant, Richard, and Herbert Robbins

1941 *What is Mathematics?* London: Oxford University Press; repr. 1961 and 1978.

Coxeter, H.S.M., M. C. Escher, et al., eds.

1986 *M. C. Escher, Art and Science: Proceedings of the International Congress on M. C. Escher.* Rome, March 26–28, 1985. Amsterdam: North-Holland.

Craig, William

1957 "Three Uses of the Herbrand-Genzen Theorem in Relating Model Theory and Proof Theory," *Journal of Symbolic Logic*, 22, pp. 269–285.

Creighton, Donald G.

1976 *The Forked Road. Canada 1939–1957.* Toronto: McClelland and Steward.

Datlin, Daniel (Jr.)

1982 *Liberal Education at Yale. The Yale College Course of Study. 1945–1978.* Washington, D.C.: University Press of America.

Dauben, Joseph W.

1979 *Georg Cantor: His Mathematics and Philosophy of the Infinite.* Cambridge, Mass.: Harvard University Press, 1979; repr. Princeton: Princeton University Press, 1990.

1988 "Abraham Robinson and Nonstandard Analysis: History, Philosophy, and Foundations of Mathematics," in *History and Philosophy of Modern Mathematics*, Minnesota Studies in the Philosophy of Science, Willaim Aspray and Philip Kitcher, eds., vol. 11, Minneapolis: University of Minnesota Press, pp. 177–200.

1989 "Abraham Robinson: Les Infinitésimals, l'analyse nonstandard, et les fondements des mathématiques," in Barreau 1989, pp. 157–184.

1990 "Abraham Robinson," *The Dictionary of Scientific Biography, Supplement,* 2, New York: Scribners, pp. 748–751.

Davidson, Carrie D.

1946 *Out of Endless Yearnings: A Memoir of Israel Davidson*. New York: Bloch Publishing Co.

Davis, Martin

1965 ed., *The Undecidable.* Hewlett, N.Y.: Raven Press.

1977 *Applied Nonstandard Analysis.* New York: John Wiley and Sons.

Davis, Martin, and Reuben Hersh

1972 "Nonstandard Analysis," *Scientific American,* 226, pp. 78–86.

Davis, Martin, Yuri V. Matiyasevich (Yu. V. Matijascvic), and Julia Robinson

1976 "Hilbert's Tenth Problem. Diophantine Equations: Positive Aspects of a Negative Solution," *Proceedings of Symposia on Pure Mathematics,* 28, Providence, R.I.: American Mathematical Society, pp. 323–378.

Dawson, John

1993 "The Compactness of First-Order Logic: From Gödel to Lindström," *History and Philosophy of Logic*, 14, pp. 14–37.

Dedekind, Richard

1888 *Was sind und was sollen die Zahlen*. Braunschweig: Vieweg, 1888. Translated by W. W. Beman as "The Nature and Meaning of Numbers" in the collection *Essays on the Theory of Numbers, Continuity and Irrational Numbers, the Nature and Meaning of Numbers*. Chicago: Open Court, 1901; repr. New York: Dover, 1963.

Deschoux, Marcel

1949 *La Philosophie de Léon Brunschvicg*. Paris: Presses universitaires de France.

Diener, Francine, and Georges Reeb

1989 *Analyse Non Standard.* Paris: Hermann.

Diener, Marc

1984 "The Canard Unchained or How Fast/Slow Dynamical Systems Bifurcate," *The Mathematical Intelligencer,* 6 (3), pp. 38–49.

Dienes, Paul

1931 *An Introduction to the Theory of Functions of a Complex Variable*, the Taylor Series. Oxford: The Clarendon Press.

1938 *Logic of Algebra*. Paris: Hermann.

Dieudonné, Jean

1970 "The Work of Nicholas Bourbaki," *American Mathematical Monthly*, 77, pp. 134–145.

1977 *Panorama des mathématiques pures: Le choix bourbachique*. Paris: Gauthier-Villars.

Dior, Christian
1957 *Dior by Dior. The Autobiography of Christian Dior*. Trans. Antonia Fraser. London: Weidenfeld and Nicolson.
Dörnte, Wilhelm
1929 "Untersuchungen über einen verallgemeinerten Gruppenbegriff," *Mathematische Zeitschrift*, 29, pp. 1–19.
Douglas, Sholto
1966 *Years of Command*. London: Collins.
Draper, Hal
1965 *Berkeley: The New Student Revolt*. New York: Grove Press.
Dubois, Diana, ed.
1982 *My Harvard, My Yale*. New York: Random House.
Dubrovsky, Diana Lydia
1971 "Computability in *P*-adically Closed Fields and Nonstandard Arithmetic," *Ph.D. Thesis*, Department of Mathematics, University of California, Los Angeles.
1974 "Some Subfields of Q_p and Their Nonstandard Analogues," *Canadian Journal of Mathematics*, 26, pp. 473–491.
Duff, George F. D.
1969 "Arthur Francis Chesterfield Stevenson," *Proceedings and Transactions of the Royal Society of Canada*, 7, pp. 103–107.
1984 "The Evolving Role of Applied Mathematics," *Applied Mathematics Notes*, 9, pp. 1–18.
Dumbrell, John, ed.
1989 *Vietnam and the Antiwar Movement*. Aldershot, England: Avebury.
Duren, Peter L., et al., eds.
1988 *A Century of Mathematics in America*. Providence, R.I.: The American Mathematical Society.
Eckmann, Benno
1942 "Systeme von Richtungsfeldern in Sphären und stetige Lösungen komplexer linearer Gleichungen," *Commentarii Mathematici Helvetici*, 15 (1942–43), pp. 1–26.
Egbert, D. D.
1970 *Social Radicalism and the Arts, Western Europe*. New York: Knopf.
Ehrenfeucht, Andrzej
1956 "Models of Axiomatic Theories Admitting Automorphisms," *Fundamenta Mathematica*, 43, pp. 50–60.
Eklof, Paul, and G. Sabbagh
1970 "Model completions and Modules," *Annals of Mathematical Logic*, 2, pp. 251–296.
Engeler, E.
1964 *Mathematical Reviews*, 27, no. 3533.
Erdélyi, Arthur
1961 "An Extension of the Concept of Real Number," *Proceedings of the Fifth Canadian Mathematics Congress. University of Montreal, 1961*. Toronto: University of Toronto Press, 1963, pp. 173–183.
1967 Review of Robinson 1966 (*Nonstandard Analysis*), *Journal of the London Mathematical Society*, 42, pp. 764–766.

Erdös, Paul, Leonard Gillman and Melvin Henriksen

1955 "An Isomorphism Theorem for Real-Closed Fields," *Annals of Mathematics*, 61, pp. 542–554.

Esnault-Pelterie, Robert

1948 *L'analyse dimensionelle*. Lausanne: Rouge.

Etherington, Ivor M. H.

1939 "Genetic Algebras," *Proceedings of the Royal Society of Edinburgh*, 59, pp. 242–258.

1939a "On Non-Associative Combinations," *Proceedings of the Royal Society of Edinburgh*, 59, pp. 153–162.

1940 "Commutative Train Algebras of Ranks 2 and 3," *Journal of the London Mathematical Society*, 15, pp. 136–149.

1941 "Non-Associative Algebra and the Symbolism of Genetics," *Proceedings of the Royal Society of Edinburgh*, Sect. B, 61, pp. 24–42.

1941a "Special Train Algebras," *Quarterly Journal of Mathematics*, 12, pp. 1–8.

Euler, Leonhard

1748 *Introductio in Analysin Infinitorum*. First ed. Lausanne.

Ewing, John H., and F. W. Gehring, eds.

1991 *Paul Halmos. Celebrating 50 Years of Mathematics*. New York: Springer-Verlag.

Feferman, Solomon, and Alfred Tarski

1953 "Review of the Paper 'A Proof of the Skolem-Löwenheim Theorem,' by H. Rasiowa and R. Sikorski," *Journal of Symbolic Logic*, 18, pp. 339–340.

Feller, William

1950 *An Introduction to Probability Theory and Its Applications*. New York: Wiley.

Ferrières, Gabrielle

1950 *Jean Cavaillès, philosophe et combattant (1903–1944)*. Paris: Presses universitaires de France.

Ferro, R., and C. Bonotto

1989 *Logic Colloquium '88, Proceedings of the Colloquium Held in Padova, Italy, August 22–31, 1988*, S. Valentini and A. Zanardo, eds. Amsterdam: North-Holland.

Fontaine, André

1968 *History of the Cold War, from the October Revolution to the Korean War, 1917–1950*. Trans. D. D. Paige. New York: Pantheon Books.

Fowell, Leonard Richard

1955 "An Exact Theory of Supersonic Flow around a Delta Wing," *Ph.D. Thesis*, University of Toronto.

Fowlie, Wallace

1960 *French Stories/Contes Français*. New York: Bantam Books.

Fraenkel, Adolph Abraham

1926 *Journal für die reine und angewandte Mathematik*, 155, pp. 129–158.

1928 *Einleitung in die Mengenlehre*. third ed. Berlin: J. Springer; repr. New York: Dover, 1946.

1953 *Abstract Set Theory*. Amsterdam: North-Holland.

1967 *Lebenskreise*. Stuttgart: Deutsche Verlags-Anstalt.

1973 Y. A. Bar-Hillel, A. Levy, and D. van Dalen, *The Foundations of Set Theory*. Second rev. ed., Amsterdam: North-Holland.

Frayne, T. E.

1956 A. C. Morel and D. S. Scott, "Reduced Direct Products," *Fundamenta Mathematica*, 51, pp. 195–218.

1958 D. S. Scott and A. Tarski, "Reduced Products," *Notices of the American Mathematical Society*, 5, pp. 673–674.

Freeman, C. Denis, and Douglas Cooper

1940 *The Road to Bordeaux*. London: Cresset Press; repr. New York: Harper and Bros., 1941.

Friedman, Joel Irwin

1966 "A Set Theory of Proper Classes," *Ph.D. Dissertation*, Department of Philosophy, University of California, Los Angeles, 1966.

Furth, Montgomery, C.-C. Chang, and Alonzo Church

1974 "Richard Montague," *In Memoriam*. Los Angeles: University of California, February, pp. 68–70.

Gal, Roger

1962 "Institut de l'UNESCO pour l'éducation à Hambourg. Dix ans d'activité," *International Review of Education*, 8, pp. 1–11; English summary, "The UNESCO Institute for Education at Hamburg. Ten Years of Activity," pp. 11–12.

Gaulle, Charles de

1959 *The Complete War Memoirs*. New York: Simon and Schuster; second ed., 1967.

Gelernter, Herbert

1957 "Theorem Proving by Machine," published in typescript as part of *Summaries of Talks, Summer Institute for Symbolic Logic. Cornell University, 1957*. Princeton: Institute for Defense Analysis, Communications Research Division, 1958; second ed. 1960, vol. 1, p. 305.

Gide, André

1946 *Journal, 1939–1942*. Paris: Gallimard; repr. with index, 1954.

Gilmore, Paul C.

1960 "A Proof Method for Quantification Theory; Its Justification and Realization," *I.B.M. Journal of Research Development*, 4, pp. 28–35.

Givant, Steven

1986 "Bibliography of Alfred Tarski," *Journal of Symbolic Logic*, 51, pp. 913–941.

Glimm, James, and Arthur Jaffe

1968 "A $(\phi^4)_2$ Quantum Field Theory without Cutoffs. I," *Physical Review*, 176, pp. 1945–1951.

1970 "The $(\phi^4)_2$ Quantum Field Theory without Cutoffs. II. The Field Operators and the Approximate Vacuum," *Annals of Mathematics*, 91, pp. 362–401.

1970a "The $(\phi^4)_2$ Quantum Field Theory without Cutoffs. III. The Physical Vacuum," *Acta Mathematica*, 125, pp. 203–267.

Glowka, Detlef

1972 "Nachruf für Saul Benjamin Robinsohn," *Neue Sammlung. Göttinger Zeitschrift für Erziehung und Gesellschaft*, 12 (May–June), pp. 180–186.

Gödel, Kurt

1940 *The Consistency of the Continuum Hypothesis*. Princeton: Princeton University Press; later printings after 1951 with corrections and notes.

1965 "Remarks before the Princeton Bicentennial Conference on Problems in Mathematics," in Davis 1965, pp. 84–88.

Goldschmidt, Dietrich

1972 "Saul Benjamin Robinsohn," *Mitteilungen aus der Max-Planck-Gesellschaft*, 2, pp. 79–85.

Goldstein, Israel

1940 *Toward a Solution*. New York: G. P. Putnam's Sons.

Goodman, Nicolas D.

1990 "Mathematics as Natural Science," *The Journal of Symbolic Logic*, 55, pp. 182–193.

Goren, Arthur A., ed.

1982 *Dissenter in Zion: From the Writings of Judah L. Magnes*. Cambridge, Mass.: Harvard University Press.

Grabiner, Judith V.

1981 *The Origins of Cauchy's Rigorous Calculus*. Cambridge, Mass.: MIT Press.

Grattan-Guinness, Ivor

1985 "Russell's Logicism versus Oxbridge Logics, 1890–1925: A Contribution to the Real History of Twentieth-Century English Philosophy," *Russell: The Journal of the Bertrand Russell Archives*, 5 (winter 1985–1986), pp. 101–131.

Green, J. J.

1970 *Aeronautics, Highway to the Future. A Study of Aeronautical Research and Development in Canada,* Background Study for the Science Council of Canada, Special Study no. 12. Ottawa: Queens Printer for Canada.

Greenberg, M. J.

1966 "Rational Points in Henselian Discrete Valuation Rings," *Publications mathématiques*, no. 31, Paris: Institut des hautes études scientifiques, pp. 59–64.

Grinnell-Milne, Duncan

1962 *The Triumph of Integrity. A Portrait of Charles de Gaulle*. New York: Macmillan.

Gunning, Robert C., and H. Rossi

1965 *Analytic Functions of Several Complex Variables*. Englewood Cliffs, N.J.: Prentice-Hall.

Hajnal, A.

1956 "On a Consistency Theorem Connected with the Generalized Continuum Problem," *Zeitschrift für Mathematische Logik und Grundlagen der Mathematik*, 2, pp. 131–136.

Halfin, Shlomo

1962 "Contributions to Differential Algebra," cited in *Abstracts of Theses Approved for the Degrees of Doctor Philosophiae (Ph.D.) and Doctor Juris (Jur.D.), in the Hebrew University of Jerusalem during the Academic Year 1961–62*, Jerusalem: The Hebrew University of Jerusalem, 1964.

Hall, A. Rupert
1980 *Philosophers at War. The Quarrel Between Newton and Leibniz.* Cambridge, England: Cambridge University Press.
Halmos, Paul R.
1955 Review of Robinson 1955, in *The Journal of Symbolic Logic*, 20, pp. 279–281.
1966 "Invariant Subspaces of Polynomially Compact Operators," *Pacific Journal of Mathematics*, 16, pp. 433–437.
1985 *I Want to be a Mathematician. An Automathography.* New York: Springer-Verlag.
Hamilton, A., and J. B. Jackson
1969 *UCLA on the Move During Fifty Golden Years 1919–1969.* Los Angeles: The Ward Kitchie Press.
Hardy, G. H., and H. Heilbronn
1938 "Edmund Landau," *Journal of the London Mathematical Society*, 13, pp. 302–310; repr. in Landau 1985, 1, pp. 15–24.
Harris, R. H.
1987 "Notes on a Century. The 'Onyxpected' Wonders of Life in Paris Cafes," *International Herald Tribune*, August 26, 1987, p. 5.
Hart, John Francis
1953 "Theory of Spin Effects in the Quantum Mechanics of a Many-Electron Atom with Special Reference to Pb^{III}," *Ph.D. Dissertation*, University of Toronto.
Heath, G. Louis
1976 *Black Panther Leaders Speak.* Metuchen, N.J.: Scarecrow Press.
The Hebrew University (Jerusalem)
1936a *Information Bulletin of the Hebrew University, Jerusalem.* Jerusalem: November.
1936b *Publicity Service Bulletin of the Hebrew University, Jerusalem.* Jerusalem: March (Adar).
1938 *Bulletin of the Hebrew University, Jerusalem*, 4 (January).
1939 *The Hebrew University. Its History and Development.* Jerusalem.
1955 "Hebrew University–Symbol of Faith in Israel's Future," *American Friends of the Hebrew University. Bulletin*, 7, p. 2.
1956 *American Friends of the Hebrew University. Bulletin* (November).
1958 *American Friends of the Hebrew University. Bulletin* (April).
1959 *American Friends of the Hebrew University. Bulletin* (January).
1961 *American Friends of the Hebrew University. Bulletin* (June).
1963 *The Hebrew University of Jerusalem*, Jerusalem: Jerusalem Post Press.
1969 *The Hebrew University of Jerusalem.* Jerusalem: Jerusalem Post Press.
Heirich, Max
1971 *The Spiral of Conflict. Berkeley 1964.* New York: Columbia University Press.
1968 *The Beginning. Berkeley 1964.* New York: Columbia University Press; repr., 1970.

Henkin, Leon

1950 "Completeness in the Theory of Types," *Journal of Symbolic Logic*, 15, pp. 81–91.

1956 "Two Concepts from the Theory of Models," *Journal of Symbolic Logic*, 21, pp. 28–32.

1962 "Are Logic and Mathematics Identical?" *Science*, 138, pp. 788–794.

Herbrand, Jacques

1930 "Recherches sur la théorie de la démonstration," *Travaux de la société des sciences et des lettres de Varsovie*, 33. This was Herbrand's Ph.D. thesis (University of Paris), chapter 5 of which was translated as "Investigations in Proof Theory: The Properties of True Propositions," van Heijenoort 1967, pp. 525–581.

Herman, Josef

1980 *The Radical Imagination: Frans Masereel, 1889–1972*. London: Journeyman Press; repr., New York: Riverrun Press, 1986.

Hermann, G.

1926 "Die Frage der endlich vielen Schritte in der Theorie der Polynomideale," *Mathematische Annalen*, 95, pp. 736–788.

Hersey, John

1970 *Letter to the Alumni*. New York: Knopf.

Hewison, Robert

1977 *Under Siege. Literary Life in London, 1939–1945*. New York: Oxford University Press.

Hilbert, David

1892 "Ueber die Irreducibilität ganzer rationaler Functionen mit ganzzahligen Coefficienten," *Journal für die reine und angewandte Mathematik*, 110, pp. 104–129.

Hirschfeld, J., and W. H. Wheeler

1975 *Forcing, Arithmetic, and Division Rings*, Springer Lecture Notes in Mathematics, vol. 454. Berlin: Springer-Verlag.

Hitti, Philip K.

1928 *The Origins of the Druze People and Religion, with Extracts from Their Sacred Writings*. New York: Columbia University Press; repr., 1966.

Hodges, W.

1986 "Alfred Tarski," *Journal of Symbolic Logic*, 51, pp. 866–868.

Holden, Reuben A.

1967 *Yale: A Pictorial History*. New Haven: Yale University Press.

Holmes, Brian

1963 "Comparative Education Society in Europe," *International Review of Education*, 8, p. 419.

1972 "Saul B. Robinsohn: In Memoriam," *International Review of Education*, 18, pp. 283–284.

Hopkins, Charles

1938 "Nilrings with Minimal Condition for Admissible Left Ideals," *Duke Mathematics Journal*, 4, pp. 664–667.

Hopkins, H.

1963 *The New Look: A Social History of the Forties and Fifties in Britain*. Secker and Warburg.

Horne, Alistaire
1969 *To Lose a Battle: France 1940.* Boston: Little, Brown and Co.
Hughes, H. Stuart
1976 *Contemporary Europe. A History*. Englewood Cliffs, N.J.: Prentice-Hall.
Humphries, Rolfe
1951 *The Aeneid of Virgil.* New York: Charles Scribner's Sons.
Hurd, A., and P. Loeb, eds.
1974 *Victoria Symposium on Nonstandard Analysis,* Springer Lecture Notes in Mathematics, vol. 369. Berlin: Springer-Verlag.
Hytier, Adrienne D.
1958 *Two Years of French Foreign Policy: Vichy, 1940–1942.* Paris–Geneva: Librairie E. Droz.
Infeld, Eryk
1970 "Leopold Infeld Bibliography," *General Relativity and Gravitation*, 1, pp. 191–208.
Infeld, Leopold
1978 "Canada," in *Why I left Canada. Reflections on Science and Politics*, by L. Infled. Trans. H. Infeld. Montreal: McGill-Queen's University Press.
Institute for Advanced Study
1980 *A Community of Scholars. The Institute for Advanced Study. Faculty and Members, 1930–1980.* Princeton, N.J.: The Institute for Advanced Study.
Jacobson, Nathan
1962 *Lie Algebras.* New York: John Wiley & Sons.
1989 *Collected Mathematical Papers.* Boston: Birkhäuser, in 3 vols.
Jech, Thomas J., ed.
1974 *Axiomatic Set Theory. Proceedings of Symposia in Pure Mathematics,* 13, Providence, R.I.: American Mathematical Society; for the first volume see Scott 1971.
Johnson, Brian
1978 *The Secret War.* New York: Methuen.
Johnson, David Randolph (Jr.)
1971 "The Construction of Elementary Extensions and the Stone-Cech Compactification," *Ph.D. Dissertation*, Yale University.
1975 with Don A. Mattson, "Some Applications of Non-Standard Analysis to Proximity Spaces," *Colloquium Mathematicum*, 34, pp. 17–24.
Jones, Reginald V.
1978 *The Wizard War. British Scientific Intelligence, 1939–1945.* New York: Coward, McCann and Geoghegan.
Kakutani, Michiko
1982 "New Haven Blues," in Dubois 1982, pp. 279–287.
Kalikstein, Kalman
1978 "Mathematical and Astronomical Commentary." New York: Post Talmudic Research Institute, 5738/1978.
Kalmár, László
1935 "Über die Axiomatisierbarkeit des Aussagenkalküls," *Acta scientiarum mathematicarum*, 7, pp. 222–243.

Keisler, H. J.

1960 "Isomorphism of Ultraproducts," *Notices of the American Mathematical Society*, 7, pp. 70–71.

1971 *Model Theory for Infinitary Logic*. Amsterdam: North-Holland.

1971a *Elementary Calculus: An Approach Using Infinitesimals*. Boston: Princle, Weber and Schmidt.

1973 "Studies in Model Theory," *Studies in Mathematics*, 8, pp. 96–133.

1977 "Introduction," Robinson 1977.

1979 "Introduction," Robinson 1979, 1, pp. xxxiii–xxxvii.

Kelemen, Peter

1972 "Quantum Mechanics, Quantum Field Theory and Hyper-Quantum Mechanics," *Victoria Symposium on Nonstandard Analysis, University of Victoria, Victoria, BC, 1972*, Lecture Notes in Mathematics, vol. 369. Berlin: Springer, 1974, pp. 116–121.

Kemeny, J. G.

1958 "Undecidable Problems of Elementary Number Theory," *Mathematische Annalen*, 135, pp. 160–169.

Kendall, David, et al.

1990 "Andrei Nikolaevich Kolmogorov (1903–1987)," *Bulletin of the London Mathematical Society*, 22, pp. 31–100.

Kent, Lenore

1951 "School Abroad," *The Varsity* (student newspaper of the University of Toronto), Thursday, November 1, 1951, p. 7.

Klein, Felix

1922 "On the Mathematical Character of Space-Intuition and the Relation of Pure Mathematics to the Applied Sciences," *Gesammelte Mathematische Abhandlungen*, R. Fricke and H. Vermeil, eds. Vol. 2, Berlin, pp. 225–231.

1926 *Vorlesungen über die Entwicklung der Mathematik im 19. Jahrhundert*. Berlin: J. Springer.

Klibansky, Raymond

1959 *Philosophy in the Mid-Century*, Firenze: La Nuova Italia Editrice, 1958–1959.

1968 ed., *La Philosophie contemporaine*. Firenze: La Nuova Italia Editrice.

Knapp, Bettina

1958 *Louis Jouvet. Man of the Theatre*. New York: Columbia University Press.

Knapp, Wilfrid

1972 *A History of War and Peace. 1939–1965*. London: Oxford University Press.

Knopp, Konrad

1951 "Edmund Landau," *Jahresbericht der Deutschen MathematikeR-Vereinigung*, 54, pp. 55–62.

Kochen, Simon

1961 "Ultraproducts in the Theory of Models," *Annals of Mathematics*, 74, pp. 221–261.

1976 "The Pure Mathematician. On Abraham Robinson's Work in Mathematical Logic," in Young 1976, pp. 312–315.

Koestler, Arthur
1980 *Bricks to Babel.* New York: Random House.
Kolchin, E. R.
1953 "Galois Theory of Differential Fields," *Journal of Mathematics,* 57, pp. 753–824.
Kolmogorov, A. N.
1933 *Grundbegriffe der Wahrscheinlichkeitsrechnung.* Berlin: Springer, 1933; Russian trans. 1936; English trans. (second ed.), N. Morrison, New York: Chelsea 1950; repr., 1956.
Körner, Stephan
1960 *The Philosophy of Mathematics.* London: Hutchinson.
1979 "Introduction to Papers on Philosophy," in Robinson 1979, 2, pp. xli–xlv.
Kozlik, Angela
1974 *Curriculumtheorie und Emanzipation. Pädagogische Reflexionen zur Curriculumtheorie Saul B. Robinsohns.* Vienna: Jugend und Volk.
Kreisel, Georg
1952 "On the Concepts of Completeness and Interpretation of Formal Systems," *Fundamenta Mathematica,* 39, pp. 103–127.
1957 "Hilbert's 17th Problem, I and II (abstracts)," *Bulletin of the American Mathematical Society,* 63, pp. 99–100.
1967 and J. L. Krivine, *Elements of Mathematical Logic.* Amsterdam: North-Holland.
1971 "Observations on Popular Discussions of Foundations," in *Axiomatic Set Theory. Proceedings of Symposia in Pure Mathematics. Part I,* D. S. Scott, ed., 13. Providence, R.I.: American Mathematical Society, pp. 189–197.
Krull, W.
1935 "Idealtheorie," *Ergebnisse der Mathematik und ihrer Grenzgebiete,* 4, Berlin: Julius Springer; repr., New York: Chelsea, 1948.
Kugler, Lawrence Dean
1966 "Nonstandard Analysis of Almost Periodic Functions," *Ph.D. Dissertation,* Department of Mathematics, University of California, Los Angeles.
1969 "Nonstandard Analysis of Almost Periodic Functions," in Luxemburg 1969, pp. 150–166.
1969a "Nonstandard Almost Periodic Functions on a Group," *Proceedings of the American Mathematical Society,* 22 (1969), pp. 527–533.
Kuhlmann, Salma
1991 "Quelques propriétés des espaces vectoriels valués en théorie des modèles," *Ph.D. Thesis,* University of Paris VII, July.
Kuhlmann, Salma, and N. L. Alling
1994 "On η_α Groups and Fields," *Order,* to appear.
Lachman, F. R.
1956 "Pilgrimage to Scopus. Isolated Campus, Vital Books, Equipment Wage Steady Battle Against Mold, Decay," *American Friends of the Hebrew University. Bulletin* (November), p. 2.

Lakatos, I., ed.

1965 *Problems in the Philosophy of Mathematics. Proceedings of the International Colloquium in the Philosophy of Science, London, 1965.* Amsterdam: North-Holland.

Lamb, Horace

1904 "On the Propagation of Tremors over the Surface of an Elastic Solid," *Philosophical Transactions of the Royal Society*, 203, pp. 1–42.

Lambert, William Miles

1965 "Effectiveness, Elementary Definability, and Prime Polynomial Ideals," *Ph.D. Thesis*, Department of Mathematics, University of California, Los Angeles.

1968 "A Notion of Effectiveness in Arbitrary Structures," *Journal of Symbolic Logic*, 33, pp. 577–602.

Landau, Edmund

1985 *Edmund Landau. Collected Works*. L. Mirsky et al., eds. Essen: Thales Verlag.

Langeron, Roger

1946 *Paris. Juin 1940.* Paris: Flammarion.

Laqueur, Walter

1976 *A History of Zionism*. New York: Schocken Books.

Laugwitz, Detlef

1959 "Eine Einführung der d-Funktionen," *Sitzungsberichte der Bayerschen Akademie der Wissenschaften*, pp. 41–59.

1961 "Anwendungen unendlich kleiner Zahlen I. Zur Theorie der Distributionen," *Journal für die reine und angewandte Mathematik*, 207, pp. 53–60.

1961a "Anwendungen unendlich kleiner Zahlen II. Ein Zugang zur Operatorenrechnung von Mikusinski," *Journal für die reine und angewandte Mathematik*, 208, pp. 22–34.

1983 "Nichstandard-Mathematik, begründet durch eine Verallgemeinerung der Körpererweiterung," *Expositiones Mathematicae*, 4, pp. 307–333.

1983a "Die Nichtstandard-Analysis: eine Wiederaufnahme der Ideen und Methoden von Leibniz und Euler," in *Leonard Euler. 1707–1783. Beiträge zu Leben und Werk.* Basel: Birkhäuser, pp. 185–197.

Leibniz, G. W.

1789 *Opera omnia*. L. Dutens, ed. Geneva: Apud fratres de Tournes.

Levensohn, Lotta

1950 *Vision and Fulfillment. The First Twenty-five Years of the Hebrew University, 1925–1950.* New York: The Greystone Press.

Levitzki, Jacob

1931 "Über nilpotente Unterringe," *Mathematische Annalen*, 105, p. 620.

1939 "On Rings which Satisfy the Minimum Condition for the Right-hand Ideals," *Compositio Mathematica*, 7, pp. 214–222.

Levy, Azriel

1957 "Indépendance conditionnelle de $V = L$ et d'axiomes qui se rattachent au système de M. Gödel," *Comptes rendus des séances de l'académie des sciences, Paris*, 245, pp. 1582–1583.

1960 "The Independence of Various Definitions of Finiteness," *Fundamenta Mathematica*, 46 (1958), pp. 1–13.

1960a "A Generalization of Gödel's Notion of Constructibility," *Journal of Symbolic Logic*, 25, pp. 147–155.

Lighthill, M. J.

1957 Review of Robinson and Laurmann 1956, in *Mathematical Reviews*, 18, p. 529.

Lightstone, Albert Harold

1955 "Contributions to the Theory of Quantification," *Ph.D. Dissertation*, University of Toronto.

Lindner, Helmut

1980 "'Deutsche' und 'gegentypische' Mathematik. Zur Begründung einer 'arteigenen' Mathematik im 'Dritten Reich' durch Ludwig Bieberbach," in *Naturwissenschaft, Technik und NS-Ideologie*. H. Mehrtens and S. Richter, eds. Frankfurt am Main: Suhrkamp, pp. 88–115.

Lipset, Seymour M., and Sheldon S. Wolin

1965 *The Berkeley Student Revolt. Facts and Interpretations*. New York: Anchor Books.

Littlefield, Joan

1946 "Covent Garden Reopens," *Theatre Arts*, 30 (4) (April).

Lobry, Claude

1989 *Et pourtant . . . ils ne remplissent pas N*. Lyon: Aleas Editeur.

Lorch, Lee

1967 "Some Recollection of the International Congress of Mathematicians," *Canadian Mathematical Bulletin*, 10, pp. 157–162; abr. in *Science*, 155 (1967), pp. 1038–1039.

Lorentz, G. G.

1957 "The Work of G. G. Lorentz," *Journal of Approximation Theory*, 13, pp. 12–16.

Łoś, Jerzy

1955 "Quelques remarques, théorèmes et problèmes sur les classes définissables d'algèbres," *Symposium on the Mathematical Interpretation of Formal Systems*. Amsterdam: North-Holland Publishing Co., 1955, pp. 98–113.

1957 and R. Suszko, "On the Extending of Models (IV)," *Fundamenta Mathematicae*, 44, pp. 52–60.

Lutz, Robert, and Michel Goze

1981 *Nonstandard Analysis. A Practical Guide with Applications*. Springer Lecture Notes in Mathematics no. 881. Berlin: Springer-Verlag.

Luxemburg, W.A.J.

1962 *Nonstandard Analysis. Lectures on A. Robinson's Theory of Infinitesimals and Infinitely Large Numbers*. Pasadena: Mathematics Department, California Institute of Technology; second corrected ed., 1964.

1969 ed., *Applications of Model Theory to Algebra, Analysis and Probability*. New York: Holt, Rinehart and Winston.

1969a "A General Theory of Monads," in Luxemburg 1969, pp. 18–86.

1976 and K. D. Stroyan, *Introduction to the Theory of Infinitesimals*. New York: Academic Press.

Lyndon, R.
1959 "An Interpolation Theorem in the Predicate Calculus," *Pacific Journal of Mathematics*, 9, pp. 129–142.
Machover, Moshé
1967 "Non-Standard Analysis without Tears: An Easy Introduction to A. Robinson's Theory of Infinitesimals," Technical Report no. 27, Office of Naval Research, Information Systems Branch. Jerusalem: The Hebrew University of Jerusalem.
1969 and J. Hirschfeld, *Lecture Notes on Nonstandard Analysis*, in the series Lecture Notes in Mathematics, 94. Berlin: Springer-Verlag.
1977 See Bell, John, and M. Machover.
Macintyre, Angus
1977 "Abraham Robinson, 1918–1974," *Bulletin of the American Mathematical Society*, 83, pp. 646–666.
Magnes, Judah L.
1939 "War and the Remnant of Israel," the President's opening address of the academic year, Hebrew University, October 19, 1939. In Goren 1982.
Mahler, K.
1934 "Über die rationalen Punkte auf Kurven vom Geschlecht Eins," *Journal für die reine und angewandte Mathematik*, 170, pp. 168–178.
Maimonides
1956 *The Code of Maimonides: Sanctification of the New Moon*. Trans. S. Gandz, intro. by J. Oberman, commentary by O. Neugebauer. New Haven: Yale University Press.
Makowski, J. A.
1979 and S. Shelah, "The Theorems of Beth and Craig in Abstract Model Theory. I. The Abstract Setting," *Transactions of the American Mathematical Society*, 256, pp. 213–239.
1981 "The Theorems of Beth and Craig in Abstract Model Theory. II. Compact Logics," *Arch. Math. Logik Grundlagen*, 21, pp. 13–35.
Mal'cev, A. I.
1968 "Some Questions Bordering on Algebra and Mathematical Logic," *American Mathematical Society Translations*, 70, pp. 89–100.
1971 *The Metamathematics of Algebraic Systems. Collected Papers: 1936–1967.* Trans. B. F. Wells, III. Amsterdam: North-Holland.
Marrus, Michael, and Robert O. Paxton
1981 *Vichy France and the Jews*. New York: Schocken, 1983.
Matiyasevich (Matijasevic), Yuri B.
1970 "Recursively Enumerable Sets are Diophantine" (in Russian), *Doklady Akademii Nauk SSSR*, 191, pp. 279–282; translated as "The Diophantineness of Enumerable Sets," *Soviet Mathematics. Doklady*, 11, pp. 354–358.
Mehrtens, Herbert, and S. Richter, eds.
1980 *Naturwissenschaft, Technik und NS-Ideologie.* Frankfurt am Main: Suhrkamp.

Meier, Amram

1962 "Analytic Continuation by Summability and Relations between Summability Methods," *Abstracts of Theses Approved for the Degrees of Doctor Philosophiae (Ph.D.) and Doctor Juris (Jur.D.), in the Hebrew University of Jerusalem during the Academic Year 1961–62*. Jerusalem: The Hebrew University of Jerusalem, 1964.

Menashe, Louis, and Ronald Radosh, eds.

1967 *Teach-Ins: USA. Reports, Opinions, Documents*. New York: Frederick A. Praeger, pp. 8–13.

Mendelson, E.

1957 "Non-standard models," in Cornell 1957, vol. 1, pp. 167–168.

Messaut, J.

1938 *La Philosophie de Léon Brunschvicg*, Paris: J. Vrin.

Miller, Michael V., and Susan Gilmore, eds.

1965 *Revolution at Berkeley*; repr. Dial Press, New York: Dell Publishing Co.

Milton, Sybil H.

1985 "Lost, Stolen, and Strayed. The Archival Heritage of Modern German-Jewish History," in *The Jewish Response to German Culture*, Jehuda Reinharz and Walter Schatzberg, eds. Hanover, N.H.: University Press of New England, for Clark University, pp. 317–335.

Mindlin, R. D.

1951 "Influence of Rotary Inertia and Shear on Flexural Motions of Isotropic Elastic Plates," *Journal of Applied Mechanics*, 18, pp. 31–38.

Mirsky, L.

1985 "In Memory of Edmund Landau. Glimpses from the Panorama of Number Theory and Analysis," in Landau 1985, 1, pp. 25–50.

Moore, Gregory H.

1980 *Zermelo's Axiom of Choice. Its Origins, Development, and Influence*. New York: Springer-Verlag.

1988 "The Origins of Forcing," *Logic Colloquium '86*, F. R. Drake and J. K. Truss, eds. Amsterdam: North-Holland, pp. 143–173.

Moore, Paul (Jr.)

1982 "A Touch of Laughter," in Dubois 1982, pp. 196–208.

Morgenstern, Christian

1965 *Gesammelte Werke*, Margareta Morgenstern, ed. München: R. Piper and Co.

Moschovakis, Yiannis N.

1988 "Mathematical Logic," commentary on "The Princeton University Bicentennial Conference on the Problems of Mathematics (1946)," in Duren 1988, vol. 2, pp. 343–346.

Mosley, Leonard

1971 *Backs to the Wall. London Under Fire, 1939–1945*. London: Weidenfeld and Nicolson.

Mostowski, Andrzej W.

1952 "Models of Axiomatic Systems," *Fundamenta Mathematica*, 39, pp. 133–158.

1955 "Quelques observations sur l'usage des méthodes non finitistes dans

la méta-mathématique," *CNRS colloque internationale, Le raisonnement en mathématiques et en sciences expérimentales (1955)*. Paris: CNRS, 1958, pp. 13–32.
1966 *Thirty Years of Foundational Studies*. New York: Barnes and Noble.

Motzkin, Theodore
1938 "Sur les arcs plans dont les courbes osculatrices ne se coupent pas," *Comptes rendus de l'académie des sciences*, 206, pp. 1700–1701.
1945 "A 5 Curve Theorem Generalizing the Theorem of Carnot," *Bulletin of the American Mathematical Society*, 51, pp. 972–975.

Müller, G. H.
1976 ed., *Sets and Classes*. Amsterdam: North-Holland.
1987 *Ω-Bibliography of Mathematical Logic*. Berlin: Springer-Verlag.

Murphy, Robert
1964 *Diplomat among Warriors*. Westport, Conn.: Greenwood Press.

Nelson, Edward
1977 "Internal Set Theory: A New Approach to Nonstandard Analyis," *Bulletin of the American Mathemaical Society*, 83, pp. 1165-1198.

Neumann, B. H.
1954 "An Embedding Theorem for Algebraic Systems," *Proceedings of the London Mathematical Society*, 4, pp. 138–153.

Noah, H. J.
1972 "Saul B. Robinsohn, 1916–1972," *International Review of Education*, 16, p. 405.

Ore, Øystein
1963 *Graphs and Their Uses*. New York: Random House.

Parikh, Rohit
1969 "A Nonstandard Theory of Topological Groups," in Luxemburg 1969, pp. 279–284.

Parzen, Herbert
1974 *The Hebrew University. 1925–1935*. New York: Ktav Publishing House.

Paxton, Robert O.
1966 *Parades and Politics at Vichy*. Princeton: Princeton University Press.
1972 *Vichy France Old Guard and New Order, 1940–1944*. New York: Columbia University Press.

Petersen, G. M.
1957 "A Tribute to G. G. Lorentz," *Journal of Approximation Theory*, 13, pp. 4–5.

Pfister, Albrecht
1976 "Hilbert's Seventeenth Problem and Related Problems on Definite Forms," in Browder 1976, pp. 483–489.

Phillips, Robert Gibson
1968 "Some contributions to nonstandard analysis," *Ph.D. Thesis*, Department of Mathematics, University of California, Los Angeles.

Pilisuk, Marc
1965 "The First Teach-In: an Insight into Professional Activism," *The Correspondent*, 34; repr. in Menashe and Radosh 1967, pp. 8–13.

Pinl, M.
1969 "Kollegen in einer dunklen Zeit," *Jahresbericht der Deutsche Mathematiker Vereinigung,*" 71, pp. 167–228, and 72 (1971), pp. 165–189.

Porter, Roy P.
1942 *Uncensored France. An Eyewitness Account of France under the Occupation.* New York: The Dial Press.

Prüfer, H.
1924 "Theorie der Abelschen Gruppen," *Mathematische Zeitschrift,* 20, pp. 165–187.
1925 "Neue Begründung der algebraischen Zahlentheorie," *Mathematische Annalen,* 94, pp. 198–243.

Pryce-Jones, David
1981 *Paris in the Third Reich. A History of the German Occupation, 1940–1944.* New York: Holt, Rinehart and Winston.

Quine, W. V.
1948 Review of Reichenbach's *Elements of Symbolic Logic* (1947), in *The Journal of Philosophy,* 45, pp. 161–166. For Reichenbach's reply to Quine, see Reichenbach 1948.

Rabin, M. O.
1959 "Arithmetical Extensions with Prescribed Cardinality," *Koninklijke Nederlandse Akademie van Wetenschappen. Proceedings,* 62, and *Indagationes Mathematicae,* 21, pp. 439–446.
1962 "Diophantine Equations in Nonstandard Models of Arithmetic," *Proceedings of the International Congress of Logic, Methodology and Philosophy of Science.* Stanford: Stanford University Press, pp. 151–158.

Rado, R.
1949 "Axiomatic Treatment of Rank in Infinite Sets," *Canadian Journal of Mathematics,* 1, pp. 337–343.

Rasiowa, Helena
1952 "A Proof of the Compactness Theorem for Arithmetical Classes," *Fundamenta Mathematica,* 39, pp. 8–14.

Reeb, Georges
1977 "Séance débat sur l'analyse nonstandard," *Gazette des mathématiciens,* 8, pp. 8–14.
1989 "Mathématique non standard," in *La Mathématique non standard,* H. Barreau and J. Harthong, eds. Paris: Éditions du CNRS, pp. 3–9.

Reichenbach, Hans
1947 *Elements of Symbolic Logic.* New York: Macmillan; repr., New York: Dover, 1980.
1948 Reply to Quine 1948. See *The Journal of Philosophy,* 45, pp. 464–467.

Richardson, A. R.
1940 "Algebras of *s* Dimensions," *Proceedings of the London Mathematical Society,* 47, pp. 38–49.

Ritt, J. F.
1950 *Differential Algebra,* American Mathematical Society, Colloquium Publications, 33.

Robinsohn, Abraham

1921 *David Wolffsohn. Ein Beitrag zur Geschichte des Zionismus*. Berlin: Jüdischer Verlag.

Robinsohn, Saul

1940 Medinat Jisrael ke-medina ledugma bekitwehem schel hogei deot politiim ba-meot ha-tet'sajin, ha-jud'sajin we-ha-jud'chet (ha-"Biblizism" be-thorat-ha-medina me-Machiavel we-ad Adam Müller, [The Israeli State as Model in the Writings of Political Thinkers of the 16th, 17th and 18th Centuries], *M.A. thesis,* Jerusalem, The Hebrew University.

1967 *Bildungsreform als Revision des Curriculum*. Berlin: Luchterhand Verlag; second and third eds. in 1970, 1971, respectively.

1968 with Helga Thomas, *Differenzierung im Sekundarschulwesen*, vol. 3 of *Gutachten und Studien der Bildungskommission des Deutschen Bildungsrates*. Stuttgart: Klett.

1969 "Das Goldene Jerusalem," *Geschichte in Wissenschaft und Unterricht*, 11/12, pp. 767–780; repr. in Robinson 1973, pp. 430–447.

1973 *Erziehung als Wissenschaft*, eds. F. Braun, D. Glowka and H. Thomas. Stuttgart: Ernst Klett, p. 1973.

Robinson, Gilbert de B.

1968 "Leopold Infeld," *Proceedings and Transactions of the Royal Society of Canada*, 6, pp. 123–125; repr. in *Canadian Mathematical Bulletin*, 14 (1971), pp. 301–302.

1971 "Samuel Beatty," *Proceedings and Transactions of the Royal Society of Canada,* 9, pp. 31–34; repr. in *Canadian Mathematical Bulletin*, 14, pp. 489–490.

1979 *The Mathematics Department in the University of Toronto, 1827–1978*. Toronto: Department of Mathematics, University of Toronto.

Robinson, Logan

1982 *An American in Leningrad*. New York: W.W. Norton and Co.

Rogosinski, W. W.

1958 "Obituary: Michael Fekete," *Journal of the London Mathematical Society*, 33, pp. 496–500.

Roquette, Peter

1975 See Roquette and Robinson 1975.

Rose, H. E., and J. C. Shepherdson, eds.

1973 *Logic Colloquium '73. Proceedings of the Logic Colloquium. Bristol, July 1973*. Amsterdam: North-Holland.

Rosenblatt, Samuel

1954 *Yossele Rosenblatt, the Story of His Life as Told by His Son*. New York: Farrar, Straus and Young.

Ross, Roderick Alexander

1955 "The Waves Produced by a Submarine Earthquake," *Ph.D. Thesis*, University of Toronto.

Rosser, J. Barkley

1952 "Message du président de l'association for symbolic logic," *Applications scientifiques de la logique mathématique; actes du 2. Colloque interna-*

tional de logique mathématique. Paris: Institut Henri Poincaré, 25–30 août. Paris: Gauthier-Villars, 1954.

Rotenstreich, N.
1966 *The Hebrew University of Jerusalem*. Jerusalem: Magnes Press.

Ryll-Nardzewski, C.
1952 "The Role of the Axiom of Induction in Elementary Arithmetic," *Fundamenta Mathematica*, 39, pp. 239–263.

Sacks, Gerald E.
1972 *Saturated Model Theory*. New York: Benjamin.

Safran, Nadav
1981 *Israel, the Embattled Ally*. Cambridge, Mass.: Belknap Press.

Samuel, Edwin
1958 "Teaching Rewarding at Hebrew University," *American Friends of the Hebrew University. Bulletin* (April 1958), p. 12.

Samuels, Gertrude
1959 "Hebrew University Rises Again," *New York Times Magazine*, Sunday, April 19, pp. 22 ff.

Saracino, D. H., and V. B. Weispfenning
1975 *Model Theory and Algebra. A Memorial Tribute to Abraham Robinson*, Lecture Notes in Mathematics, vol. 498. Berlin: Springer-Verlag.

Sartre, Jean-Paul
1984 *Les Carnets de la drôle de guerre: Novembre 1939–Mars 1940*. Paris: Éditions Gallimard, 1983. English trans. Q. Hoare. *The War Diaries. November 1939–March 1940*. New York: Pantheon Books.

Schiff, Ze'ev
1974 *A History of the Israeli Army*. Trans. R. Rothstein. San Francisco: Straight Arrow Books.

Schleicher, Klaus
1971 Review of Robinsohn and Thomas 1968, *International Review of Education*, 17, pp. 119–122.

Schmeiden, C., and D. Laugwitz
1958 "Eine Erweiterung der Infinitesimalrechnung," *Mathematische Zeitschrift*, 69, pp. 1–39.

Schreiber, Shmuel
1963 "Loop Rings," *Abstracts of Theses Approved for the Degrees of Doctor Philosophiae (Ph.D.) and Doctor Juris (Jur.D.), in the Hebrew University of Jerusalem during the Academic Year 1963–64*. Jerusalem: The Hebrew University of Jerusalem, 1965.

Scofield, John
1959 "Jerusalem, the Divided City," *National Geographic Magazine*, 115 (April), pp. 492–531.

Scott, Dana S., ed.
1971 *Axiomatic Set Theory. Proceedings of Symposia in Pure Mathematics*, 13, Providence, R.I.: American Mathematical Society; for the second vol. see Jech 1974.

Scott-James, Anne
1953 *In the Mink*. New York: Dutton.

Searle, John

1965 "The Faculty Resolution," in Miller and Gilmore 1965, pp. 92–104.

Seidenberg, A. J.

1956 "An Elimination Theory for Differential Algebra," *University of California Publications in Mathematics*, 3, pp. 31–66.

Seligman, George

1979 "Biography of Abraham Robinson," in Robinson 1979, 1, pp. xiii–xxxii.

Shoenfield, J. R.

1957 "Constructible Sets (abstract)," in Cornell 1957, 2, p. 214.

1959 "On the Independence of the Axiom of Constructibility," *American Journal of Mathematics*, 81, pp. 537–540.

Siegel, Carl Ludwig

1929 "Über einige Anwendungen diophantischer Approximationen," *Abhandlungen der Preussischen Akademie der Wissenschaften, Physikalisch-Mathematische Klasse*, 1, in Siegel, *Gesammelte Abhandlungen*, 1. Berlin: Springer-Verlag, 1966, pp. 209–266.

Simmons, H.

1972 "Existentially Closed Structures," *The Journal of Symbolic Logic*, 32, pp. 293–310.

Sinaceur, Hourya

1984 "De D. Hilbert à E. Artin: Les Différents aspects du dix-septième problème et les filiations conceptuelles de la théorie des corps réels clos," *Archive for History of Exact Sciences*, 29, pp. 267–287.

1985 "La théorie d'Artin et Schreier et l'analyse non-standard d'Abraham Robinson," *Archive for History of Exact Sciences*, 34, pp. 257–264.

1989 "Une origine du concept d'analyse non-standard," in Barreau 1989, pp. 143–156.

1991 *Corps et modèles, essai sur l'histoire de l'algèbre réelle*. Paris: Vrin.

1993 "Du Formalisme à la constructivité: le finitisme," *Revue internationale de philosophie*, 4, Paris: Presses universitaires de France, pp. 251–283.

1994 *Jean Cavaillès. Philosophie mathématique*. Paris: Presses universitaires de France.

Sissons, Michael, and Philip French, eds.

1964 *The Age of Austerity*. London: Penguin, 1964, pp. 35–57.

Skolem, T.

1934 "Über die Nichtcharakterisierbarkeit der Zahlenreihe mittels endlich oder abzählbar unendlich vieler Aussagen mit ausschliesslich Zahlenvariablen," *Fundamenta Mathematicae*, 23, pp. 150–161.

1955 "Peano's Axioms and Models of Arithmetic," *Symposium on the Mathematical Interpretation of Formal Systems*. Amsterdam: North-Holland, pp. 1–14.

Spears, Edward Louis

1954 *Assignment to Catastrophe*. London: Heinemann.

Stadtman, V. A., ed.

1967 *The Centennial Record of the University of California*. Berkeley: University of California Printing Department.

Stapledon, Olaf

1948 "Interplanetary Man," *Journal of the British Interplanetary Society*, 7 (6) (November), pp. 213–233.

Steinitz, E.

1910 "Algebraische Theorie der Körper," *Journal für die reine und angewandte Mathematik*, 137, pp. 167–309.

Steketee, Jakob Abraham

1955 "Some Problems in Boundary-Layer Transition," *Ph.D. Thesis*, University of Toronto.

Stewart, Ian

1991 *The New York Review of Books*, 38 (December 5), p. 75.

Stroyan, K. D., and José Manuel Bayod

1986 *Foundations of Infinitesimal Stochastic Analysis.* Amsterdam: North-Holland.

Super, R. H.

1962 *Matthew Arnold. Lectures and Essays in Criticism.* Ann Arbor: University of Michigan Press.

Sutton, John L.

1973 *Black Panther Banner.* Metuchen, N.J.: Scarecrow Press.

Synge, J. L.

1972 "Curriculum Vitae of J. L. Synge," in *General Relativity (Papers in Honor of J. L. Synge).* Oxford: Clarendon Press.

Szmielew, W.

1955 "Elementary Properties of Abelian Groups," *Fundamenta Mathematicae*, 41, pp. 203–271.

Tamari, D.

1953 "On the Embedding of Birkhoff-Witt Rings in Quotient Fields," *Proceedings of the American Mathematical Society*, 4, pp. 197–202.

Tarcici, Adnan

1971 *The Queen of Sheba's Land: Yemen (Arabia Felix).* Beirut: Nowfel.

Tarski, Alfred

1939 "New Investigations on the Completeness of Deductive Theories" (abstract), *Journal of Symbolic Logic*, 4, p. 176.

1948 *A Decision Method for Elementary Algebra and Geometry*. Santa Monica, Calif.: The Rand Corporation.

1948 and J.C.C. McKinsey, *A Decision Method for Elementary Algebra and Geometry*. Berkeley: University of California Press; second ed. 1951; repr., Ann Arbor: University Microfilms, 1979.

1953 Andrzej Mostowski and R. M. Robinson, *Undecidable Theories.* Amsterdam: North-Holland; repr., 1968.

1957 and Robert L. Vaught, "Arithmetical Extensions of Relational Systems," *Compositio Mathematica*, 13, pp. 81–102.

1960 "Some Problems and Results Relevant to the Foundations of Set Theory," in *Logic, Methodology and Philosophy of Science. Proceedings of the 1960 International Congress for Logic, Methodology and Philosophy of Science*, E. Nagel, P. Suppes, and A. Tarski, eds. Stanford: Stanford University Press, 1962, pp. 125–135.

1965 *Introduction to Logic and to the Methodology of Deductive Sciences*. New York: Oxford University Press.

Tartakower, Chaim

1918 "Dr. Robinsohn plötzlich gestorben," *Jüdische Rundschau (Beiträge zur Jüdischen Zeitung der Jüdische Nationalfond)*, 19 (May 10).

Thackrah, John R.

1981 *The University and Colleges of Oxford*. Lavenham: Terence Dalton Ltd.

Thomas, I.

1964 "Medieval Aftermath: Oxford Logic and Logicians of the Seventeenth Century," *Oxford Studies Presented to Daniel Callus*, Oxford Historical Society, New Series, vol. 16. Oxford: Clarendon Press, pp. 297–311.

Tolstoy, Nikolai

1972 *Night of the Long Knives*. New York: Ballantine Books.

Traubel, Helen

1959 *St. Louis Woman*. New York: Duell, Sloan and Pearce.

Trautman, Andrzej

1973 "Leopold Infeld," *Dictionary of Scientific Biography*, 7, pp. 10–11.

Travis, Larry Evan

1966 "A Logical Analysis of the Concept of Stored Program: A Step Toward A Possible Theory of Rational Learning," *Ph.D. Dissertation*, Department of Philosophy, University of California, Los Angeles, 1966.

Trillin, Calvin

1965 "Letter from Berkeley," *The New Yorker*, March 3, 1965; in Miller and Gilmore 1965, pp. 253–284.

Tripodes, Peter Gust

1968 "Structural Properties of Certain Classes of Sentences," *Ph.D. Thesis*, Department of Mathematics, University of California, Los Angeles, 1968.

UCLA

1962 *General Catalogue Issue. Fall and Spring Semesters. 1962–1963*. Los Angeles: University of California.

1963 *General Catalogue Issue. Fall and Spring Semesters. 1963–1964*. Los Angeles: University of California.

1964 *General Catalogue Issue. Fall and Spring Semesters. 1964–1965*. Los Angeles: University of California.

1966 *UCLA General Catalog for 1966–1967*. Los Angeles: University of California.

Vaihinger, Hans

1911 *Die Philosophie des Als Ob*. Berlin: Reuther & Reichard; Leipzig, F. Meiner, 1920. Trans. C. K. Ogden, *The Philosophy of 'As If.'* New York: Harcourt, Brace & Co., 1924.

Van der Waerden, B. L.

1985 *A History of Algebra from al-Khwa¯rizmı¯ to Emmy Noether*. Berlin: Springer-Verlag.

van Heijenoort, J., ed.

1967 *From Frege to Gödel: a Source Book in Mathematical Logic*. Cambridge, Mass.: Harvard University Press.

van Rootselaar, B.
1972 "Adolf Abraham Fraenkel," *Dictionary of Scientific Biography*, C. C. Gillispie, ed., 5, pp. 107–109.
Vaught, Robert L.
1954 "Applications of the Löwenheim-Skolem-Tarski Theorem," *Indagiones Mathematicae*, 16, pp. 467–472.
1956 "On the Axiom of Choice and Some Metamathematical Theorems," *Bulletin of the American Mathematical Society*, 62, pp. 262–263.
1960 Review of Robinson 1956e, *Complete Theories,* in *Journal of Symbolic Logic*, 25, pp. 174–176.
Veblen, Oswald
1950 "Opening Address," in Congress 1950, pp. 124–125.
Vermes, P.
1967 "Richard George Cooke," *Journal of the London Mathematical Society*, 42, pp. 553–558.
von Neumann, J.
1926 "Zur Prüferschen Theorie der idealen Zahlen," *Acta Szeged*, 2, pp. 193–227.
Wang, Hao
1954 "The Formalization of Mathematics," *The Journal of Symbolic Logic*, 19, pp. 241–242.
Ward, G. N.
1952 "On the Integration of Some Vector Differential Equations," *Quarterly Journal of Mechanics and Applied Mathematics*, 5, p. 446.
Warmington, E. H.
1954 *A History of Birkbeck College, University of London, During the Second World War 1939–1945*. London: Birkbeck College.
Warner, Geoffrey
1968 *Pierre Laval and the Eclipse of France*. New York: Macmillan.
Wasiutyński, Bohdan
1930 *Ludnośč Zydowska w Polsce w Wiekach XIX i XX*, Warsaw: Wydawn. Kasy im. Mianowskiego.
Weil, André
1928 "L'Arithmétique sur les courbes algébriques," *Acta mathematica*, 52, pp. 281–315.
Weizmann, Chaim
1949 *Trial and Error, the Autobiography of Chaim Weizmann*, 1949; repr., 1977.
Werth, Alexander
1940 *The Last Days of Paris: A Journalist's Diary*. London: H. Hamilton.
West, Bruce
1967 *Toronto*. Toronto: Doubleday Canada Ltd.
Wigner, E. P.
1960 "The Unreasonable Effectiveness of Mathematics in the Natural Sciences," *Communications on Pure and Applied Mathematics*, 13, pp. 1–14.
Wolin, Sheldon S., and John H. Schaar
1970 *The Berkeley Rebellion and Beyond. Essays on Politics and Education in the Technological Society*. New York: Vintage Books.

Wood, Carol

1971 "Forcing for Infinitary Languages," *Ph.D. Dissertation*, Yale University.

1972 "Forcing for Infinitary Languages," *Zeitschrift für mathematische Logik und Grundlagen der Mathematik*, 18, pp. 385–402.

1973 "The Model Theory of Differential Fields of Characteristic 0," *Proceedings of the American Mathematical Society*, 40, pp. 577–584.

1974 "Prime Model Extensions for Differential Fields of Characteristic 0," *Journal of Symbolic Logic*, 39, pp. 469–477.

1976 "The Model Theory of Differential Fields Revisited," *Israel Journal of Mathematics*, 25, pp. 331–352.

Wooley, A. R.

1972 *The Clarendon Guide to Oxford.* Oxford: Oxford Univeristy Press.

Woolf, Virginia

1984 *The Diary of Virginia Woolf*, A. O. Bell, ed., vol. 5 (1936–1940). San Diego: Harcourt Brace Jovanovich.

Young, A. D.

1976 "The Applied Mathematician," in Young et al., "Abraham Robinson," *Bulletin of the London Mathematical Society*, 8, pp. 307–323.

1979 "Introduction," in Robinson 1979, 3, pp. xxix–xxxii.

Zakon, Elias

1953 "Left Side Distributive Law of the Multiplication of Transfinite Numbers," *Riveon Lematematika*, 6, pp. 28–32 (in Hebrew).

1954 "On the Relation of 'Similarity' between Ordinal Numbers," *Riveon Lematematika*, 7, pp. 44–49 (in Hebrew).

1954a *Collection of Exercises in Higher Mathematics*, Haifa: Technion Edition, 1954–1955.

1955 "On Fractions of Ordinal Numbers," *Scientific Publications of the Haifa Institute of Technology*, 6 (1955).

1961 "Generalized Archimedean Groups," *Transactions of the American Mathematical Society*, 99, pp. 21–40.

INDEX

Numbers in *italic print* indicate pages on which either photographs or illustrations related to the subject matter may be found. Interviews and manuscripts as listed on pp. 495–498 of the bibliography have also been indexed (in **boldface**); Robinson's papers are included under "Publications, Abraham Robinson," listed by years keyed to the bibliography as well. Cities may be found according to their English spellings, and names are transliterated in their usual or preferred forms with alternatives in parentheses. Thus Köln is indexed under Cologne; Matiyasevich (Matijasevic) and Mal'cev (Malcev, Maltsev) indicate alternative transliterations. Institutions are usually listed by city, sometimes by county; for example, Harvard University is found under Cambridge (USA); where locations are ambiguous or not well known, the institution is indexed directly (see, for example, IBM). Named mathematical theorems are indexed together under Theorems. Readers interested in Hilbert's *Nullstellensatz*, for example, will find it listed there (along with Ritt's and Rückert's).

JOSEPH WARREN DAUBEN is Professor of History and the History of Science at Herbert H. Lehman College and at the Graduate Center of the City University of New York. Among his works is *Georg Cantor: His Mathematics and Philosophy of the Infinite* (Princeton).